Handbook of Optical Dimensional Metrology

SERIES IN OPTICS AND OPTOELECTRONICS

Recent titles in the series

Handbook of 3D Machine Vision: Optical Metrology and Imaging
Song Zhang

Handbook of Optical Dimensional Metrology
Kevin Harding

Laser-Based Measurements for Time and Frequency Domain Applications: A Handbook
Pasquale Maddaloni, Marco Bellini, and Paolo De Natale

Handbook of Silicon Photonics
Laurent Vivien and Lorenzo Pavesi

Biomimetics in Photonics
Olaf Karthaus

Optical Properties of Photonic Structures: Interplay of Order and Disorder
Mikhail F Limonov and Richard De La Rue (Eds.)

Nitride Phosphors and Solid-State Lighting
Rong-Jun Xie, Yuan Qiang Li, Naoto Hirosaki, and Hajime Yamamoto

Molded Optics: Design and Manufacture
Michael Schaub, Jim Schwiegerling, Eric Fest, R Hamilton Shepard, and Alan Symmons

An Introduction to Quantum Optics: Photon and Biphoton Physics
Yanhua Shih

Principles of Adaptive Optics, Third Edition
Robert Tyson

Optical Tweezers: Methods and Applications
Miles J Padgett, Justin Molloy, and David McGloin (Eds.)

Thin-Film Optical Filters, Fourth Edition
H Angus Macleod

Laser Induced Damage of Optical Materials
R M Wood

Principles of Nanophotonics
Motoichi Ohtsu, Kiyoshi Kobayashi, Tadashi Kawazoe, Tadashi Yatsui, and Makoto Naruse

Handbook of Optical Dimensional Metrology

Edited by
Kevin Harding

CRC Press
Taylor & Francis Group
Boca Raton London New York

CRC Press is an imprint of the
Taylor & Francis Group, an **informa** business
A TAYLOR & FRANCIS BOOK

CRC Press
Taylor & Francis Group
6000 Broken Sound Parkway NW, Suite 300
Boca Raton, FL 33487-2742

Version Date: 20121210

International Standard Book Number: 978-1-4398-5481-5 (Hardback)

Library of Congress Cataloging-in-Publication Data

Handbook of optical dimensional metrology / editor, Kevin Harding.
pages cm. -- (Series in optics and optoelectronics)
Summary: "The need for faster and more accurate dimensional measurements has sparked the development of a range of laser types and methods with greater computational power. Featuring contributors from industry and academia, this handbook reviews modern optical dimensional measurement technologies and their applications. Covering scales ranging from nanometers to kilometers, it presents the principles, techniques, and devices used in modern optical dimensional metrology. An essential desk reference for metrologists and engineers, the book incorporates data tables and a multitude of available techniques"-- Provided by publisher.
Includes bibliographical references and index.
ISBN 978-1-4398-5481-5 (hardback)
1. Measurement--Handbooks, manuals, etc. 2. Optical measurements--Handbooks, manuals, etc. 3. Dimensional analysis--Handbooks, manuals, etc. I. Harding, Kevin G.

T50.H268 2013
530.8--dc23 2012038080

Contents

Part V Advanced Optical Micro-Metrology Methods

Preface

The field of optical metrology has seen significant growth in the last 30 years. Tools that at one time were limited to special applications as means to measure optical components are now seeing widespread use in all phases of manufacturing. This change for optical metrology has been driven by two primary factors. First, computing power improvements have moved the analysis of optical metrology data from long, painstaking hours of skilled interpreters looking at patterns only a select few understood to near instantaneous interpretations ready for use by the manufacturer. The other factor driving the change in optical metrology is the needs of manufacturing.

Modern day production lines are making and moving parts at speeds much faster than at any other time in history. The standards of six sigma quality have demanded much better control than ever over even small, cosmetic defects. Industries such as primary metals, automotive, textiles, and even plastic extruders have found that having the right dimensions and being "functional" is just not enough. Manufacturers are finding that any appearance of quality problems, be it pits and scratches or a bad overall shape, can mean rejections of full lots of product, costing millions of dollars to a company and affecting their bottom line. In the most critical industries, over 80% of customer rejects are often due to cosmetic defects that do not impact the function of the part but can easily be on the micron level in size. At modern speeds of production and tight dimensional tolerances, human inspectors have trouble keeping up to production. Studies have shown that even after 2 hours of such work, the human inspector becomes distracted. The same mind that provides for high defect discrimination can "fill in" missing pieces, even when they are not present. After seeing 1000 parts with a hole in the center, part 1001 will appear as though it has a hole, whether it does or not.

Computers and the Internet have provided the tools to deal with large amounts of information very quickly. The same limitation that requires that a task be completely spelled out for a computer ensures that it will find that missing hole in part 1001 as consistently as in part 1,000,001. In addition, the simple act of reporting a dimensional variation, inherent to the philosophy of statistical process control, becomes a quick transfer of data over Internet lines in the digital form needed for statistical process control (SPC) software. So, computer-based inspection and metrology not only affords the programmable flexibility demanded by flexible manufacturing but also provides the quick data collecting and tracking abilities needed for high-speed repetitive operations.

But why use optical-based inspection and gaging? Manufacturing has employed contact probes and gages in regular use since the turn of the twentieth century. Coordinate measurement machines (CMMs) have gone from slow laboratory systems to automated factory floor systems. But even with those improvements, 100% inspection is rarely feasible with CMMs alone. Many fixed gages have now become computerized as well, providing a dedicated part gage, with computer output at the speeds needed. For loading these gages, robotic systems are able to load and unload parts in a highly repeatable manner so well that they have revolutionized the electronics fabrication industry. But this option means a dedicated set of gages for each part, demanding rooms full of gages and billions of dollars in expenses each year. At these costs, the small batch run envisioned as the main tool of flexible manufacturing systems is just not economically feasible.

Even with these computerized advances, the high speed and high tolerances of new parts have pushed past the limits of these more traditional sensors. The combination of speed and resolution has hit mechanical limits that even with new lightweight materials have not yet been overcome. The flexibility of optical metrology methods to check hundreds of points on one part, then a different set of points on the next part, all in a matter of seconds, has provided a capability not before available with traditional fixed gages.

The progression of optical metrology as a tool in manufacturing has not been an overnight phenomenon. Early applications of optical methods as a sorting tool and part ID aid were little more than $100,000 bar code scanners. High speed, low cost, and flexible change over in the fast-moving computer and semiconductor industries have acted as a catalyst to increase the speed of these metrology systems as well as increasing the sophistication from sorters to high-precision measurement tools working in the micron regime.

Early automated optical metrology systems using simple processor chips progressed to dedicated ICs, gate arrays, digital signal processing (DSP) chips, and now integrated Internet devices. The dynamic nature of the electronics and semiconductor market segment has kept these areas as the largest current application of optical metrology, still accounting for over 50% of sales in a multibillion dollar worldwide market just for the area of machine vision today. Even 30 years after the early factory applications of optical metrology, many durable goods manufacturing plants have not embraced the technology. But the competition for tighter quality control may push metrology technology in to even the most conservative metal cutting and forming operations.

For these later types of applications that may involve single part integrity and process control in a machine tool operation, more data at much higher speeds is often desired if the measurement is not to be a time constraint on the manufacturing process. This area of durable goods manufacturing, long dominated by mechanical CMMs using mechanical touch probes, has hit the limit where this dimensional probing is not a significant part of the manufacturing process. The machine tools themselves have become much faster, requiring greater speed of the measurement tools. This higher speed is currently the domain of optical 3D scanners. Optical 3D systems come in a few different types, ranging from fast point scanning of a laser spot to full area coverage that is electronically scanned by video sensors. Speeds in the range of 20,000 to over 1,000,000 points per second are not unreasonable to expect from such systems. Moving a gantry or arm-type structure of a CMM at these types of speeds is just not practical due to simple momentum of the systems.

Speed on optical scanners, however, typically comes at some price of flexibility. In the simplest case, this may just mean that the sensor scans from a particular point of perspective; thus, seeing multiple sides of the part means either movement of the part or movement of (or multiple) sensors. In other cases, the performance of a sensor may be optimized over a very limited range of measurement to simplify the data handling of the potential millions of data points. For example, in looking at the coplanarity of chip leads, it is certainly expected that all the leads will be very close to being in a single plane. There is no need in this case to measure large volumes, since anything beyond this small range is really not of interest to the measurement. In fact, the sensor may only measure select areas of interest and not provide a full volumetric mapping at all. In this respect, such 3D optical sensors are closer to a hard gauge than a flexible one like a CMM.

There have been quite a few other applications that have proven themselves for optical metrology tools in the industry, as well as those that have not panned out.

The measurement of primary metal products using laser-based metrology gages has removed some of the most dangerous measurement tasks from the operator. Measuring bearings for dimensional accuracy has saved this industry millions in failures. Some of the tougher problems such as quantifying small dimensional features, for example, critical edges or surface flaws, are only today starting to see success, requiring a degree of reasoning not as easily provided by a computer. The high speeds of computers today coupled with extreme competition have been pushing optical methods into even these areas. The alternative has proven to be loss of business or even bankruptcy for those who have not made the transition.

So, just what is this field of optical metrology? Optical metrology takes on many forms. The optical part may be a laser, a white light source, and a fast sensor or a camera. In each case, some change in how light reflects or otherwise interacts is used to make the measurement. The basic mechanisms of optical metrology include the following:

1. A change in the amount of light that is reflected or transmitted
2. A change in the direction of the light
3. A change in the nature of light such as phase, coherence, or polarization
4. A change in the distribution of the light returned (such as focus)

These simple mechanisms provide a wide range of optical tools that allow the measurement of a wide range of subjects from liquids to mountains.

Optical metrology tools can be categorized based upon how they are going to be used. There are applications that require the measurement of large subjects, typically requiring measurements in the millimeter range over meters of area. The subject might be buildings, bridges, ships, planes, roads, or any range of larger machinery such as wind turbines today. The optical metrology tools that work in this domain include the following:

- Laser trackers that use interferometry to make point-to-point measurements
- Photogrammetry-based trackers that use multiple views to map targets
- Laser radar systems that use the time of flight of a light beam (using phase)
- Camera-based machine vision systems that measure using 2D images or focus

On a finer scale are most of the world of durable manufactured goods such as car parts, aircraft engines, appliances, or machinery. This realm works in sizes of a few meters down to millimeters with measurements in the tens of microns. The optical metrology tools that work in this domain include the following:

- Laser gages used to measure a point at a time
- Laser line systems that provide profiles along a line
- 3D scanners that collect a volume of data often in mere seconds
- Dedicated machine vision tools that use microscope level imaging

The final realm of measurement is below the range we can experience directly, measuring on the micron and nanometer scale of subjects that are so precise as to be almost invisible to us such as the precision surfaces of optical components and to nanostructures viewable only under high magnification such as surface finishes and biological samples. The optical metrology tools that work in this domain include the following:

- Interferometry ranging from classical optical measurements to white light–based mappers
- Optical microscopy that uses the change in the coherence of light
- Field-based systems that look at how light interferes with edges or structures

In this book, we will look at each of these areas, starting with the basic tools for large parts, all the way to the nanomeasurement area. As an introduction, we will first explore the many terms of metrology as they will be used in this book to establish a basic vocabulary. This will be followed by a comparison of optical metrology to other forms of metrology, most notably mechanical gaging, with specific attention to the limitation and errors associated with each mode of measurement on a very general level. This comparison will be particularly helpful to current industry users of metrology who use the most widely applied mechanical tools today. With this background, the reader will have a starting point on how to evaluate the optical metrology methods that follow so as to understand how to apply and use these technologies on a general level.

The technical discussions that follow the introduction are intended to define the current state of the art of optical metrology for large area, medium-part-sized, and submicron measurements. The objective is to provide the reader with the background to effectively evaluate the application of optical metrology to current problems.

Kevin Harding

Editor

Kevin Harding is a principal scientist at GE Research, Niskayuna, New York. He leads the work in optical metrology at the R&D center and provides guidance to a wide range of optical technology projects, working for the many businesses of GE. Before joining GE, he was director of the Electro-Optics Lab at the Industrial Technology Institute for 14 years and built the electro-optics business, working on over 200 projects and spinning off a number (six) of commercial products. He has been in the technical community for over 25 years and has developed and chaired technical conferences and workshops, including the industry-recognized Optics and Lighting for Machine Vision, which he has taught for 15 years. Harding is internationally recognized for his expertise in 3D measurement technology and has been recognized for his work by many organizations. He has received the Outstanding Young Engineer Award from the Society of Manufacturing Engineers (SME) in 1989, the Engineering Society of Detroit (ESD) Leadership Award in 1990, the Automated Imaging Association Leadership Award in 1994, and the SME Eli Whitney Productivity Award in 1997.

Harding has published over 120 technical papers, taught more than 60 short courses and tutorials to industrial and academic audiences, as well as video course on optical metrology, contributed sections to 6 books, and received over 55 patents. He has also served as association chair, society committee chair, and conference chair for over 20 years, working with the International Society for Optics and Photonics (SPIE), Laser Institute of America (LIA), ESD, SME, and Optical Society of America (OSA). Kevin was the president of the SPIE in 2008 and currently serves as a fellow.

Contributors

Gil Abramovich
GE Global Research
Niskayuna, New York

Vivek G. Badami
Zygo Corporation
Middlefield, Connecticut

Daniel Brown
Creaform
Lévis, Québec, Canada

Peter J. de Groot
Zygo Corporation
Middlefield, Connecticut

Bryan Guenther
Bruker Nano Surfaces Division
Tucson, Arizona

Kevin Harding
GE Global Research
Niskayuna, New York

Qingying Hu
Quest Integrated, Inc
Kent, Washington

Lianhua Jin
Department of Mechatronics
University of Yamanashi
Kofu, Japan

Michał Józwik
Institute of Micromechanics and Photonics
Warsaw University of Technology
Warsaw, Poland

Małgorzata Kujawińska
Institute of Micromechanics and Photonics
Warsaw University of Technology
Warsaw, Poland

Stephen Kyle
University College London
London, United Kingdom

Jean-Francois Larue
Creaform
Grenoble, France

Erik Novak
Bruker Nano Surfaces Division
Tucson, Arizona

Scott Sandwith
New River Kinematics
Williamsburg, Virginia

Joanna Schmit
Bruker Nano Surfaces Division
Tucson, Arizona

H. Philip Stahl
Marshall Space Flight Center
National Aeronautics and Space Administration
Huntsville, Alabama

Adam Styk
Institute of Micromechanics and Photonics
Warsaw University of Technology
Warsaw, Poland

Toru Yoshizawa
Department of Biomedical Engineering
Saitama Medical University
Saitama, Japan

Part I

Optical Metrology: Introduction

1

Optical Metrology Overview

Kevin Harding

CONTENTS

1.1 Introduction

Modern tools of manufacturing add new flexibility to how parts can be made. Multiple axes of motion, multi-pass operations, fine control in some areas, and fast sweeps in others are all means to improve the speed, quality, and flexibility of manufacturing. A key set of tools that is needed to work within this new multidimensional environment is metrology, and this metrology tool set must be up to the task of providing the type of information needed to control manufacturing systems.

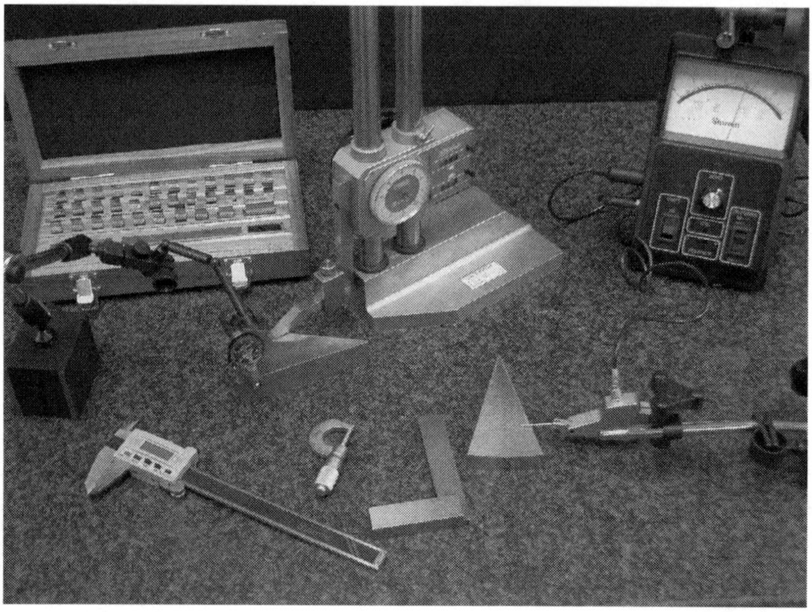

FIGURE 1.1
Mechanical gages are traditionally used for measurements of manufactured parts.

In old times of manufacturing, metrology was often left as a last step in the manufacturing process. The part was designed based upon two-dimensional (2D) views and a fixed set of primitive features such as holes, flat surface, or edges. As each feature was made, there might be a go, no-go check such as using a plug gage to verify if a drilled hole was of the right diameter, but little other in-process measurements were done. When the part was complete, a limited set of key parameters might be checked using micrometers or surface plate tools such as mechanical gages (see Figure 1.1). But ultimately, the check of the correctness of the part was purely functional. Did the part fit where it was supposed to fit, and if not, could we do minor adjustments (without measurement) to make it fit?

For many years, many automotive parts would be sorted into large, medium, and small bins. As a system like an engine was assembled, parts would be tried out. If a cylinder is a little large for the bored hole that was made, try the smaller size. This fitting process was commonplace and accommodated the many manual operations and variability such as tool wear that would lead to small part variations.

This type of metrology began to change with the introduction of more automated processes such as computer numerical controlled (CNC) machining and robotic assembly. With CNC machining, it was possible to make parts in a much more repeatable manner. To accomplish this repeatability, touch probes, probes that determine a part location by touching the part, were added to many CNC machines to check for such things as part setup position and tool offsets due to either the mounting of the cutting tool or wear on the tool.

The touch probe works by using the actual machine tool's electronic scales that are used by the machine to position cutting tools (see Figure 1.2). The probe is loaded into the spindle or tool holder of the machine tool just like any cutting tool. However, in this case, the machine slowly moves the probe toward the part surface until the probe just touches the surface. The probe acts like a switch. As soon as the probe tip is displaced slightly

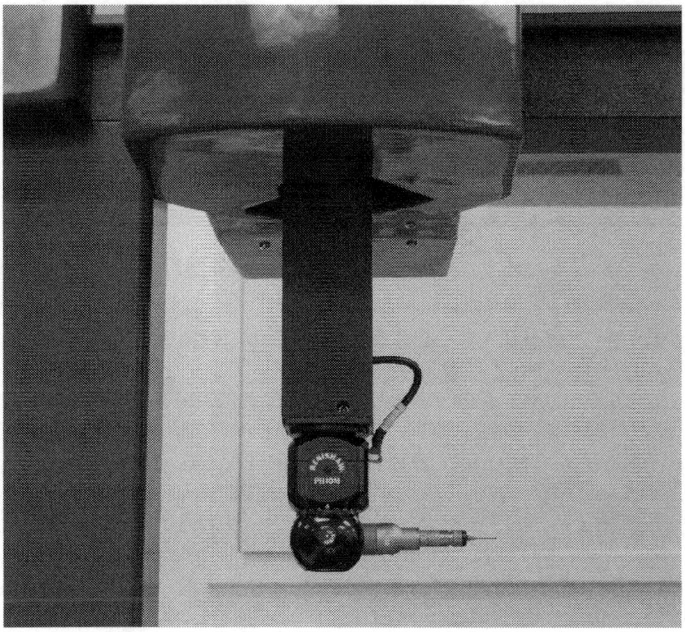

FIGURE 1.2
A touch probe used to set the offsets on a machine tool.

by the touch on the part, it sends a signal telling the machine tool to stop. The machine tool then reads out the position of the touch probe using the built-in position scales needed by the machine to do automated machining.

This type of touch probe check allows the CNC machine to verify the position of a feature on the part, and to use any changes from the ideal location to correct or offset the path, the actual cutting tool will need to take to do the desired machining. This process can be slow. On a high value part, such as a critical part in an aircraft engine, where a small mistake may mean the part cannot be used, costing the manufacturer thousands of dollars, it is not unusual for the CNC machine to spend 10%–20% of the machining time checking features or positions with touch probes.

On a CNC drill, manufacturers have learned to use power monitoring to verify the drill is actually cutting something and may even look for a characteristic signature of how the power to a motor should change during a processing operation. With a modern manufacturing process such as a laser material processing, this type of monitoring based upon force or vibration feedback, resulting from the physical interaction of the tool and the part, may not be possible as there may not be any such physical interaction. Different interactions, not involving contact, may be needed to monitor the process.

With the right information, the flexibility of modern manufacturing can offer many advantages to correct small problems with a part during processing, providing a high-quality part every time. In many cases, even issues of tool wear become irrelevant with modern tools such as laser or electro-discharge machining (EDM). Making sure the process is done right makes possible the opportunity of highly repeatable manufacturing results.

Fortunately, there is a wide range of metrology tools capable of measuring points, lines, or surfaces at speeds thousands of times faster than a touch probe or manual operation that can be integrated into these new energy field manufacturing systems. The rest of this chapter will review these metrology tools, including the pros and cons of each of them. Finally, we will look at how new capabilities being developed today may provide even more options for the future of manufacturing that may provide the means to completely rethink manufacturing methods and strategies.

1.1.1 Sensor Technology Justification

The advent of automated manufacturing processes has placed new demands on the controls to those processes. In the past, the human machine operator was expected to monitor the manufacturing process and insure that the finished product was of high quality. High-quality products have long been associated with the skilled craftsman. Now, after a period of growth in automation that often compromised quality for volume, there is a new emphasis in industry on the production of "quality" product. To be competitive in today's marketplace requires not only that you make your product cheaper but that you must also make it "better" than ever before. The drive toward quality has forced a rethinking of the role of sensors in manufacturing and how the results of those sensors are used.[1–6] The days of the skilled craftsman with the caliper in his back pocket are giving way to untended machines which must perform all of the tasks formerly done by the craftsman that were taken for granted.

Machines may be getting smarter, but they are still a long way from the sophistication of the skilled craftsman. When a person looks out the window and sees a tree, they recognize it as being a tree no matter whether it is a pine or an apple tree, in full bloom or dead. That person has used a variety of sensors and knowledge to recognize the tree. He may have used stereo perception to estimate its size and distinguish it from a painting, he may have heard leaves rustling in the breeze, or he may have caught a whiff of apple blossom. The actual interpretation of these data about the tree has drawn upon many years of experience of seeing other trees, smelling flowers, or listening to noises in the woods. What actually distinguishes the sound that leaves make in the breeze from that of a babbling brook or a slow-moving freight train? These may seem obvious questions to you or me, but a computer has no such experience base to draw upon. The sensory data received by a machine must be of a very succinct nature. The data must be unambiguous in what it means and there needs to be a clear understanding of what the machine must do with that information.

1.2 Understanding the Problem

For a sensor to be effective as a tool for controlling quality, the implementation of the sensor must be right. At first glance, we may say we want to measure the wear of the cutting tool, but is that really what we are interested in measuring, or is it the part surface finish or shape we want to measure? A dimensional measure of a diameter may seem an obvious application for a micrometer but what of the environment and materials handling in the system? Should the micrometer become broken or bent, we will receive incorrect data. The error may be obvious to the operator, but will not be obvious to a deaf, dumb, and blind machine. The right technology must be matched to the task. There are many ways to make a measurement, but only one of them will likely be the best, and even then may not be optimum. Beyond the technology, implementation of sensors requires

- An organizational strategy, incorporating such points as management acceptance and cost justification
- Training of and understanding by the operators who must maintain the equipment
- Interfacing to the environment of the physical plant, users, equipment, etc.
- Some means of using the information provided by the sensor

A sensor without a useful "receptacle" for the sensed data is like a leaky faucet, at best an annoyance and at worst a waste of money.

The purpose of sensing and metrology is to measure some parameters which will help the manufacturing process, either by keeping the machines at their peak through machine monitoring or by verifying the quality of the finished product at each step of operation to minimize the cost of a mistake. It has been said that "any good inspection system should be self-obsoleting." Throwing away the bad parts is at best a stop gap measure in most cases. To insure quality, we would like to improve the process so that is does not make bad parts in the first place! Once we no longer make bad parts, we should no longer need to sort the parts.

1.2.1 Basic Terms for Sensor Technology

The first step in applying sensors is to understand the language. There are many good references that describe these terms in more detail, so only a general review will be given here.[1,2,7]

1.2.1.1 Repeatability

Of primary interest in an automated process is the issue of repeatability. A sensor can have a high precision, that is, output very small numbers or many decimal places, but if the same physical quantity gives rise to a different number each time, the output cannot be used to control the process or insure quality. Repeatability is effectively a measure of how reliable the results are over the long haul. To repeat a number does not insure that it is correct in the eyes of the technical community at large, but at least the number is consistent.

Example 1—Photoelectric Proximity: A typical photoelectric proximity sensor has a repeatability of 0.001 in. This means that if a particular part is brought in proximity of the sensor in a consistent manner over and over, the sensor will produce a particular signal, typically a simple switch closure. The switch will close at the same part position each time. However, if the part is brought to the sensor from a different direction, the sensor switch will likely close at an entirely different part location. The sensor was repeatable, but does not alone tell whether the part is in the correct place.

Example 2—Electrical Scales: Electrical scales are used on many systems to measure linear distances. Such a scale may have a repeatability of 1 μin. but an accuracy of 50–80 μin. (2 μm). At a particular location on the scale, the sensor will produce the same reading very consistently, but the relation between that point and some other point on the scale is only correct, by conventional standards, to the 0.00008 in. In this case, the repeatability alone is not sufficient to provide the information we want.

1.2.1.2 Resolution

An often quoted number as related to measurement is resolution. In the terms of metrology, the resolution is the ability of the system to distinguish two closely spaced measurement points. In simple terms, resolution is the smallest change you can reliably measure. What prevents this measure from being reliable is typically noise. If the signal associated with a small change in the measurement is overshadowed by noise, then sometimes we will measure the change, sometimes we will not measure the change, and sometimes we will measure the noise as being a change in the measurement, and therefore, it will not be repeatable.

Example 1—Photoelectric Proximity: A rating of resolution for a photoelectric proximity sensor might be 0.01 in. but still have a repeatability of 0.001 in. In the case of a proximity sensor, the resolution indicates to how small of a change of part position the "switch closure" of the sensor can be adjusted. This does not mean the sensor will actually measure the change, just as the human eye cannot measure stars in the sky, but the sensor will detect it.

Example 2—Electrical Scales: The resolution of an electrical scale is generally set by the counting mechanism used to read the scale. In this case, one count may be on the order of 20 millionths of an inch (20 μin.), but four counts would be needed to make a reliable reading. Therefore, the resolution is necessarily better than the actual usable measurement obtained from the sensor.

1.2.1.3 Accuracy

The issue of accuracy is an even more difficult one to address. To metrology, accuracy requires that the number be traceable to some primary standard, accepted by the industry and justifiable by the laws of physics. Accuracy is the means to insure that two different sensors provide numbers which relate to each other in a "known" manner. When the supplier makes a part to some dimension and tolerance, the original equipment manufacture (OEM) builder wants to be able to measure that part and get the same results, otherwise the part may not fit mating parts made by other suppliers.

Example—Electrical Scale: The accuracy of the scale was given before as around 0.00008 in. If we have two scales with this accuracy and we measure a common displacement, they should both provide the same reading. In fact, if we compare the reading of the scale for any displacement it can measure, we should be able to compare that number against any other sensor of the same or better accuracy, such as a laser interferometer, and get the same reading. Accuracy provides the only common ground for comparing the measurements across many sensors and from company to company. In comparison, the optical proximity is, for this reason, not accurate at all, but rather just self-consistent.

The need for common numbers is the reason for industry-wide "standards" of measurement. When the woodworker is making that cabinet you ordered for your dining room, he can make the door fit just right and not need to know exactly the size of the door. The woodworker is using the same measures for the door and the opening, even if it is just a piece of cut wood, so it does not matter if his measures do not match anyone else's. He needs resolution, but not accuracy. He is effectively inspecting the part to fit, not to tolerance. When a similar situation occurs between a supplier of car doors and the auto manufacturer such that the doors are made to one measure and the door openings to another, it requires the time and expense of a worker with a "big hammer" to bring the two measures into agreement. A common standard of measure is not being used in this example, so the measurements are not accurately related.

Obviously, if a number is not repeatable or resolvable, it cannot really be proven to be accurate. A popular rule of thumb is to use the "rule of ten" or what I call the "wooden ruler rule." That is, if you need to know that a dimension is good to a certain number, you need to measure it to 10 times better than the resolution of that number to insure that it is accurate. I call this the wooden ruler rule because the number 10 seems to relate more to the resolution of divisions on a wooden ruler or the number of fingers of the metrologist than any statistical significance.

A more statistical rule of measurement is the 40% rule (ala Nyquist sampling) which says that you must sample the number to within 40% to know which way to round it for the final answer. The 40% rule seems to inherently imply a rule of ten in any case, but it can slow down the "runaway specification." When these rules of ten start getting piled on top of each other, a measure can easily become over specified to the extent that you may be measuring to a factor of a hundred times more accuracy than the process can manufacture to in any case, leading to the leaky faucet of information.

As an example of how rules of 10 pile up, consider a part that must be correct to 0.01 in. A tolerance of 0.001 in. would be placed on the part's dimensions to insure a 0.01 accuracy. To insure meeting the 0.001 in., that part is measured to an accuracy of 0.0001 in. In order to assure the 0.0001 in. accuracy, the sensor is required to have a resolution of 0.00001 in. or 10 μin., to measure a part whose dimension is important to 0.01 in., a factor of a thousand times coarser. The actual measurement tolerance, and not the rule of thumb, is what we must keep in mind when specifying the sensor needed.

An interesting question arises when a surface dimension is specified to be measured to an accuracy which is much finer than the surface finish of the surface. Since we want the number to be repeatable by anyone, perhaps we must ask whether we measure the top of the surface finish "hills," the bottom of the surface finish "valleys," or perhaps "which" hill or valley we should measure. It is for this reason that a location relative to a common datum (e.g., 1 in. from the leading edge) should be specified for the measurement for the tolerance to be meaningful.

1.2.1.4 Dynamic Range

The dynamic range relates to the "range of measurements" that can be made by the sensor. There is often also a standoff (the physical dimension from the sensor to the part), which is not part of dynamic range, and a working range which is the high and the low value of the measurements. The working range of a sensor divided by the resolution, the smallest change that can be measured, gives an indication of the dynamic range. If the dynamic range is 4000–1, this implies there are 4000 resolvable elements that can be distinguished by the sensor. If this is now read out as 8 bits of information, which only describes 256 numbers, the significance of the 4000 elements is moot unless only a limited part of the entire range is used at a time.

Example 1—Optical Proximity: The working range of a typical proximity sensor might be 3 in. This means the sensor will detect a part as far away as 3 in., as well as closer. However, once set to a particular detection level, the proximity sensor tells nothing about where the part is within that range. Proximity sensors are inherently on–off devices and as such do not have a range of measurements or dynamic range to speak of but rather only a static standoff range.

Example 2—Electrical Scale: Scales typically produce a measurement along their entire distance of use. So if we have a measurement resolution of 0.00004 in., and a working length of 4 in., the scale would have a dynamic range of 100,000–1. Since with a scale we are concerned with actually accuracy, it may be more relevant to consider the dynamic range with respect to the usable number produced. So, if we have an accuracy of 0.00008 in., and a range of 4 in., the dynamic range is just 50,000–1 (about 16 bits of information). The dynamic range of measurement sensors is typically very important in considering how good of a measure you can obtain over some range. With many modern sensors, this range is in fact limited to the digital data that can be used, so a 16 bit sensor can only describe 64,000 measurements over whatever range you chose to measure. Beyond the basic range, sensors such as scales are often cascaded together to obtain a larger dynamic range.

1.2.1.5 Speed of Measurement versus Bandwidth

A similar question of dynamic range arises with respect to speed of measurement. When we speak of the speed of measurement or the rate at which we can make a measurement, we are referring to the rate at which actual data points can be completely obtained to the extent that they are usable as a measurement of the part. The bandwidth of the sensor is not necessarily the speed at which data can be obtained but relates to the electrical or other operating frequency of the detector.

For example, a 2000 element array of detectors can have a bandwidth of 5 MHz but to get the measurement requires that the 2000 elements be read out, each in 200 ns, sequentially, and before that happens, the detectors may require some integration time to obtain the energy or force they are sensing. The result would be a detector which can be read out every 1 or 2 ms, with a bandwidth of 5 MHz. The bandwidth will generally limit the signal to noise ratio that can be expected. For optical detectors, the response is actually specified for a specific bandwidth and changes as the square root of the bandwidth. If you want to know the speed of the data output, ask for the data output rate and not the bandwidth. Speed can be over specified by looking at the wrong number. Many controllers can only respond on multi-millisecond or multi-second time frames, so using a detector which tells you of impending disaster a millisecond before it happens becomes hindsight in reality.

1.3 Process Control Sensors: Background

Modern production lines are making and moving parts at speeds much faster than any other time in history. The standards of six sigma quality have demanded much better control than ever over even small, cosmetic defects. Industries such as primary metals, automotive, textiles, and even plastic extruders have found that having about the right dimensions and being "functional" just isn't enough. Manufacturers are finding that any appearance of quality problems, be it pits and scratches or a bad overall appearance, can mean rejections of full lots of product, costing millions of dollars to a company, and affecting their bottom line. At modern speeds of production and tight defect tolerances, human inspectors have trouble keeping up to production. Studies have shown that even after 2 h of such work, the human inspector becomes distracted. The same mind that provides for high-defect discrimination can "fill in" missing pieces, even when they are not present.

After seeing 1000 parts with a hole in the center, part 1001 will appear as though it has a hole, whether it does or not.

Computers and the Internet have provided the tools to deal with large amounts of information very quickly. The same limitation that requires that a task be completely spelled out for a computer insures that it will find that missing hole in part 1001 as consistently as in part 1,000,001. In addition, the simple act of reporting a variation, inherent to the philosophy of statistical process control, becomes a quick transfer of data over Internet lines, in the digital form needed for SPC software. So, computer-based inspection and monitoring not only affords the programmable flexibility demanded by flexible modern manufacturing but also provides the quick data collecting and tracking abilities needed for high-speed repetitive operations.

Simple sensors such as touch probes have been used in traditional metal-cutting machines for some years. There are many instances where sparse occasional data are all that is needed, and as such, touch probes are a reasonable tool to use. A touch probe, however, does not provide any measurement itself; it is merely a switch that says "I have touched something." The measurement actually comes from a machine axis, such as on a traditional milling machine. With the advent of energy field manufacturing, the machines often do not have the traditional tool holder and may have a much different type of axis system than is needed to slowly approach and touch a part with a touch probe. So, although touch probes can still be a viable tool, the flexibility and speed of noncontact optical metrology probes will generally be a better fit with the demands of flexible manufacturing methods.

Optical noncontact sensors made for large standoffs include optical systems such as machine vision, laser-based probes, and three dimensional (3D) mapping systems.[3,4,8] We will review the details of these optical-based measurement systems within the context of fast, flexible manufacturing methods, then contrast some of the application challenges and errors with the contact-based systems.

1.3.1 Machine Vision Sensors Overview

Manufacturing has employed contact probes and gages in regular use since the turn of the twentieth century. Coordinate measurement machines (CMMs) have gone from slow, laboratory systems to automated factory floor systems. But even with those improvements, 100% inspection is rarely feasible with CMMs alone. Many fixed gages have now become computerized as well, providing a dedicated part gage, with computer output at the speeds needed. For loading these gages, robotic systems are able to load and unload parts in a highly repeatable manner, so good that they have revolutionized the electronics fabrication industry. But this option means a dedicated set of gages for each part, demanding rooms full of gages and billions of dollars in expenses each year.

At the billion dollar costs of fixed electronic gages, the small batch run envisioned as the main tool of flexible manufacturing systems just is not economically feasible. Even with these computerized advances, the high speed and high tolerances of new parts have pushed past the limits of these more traditional sensors. The flexibility of machine vision to check hundreds of points on one part, then a different set of points on the next part, all in a matter of seconds has provided a capability not before available with traditional fixed gages.

The progression of machine vision as a tool in process control and metrology within the manufacturing process has not been an overnight occurrence.[9] Early applications of machine vision as a sorting tool and part ID aid were little more than a hundred thousand dollar bar code scanners. High-speed, low-cost, and flexible changeover in the fast-moving computer and semiconductor industries has acted as a catalyst to increase the speed of

FIGURE 1.3
The electronics industry has made extensive use of machine vision for part inspection to allow automated processing and assembly. Verifying all the leads are in the right place on a chip and the chip number allows a robot to automatically place it on a PC board.

these machine vision systems. Early machine vision systems using simple processor chips progressed to dedicated integrated circuits (ICs), gate arrays, digital signal processing (DSP) chips, and now integrated internet devices. The dynamic nature of the electronics and semiconductor market segment has kept these areas as the largest current application of machine vision, still accounting for over 50% of sales in a multibillion dollar worldwide machine vision market today (see Figure 1.3). New processors, special lighting and cameras, and advanced algorithms have all greatly improved the capabilities of machine vision.[10] The competition for tighter quality control will push vision technology into even the most conservative metal-cutting and metal-forming operations and when applied to flexible manufacturing may offer a natural marriage of fast, new manufacturing technologies. Machine vision will be discussed in more detail in Chapter 2.

1.4 3D Sensors: Overview

Just as we can now easily scan a 2D image into computer memory, the tools are commercially available to do the same with a 3D part. There are many tools available for digitizing 3D shapes. Some applications may require a very high density of data over very small areas to capture a complex shape. A quarter or dime would be an example of such a part. For other applications, the sizes in question may be quite large, with only minimal variations from one area to another (see examples in Figure 1.4). There are systems available on both ends of this spectrum and at many points in between. Choosing the best tool for a particular job is the challenge to be met by the designer. Many of these systems have been made to address a range of applications from robot guidance to surface structure analysis.[11,12] No single system is likely to ever address all the possible applications for 3D contouring in the near future.

For example, a system capable of describing the work area of a robot doing welding may be looking at an area of a few square meters to a resolution of a few millimeters, while a surface laser treatment system may be looking at a few millimeters to submicron levels. The density of data is not the same for all applications either. If the concern is the presence of high spots on a part that may lead to cracking, then the sensor cannot skip points. In the

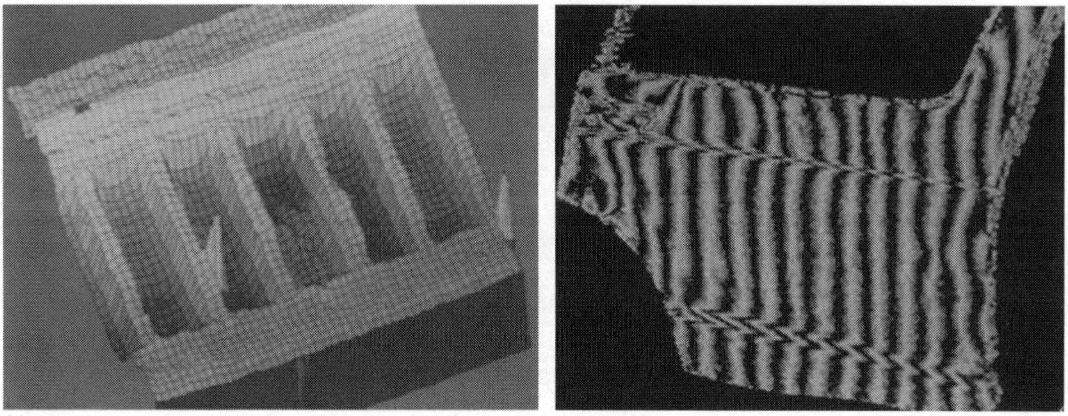

(a) (b)

FIGURE 1.4
For small features, very high density of 3D data may be needed (back of a penny: a), while on larger parts, high resolution may be needed (a car panel: b), but not as much area resolution.

case of many robotic manufacturing applications, only the distance to the part and where one or two edges are located are important. In these latter cases, perhaps just a line or a small array of a dozen points will be sufficient.

One of the early applications of using 3D information was the Consight system developed by General Motors.[13] The purpose of this system was not to measure the 3D shape of the part, but rather to take advantage of the known difference in the 3D shape to sort the parts. The parts in this application were gray metal castings on a gray conveyor belt. These parts were hard to distinguish using only 2D images. The 2D silhouette was not necessarily different, and the features of the gray, cast parts were too low in contrast to pick out of a typical 2D view. In this case, the density of data needed was small. A single white line projected from an angle provided a changing cross section silhouette of the part shape. This information was sufficient for the task of sorting the parts.

In some cases, a sensor made for low data density can be used to build up the data. Scanning a sensor, which measures one point or a line of points, can be used to build up a full, 3D shape. In the case of a complicated shape like an airfoil surface or plastic molding, building up the shape may be a long process one point at a time, suggesting the need for more of a full-field data collection sensor if real-time data are needed to control the shaping process. This does not mean it is necessarily desirable to work with the maximum number of points at all times. A typical video frame has a quarter of a million data points. If there is a depth associated with each data point in such an image, there is, indeed, a large amount of data, more than may be practical to handle in the time available in a production operation.

Because of the variety of applications for 3D sensing, there are a variety of systems available.[11] These sensors can perhaps be broken into a few basic types:

- Point-scanning sensors measure only the specific points of interest, typically in a serial fashion.
- Line sensors provide a single line of points in the form of a cross section of the contour of interest.
- Full-field sensors provide an X, Y, Z map of all the points in the scene, which must then be analyzed down to the information of interest.

Each of these types of sensors has been developed using technology that is suited to the application. In some cases, the technology is capable of multiple modes of operation (finding a point on a surface, or defining the full surface) as well, but this often stretches the technology into a field overlapping other technologies. There has not to date been any single sensor in industrial applications which does everything. The result has been an assortment of sensors finding their best fit to specific applications.

1.4.1 Discussion of 3D Technologies

Before we address the performance of specific sensors, it is useful to establish the basic technologies in use. There are methods that can be used to find the distance to an object.[13–27] A simple version is to focus a beam of light on the object at a given distance. As the object surface moves closer or more distant, the spot on the object surface will enlarge, with the size of the spot being directly proportional to the change in surface height. This method has not seen much industrial use, so will not be further explored at this time. Some of the other methods, such as the scanning and full-field methods, have seen commercial success and have the potential to be used in process control as well as detailed gaging functions.

1.4.2 Point Triangulation

The most popular commercial versions of range finding use the triangulation method where a beam of light is projected onto the object's surface at some angle and the image of this spot or line of light is viewed at some other angle (see Figure 1.5). As the object distance changes, a spot of light on the surface will move along the surface by

$$\text{Change in spot position} = \text{Change in distance}/\big(\tan(\text{incident angle}) + \tan(\text{viewing angle})\big)$$

A wide range of commercial gages exist which can provide a single point of measurement based upon this triangulation principle. To make a discrete point measurement as a process control tool, such a sensor can be directed at the location of interest, with a wide range of possible standoff distances and send data at thousands of points per second in most cases. In order to obtain a contour map, these systems are typically

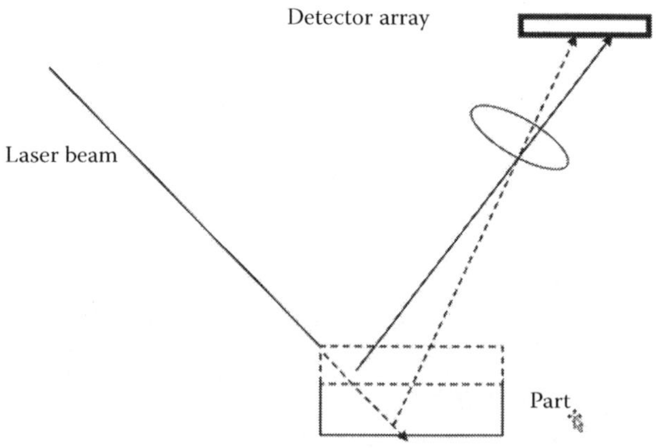

FIGURE 1.5
A triangulation-based system using a point of light to obtain distance.

scanned across the part.[27–29] The scanning has been addressed both by scanning the entire sensor head in a mechanical manner and by using scanning mirrors. Some of the mirror-based systems can collect a full field of data at nearly the rate of data of a video camera. Resolution of a few microns to tens of microns has been realized with point base triangulation sensors.

Most triangulation gages today use laser light. When a laser beam is incident on an opaque, rough surface, the microstructure of the surface can act as though it is made of a range of small mirrors, pointing in numerous directions. These micro-mirrors may reflect the light off in a particular direction or may direct the light along the surface of the part. Depending on how random or directional the pointing of these micro-mirrors may be, the apparent spot seen on the surface will not be a direct representation of the light beam incident on the part surface. The effects of a laser beam reflected off a rough surface include[28]

- Directional reflection due to surface ridges
- Skewing of the apparent light distribution due to highlights
- Expansion of the incident laser spot due to micro surface piping

The result of this type of laser reflection or "speckle" is a noisy signal from some surfaces such as shown in Figure 1.6. Trying to determine the centroid of such a signal will likely lead to some errors in the measurement. In like manner, there can be a problem with translucent surfaces such as plastics or electronics circuit boards. For translucent surfaces, the laser light will scatter through the medium and produce a false return signal. For a laser-based sensor, a smooth, non-mirrorlike, opaque surface produces the best results. Just as a contact probe has problems measuring a soft or delicate part (such as a gasket of a metal foil part), laser probes must be adapted to measure optically unfriendly parts. There have been a number of methods developed for dealing with such parts with laser gages, which are typically based upon restricting the view of the surface to only those areas where the laser beam should be seen and using smart data processing. Restricting the view is perfectly reasonable since the laser probe is only measuring a specific point on the part.

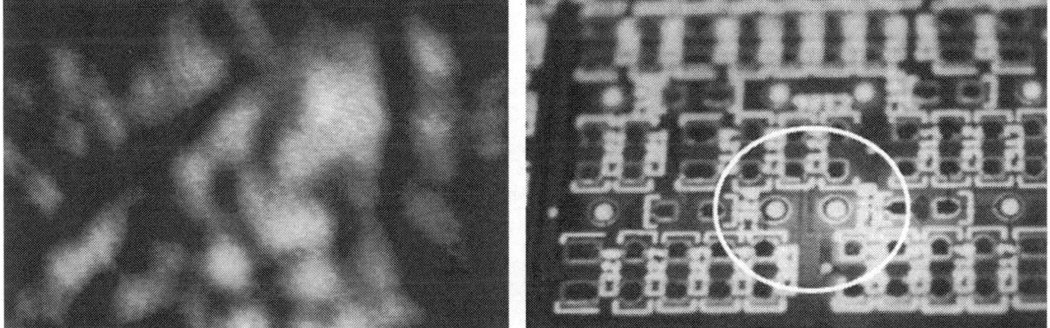

(a) (b)

FIGURE 1.6
Laser light does not always provide a clean spot to use for measurement (a). Scattering surfaces or translucent surfaces (b) can provide an uncertain spot location.

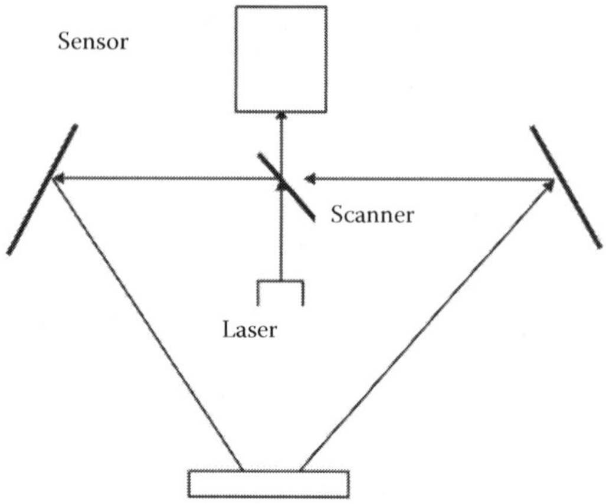

FIGURE 1.7
A synchronized scanning system with a limited range of view.

An active variation of restricting the view uses synchronized scanning.[29] In the synchronized scanning approach (see Figure 1.7), both the laser beam and the viewing point are scanned across the field. In this manner, the detector only looks at where the laser is going. This method does require an active scan but can be made more selective to what view the detector sees. The view cannot be completely restricted with synchronized scanning if an array or a lateral effect photodiode is used.

1.4.3 Line Triangulation

In contrast with a single spot of light, if a line is projected onto the surface by imaging or by scanning a beam, as shown in Figure 1.8, the line will deform as it moves across a contoured surface as each point of the line moves as described earlier.[18–23] The effect is to provide an image of a single profile of the part (see Figure 1.9). In applications requiring only a profile measurement, these techniques offer good speed of measurement. If the full contour is of interest, then the line is scanned over the part, requiring a video frame of data for each profile line of interest.

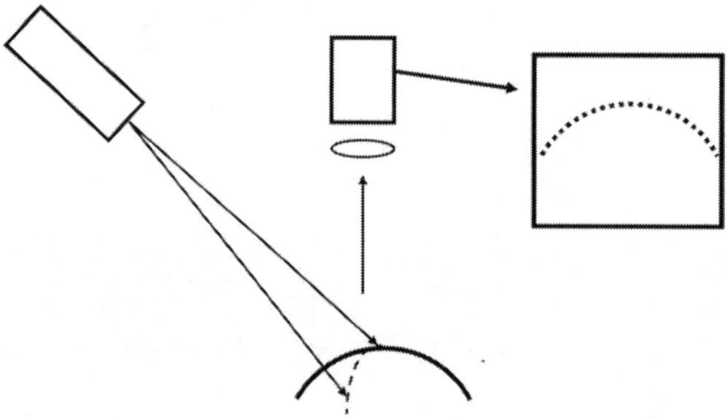

FIGURE 1.8
A line of light-based sensor showing the surface profile.

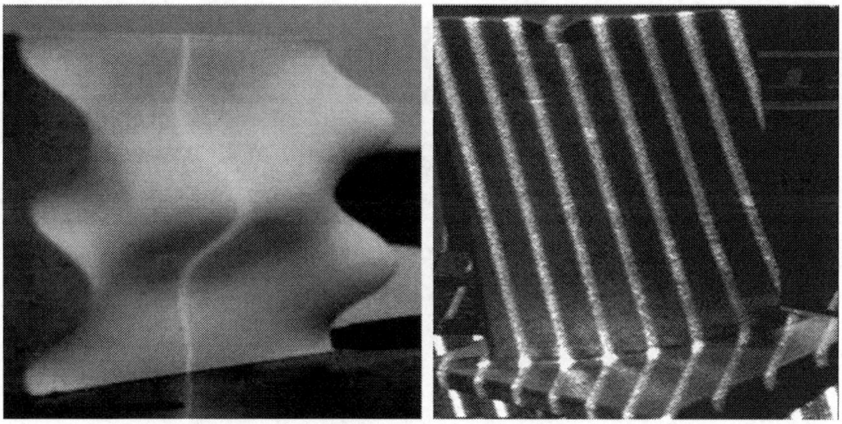

FIGURE 1.9
One or more lines of light on a part provide cross sections of the shape.

1.4.4 Area Triangulation and Moire Contouring

A popular extension in industry for the individual line of light system has been the use of multiple lines or patterns such as reticles, to cover more area at a time.[19,20] These patterns can take the form of encoded dot patterns, distorted grid lines, or simple gratings. Today many of these systems use white-light sources rather than lasers to reduce the noise associated with laser light. The structured light patterns are often analyzed with a technology known as phase shifting, which allows the system to produce an X, Y, Z point at every pixel (picture element) within the image (see Figure 1.10). More on this technology and the analysis will be covered in Chapter 7.

One special case of structured lighting using simple gratings is moire contouring.[11,30,31] In the case of moire contouring, it is not the grating lines that are analyzed directly, but rather the result when the initial grating as seen on the part is beat against a secondary or submaster grating. The resulting beat pattern or moire pattern creates lines of constant height that will delineate the surface the same way that a topographic map delineates the land (see Figure 1.11). This beat effect provides an extra leverage, since the grating line changes do need not to be directly detected and data are available at every point in the field to be captured within a single video image. This leverage can be useful in special applications such as flatness monitoring, as the depth resolution can be made much finer than in the X–Y plane.

The optical system for a moire system is more complicated than that of simple structured light (see Figure 1.12). So, only in some specific applications where very high-depth resolution is needed, such as sheet metal flatness as shown in Figure 1.13, has this technology been used. The other drawbacks of a moire contour include the difficulty in distinguishing a peak from a valley, ambiguity over steps, and the large amount of data generated. With the current commercial systems and computing technology, most of these issues regarding moire and structured light in general have been addressed. The methods of analyzing such patterns have been well established.[32–42] In fact, many commercial structured light systems, which directly analyze a projected grid pattern, use the same type of analysis as is used in interferometry in the optics industry. Interferometry provides nanometer level resolutions, which are typically beyond most applications in manufacturing, so will not be further explored here.

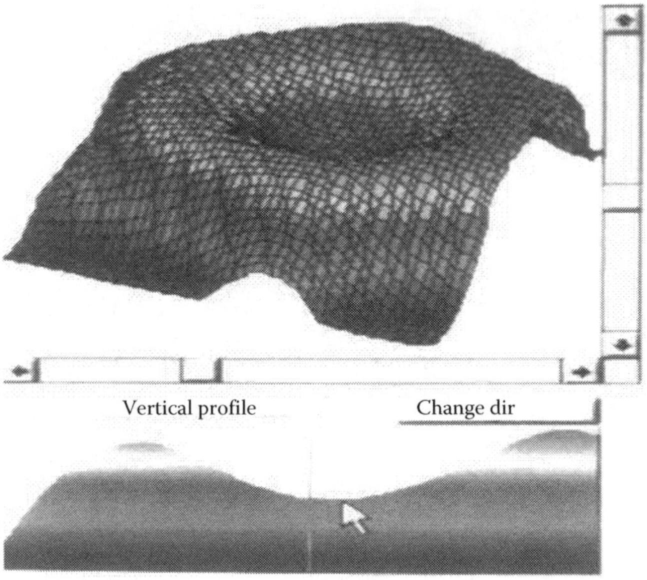

FIGURE 1.10
Three-dimensional data taken with a system using a structure light pattern, generating an X, Y, Z measurement at each picture element in the image.

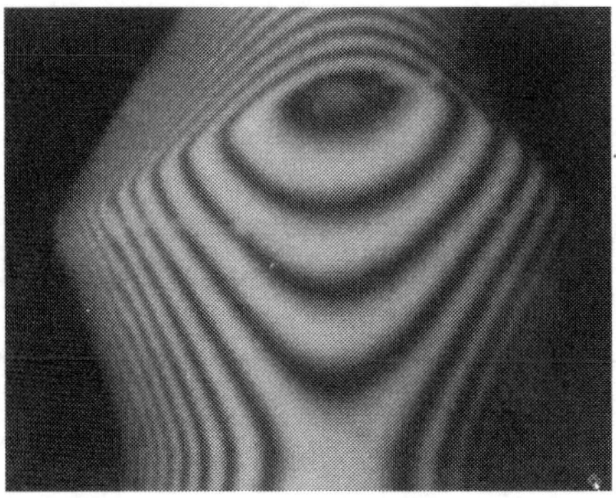

FIGURE 1.11
The moire beat lines delineate the shape of a plastic soap bottle.

1.4.5 Current Applications of Laser Probes

Triangulation-based distance sensors have been around since the time of ancient Egypt. Modern sensors have resolutions approaching a few microns. The most common industrial uses are in semifixed sensing operations where a fixed set or a few fixed sets of points are measured in a fixture. Entire car bodies, engine blocks, or other machined parts are measured by this means. For the purpose of reverse engineering metrology, the flexibility of the "scanning" triangulation sensor offers some attractive capabilities.

The individual laser probes have seen nearly 10-fold improvement in resolution in the past few years. The application of such probes in energy-based manufacturing has been a great benefit as a feedback control in systems where monitoring the force of a mechanical contact may not be possible. Scanning and fixed triangulation systems have been used to contour

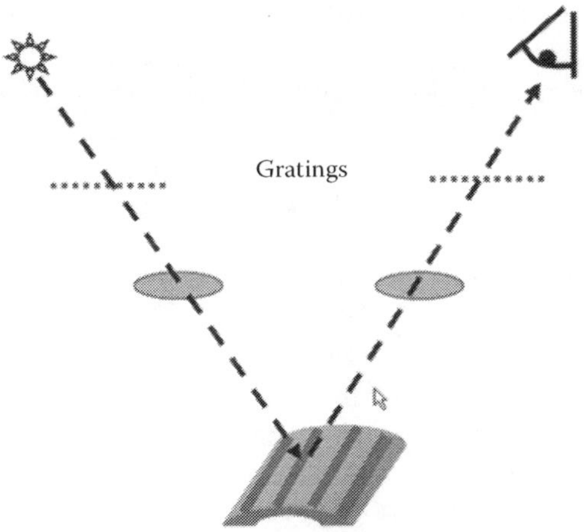

FIGURE 1.12
Moire setup showing two gratings used to create the beat pattern.

FIGURE 1.13
Moire systems have been used for online steel flatness monitoring.

large structures such as airplanes, airfoil shapes, and flatness of rolled metal. The large area systems primarily have used one or multiple lines of light to obtain a cross-sectional profile at a time. In many cases, these line sensors are connected with a machine tool axis of motion to extend the working range of the sensor. The resolutions of such systems need typically be less than a millimeter and more typically are around 2.5 μm (0.0001 in.).

Because of the long time the triangulation-based systems have been around, and the well-ordered nature of the line profile, there has been very good progress in adapting this technology to the needs of energy field manufacturing such as welding. A number of systems are available with direct CAD interface capabilities and would be capable of generating CAM type data as well. There is extensive second source software available that permits the large "clouds" of data to be reduced to CAD type of information for direct comparison to the computer data of the part. Most such comparisons have been largely specialized in nature, but as computer power increases, the user friendliness of such software is increasing.

Scanning triangulation sensors have been used in the manufacturing of small parts, such as precision parts made by laser machining.[43,44] The resolutions for these smaller sensors have been in the micron range, over distances of a few millimeters at a time, with data rates approaching a Megahertz. Dedicated inspection systems which work like full-field coordinate measurement systems for small parts are commercially available, gaining wide use particularly in the electronics industry (see Figure 1.14). The use of these sensors in manufacturing has been a significant tool in the electronic data transfer of dimensional information.

Full-field structured light systems, based upon projected grids by direct sensing of the grid or related to moire are also commercially available. The primary application of this type of sensor has been the contouring of continuously curved, non-prismatic parts such as turbine airfoils, sheet metal, clay models, and similar shaped parts, as shown in Figure 1.15.

Special compact sensors for use on machine tools are also available with this technology. Most of the applications of this technology have been on applications requiring dense data but have also been engineered to enhance video data for the purpose of 3D "comparator" type measurements on objects ranging from rolled sheet metal (for flatness) to car bodies.

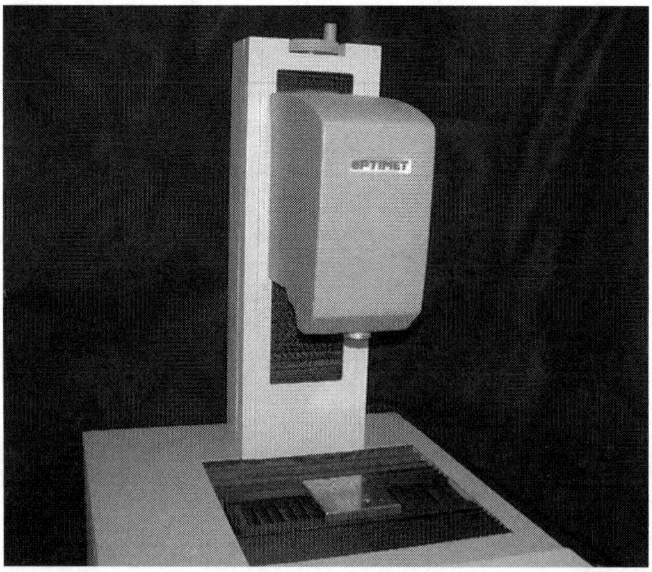

FIGURE 1.14
Point-scanning system with small X–Y table serves as a noncontact coordinate measuring system for parts such as circuit boards.

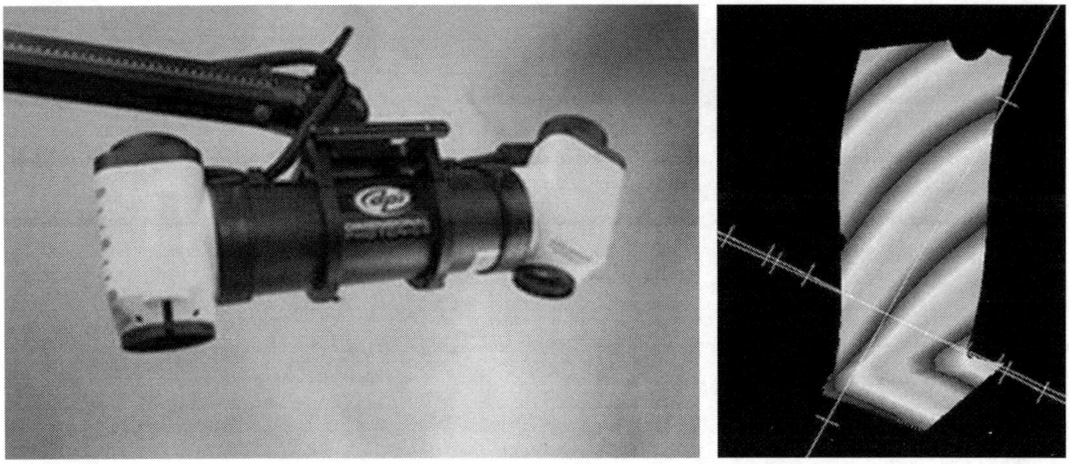

FIGURE 1.15
Structured light 3D systems are available and are used for mapping continuous shapes such as a compressor blade.

Coverage of 2 m^2 areas at a time has permitted very high-speed relative measurements of large structures to several micron resolutions. Typical resolutions of commercial full-field structured light systems are in the range of submillimeter to several microns (down to 0.0001 in.), with data collection taking from one to five frames of video in a few seconds or less. This technology has benefited greatly from the advances in computing power due to the large amounts of data involved (up to a quarter of a million data points in a few seconds). These systems have also been interfaced to provide direct CAD data inputs. Area-based structured light systems offer better speed in applications requiring dense data over complex shapes as opposed to selected regions of a part.

1.5 Application Error Considerations

As discussed previously, noncontact probes may be better suited for energy field manufacturing than touch probes. Touch probes are used in traditional machining for process control. The tool holders and mechanical scales are not adaptable enough to use touch probes in manufacturing system. However, to best understand the errors that may be encountered in any process control, we will examine both contact and noncontact probes.

Both touch-based probes and optical probes have certain errors associated with their operation. The errors tend to be inherent in the nature of the sensor. Each sensor technology has operations it is good at measuring, while with other operations, it has problems. In the case of touch probes, measuring any feature on a sharply curved surface, be it the diameter of a hole or going around a corner edge, requires more points to compensate for how the touch probe makes measurements. In the case of optical probes, the biggest errors tend to come from the edges of parts, either because the probe cannot see past the edge or because the measurement point is larger than the edge. Understanding what these basic errors are with the probes used for control of a process is an important step in correctly applying the technology and getting useful data to control the process.

1.5.1 Contact Probe Errors

In the case of touch probes, the operations are geared toward the types of potential errors specifically found in these sensors. Touch probes used on machine tools have errors associated with the direction of touch.[45–47] These errors fall into two categories. First, since the end of the touch probe is a ball of finite size, the measurement that the machine tool axis provides must be combined with the offset of the radius of the ball and added to the measurement to offset the measure in the direction of the normal of touch to the ball. Of course, knowing precisely what the angle of the normal of touch can be a difficult question. As a sphere, the touch can be in any direction over nearly 360°. Therefore, in operation, additional points are taken around the first touch point to try to establish the local plane of the object. The orientation of the plane of the object is used to determine the direction of offset of the measured values.

Much work has been done to minimize these touch offset errors, both in determining the minimum number of points needed to establish the direction of touch as well as the means to devise durable small point touch probes to reduce this potential of error on high-precision machines. However, as can be seen in Figure 1.16, there remain many error conditions that may still provide an erroneous reading. Features such as corners and small holes necessarily remain a problem for touch probes. A sharp corner location is typically inferred from the intersecting surfaces forming the corner (see Figure 1.17).

The second type of error associated with direction of touch is the so-called lobing errors present in many touch probes (see Figure 1.18). The lobing error is the result of the design and operation of the probe. The probe responds faster to the touch in some directions than in other directions. The result is an additional error that is systematic and consistent with respect to the orientation of the touch probe. Any calibration test must map the response of

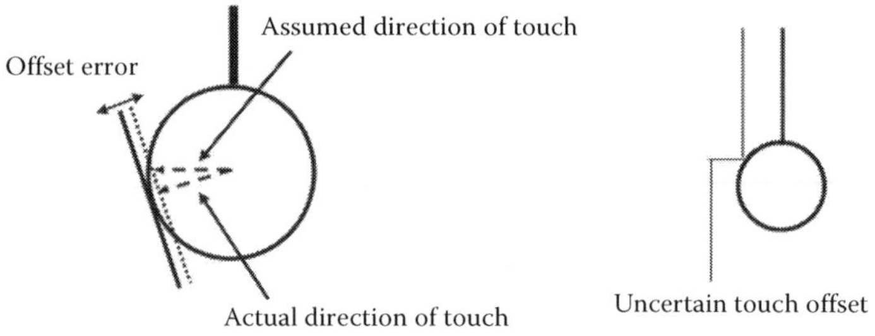

FIGURE 1.16
Errors from touch direction on contact probes due to an uncertainty in the direction of touch on the ball tip.

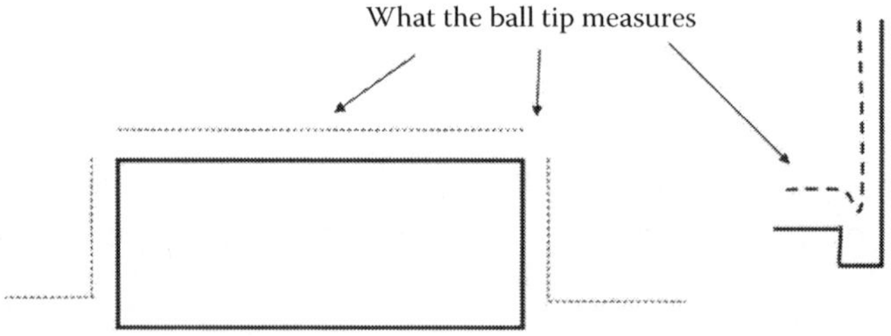

FIGURE 1.17
Actual shapes versus what is measured by a touch probe where the ball can not follow the corner due to touch direction (left) or may not fit into a feature (right).

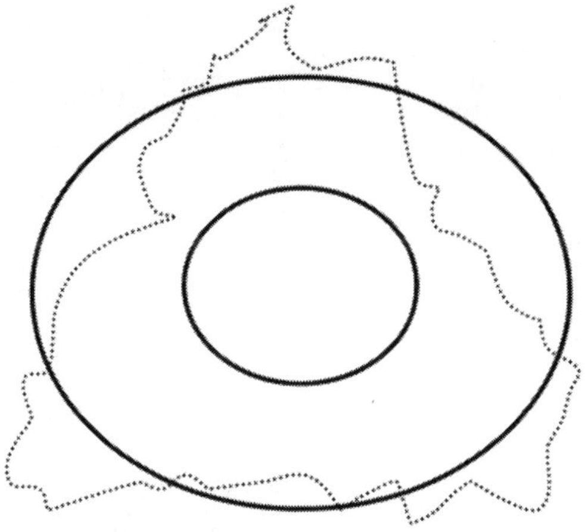

FIGURE 1.18
Typical lobing errors from a touch probe caused by the directional response of the electronics in the probe. These errors are canceled out by probing in multiple directions, such as on a sphere or hole, or by adding a correction factor.

the probe at a full range of angles and approach speeds. Touch probes are typically tested using a sphere of known size. By finding the center of the sphere, the errors associated with lobing and ball touch angle can be corrected.

For most point probes used on a machine tool, the measurement is actually being made using the scales on the machine. A touch probe itself does not really provide any measurement directly; it only acts as a switch to indicate when to take a measurement. There are available analog touch probes that provide some small measurement range directly. If part of the measurement comes from the movement of a machine tool and some from a sensor, the alignment and calibration of one source of measurement to the other is very important to the overall performance of the measurement. In either case, the machine scales are playing a significant role in the measurement of the part. The machine axes themselves are what is often used to do material processing, and as such, any measurement made with them will be self-consistent, whether they are right or wrong.

1.5.2 Machine Axis Errors

The errors associated with the linear axes of the machine include errors in the read out of the stages, as linear errors in X, Y, and Z, as well as the squareness of these three axes. The specific nature of these errors is unique to the machine tool operation. Scale errors tend to be linear, often as a result of the axis not being in line with the assumed direction, but rather at a small angle. The result of a small angle in the axis is referred to as the cosine error effect (see Figure 1.19).

The straightness of the Cartesian motion axis of machine tool can also contribute to the cosine error. However, the motion axis alignment is more a design parameter than something that can be fixed by some user alignment. That is, the axis may actually be slightly bowed or twisted, due to mechanical sagging of the beam carrying the cutting head or tool holder. In addition, the initial straightness of the ways used to build the machine may not be perfect. Because it is the composite performance that is important, touch probe positions on machine tools are usually calibrated using a ball bar. The balls at the end of the bars

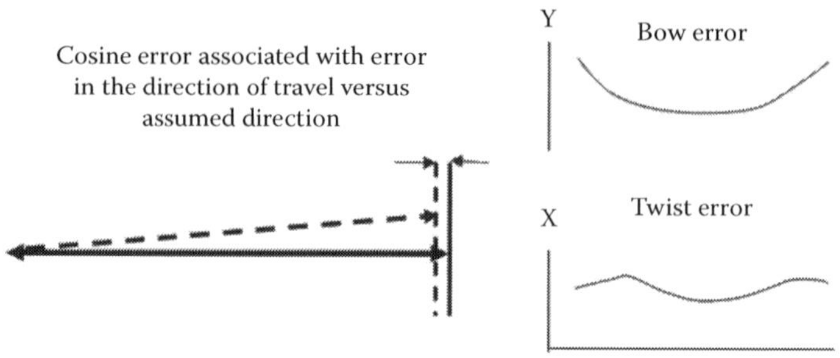

FIGURE 1.19
Cosine errors due to axis alignment errors will cause the measurement to be larger than the actual motion.

permit the errors of the touch probe to be taken into account, while the length and various angle positions of the bar tests out the machine scale accuracies and the squareness of the movements of the machine axis.

1.5.3 Noncontact Probe Errors

In the case of noncontact gaging systems, the potential causes of errors are different, requiring different types of tests to isolate. Unlike the touch probe on a machine tool, whose variations tend to be not in what movements or errors it may poses, but rather the particulars of how it makes these movements. Optical measurement systems are much more varied in the basics of what they do. The variety of 3D optical systems might be classified into three basic areas:[48]

1. Radial scanners measure the distance along a line of sight from some central location, such as laser radar or conoscopic systems. Errors in these systems are related to errors in scan angles.[49] The base coordinate system is typically R theta in this case (Figure 1.20a). A special case of such a scanner would be when the scan center is at infinity. In this case, the scan is telecentric or parallel. There is no angle effect if the scan is parallel, but the linear translation may have a small error.

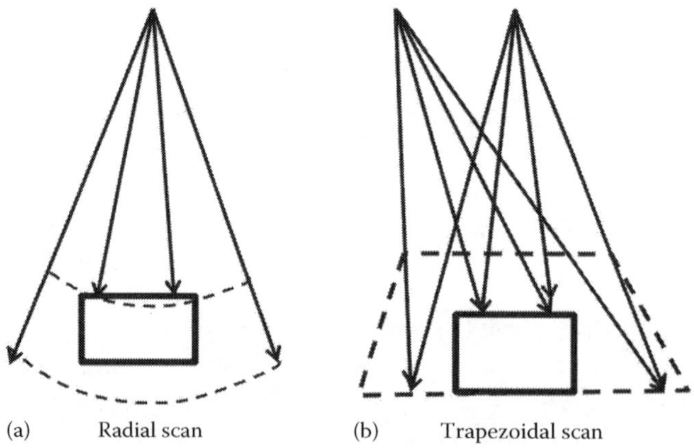

FIGURE 1.20
Coordinate systems formed by different 3D optical scanning methods create either a (a) radial or (b) trapezoid shaped measurement format.

2. Triangulation-based systems, such as point laser probes and structured light probes, rely on obtaining information from two angles of view. Both views can be passive as with a stereo view, or one view can be active in the form of a projected dot or pattern of some sort. The coordinate system with triangulation systems is typically taken as Cartesian but in fact is at best trapezoidal (see Figure 1.20b). The errors associated with triangulation-based systems tend to produce a field that is curved or saddle shaped. The errors in a curved field include magnification effects and the change in the triangulation angle. Both the magnification and angle can change with position across the field and with changes in distance.[50,51] The interaction of the two or more optical systems must be taken into account when addressing the actual calibration.
3. Interferometric-based systems, such as classic interferometry or so-called phase shift structured light systems, make measurements based upon the distance light travels relative to some reference surfaces (real or virtual). In this case, the calibration is tied to the real or effective wavelength being used to measure this difference. Moire contouring is an example that can be analyzed using interferometric analysis based on an effective period of light (typically much longer than the optical wavelength). However, moire is also a triangulation method and therefore subject to the variations and constraints of magnification changes and angle of view.

Clearly when applying a noncontact 3D system to an application currently done by machine tools, the very basic question of what coordinate system is being used must be answered. Machine tools typically are built around three axes all perpendicular to each other. An optical 3D system may have a curved measurement area, one that is trapezoidal or even spherical. Much of this variation in coordinate system is accommodated for in the calibration routines of the sensor. It is not necessarily the case that a spherical coordinate system is incorrect, but typically, parts are specified in square Cartesian coordinates.

In order to apply optical methods, the coordinates are translated from the inherent system coordinates of the optical sensor into the equivalent Cartesian space native to the machining operation. Such transformations always have their errors and approximations. In the case of a trapezoidal or spherical measurement, this may mean reducing the accuracy of the measurement to that obtained in the worst area of the measured volume. This worst area is typically the points furthest or most off-center from the sensor. If machines were initially made with spherical coordinate geometries, then the transition to some types of optical-based measurement tools might be a simpler task. For some manufacturing systems, this might be an option. One type of coordinate system is not necessarily superior to the other; it is just a matter of what is being used.

In applying optical-based measurement systems to on-machine operations, the other primary issue is how optical-based measurements handle edges. We have already described the potential errors that occur when a touch probe goes over an edge and the uncertainty in offsets that can arise depending on the angle of attack to the surface. Optical probes that are based upon finding the center of a laser spot typically have just the opposite problem from a touch probe. As the laser spot goes over the edge (see Figures 1.21 and 1.22), part of the spot is no longer seen by the sensor. The center of the spot actually seen is not in the same location as it would be if the whole spot were visible. The result is a measurement suggesting that there is a raised lip on the edge that is not really there.

Typically, a laser spot in a triangulation sensor is less than 50 μm and perhaps only a few microns in size. Even so, this finite spot size produces an offset error that increases

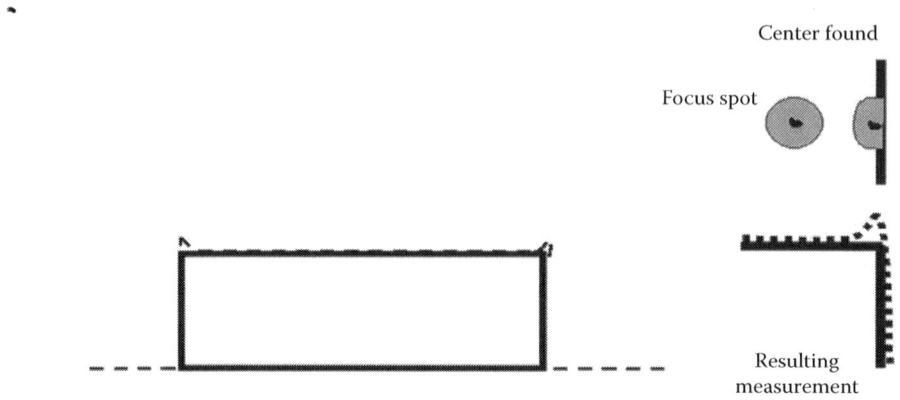

FIGURE 1.21
Edge error associated with many laser probes causes an apparent liftoff at the corner, due to part of the laser spot being lost over the edge.

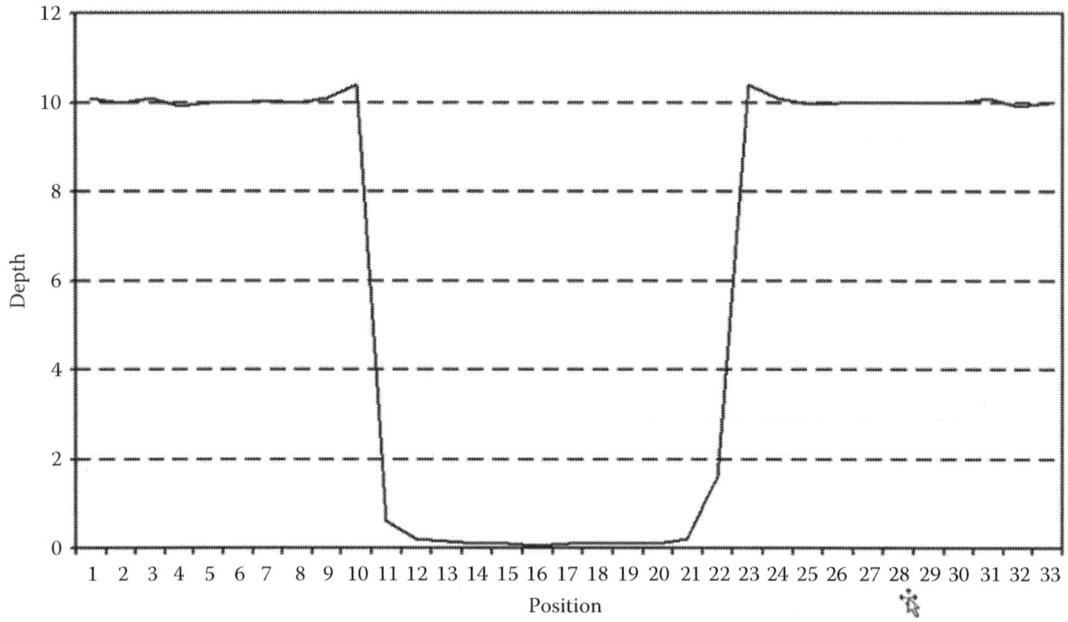

FIGURE 1.22
Graph of laser point sensor performance at an edge showing upturns at the top and round offs of lower inside corners.

as the spot goes across the edge. The actual centroid calculation may depend on the intensity of the spot, the surface finish, the shape of the spot, and the algorithm used to estimate the center. For many optical-based systems (other than interferometric or laser radar), this edge liftoff exists whether there is a real edge or just a transition from a bright to a dark area.

Clearly, 3D systems that rely more on area-based averaging will have more of a problem with how close they can measure an edge before errors start to come into play.

Some methods, such as the interferometric or phase step-based systems that calculate a range at each pixel, can typically measure closer to the edge than a system that uses a spot or a line that may be many pixels wide. Such differences in offset errors and how systems see edges often mean one type of system is seen as superior over the other. A sensor that can measure closer to a physical edge may be judged better than one that can only measure to within a millimeter of the edge. That fact is, just as touch probes can be used around an edge and the offset compensated for in the analysis, the same can be said about the optical probes. The correction for the edge offset is different for optical versus touch probes, but not less predictable for either method of measurement.

For example, in the case shown in Figure 1.21, the laser spot is shown as being a round spot, with the measurement based upon finding the centroid of that spot. Therefore, we can predict that the spot centroid error will change as a quadratic function of the form:

$$\text{delta}(Z) = Z + \frac{P \times X^2}{R^2}$$

where

P is the triangulation factor of Z(X)
R is the spot radius
X is the displacement past the edge

As the laser spot hits the bottom of an edge, some of the spot will highlight the side wall. Depending on the steepness of the wall, this may then lead to the complimentary effect of a rounded bottom corner that follows the form:

$$\text{delta}(Z) = Z - \frac{P \times R^2}{X^2}$$

which is just the inverse as seen on the top of the edge (see Figure 1.21). This basic form agrees fairly well with what is typically seen from experimental data of this type, as shown in Figure 1.22. Once the laser spot center has moved one half of the spot diameter from the wall, then the spot is completely on the bottom, and a correct measurement is available directly. This correction to triangulation sensors assumes that the triangulation angle has not been occluded by the wall, which would block the beam. Occlusions going past an edge are really more of a problem than the liftoff, since there are no data to correct. For this reason, many triangulation sensors that are used for this type of application will view the laser spot from two or more perpendicular directions to avoid occlusion issues.

The point of this discussion is to show that the errors from optical sensors near the edge are both understandable and predictable and can be corrected in the same manner as touch probes accommodating the ball radius. As an additional complication, if the edge causes a bright glint of light, the error in a standard centroid-based triangulation system can be compounded. Some manufacturers monitor the change in light level to recognize such glint conditions, either to reject the data or to attempt to correct that spot.

In like manner, if the side of a step is not steep, then light may be seen from the detector on the side wall, as shown in the left image in Figure 1.23, creating a very elongated spot and again increasing the error. A groove may appear to be two spots, as shown on the right side of Figure 1.23, confusing the interpretation of the centroid.

These reflection problems are a function of the surface finish and the geometry of the edge, so are more difficult to predict. For phase- and frequency-based sensors, a bright

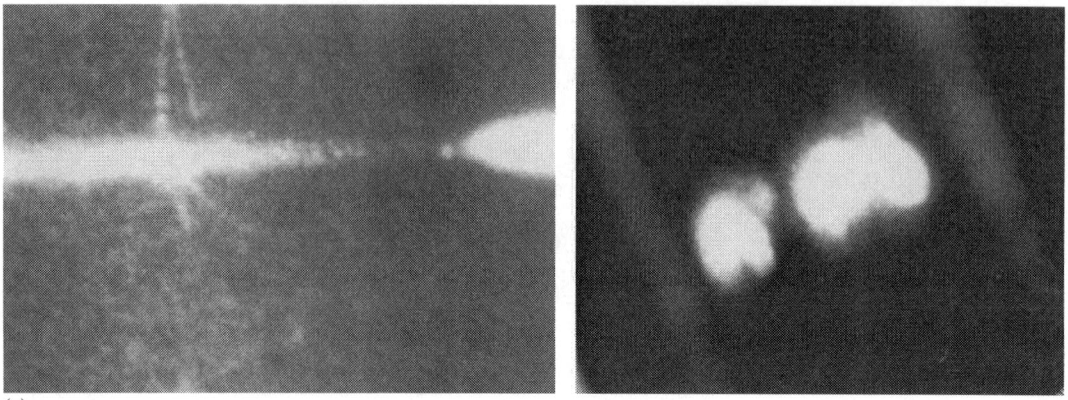

(a) (b)

FIGURE 1.23
Two laser spots showing part of the spot bouncing off a side wall (a) and one being broken into two bright but irregular spots (b) as the laser spot seen in a groove.

glint is typically not a problem assuming the sensor has the dynamic light range to accommodate the extra light. For that matter, if the triangulation sensor can be used with the plane of triangulation along the edge, then relatively little offset would be seen. However, depending on how the spot is sensed, an area-based system can still misinterpret the Y displacement as a change in Z.

1.5.4 3D Probing Errors

In the case of 3D measurements, the issues described earlier can produce a range or errors that are inherently different from the ones encountered in a mechanical measurement system when measuring real parts.[52,53] The issues discussed, such as radial coordinates, edge effects, and even such effects as light source variations and optical aberrations can warp and displace the measurements made with an optical system of any type.[54] Following the example of the calibration done for touch probes and CMMs, there are similar tests that can be done to detail the performance of an optical metrology system. We will discuss some example potential tests that can be used as a starting point. These tests are intended to both highlight the likely errors with optical metrology systems but also exercise the optical systems in a manner appropriate to the preferred means of use. Given the sensitivity of optical system to such a wide range of factors from surface finish and reflectivity to surface angles, no performance test is likely to be completely inclusive of all factors. For any specific application, specific artifact tests may prove to be the best tool to define the performance of the system as the user wants to employ the system.

The 3D optical system equivalent to a touch probe sphere test would be to measure the angle between two flat surfaces, oriented with the edge horizontal and then vertical, or by viewing the apex of a cube or pyramid, using data away from the edges to define the surfaces then calculating the intersection of those surfaces (see Figure 1.24). This measure provides a separable quantification of the in-plane and depth scale accuracy over a local region, which is the purpose of the measurement. Adding in surfaces onto the measurement object at more than one angle allows a determination of the angle sensitivity of the system as well. Measuring beyond the angle where the measurement points are acceptable

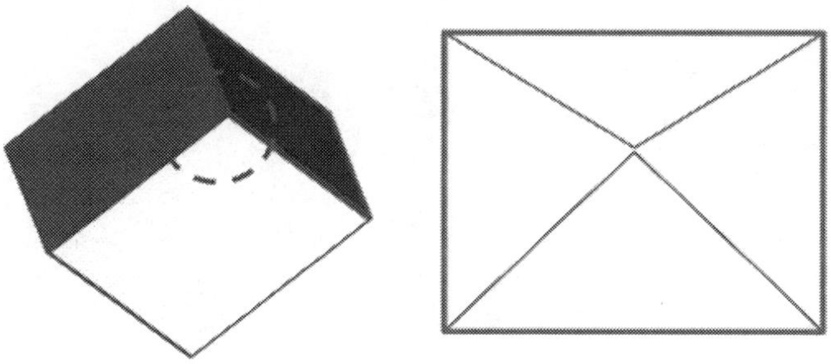

FIGURE 1.24
Fitting planes to the surface of a cube or the surfaces of a four-sided pyramid, then using the intersections to define the apex provide a consistent reference point that uses the best sensor data.

is the equivalent of trying to reach too far around a corner with a touch probe and hitting the shaft of the probe, rather than the tip.

A common test used on CMMs to determine volumetric errors in measuring point to point distances is the use of a ball bar. A ball bar is a long bar with a sphere on each end that can be moved around the measurement volume of a system. The length of the bar should always come out to be the same anywhere within the measurement volume. To do the equivalent ball bar test for a 3D optical system requires measuring from multiple approach directions with the optical system, just as is done with a touch probe to define the sphere. However, rather than compensating for errors relating to direction of touch, the method as described using sphere diameters compounds the long-distance calibration of the 3D sensor with variations due to angle of view, which should be measured separately as described earlier. To avoid combining the angle sensitivity and distance calibrations, the measurement should be made based upon the calculated center of the spheres using the limitation of the data used from the sphere as described earlier, as this separates the measurement of the sphere spacing from a measurement of the sphere diameter.

In order to be sure, sufficient points are used to find a sphere center, typically a minimum of 10,000 points should be used. As stated before, if sphere centers are used, no points should be used that are further around the sphere than 90° minus 30° minus the triangulation angle from normal for triangulation-based systems or the point at which liftoff exceeds the process tolerance should be used in defining the sphere center. If insufficient points are not available over the usable surface, a larger sphere should be used. Alternately, an end artifact, which contains intersection points of flat planes (cube, tetrahedron) whereby an area-based sensor can well define the intersection of three surfaces to define a point, uses the strength of the 3D optical sensor to provide a higher confidence local measure and focuses the test on the long-distance, volume calibration, which is the point of the ball bar test, such as suggested in Figure 1.25. These corner points can then be used to measure the separations of the cubes.

When making this type of measurement with a CMM, the measures are made a point at a time. As such, on a CMM, it takes little more time to use one ball bar and move it around the volume as it would to use multiple ball bars. In the case of 3D optical systems, they have the capability to measure multiple points in parallel. The time spent moving the bar around, particularly if handled by a person, can cause changes due to thermal expansion and drift which can be confounded with the intended measurement of volume accuracy.

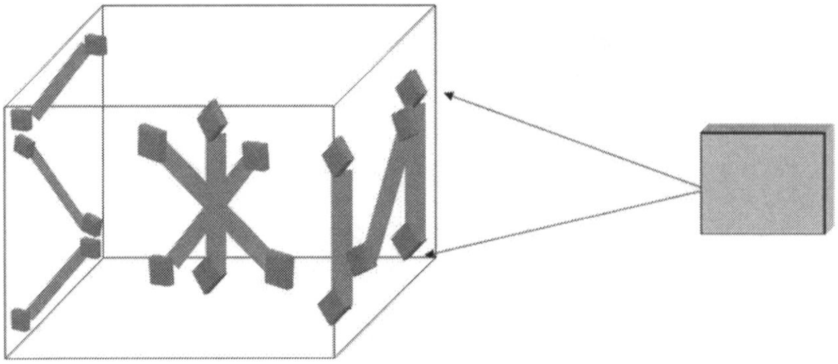

FIGURE 1.25
A bar with a ball or cube (better for optical systems) on each end is positioned to check the volumetric errors of the 3D optical system. Redundant orientations recognize the interdependencies of the position and the axis.

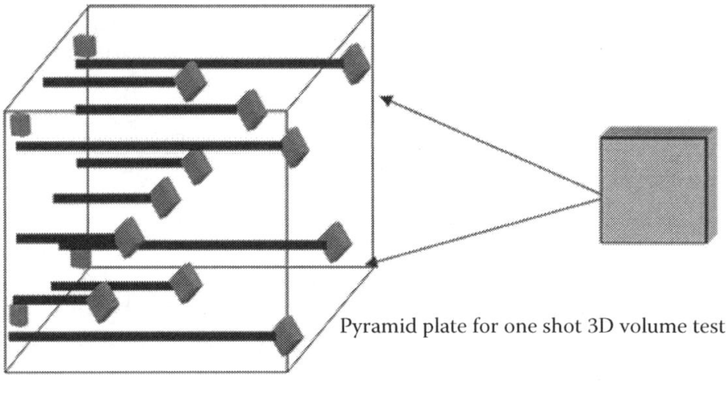

FIGURE 1.26
A ball plate (vertical at left) provides a means to test the volumetric accuracy of a 3D optical system using one data set, without changing angles of view.

A much more commonly used test for 3D optical volumetric systems is to use a ball plate or similar plate with pyramid targets such as shown in Figure 1.26. The objective of the test is to determine spatial accuracy capability, not the ability to measure a sphere in many locations. So, using a fixed plate with all points of interest determined by the intersection of planes uses the optical system to its best capability locally, thereby separating out local noise from volumetric measures (just as the balls do for a CMM) and permits the test to be done efficiently and quickly to avoid any thermal or drift effects.

In any measurement system, the simplest type of measurement is to measure a plane. Ideally, anywhere in the volume of the system, the plane should show as being flat. 3D optical systems can have localized errors such as waves that would not be picked up by a narrow bar. Since one advantage of the optical system is often the ability to take many points quickly, it makes sense to look at a large surface which provides a clear picture of the local as well as global variations over the working field of the system.[55] A normal view of near, middle, and far positions is a good start (vertical lines in Figure 1.27). However, as the performance of optical sensors is a function of the angle of view on a surface, two tilt angles should be used that are different angles tilted in the same plane.[56] Without multiple angle information, significant errors relating to phase approximations (for phase shift–based systems) can be missed.

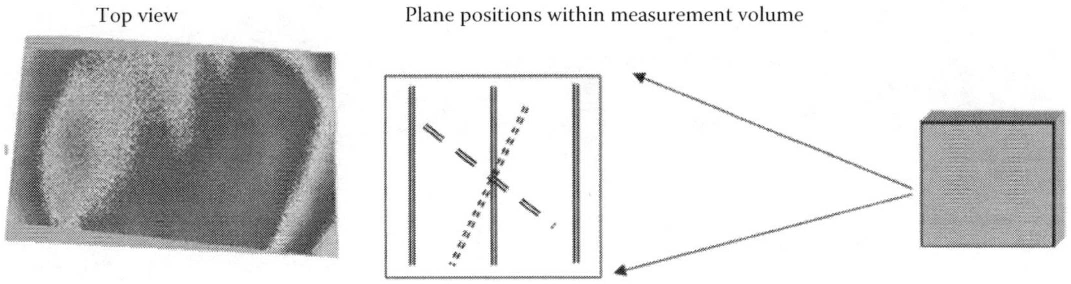

FIGURE 1.27
Positions needed for a test plate to obtain flat plane errors in an optical 3D system.

A diagonal measurement is redundant, as is a vertical angle. When using a larger plate surface, rather than a narrow bar, both the symmetric and asymmetric variations are already measured. These skewed angles are both positions difficult to accomplish and really provide no additional information relative to 3D optical system errors. The types of errors possible with a 3D optical system are not necessarily only spherical in nature and can include saddle points and zonal waves that would tend to be ignored by a spherical (squared) fit, so higher order fitting is in order. Comparison of the measured surface to a plane, considering zonal deviations, can provide greater insights to any errors of on optical metrology system.

The individual tests we have discussed are primarily made using just one view of the part. In reality, a part has many sides and may need to be measured using multiple viewing angles that are then put together. A useful means to consider the overall accuracy and stability of a system is to use some type of golden part or artifact. As we have seen, in an optical system, many errors are interdependent. So, as a test of overall performance, a known part with a shape similar to the part to be tested, with key gage values defined, can be characterized by some other means accepted by the shop. This test can provide local, multisided performance measurement (e.g., repeatability of thickness) appropriate to the final part measurement.

The suggested test procedure could be as follows (see Figure 1.28):

1. Use a reference artifact made to be similar to parts with optically dull surface, on which dense CMM data have been taken, and at least five profiles have been defined with thickness values defined at 20% from the edges of the part and at the maximum thickness location. At least 20,000 points should be used per side of the artifact.
2. Measure the part on both sides at center of volume at best angles and calculate contour and thickness values.
3. Move the part to the top of the volume and repeat tests.
4. Move the part to one side of the volume and repeat tests.
5. Move the part to the rear of the range of the volume at the center and repeat the test.
6. Move the part to the front of the range of the volume at the center and repeat the test.
7. Move the part to one far corner of volume and repeat the test.

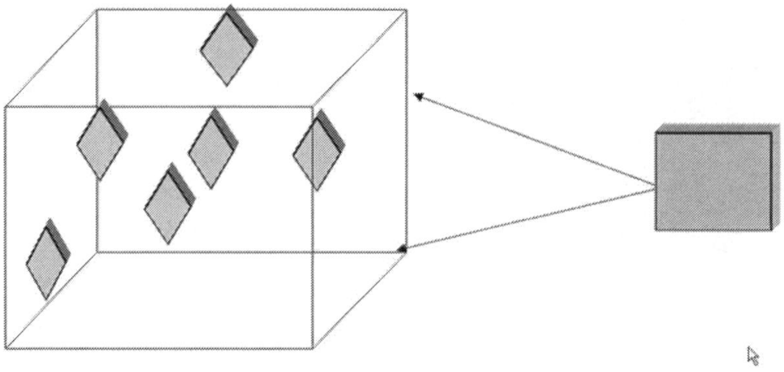

FIGURE 1.28
Positioning of an artifact within the measurement volume used to check overall performance.

Procedures such as described would allow a user to determine the errors inherent within an optical-based measurement system, be it ranging, triangulation, point scanning, or structured pattern based. Such tests can also be used as a means to compare different systems. However, many other considerations must go into a decision on using a particular optical measurement system for use as a production tool. Carefully considering how the part interacts with light, how the total system operates, and how the data from the measurement system are to be used is an important next step in employing these technologies.

1.5.5 Measuring to Datum

When any feature on a part is measured, it is measured at a location and relative to some predefined references or datum points. The datum point may be on the part, on the fixture holding the part, or somehow defined by the shape or fit of the part. Machine-based measurements commonly start a measurement by locating a few key datum features such as planes or holes in the part. These are features a simple point sensor can define with a minimal number of points. For production applications, the part is typically referenced off a fixed position fixture, pre-located on the processing machine tool.

To the extent that the fixture is repeatable and kinematic, the machine axis or other measurement device, such as a hard gage, can expect to make measurements in the correct, prescribed positions. If the fixture is off or the part does not sit right in the fixture, say due to a slight error in the edge geometry of the part, the system will make a good measurement, just not in the right place. Clearly, from the discussion previously, the way in which a measurement can be mispositioned can be very different when using a contact sensor versus a 3D optical gage. In both cases, understanding how to correct these errors is key to making good measurements.

1.6 Summary and the Future

We have discussed a wide range of possible sensors for use with flexible manufacturing operations. The intent of the use of any of these sensors would be to control the process in a situation where there is no traditional contact with the part being machined or formed.

TABLE 1.1

Overview of Primary Sensor Types of On-Machine Monitoring

Sensor	Data Rate (pts/s)	Resolution	Issues
Touch probe	1/s	na	Uses scales to measure
Point optical probe	500–20,000 Hz	1–3 μm	Reflections, edges
Laser line probe	20,000 Hz	10–50 μm	Laser speckle
Machine vision	To 50,000	5–100 μm	Resolution depends on FOV
3D structured light	>1 million (large data sets)	10–50 μm	Resolution depends on FOV

The possible sensors range from point contact probes, currently in wide use on traditional metal-cutting machines, noncontact point laser probes, 2D machine vision camera-based measurement systems, and full-3D mapping systems. A summary of these methods and typical capabilities is shown in Table 1.1.

The right sensor for an application is very dependent on the nature and amount of data needed to provide feedback to the process. To monitor a few key points, a touch or point laser probe can provide sufficient feedback and is in wide use in many industries today as a process control tool.

Laser line probes are typically used in continuous process applications such as extrusions where only a contour section really matters to the process control. One wide use of these sensors is to monitor welds as they are being formed.

Machine vision is widely used as a feature inspection tool, including applications such as aligning and verifying holes made by EDM and laser drilling. The full-field, structured light 3D systems are still new on the market and are primarily being used to verify only the first parts made in production. However, the speed of 3D systems is such that monitoring a fast manufacturing operation is practical.

The processing capabilities of computers will continue to make any of these sensors faster, easier to interface to manufacturing systems, and easier to interpret. The combination of fast 3D sensors with energy field manufacturing has the potential to enable completely automated processes that go from drawing to finished product. The capability exists today to make a 3D copier machine that would work as easily as a 2D document copier. Such a device could completely change the way we do manufacturing in the future.

References

1. C. W. Kennedy and D. E. Andrews, *Inspection and Gaging*, Industrial Press, New York (1977).
2. E. O. Doebelin, *Measurement Systems Application and Design*, McGraw-Hill Book Company, New York (1983).
3. P. Cielo, *Optical Techniques for Industrial Inspection*, Academic Press, Boston, MA (1988).
4. A. R. Luxmore, Ed., *Optical Transducers and Techniques in Engineering Measurement*, Applied Science Publishers, London, U.K. (1983).
5. R. G. Seippel, *Transducers, Sensors, and Detectors*, Prentice-Hall, Reston, VA (1983).
6. R. P. Hunter, *Automated Process Control Systems*, Prentice-Hall, Englewood Cliffs, NJ (1978).
7. L. Walsh, R. Wurster, and R. J. Kimber, Eds., *Quality Management Handbook*, Marcel Dekker, Inc., New York (1986).

8. N. Zuech, *Applying Machine Vision*, John Wiley & Sons, New York (1988).
9. K. G. Harding, The promise and payoff of 2D and 3D machine vision: Where are we today? in *Proceedings of SPIE, Two- and Three-Dimensional Vision Systems for Inspection, Control, and Metrology*, B. G. Batchelor and H. Hugli, Eds., Vol. 5265, pp. 1–15 (2004).
10. K. Harding, Machine vision lighting, in *The Encyclopedia of Optical Engineering*, Marcel Dekker, New York (2000).

Sensor References

11. K. Harding, Overview of non-contact 3D sensing methods, in *The Encyclopedia of Optical Engineering*, Marcel Dekker, New York (2000).
12. E. L. Hall and C. A. McPherson, Three dimensional perception for robot vision, *SPIE Proc.* 442, 117 (1983).
13. M. R. Ward, D. P. Rheaume, and S. W. Holland, Production plant CONSIGHT installations, *SPIE Proc.* 360, 297 (1982).
14. G. J. Agin and P. T. Highnam, Movable light-stripe sensor for obtaining three-dimensional coordinate measurements, *SPIE Proc.* 360, 326 (1983).
15. K. Melchior, U. Ahrens, and M. Rueff, Sensors and flexible production, *SPIE Proc.* 449, 127 (1983).
16. C. G. Morgan, J. S. E. Bromley, P. G. Davey, and A. R. Vidler, Visual guidance techniques for robot arc-welding, *SPIE Proc.* 449, 390 (1983).
17. G. L. Oomen and W. J. P. A. Verbeck, A real-time optical profile sensor for robot arc welding, *SPIE Proc.* 449, 62 (1983).
18. K. Harding and D. Markham, Improved optical design for light stripe gages, *SME Sensor '86*, pp. 26–34, Detroit, MI (1986).
19. B. F. Alexander and K. C. Ng, 3-D shape measurement by active triangulation using an array of coded light stripes, *SPIE Proc.* 850, 199 (1987).
20. M. C. Chiang, J. B. K. Tio, and E. L. Hall, Robot vision using a projection method, *SPIE Proc.* 449, 74 (1983).
21. J. Y. S. Luh and J. A. Klaasen, A real-time 3-D multi-camera vision system, *SPIE Proc.* 449, 400 (1983).
22. G. Hobrough and T. Hobrough, Stereopsis for robots by iterative stereo image matching, *SPIE Proc.* 449, 94 (1983).
23. N. Kerkeni, M. Leroi, and M. Bourton, Image analysis and three-dimensional object recognition, *SPIE Proc.* 449, 426 (1983).
24. C. A. McPhenson, Three-dimensional robot vision, *SPIE Proc.* 449, 116 (1983).
25. J. A. Beraldin, F. Blais, M. Rioux, and J. Domey, Signal processing requirements for a video rate laser range finder based upon the synchronized scanner approach, *SPIE Proc.* 850, 189 (1987).
26. F. Blais, M. Rioux, J. R. Domey, and J. A. Baraldin, Very compact real time 3-D range sensor for mobile robot applications, *SPIE Proc.* 1007, 330 (1988).
27. D. J. Svetkoff, D. K. Rohrer, B. L. Doss, R. W. Kelley, A. A. Jakincius, A high-speed 3-D imager for inspection and measurement of miniature industrial parts, *SME Vision '89 Proceedings*, pp. 95–106, Chicago, IL (1989).
28. K. Harding and D. Svetkoff, 3D Laser measurements on scattering and translucent surfaces, *SPIE Proc.* 2599, 2599 (1995).
29. M. Rioux, Laser range finder based on synchronized scanners, *Appl. Opt.* 23(21), 3827–3836 (1984).
30. K. Harding, Moire interferometry for industrial inspection, *Lasers Appl.* November, 73 (1983).
31. K. Harding, Moire techniques applied to automated inspection of machined parts, *SME Vision '86*, pp. 2–15, Detroit, MI (June 1986).
32. A. J. Boehnlein and K. G. Harding, Adaption of a parallel architecture computer to phase shifted moire interferometry, *SPIE Proc.* 728, 183 (1986).
33. H. E. Cline, A. S. Holik, and W. E. Lorensen, Computer-aided surface reconstruction of interference contours, *Appl. Opt.* 21(24), 4481 (1982).
34. W. W. Macy, Jr., Two-dimensional fringe-pattern analysis, *Appl. Opt.* 22(22), 3898 (1983).

35. L. Mertz, Real-time fringe pattern analysis, *Appl. Opt.* 22(10), 1535 (1983).
36. L. Bieman, K. Harding, and A. Boehnlein, Absolute measurement using field shifted moire, *Proceedings of SPIE, Optics, Illumination and Image Sensing for Machine Vision*, D. Svetkoff, Ed., Boston, MA, Vol. 1614, p. 259 (1991).
37. M. Idesawa, T. Yatagai, and T. Soma, Scanning moire method and automatic measurement of 3-D shapes, *Appl. Opt.* 16(8), 2152 (1977).
38. G. Indebetouw, Profile measurement using projection of running fringes, *Appl. Opt.* 17(18), 2930 (1978).
39. D. T. Moore and B. E. Truax, Phase-locked moire fringe analysis for automated contouring of diffuse surfaces, *Appl. Opt.* 18(1), 91 (1979).
40. R. N. Shagam, Heterodyne interferometric method for profiling recorded moire interferograms, *Opt. Eng.* 19(6), 806 (1980).
41. M. Halioua, R. S. Krishnamurthy, H. Liu, and F. P. Chiang, Projection moire with moving gratings for automated 3-D topography, *Appl. Opt.* 22(6), 850 (1983).
42. K. Harding and L. Bieman, Moire interferometry gives machine vision a third dimensional, *Sensors* October, 24 (1989).
43. K. G. Harding and S.-G. G. Tang, Machine vision method for small feature measurements, *Proceedings of SPIE, Two- and Three-Dimensional Vision Systems for Inspection, Control, and Metrology II*, K. G. Harding, Ed., Vol. 5606, pp. 153–160 (2004).
44. K. G. Harding and K. Goodson, Hybrid, high accuracy structured light profiler, *SPIE Proc.* 728, 132 (1986).

Error References

45. S. D. Phillips, Performance evaluations, in *CMMs & Systems*, J. A. Bosch, Ed., Marcel Dekker, Inc., New York, pp. 137–226, Chapter 7 (1995).
46. S. D. Phillips, B. R. Borchardt, G. W. Caskey, D. Ward, B. S. Faust, and S. Sawyer, A novel CMM interim testing artifact, *CAL LAB* 1(5), 7 (1994); also in *Proceedings of the Measurement Science Conference*, Pasadena, CA (1994).
47. S. D. Phillips and W. T. Estler, Improving kinematic touch trigger probe performance, *Qual. Mag.* April, 72–74 (1999).
48. K. G. Harding, Current state of the art of contouring techniques in manufacturing, *J. Laser Appl.* 2(2–3), 41–48 (1990).
49. I. Moring, H. Ailisto, T. Heikkinen, A. Kilpela, R. Myllya, and M. Pietikainen, Acquisition and processing of range data using a laser scanner-based 3-D vision system, in *Proceedings of SPIE, Optics, Illumination, and Image Sensing for Machine Vision II*, D. J. Svetkoff, Ed., Vol. 850, pp. 174–184 (1987).
50. K. G. Harding, Calibration methods for 3D measurement systems, in *Proceedings of SPIE, Machine Vision and Three-Dimensional Imaging Systems for Inspection and Metrology*, K. G. Harding, Ed., Vol. 4189, p. 239 (2000).
51. K. G. Harding, Sine wave artifact as a means of calibrating structured light systems, in *Proceedings of SPIE, Machine Vision and Three-Dimensional Imaging Systems for Inspection and Metrology*, K. Harding, Ed., Vol. 3835, pp. 192–202 (1999).
52. K. Harding, Optical metrology for aircraft engine blades, *Nat. Photonics—Ind. Perspect.* Vol. 2, pp. 667–669 (2008).
53. K. Harding, Challenges and opportunities for 3D optical metrology: What is needed today from an industry perspective, *SPIE Proc.* 7066, 706603 (2008).
54. K. Harding, Hardware based error compensation in 3D optical metrology systems, *ICIEA Conference, SPIE*, Singapore, pp. 71550–1—715505–9 (2008).
55. Q. Hu, K. G. Harding, and D. Hamilton, Image bias correction in structured light sensor, *SPIE Proc.* 5606, 117–123 (2004).
56. X. Qian and K. G. Harding, A computational approach for optimal sensor setup, *Opt. Eng.* 42(5), 1238–1248 (2003).

2

Machine Vision for Metrology

Kevin Harding and Gil Abramovich

CONTENTS

2.1 Machine Vision Operation Technology

Machine vision can actually be any system where visual information is sensed and analyzed by a computer system for the purpose of determining something about the manufacturing process. Such a system typically consists of a light source to highlight the features, some optics to image, a video camera to sense the scene, a digitizer to move the video into the computer format, and a computer system with analysis software (see diagram in Figure 2.1). By industry consensus, machine vision strictly relates only to the application of this technology to the manufacturing environment, which has been the mainstay of vision technology for the past 25 plus years.[1–6] However, it is very telling of the maturity of the technology that applications in other areas such as medical image analysis, transportation, and security are being seriously pursued today.

In operation, the lighting and optics system creates an image on the video camera of some region of interest (ROI) or field of view (FOV). The image is recorded by the video camera and digitizer to create an array of picture elements or "pixels," each of which denotes one point in the image. The image might then be smoothed by means of averaging or filtering, then segmented to isolate key "features" of interest.

The value of each pixel may be a gray level of light or may be binarized to be either dark or light such as shown in Figure 2.2. The analysis then typically uses simple tools that relate the pixels to each other by identifying edges, counting the number of dark or light pixels in an area, or looking for certain patterns in the arrangement and values of the pixels.

There have been many advances made in dedicated computer systems and user-friendly software over the years. Easy-to-program icon-based systems for computers have also helped to provide much easier-to-use software for machine vision. Some of the new vision algorithms have just been refinements of old ideas on new, faster computer platforms, where others are new ways of doing things. Camera systems have also improved over recent years, fueled by the use of home digital cameras. The combination of camera technology and computers has facilitated a whole new area of machine vision referred to as smart cameras. These smart cameras have onboard processors which can perform a powerful, if limited, set of operations for such defined operations as part ID, location, and simple measurements, then communicate the results over an internet connection. Let us consider where the technology of machine vision is today and where that technology may apply to flexible manufacturing in the future.

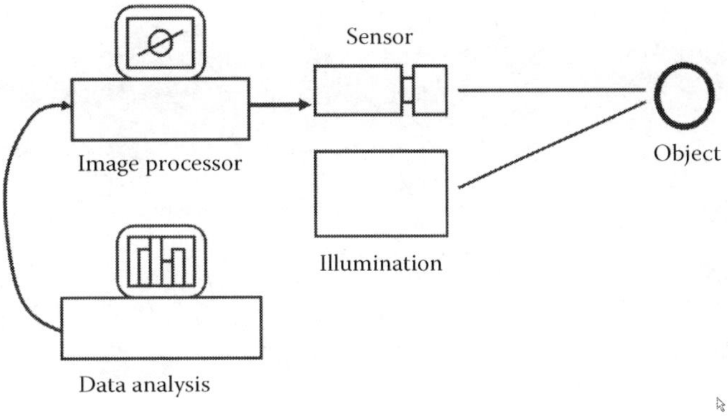

FIGURE 2.1
The components of a machine vision system.

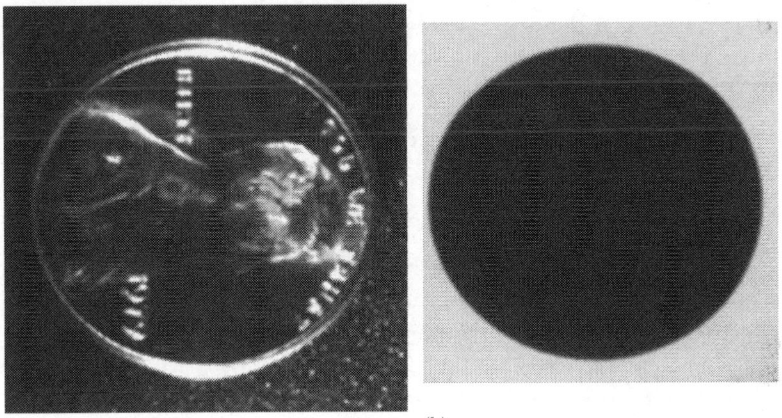

(a) (b)

FIGURE 2.2
A part image (a) which had been binarized, (b) showing only black-and-white pixels.

2.2 Making Machine Vision Work

The area of machine vision has seen quite a few advances in the past 25 years.[7] The first step, and perhaps the least developed today, is getting a good image to analyze. The problem is not that the technology for producing high-quality imaging does not exist; it is more that the machine vision market has not as yet reached the type of commodity volumes, such as home cameras, that make investing in making many products for the machine vision market profitable. However, recent years has seen a few notable exceptions.

The lighting and optics have long been widely recognized as an essential first step to a successful vision application. A clean image can make the difference between an easy-to-analyze scene and an unreliable failure (see Figure 2.3). Parts are not designed like optical elements, with well-defined optical characteristics. They are designed as bearings, pumps, fasteners, and any number of other items, the appearance of which is

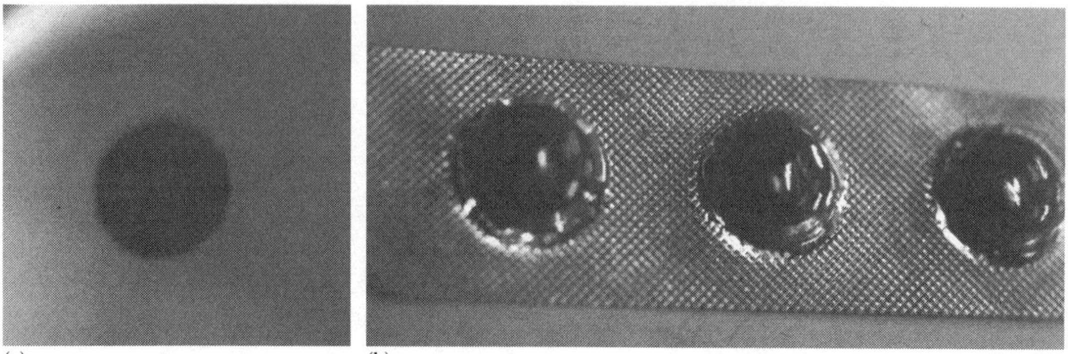
(a) (b)

FIGURE 2.3
An example of how a good optical system can produce a usable image (a) out of one that is not usable (b) to do a measurement.

at best secondary. However, the optical performance of the part has a major impact on the ultimate performance of the machine vision system.

2.3 Machine Vision Lighting

There are many lighting techniques that have been developed both by accident and by design that can be used in a vision system.[8,9] The objective of the lighting is to make the features of interest stand out in the image. Typically, this means the features of interest must be evident in a black-and-white image. Black-and-white cameras still provide the best resolution, cost, and flexibility for machine vision. Even for colored parts, separating colors with specific color filters typically provides better control of the color of a part than what can be realized with a color video camera.

Determining what is required in the image outlines the task and defines the limitation of performance that can be expected from the viewing system. A simple shape identification task may not need the high resolution needed to accurately gage a small feature to small tolerances. As the resolution performance of the lens system is approached, the lens degrades the contrast of the image until the small dimensional changes are washed out of the image. An initially low-contrast image produced by the lighting further degrades the limiting resolution of the viewing system.

There are many other considerations in a machine vision application such as mechanical vibrations, fixturing, and space limitations. However, getting the lighting and optics right goes a long way toward a successful application. To facilitate getting the right image for machine vision, there have been a wide range of tools that have appeared on the market in the past few years. On the lighting side, some of these tools have included (see Figure 2.4)

- Diffuse lighting modules, both on axis and surrounding like a tent that help to decrease the sensitivity to local shiny spots or surface irregularities
- Directional lighting modules, including line illuminators, dark field illuminators, and collimated illuminators that highlight surface textures, some point defects such as scratches, and surface irregularities such as flatness issues

FIGURE 2.4
New LED lighting modules that provide diffuse of directional lighting.

- High-frequency lights to permit asynchronous image capturing
- Highly stabilized lights and high light uniformity, both of which provide for a more repeatable image

Perhaps the most useful advance as can be applied to on-machine monitoring and sensor for energy-based manufacturing is the use of light emitting diodes (LEDs). The improved brightness and longer operation life of LEDs in the past years have made them a good alternative to halogen lamps that put out heat and have limited lifetimes or fluorescent lamps that remained too bulky to fit in a machine environment and expensive to customize. Although not yet powerful enough to be used in many larger FOV applications, and still a more expensive option than incandescent lights, LEDs offer a degree of flexibility of design and a potential ruggedness of manufacture that is needed for on-machine types of applications. Making an odd-shaped line, a spot light, or a surrounding light that can be adjusted by quadrants are all tools now in use in machine vision using LEDs as the

(a) (b)

FIGURE 2.5
Special purpose line (a) or surrounding cylinder lighting (b).

light source (see Figure 2.5). To illuminate a large part with a thousand LEDs may not be desirable. However, as many applications in manufacturing often imply looking on a more detailed scale, the size issue has not been a big limitation to the use of LED. As LEDs make their way into more consumer products such as cars and appliances, this market is likely to both consolidate to stronger companies and likewise grow and become more cost competitive.

Using the new lights made especially for machine vision, it is not difficult to realize a number of basic lighting configurations that will help highlight features of interest while suppressing those features not of interest. The most common forms of lighting for machine vision include

- Diffuse backlighting or directional backlighting
- Diffuse frontlighting or directional frontlighting (typically from the side)
- Structured lighting

Diffuse backlighting is often used when only the silhouette of a part is needed to do a measurement. This might be a simple diameter, general sizing, or cross-sectional measurement. A basic diffuse backlighting arrangement is shown in Figure 2.6. In this simple example, a cylinder shows up as a square box. A flat object, such as an integrated circuit (IC) chip with leads coming out to the side, will be well represented by the diffuse backlit image as shown. A disadvantage of a diffuse backlight is if the part has height, what is seen is the projection of the upper edge, rather than necessarily the "widest" cross section of the part. For obtaining an accurate measurement with a taller part, it may be desirable to use a highly direction backlight as shown in Figure 2.7.

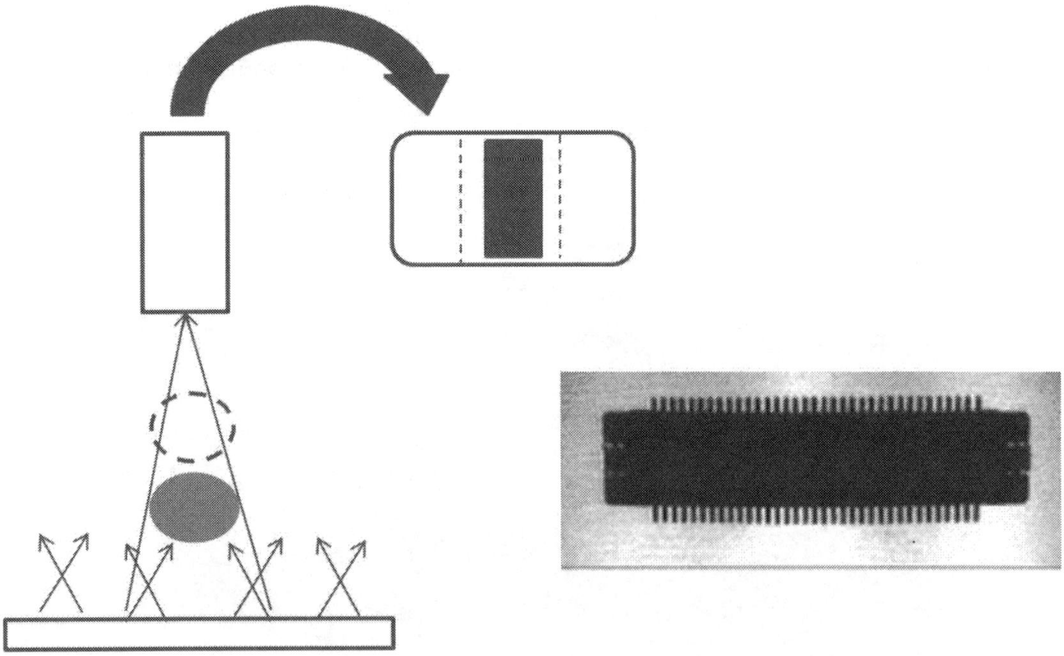

FIGURE 2.6
Diffuse backlight provides an outline image of the part. This method might be good for finding the leads on an IC chip (right). However, if the part moves closer (dashed outline), then the part appears to become larger.

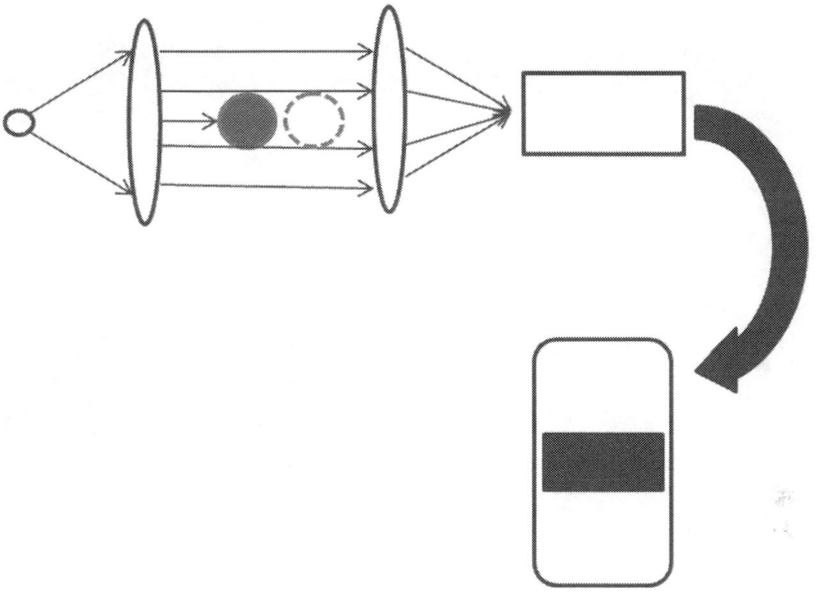

FIGURE 2.7
With a directional backlight, the outline of the part remains constant and represents the maximum width of the part.

The other approach to lighting is to light up the side the camera is viewing or front-lighting. Again, the light can be diffused, coming from many different directions, or it can be very directional. The way to think about such lighting is to consider the light in a room. It is very common to light a room using indirect lighting that bounces off the ceiling or walls. In doing so, the light becomes diffuse and illuminates everything in the room from multiple directions. The result is very few shadows and typically uniform light on everything you see. If the light is coming in the window from the sun, then you see distinct shadows. In machine vision, there are reasons for doing each of these lighting methods. Figure 2.8 shows a diffuse lighting either from a couple of panels or by completely surrounding the part with indirect lighting as discussed. In this case, there is going to be no shadow or bright glints off of any surfaces. So, a part that might otherwise have very bright and dark areas becomes uniformly lit as shown in the picture in Figure 2.8.

Other times, it may be desirable to encourage the shadows to bring out a texture or shape. In Figure 2.8, it is evident that the foil is crumpled in the left image, but not in the right one under diffuse lighting. Figure 2.9 shows a directional lighting system, which is able to highlight even small texture on a part. The lower the angle of the light relative to the surface, the longer the shadows created and more evident small bumps may appear.

Perhaps a special case of directional lighting is to put a structured pattern into the light field and observe how that pattern follows the surface of the part (shown in Figure 2.10). The effect seen is due to what is called triangulation. Effectively, the light pattern encodes the three-dimensional (3D) shape into a pattern that can be seen with a 3D camera. The use of this technology to produce detailed 3D measurements in covered in Chapters 1 and 7, but as a tool to enhance and bring out key features such as the height of curvature of a part is well within the analysis tools of standard machine vision practices.

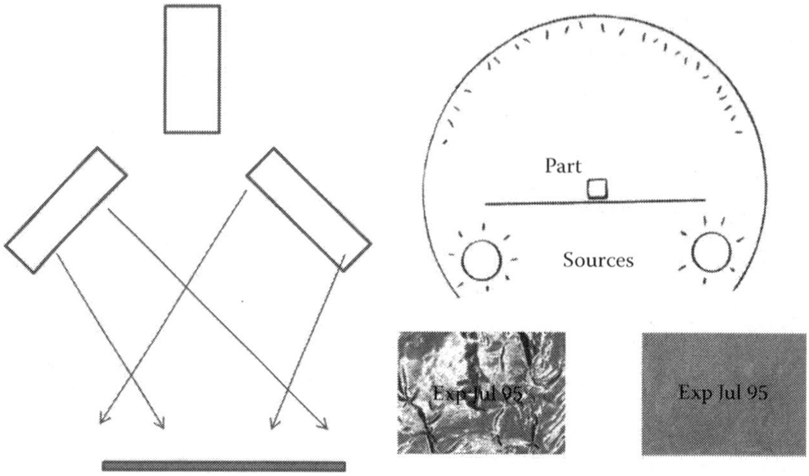

FIGURE 2.8
Diffuse frontlighting can be formed by means of an extended source or a "tent" formed over the part. Lighting from many directions makes the printing on the foil readable by removing all shadows and glints (right image).

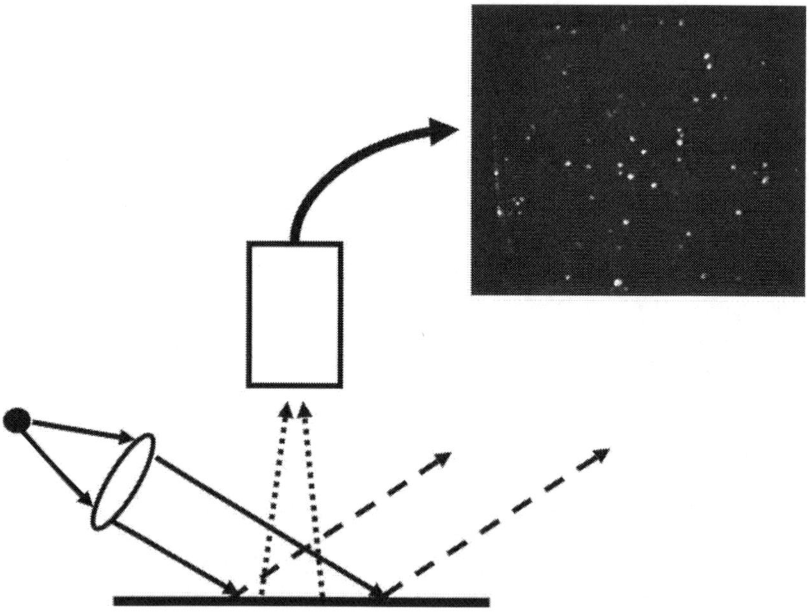

FIGURE 2.9
Directional lighting from a spotlight off to the side highlights the surface texture of a painted surface.

2.4 Machine Vision Optics: Imaging

Before an image ever gets into the machine vision processor or even turns into a video signal, there are a large range of operations performed which serve to form, process, transform, or just relay that image. To extract the ultimate inspection data required from an

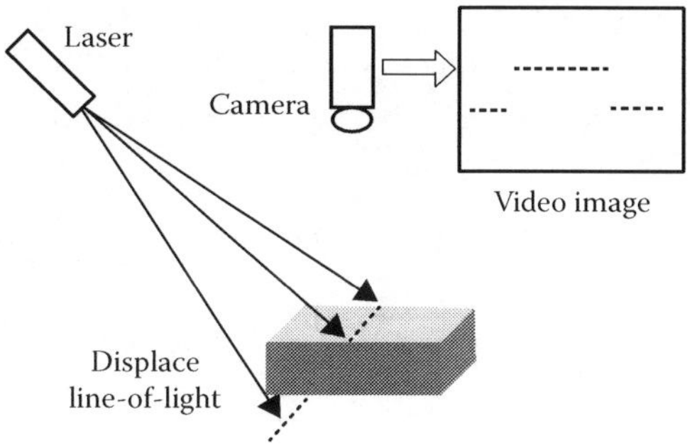

FIGURE 2.10
Structured lighting provides a means to enhance the shape of subjects in the 2D image.

image, that data must first exist in a form usable to the machine vision processor. The "front end" optical components, the lighting and imaging optics, can be used to manipulate the information contained in the light reflected from the part under inspection to optimize the information content and in many cases separate the desired information from the noise.

The components of the machine vision optical front end can be made to work in harmony and must be considered as a whole to obtain the best possible result. These components may include lensing, filters, mirrors and prisms, gratings, apertures, light sources, diffusers, backgrounds, and, not to be overlooked, the part under test itself. Though often seen as only an end in itself, the part should be considered as part of the whole optical train as it can have as much or more influence on the perturbations imposed on the light as any lens or other element in the system.

To properly design the best optical system requires the consideration of the parameters of the light field at each stage in the optical train so as to gain an understanding of the effect each component will have on the final product of this optical system. Parameters to be considered in analyzing any optical system include aberrations, divergence, light-ray directions (direction cosines), polarization, color, spatial frequency content (how shapes affect the light), and losses (a much neglected factor). The effect of each component in the optical train on these parameters will finally accumulate to a total which is greater than any individual perturbation and (according to Murphy) in a manner so as to give the most devastating and typically least expected result.

This discussion will first review some of the details of the optical component specifications which we will need throughout the remainder of the discussion. We will then discuss both the conventional and less conventional components at our disposal for our optical train. As an illustration of the cumulative effects involved, we will then trace through the effects of losses and some of the other parameters in typical example cases.

We will just briefly review the basis of imaging. There are many good books on the subject.[10–13] Figure 2.11 illustrates a few basic terms regarding lenses. The focal point of a lens is the point at which the lens will cause a set of parallel rays (rays coming from infinity) to converge to a point. The plane at which the parallel rays appear to bend from being parallel toward the focal point is called the principal plane, or for a lens in air, this plane may be called the nodal plane. A lens which is other than a theoretical "thin lens" will typically have multiple principal planes. The separation of the principal planes and their positioning can be used to determine the optimum use of a lens (the best object to image distance ratio).

An experimental method for locating approximately where the rear principal plane is located is to tilt the lens around various lines which are perpendicular to the optical axis

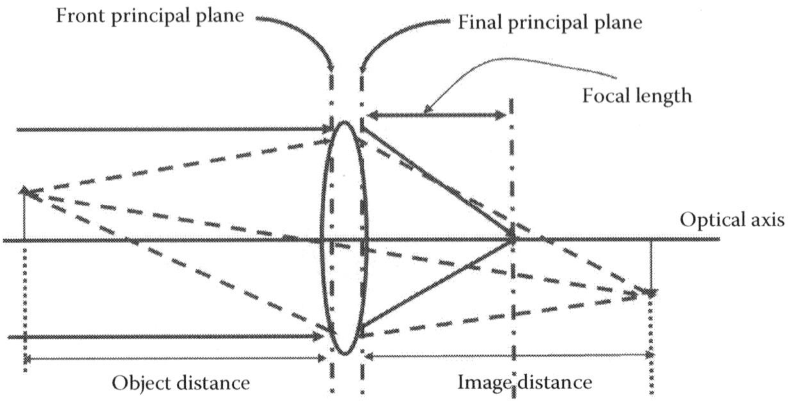

FIGURE 2.11
Parameters of a simple lens.

while viewing the image of a very distant (infinity if practical) object. When the lens is tilted about the principal or nodal plane, the image will not move. Alternately, since the focal length is generally given, a collimated beam can be focused to a point, which will be the focal point of the lens. The distance from the final principal plane, which is not necessarily located at the back of the lens (nor is it necessarily behind the first principal plane), to the focal point is the focal length of the lens. Often a back focal length of a lens is specified, which is the distance from the rear element of the lens to the focal point.

The f-number of a lens is the focal length divided by the diameter of the aperture (the limiting aperture) of the lens. For example, a lens with a 50 mm focal length and a 10 mm aperture diameter will be an f/5 lens. The f-number is useful in determining the relative illuminance at the image. The illuminance is actually a function of the cone of light which converges to a point in the image (mathematically as the square of the sine of the half angle of the cone of light, so we are making a small-angle approximation). This function changes as the area of the aperture, generally indicated on photographic lenses as f-stops for changes of a factor of two in illuminance. Some standard f-stops (often mechanical click stops) on photographic lenses are for f-numbers of 1.4, 2, 2.8, 4, 5.6, 8, 11, 16, and so on. Since the f-number is a function of the inverse of the lens aperture diameter, the illuminance will change as the inverse of the f-number squared. Therefore, there is a change of about a factor of 2 from f/8 (squared is 64) to f/11 (squared is 121 or about 64×2).

The focal length and f-number of any commercial lens is generally given right on the lens barrel. Given this information, the first step to presenting a useable image is to apply the basic imaging equations:

$$\frac{1}{(\text{Lens focal length})} = \frac{1}{(\text{Image distance})} + \frac{1}{(\text{Object distance})}$$

(This assumes a theoretical lens of no thickness but is sufficient for setup calculations.)

$$\text{Lateral magnification} = \frac{\text{Image distance}}{\text{Object distance}}$$

$$\text{Longitudinal magnification} = \text{Lateral magnification squared}$$

The object to image distance ratio is the image conjugates of the system (e.g., a magnification of 1/10 is 10:1 image conjugates).

With these simple equations, it is possible to follow the image through the optical system and determine its location and size. By doing this, one can also determine what the FOV of a given system will be by calculating what the image size of the sensor would be at the subject plane. For example, if we have a sensor which is 10 mm across, and we place a lens with a focal length of 50 mm, 70 mm in front of the sensor, then the subject distance is

$$\left(\frac{1}{50}\right)-\left(\frac{1}{70}\right)=\left(\frac{1}{\text{Object distance}}\right)=\left(\frac{1}{175}\right)$$

and magnification is 70/175 = 0.4, (2.5 demagnification) giving a 25 mm FOV.

If our image distance was only 30 mm, then the required subject distance would be (−)75 or 45 mm behind the sensor. The image of the sensor in this case would be a virtual image (as opposed to a real image) and as such would not be useful to us.

An important note when locating the image plane is that the orientation of the lens affects the orientation of the image in terms of tilt. The easiest way to conceptualize how the image plane is tilted (see Figure 2.12) is to extend the plane that the object lies in and to intersect a plane which goes through the center of the lens (actually this should be the principal plane of the lens) and is perpendicular to the lens optical axis (the lens optical axis is the line which goes through the center of the lens such that it would be undeviated by the lens by being normal to all surfaces). The intersection of these two planes forms a line. If the optical axis of the lens was perpendicular to the subject plane, then this intersection is at infinity. The best image plane will now be within that plane which will contain both the line of intersection just created and the calculated image point, behind the lens, on the lens optical axis. This condition is called the Scheimpflug condition.

For example, let's consider a case where the optical axis of a lens is set at 45° to the average normal to the surface being viewed and the magnification is set to one. The image plane will actually be perpendicular to the subject plane in this case. On the other hand,

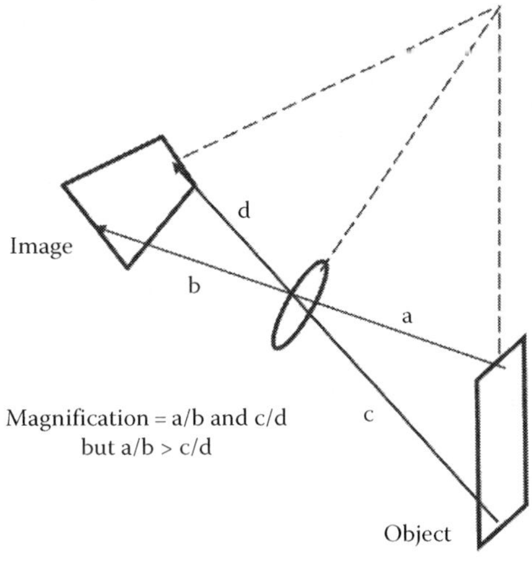

FIGURE 2.12
The effects of a tilted lens create tilt of the image plane and a change in magnification across the image creating a keystone shape.

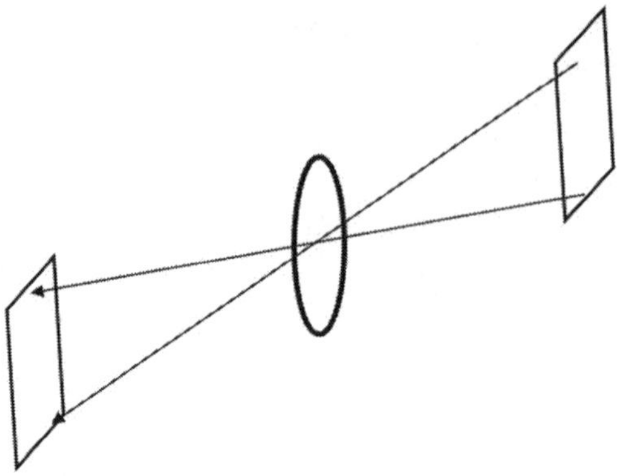

FIGURE 2.13
Placing the lens and image plane parallel to the subject plane, a uniform magnification can be maintained. This is a useful configuration for gaging applications.

if the lens is viewing the surface at normal incidence, then all three planes would intersect at infinity, and the image plane would be parallel to the subject plane.

It is important to note that even if the subject is properly imaged to the sensor by tilting the lens and camera, the image may not be acceptable because of perspective errors. Unless the image plane, lens plane, and subject plane are maintained parallel as in Figure 2.13, there will be a variation in magnification across the field. This magnification error will make a rectangle appear wider at one end than the other, creating the so-called keystone effect as shown in Figure 2.12. If a calibrated measurement or position is to be determined, the keystone effect will change the measurement according to where in the FOV the subject is located. This correction can be a very laborious task which can be usually be corrected optically very easily.

This does not necessarily mean you can't still screw your lens onto the front of the camera, but if the error across the image as to the image distance versus the actual sensor distance is larger than the depth of focus of the system, you will be looking at a blurred image that will change greatly in quality across the scene. This brings up the question of what the depth of focus actually is in the imaging system.

The depth of field is the distance along the optical axis (the center line through the lens) at the subject through which the object can be located and still be properly imaged. The depth of focus is the distance along the optical axis at the sensor through which the image will be in focus. To determine these depths, we must first introduce the concept of circle of confusion. The circle of confusion is the small amount of blur which can be tolerated by the sensor without degradation to the image. In the case of many vision systems, this allowable blur is one pixel size. The depth, at the image, through which this blur will stay within this limit is the f-number of the light (the focal distance of the lens divided by the diameter of the aperture) times the allowable blur circle.

Determining the depth of field in the subject plane is not quite as simple. Referring to Figure 2.14, there are two depths to consider:

$$\text{Far depth of field} = \frac{\text{co}}{(\text{A} - \text{c})}$$

$$\text{Near depth of field} = \frac{\text{co}}{(\text{A} + \text{c})}$$

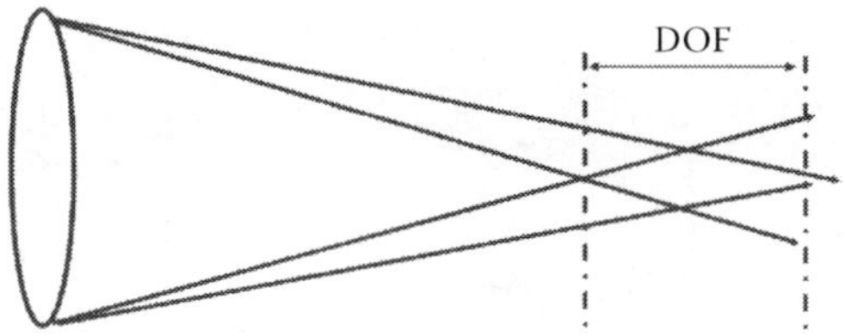

FIGURE 2.14
The depth of field of an image is defined as the region over which the size of a chosen feature does not appear to change.

such that

$$\text{Depth of field} = \text{far depth} + \text{near depth}$$

where
c is the blur circle diameter at subject
o is the object distance
A is the lens aperture diameter

The distance from the image plane to the minimum focus distance (the closest distance from the lens a feature will stay in focus) is the near depth of field. The distance from the image plane to the maximum focus distance (the furthest from the lens a feature will stay in focus) is the far depth of field. The sum of the near and far depth of field is the total useable depth of field of the viewing system for the specified feature size. This means that points on the object, which are within this depth range of distances from the camera, will be in focus.

For a feature size much smaller than the lens aperture, the focus range again reduces to the blur circle times the effective f-number at the subject (the object distance divided by the lens aperture). If a particular size feature is to be resolved on the object, that means the range through which the feature can move is limited by the effective f-number of the system. It is interesting to note that the feature will go out of focus at a different rate on one side of best focus versus the other. In fact, if the feature is as large as the lens aperture, it will remain in focus all the way out to infinity (this does not mean it is resolved, just that the focus doesn't degrade). The depth of field can obviously be increased by stopping down the lens, but this will also decrease the resolution (increasing the allowable circle of confusion) and the light level available at the image.

As a good rule of thumb, the acceptable blur circle at the subject (which will be the smallest resolvable feature as well) will be the magnification (here object distance divided by the image distance) times the pixel size. The depth of field will then be the effective f-number times this blur size or, alternately, the square of the magnification times the depth of focus at the image (longitudinal magnification goes as the square of lateral magnification).

2.4.1 Resolution

Given a basic understanding of the terms and method a lens forms an image, we can now address the question of the performance of a lens. The first parameter in defining the quality of an optical image is the resolution. In optical terms, the resolution in an image is the ability to distinguish two closely spaced points. As this definition is much older than

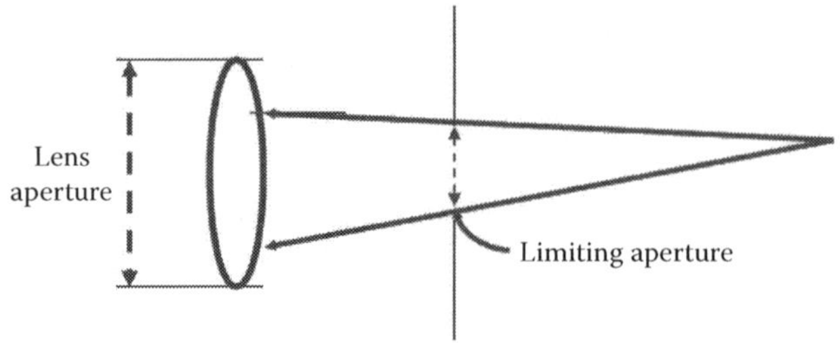

FIGURE 2.15
The smaller aperture in an imaging system limits the resolution of the image collected by a lens.

machine vision, we will stick to it for our discussion. The inherent limiting factor in the resolution of a lens is what is referred to as the diffraction limit. This is the inherent cut off of the frequency of information that is collected by the lens, before consideration of aberrations or other factors that might limit the image quality.

The diffraction limit is given by the following:

$$\text{(Diameter feature which can be resolved is limited to)} = 2.44\ \text{(wavelength of the light)} \times$$
$$\text{(the distance to the focus)} \times \text{(the aperture or size of the beam on the optic)}$$

What this equation means, however, is that any aperture which restricts or transfers the light, unless positioned at an image (distance to focus goes to zero), will limit the ultimate resolution of the system as shown in Figure 2.15. Therefore, to analyze the optical system properly, the light path must be traced through the optical system, adding on aberrations introduced and adding limitations imposed by restricting (actually, "the" limiting) apertures.

For example, if we were looking at a feature at the bottom of a hole which is 10 times deeper than its opening, the diffraction limit imposed by that hole entrance is about 12 μm or half a thousandth of an inch. Now we will reflect off two plate glass mirrors, pass through a 5 mm thick color glass filter and then a plate glass window also about 5 mm thick. Our diffraction-limited performance of 12 μm of theoretical limit has been degraded to about 25+ μm or about one thousandth of an inch, and we haven't even considered the imaging lens yet.

The concern of most of the components mentioned in this example is usually more a concern in considering light losses, but they affect the imaging performance directly as well. In fact, the larger the field angle of the lens (the angle of view from the lens to the subject) and the size or light collection cone of the lens (working f-number), the greater the effect of components in the path. Often a larger lens with a lower f-number is used to collect more light, however, since aberrations like spherical increase as f-number decreases (spherical is a focus blur that decreases as the cube of f-number increase), the extra light gathered to compensation for losses in filters, mirrors, or windows in the field may simply lead to a greater degradation of the image quality or modulation transfer function (MTF) by the lens.

Some lenses, such as doublets, can only be practically designed to cover a few degrees of field angle. When more complex lenses are considered, a field angle of 30°–50° is typical (such as in normal photographic lenses). Lenses with field angles of 80°–100° are wide angle, requiring special designs which often sacrifice MTF for a larger FOV.

Trying to use a lens beyond its designed field angle will result in a degraded and often unusable image.

The earlier definition for resolution does not hold for coherent light, where interference effects come into play. In coherent light, there are unique sets of rays associated with each feature on the subject. In fact, the coherent transfer function can be almost binary, either producing high contrast of a feature or not passing the information at all. The Abbe criterion for coherent resolution as applied to microscopy gives the limiting radius or a feature diameter of about 1/(twice the lens f-number). which can be transferred or formed in the image as

$$Y = 0.61\left(\frac{\text{Wavelength of light}}{\text{Lens numerical aperture}}\right)$$

remembering that the lens numerical aperture is given by the sine of the half angle of light times the index of refraction of the medium, or about 1/(twice the lens f-number). Typically, the direction and transfer of individual cones of light must be considered to analyze the resolution and performance of a coherent optical system. In general, the resolution and contrast for a coherent imaging system will be consistently better than for a comparable incoherent system, but the limiting resolution will be not as good. Theoretically, the coherent transfer function resolution cutoff is at half the spatial frequency of the incoherent case, but in the incoherent case the contrast will steadily decrease with increasing spatial frequency (decreasing feature size) rather than stay high until the cutoff as in the coherent case. In general, the contrast for a coherent imaging system will be consistently better than for a comparable incoherent system, but the limiting resolution will not be as good.

2.5 Optical Specifications

Aberrations, reflection losses, glare, and resolution are all typical measures of the performance of a lens or any other optical component. Not all of these factors are of equal importance to a particular application. For example, in a part identification or sorting operation, unwanted glare or ghost images could make a part appear to be other than it actually is shaped, by superimposing light from another part in the field. However, aberrations such as spherical, coma, or chromatic, which will reduce the definition of features everywhere in the field, would be of little concern when simply identifying what part is under view. If we were trying to measure the part dimensions between two sharp edges on the part, the "blurring aberrations" would greatly complicate our task by discarding the fine detail we want to reference our measurement against. A glare or ghost somewhere in an image to be measured, though undesirable, may not be a problem as it would not be of the same character as the features we are using. Detailed discussions of optical parameters can be found in a number of text and so will not be covered here.[12–14]

2.5.1 Lenses

It is easy to say "I want a perfect lens," but it is very difficult to pay for it. For this reason, we want to carefully consider the constraints of our system to derive a reasonable specification. One of the primary measures of performance of a lens is the MTF. To review, the

MTF is a measure of the contrast of a given spatial frequency content that comes out of the lens versus what went into the lens. In other words, for a given size feature or edge transition, how sharp or clear is that feature in the image. An MTF of 0.5 at 20 lines/mm means that if we have a feature of 1/20 of a millimeter in the image, the contrast of this feature will only be 50%. Remember that with typical demagnifications from say a 12 in. field, this is a feature size of about 2 mm or a tenth of an inch on the part.

Often, only a limiting resolution is provided in the lens spec (the MTF curves are typically available on request). This limiting performance is not at 100% contrast, but is typically set closer to 20% contrast. A contrast of about 20% was chosen about a hundred years ago by Lord Rayleigh as the lowest contrast at which two closely spaced points can be distinguished. A contrast of 20% is not typically acceptable for machine vision systems (they have poor imagination). A lens specification of twice the lines per millimeter limiting MTF than what you ultimately need is typically a minimum safety factor. If the curves are available, a contrast of 80% (MTF of 0.8) for the resolution required is a good benchmark to use. A contrast of 100% is not realistic for any imaging system, where a contrast of 50%–60% is reasonable to expect from any simple imaging optic, and may often be acceptable.

When leaning toward the lower MTF, the MTF of other components in the system should be considered as well. As one would expect, the fidelity of the image will usually always be degraded by each optic's MTF (multiplicative in contrast but additive in aberrations). There are some cases where one lens's aberrations may tend to cancel out the aberrations of another. This cancelation can happen with spherical aberration and is actually the tool used to reduce chromatic aberrations. An example of such a case where the spherical aberrations can be made to cancel is in image relay systems. Two plano-convex lenses oriented so that the curved faces face each other, with appropriate spacing, can yield near-zero spherical aberration, even though alone they exhibit quite a lot. In fact, such an arrangement of doublet lenses is often the basis of more sophisticated imaging lenses.

In the optics area, manufacturers have also found it financially beneficial to address the machine vision needs, primarily on two fronts. The first area is just the availability of new high-resolution lenses with better corrected fields as needed for high-accuracy inspections in a machine environment. In the earlier days of machine vision, it was often argued that the "software" could "correct" the image. However, these corrections are something that takes up processing time and computing power and in many cases are only partially successful. Now, several companies have introduced lines of small, highly corrected, C-mount lenses that reduce aberrations such as field curvature and distortion. Other companies have made adapters for existing lenses such as the wide selection of enlarger lenses on the market today. Enlarger lenses are typically made for good geometric reproduction and a flat, uniform image plane, both of which are good characteristics for using the image to gage features on a part. Optical errors in standard security-type lenses made for camera C-mounts are often up to 10% of the field in security camera lenses. New machine vision lenses make measurements at thousandths of an inch level both meaningful and cost effective with a vision system.

The other advance in machine vision as a metrology tool has been the availability of telecentric optics. Telecentric lenses provide a means first to present a uniform intensity image onto the camera if the subject is uniformly lit. These lenses provide uniform collection by using a telecentric stop to provide a uniform cone angle of collection of light as shown in Figure 2.16. This may not sound like much, but in fact most common lenses vignette the light that is not in the middle of the image field, causing the light at the edges to be reduced. Just looking at an image, most people might not even see this effect as people are very tolerant to light level changes. However, given a vision system that is thresholding

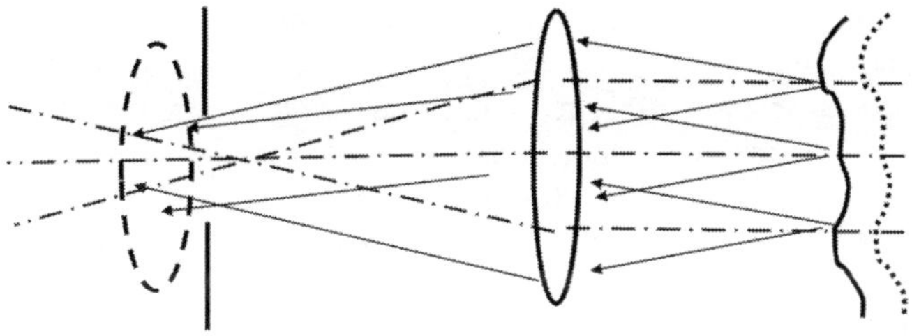

FIGURE 2.16
A telecentric lens with a fixed stop (left) to provide uniform lighting.

the light at some level in order to do the inspection, this light level variation is at least a nuisance. Even with adaptive thresholding, the limited dynamic range of cameras and the processing time taken for such filtering are both things that would be nice to not use up on something like light uniformity, given that the light sources themselves have become much better at presenting a uniformly lit field.

The second benefit telecentric optics provides (the ones telecentric in object space) is to produce an image that does not change in magnification for small shifts in the object distance (shown in Figure 2.17). This means the system can be much more tolerant to shifts in the position of the part under inspection, both within the field and in distance, without the character of the image changing, either in terms of perspective view or magnification. This means a measurement made by a machine vision system can be tolerant to changes in distance to the part, without losing measurement accuracy. Telecentric optics does have its

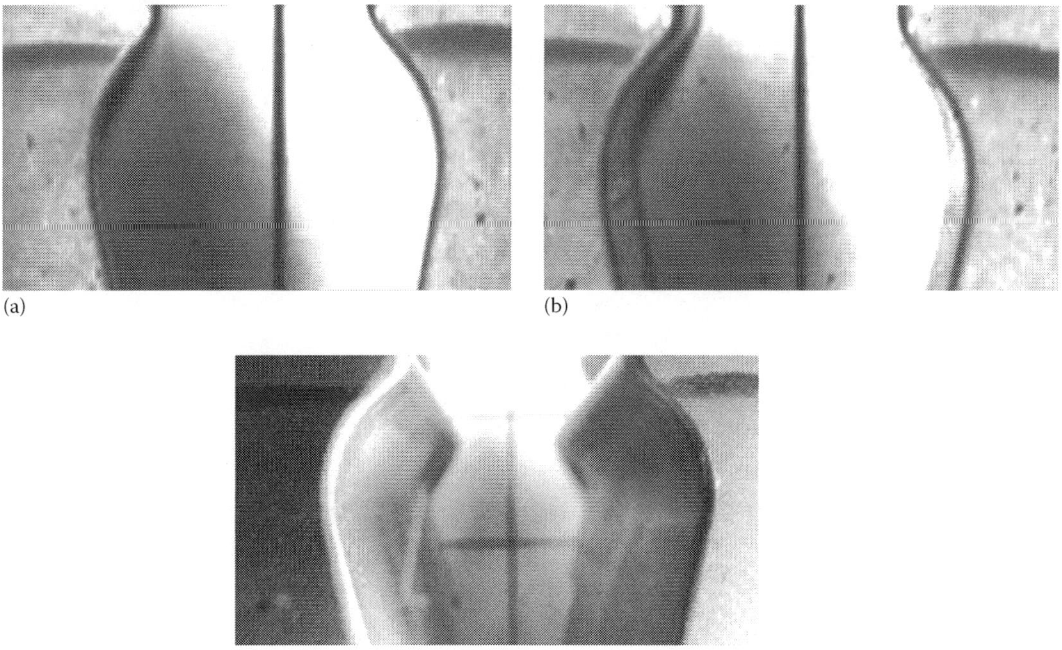

(a) (b) (c)

FIGURE 2.17
A telecentric view (a) looking into a gap versus a normal lens view (b) or a wide angle lens view (c) where the sides of the gap can be seen, with different magnification from top to bottom of the gap.

limitations. To be completely effective, there is some loss of light due to larger f-numbers, and the lens system does need to be larger than the part under inspection. But with the great majority of precision metrology applications being the inspection of small part areas or features, telecentric optics has become a commonly used tool today.

2.5.2 Windows

More than only the lensing must be considered in the ultimate MTF of a system. Other components such as the camera sensor performance, any filters, mirrors or prisms, and even the light pattern put out by the source can have an influence on the MTF of the imaging system. Passing through any glass will introduce about 5 μm or 0.0002 in. of spherical aberrations for each centimeter of glass the light passes through (this assumes roughly an f/5 cone of light for the purists). This occurs because the light going through the glass at steep angles will be deviated or translated relative to their original path, causing them to ultimately focus further from the lens than light rays going through the glass near normally (see Figure 2.18).

An irregular piece of glass can introduce irregularities and degrade the image severely. A window made of regular plate glass can often present this problem. For most applications, a window made of a float glass will introduce only a few waves (and half as many microns) of aberrations into the image. For critical measurement or high-resolution cases, windows are available, for a price (say $70 to a few hundred dollars for a few inch diameter), which are rated for transmissions of 1/4 wavelength of aberration to the light or less. These optical quality windows will typically have a minimal effect on the image, though even these will exhibit the spherical aberrations described earlier (not due to quality, but to physics).

2.5.3 Mirrors

A mirror can introduce aberrations or irregularities into an image, even though it is considered to be a flat mirror. A typical second-surface mirror (like a bathroom mirror) of thickness of 6 mm actually presents a glass path length for the light at 45° of about 22 mm, yielding about 12 μm or 0.0005 in. of just spherical (there is astigmatism, distortion, and chromatic as well of course). This means the limiting resolution you started with

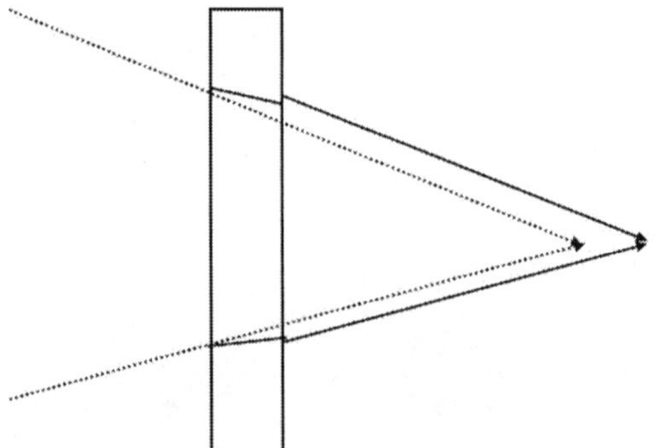

FIGURE 2.18
Focus displacement due to the thickness of a glass window.

is degraded by 12 μm of spherical aberrations (plus about 20 μm of others) just by reflecting off a mirror. A first-surface, plate glass mirror presents less problems than a second-surface mirror, typically introducing about half this value (about 6 μm or 0.0003 in.) in aberration, due mostly to surface figure irregularities, degrading the contrast (the MTF) of the image by a few percent per reflection. Float glass goes down in effects introduced by about another half (2–3 μm). A ground and polished mirror specified as flat peak to valley by 1/4 wave is considered diffraction limited, meaning it will not introduce any additional aberrations to an image beyond the degradation associated with the physics of light reflecting off a physical surface.

When specifying windows or mirrors, as well as single lenses, it should be noted that there is a mounting area assumed around the edge of the optic. By this convention, optical specifications refer only to about the central 80% of the physical aperture of the optic. Making an optic to work out to the edges is actually very difficult, just because of the processes used to make optical components.

2.5.4 Surface Quality

An often overlooked parameter of lenses as well as mirrors, windows, and prisms is the surface quality. An industrial optical grade of surface finish would be a scratch/dig specification of 80/50. Scratch and dig is specified on optical components because the polishing process does not leave very much in terms of surface roughness. The actual RMS surface roughness is typically under a microinch (actually 100–200 Å, there being 250 Å to a microinch). The scratches and digs are the leftover scratches and digs on the surface which began much bigger and were not completely polished out of the surface. The first number in the specification is the "apparent width" of a scratch on the surface. The apparent width of a scratch on the surface of an optic can be affected by the sharpness of the edges of the scratch, unpolished area around the scratch, and even the power of the surface. The second number, under like conditions, refers to the permitted diameter of a dig. These parameters are not dependent on the surface figure accuracy, which is how accurately the surface shape matches a flat, a particular sphere, or some other conic section. Surface accuracy leads to aberrations, while surface quality factors such as scratch and dig will introduce scattered light and hence noise into the system.

Along with scratch and dig, there are possible inclusions such as bubbles, striate (actually local variations in the index of refraction of the material), or even physical inclusions such as dirt in the glass material. Striate will appear as local streaks where the image will appear to "wiggle" or deform. These local glass irregularities are actually common in window glass and other glass items where large batches are made "in a hurry." An industrial quality-rated glass can have a variety of these noise-contributing factors. A bad surface may scatter 1%–2% of the light. The scattering begins to degrade the contrast of the image by making the image appear as though it is being viewed through a dirty window or other diffuser (which it is). Noise from a series of mirrors, widows, and lenses may reduce the overall contrast of an image by a few percent.

The typical specifications for surface and glass quality fall into three major categories with variations depending on the vendor.

Industrial quality or class III—scratch/dig 80/50, may have comparable size bubbles and striate present. Cost example, a 2 in. optical window would cost a few dollars (mostly due to the cost of cutting the glass). The scatter can be 1%–2%.

Grade A optical quality or class II—scratch/dig 60/40 to 40/20—glass essentially free of bubbles and inclusions may have some striate. Cost example, expect to pay $20–$50 dollars for the 2 in. window, before antireflection (AR) coating. Most standard lenses are of this quality. Scatter is typically less than 0.1%.

Research, laser, Schlieren grade or class I will have a scratch/dig of 40/20 (actually should be 20/10) to 10/5, and be free of inclusions and striate. (This is the really good stuff, but you pay for it.) Cost example, the 2 in. window indicated earlier in glass would run about $150–$300, with a good surface figure, before coating. Scatter from these is so low that it is often hard to tell where the surface is located, the scatter being less than about 0.001% typically.

The surface quality particularly affects the image if it is a window or even a filter which may be close to the image plane. Noise factors such as scatter can also be introduced by coatings put on the surface to produce mirrors, reduce reflections, or filter the light passing through the optics. A good-quality surface with a bad coating is still a bad surface. Therefore, next we will consider what coatings can do for and to the performance of the optical system.

2.6 Filters

The topic of optical properties such as polarization and color is beyond the scope of this discussion but is well covered in other text.[15–18] Filters that created polarized light are a valuable tool in generating the image desired for effective machine vision application. Although not always thought of as such, filters are devices which sort the properties of light, to direct different properties in a different manner. We have already talked about the use of a variety of filters, so now I will try to present categories for the types of filters available. Filters can take the form of

- Thin films using the interference of standing waves described before; these are called interference filters.
- Absorption by bulk materials, color glass filters, as well as plastic polarizers fall into this category.
- Selective blocking or reflecting of light of a certain character, in which the polarizing prisms described fall, as well as glare or selective angle blocks (which we will discuss in more detail).

Thin film interference filters separate light of different properties simply by acting on some of the light, while not acting on the rest. One type of component that does this is the beam splitter used to split or combine different beams of light as introduced earlier. In some cases, the ability of the coatings to separate polarizations is used to advantage to create polarizing beam splitters. These devices will reflect one polarization while transmitting the other. In this manner, one can observe specular glints in one view and view only diffuse reflections, glint-free, in another.

In some cases, a polarizing beam splitter can be used for high light efficiency, such as to illuminate with one polarization and view the other reflected off the same beam splitter. A quarter wave plate will change the linearly polarized light into circularly polarized light. After specularly reflecting off the part, the light goes back through the quarter wave plate which now turns it back into linearly polarized light, but now rotated in polarization by 90° (i.e., the horizontal or "p" polarized light turns into vertical or "s" polarized light). The return beam will now go to the opposite leg of the beam splitter that it came from, that is, if the light was originally transmitted by the beam splitter, it will now be reflected completely. This arrangement is also used as an optical isolator to prevent specularly reflected light from going back to the source. If a line of sight illumination is used (the light is nearly colinear with the line of sight of the camera), this prevents the specular light from going back to the camera. Plastic sheet circular polarizers can also be used in this fashion.

Polarizing beam splitters are typically in the form of a beam-splitting cube of 1–2 in. aperture and can split the beam with an extinction of 400–1 or better (the ratio of properly polarized light to unpolarized), with light losses of less than a percent. There are some such devices available to cover most of the visible light spectrum at once (for a cost of a few hundred), but typically a polarizing beam splitter will be limited to a narrow wavelength region for the reasons discussed before.

2.6.1 Interference Filters

Many of the color filters that provide the best control of light sorting are of the thin film coating interference type. These filters include

- Band-pass filters which pass only a limited range of wavelengths
- Long-pass filters which pass only the longer toward the infrared wavelengths
- Short-pass filters which pass the shorter toward the blue wavelengths

Interference filters can work in either reflection or transmission, often with both reflection and transmission being usable (as in the case of the polarizers). A special class of color filters of the interference type in which both reflected and transmitted light is used is the dichroic mirror or beam splitter. A dichroic mirror will reflect a limited color band of light while transmitting the rest. Some projection color televisions use this principle to combine the three colors (red-green-blue) so as to travel along a common path. By combining the three colors along a common path in a projection TV system, the color ghost which is often seen due to slight misalignments of the projector or an error in the overlap distance (per best focus) is not a problem. In reversing this design, an optical system can look along a common line of sight, yet the light be split to view, for example, a grayscale image by one camera, and a laser illuminated structured or directionally lit (shadowed) image with another camera.

As these interference filters rely on a standing wave of light, the optical thickness of the coating seen by the light is critical. The sensitivity of interference filters to changes in path length makes them sensitive to the angle of incidence on the filter. For this reason, interference filters typically have a limited angle over which they will work properly, typically in the 10°–20° range. Tilting of an interference filter will cause the wavelength to shift, as shown in Figure 2.19.

The performance of interference filters can be very high. There is typically a transmission loss of 20%–50% of the light (depending on how much you want to spend), but the signal to noise is very high because they can reject the unwanted or redirected light by a ratio of better than 10,000–1. As with any performance parameter, you must pay for what you get. Common interference filters to pass a bandwidth of light can cost only $25 for a 25 mm (1 in.) aperture. More specialized filters and larger apertures can cost a few hundred to a few thousand dollars. A good minimum cost to pay for a specially made filter is $600–$800, depending on complexity. Over most of the visible spectrum, the filter vendors have standard products which cover almost any combination of specifications that most people would need, for under $100 a piece.

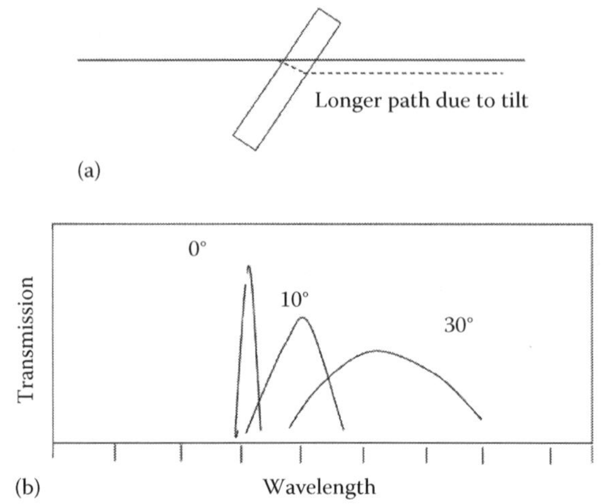

FIGURE 2.19
Tilting an interference filter changes the path length through the thin film of the filter (a), causing the wavelength and characteristics to change (b).

2.6.2 Color Glass Filters

An alternative to color interference filters are the glass absorption filters. Actually, both glass and plastic films are available, such as the polarizing sheet plastic already discussed. The mechanical stability and optical quality of the plastics is typically well below that of the glass (they vibrate), but the plastic sheets cost less in most cases. The spectral transmission properties of these filters are not as clean or symmetric as the interference filters, as they rely on available dyes to absorb unwanted colors. The color glass filters also can serve as band-pass, long-pass, or short-pass filters. Color glass (or the plastic equivalent) filters are more likely than the interference filters to exhibit multiple regions of the visible spectrum where they transmit light. Care must also be taken to consider what happens to the absorbed light. Heat-absorbing glass filters, for example, are actually just short-pass filters which absorb infrared light. The absorbed energy must be taken away from the glass, usually by air cooling, or the glass may break. In the case where the sensor is responsive to infrared light, a hot filter will actually radiate light in the infrared which the sensor may see, producing a general background noise. A similar factor can occur if ultraviolet light is used in the system. Many color glass filters will actually fluoresce, producing visible light from the ultraviolet. It is relevant to consider, not only what happens to the light being transmitted when using absorption filters, but also what happens to the absorbed light as well.

The transmission of absorption filters can be very high in selected regions, approaching better than 95%. In other regions of the spectrum, the transmission may only be reduced by 10% or 20%. In application, long-pass or short-pass filters made of color glass are often the best option, for cost and performance, perhaps with just a less steep cutoff point than possible with thin film technology. In fact, many filters of this type use a thin film coating to create a sharper cutoff wavelength (beyond which the filter does not transmit light), on top of the color glass to give the broad light rejection over a wide color region at a lower cost.

Color glass filters are available in a variety of sizes, ranging from 25 mm (1 in.) sizes for just a few dollars to many inches for a few tens of dollars. There is a wide selection of photographic color filters, some in glass that screw onto the front of 35 mm camera lenses and some in plastic which fit into holders which screw onto 35 mm camera lenses. The camera filters usually cost more than the stock color glass of plastics but come with their own mounts if the matching lensing is used (however, this is no basis for selecting a lens). When using a thick color glass filter to get good signal to noise, the thickness of the glass will affect the aberrations and resolution of the optical system as discussed in the specification section.

2.7 Coatings

Image geometry is only one optical performance specification to consider. Equally important can be reflections that create ghost images, light loss and scatter on optical surfaces, spectral transmission properties, polarization changes, and robustness to environment such as large temperature changes or vibrations. A key means of controlling this latter set of parameters is through proper optical coatings. Optical coatings can control reflectivities, polarization, spectral transmission, and even the robustness to some environmental issues like dirt, temperature, and moisture.

2.7.1 Antireflection

One of the more common types of coatings used on optical elements is the antireflection or AR coating. AR coatings are composed of either single or multiple layers of dielectric materials, such as magnesium fluoride, lithium fluoride, or silicon oxides, made in layers which are a fraction of a wavelength thick. The thickness of the coatings is carefully controlled such as to create standing waves of optical interference between the surface of the coating and where the coating meets the optical lens material. These interference waves act to null out a reflection from the surface of the optical component, actually yielding higher light transmission as show in Figure 2.20.

A stray reflection off one surface in a lens can wind up being reflected off other surfaces in a manner to actually create an image of low intensity (a so-called ghost image). This same ghosting occurs from uncoated windows. The light going through the window can reflect off the back surface of the window, then off the front surface so as to be directed into the imaging lens, except now the light has traveled a longer path than the unreflected light, so can create an image slightly closer to the window than the primary image. If the view through the window is at an angle, the ghost image will be slightly displaced from the

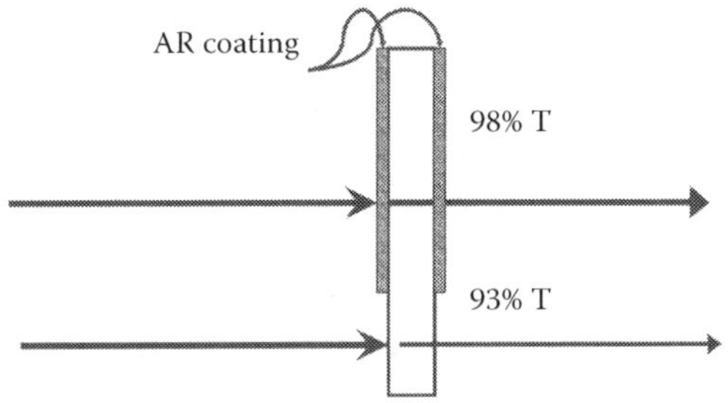

FIGURE 2.20
Antireflecting coatings null out the reflection from a piece of glass.

primary image, an effect which could bias any measurements made of the composite image. If there is no tilt of the window, the ghost may not even be noticed, yet exist as a slightly out-of-focus image which will produce a slight blur or haze to the primary image. These are problems which can be traced and removed.

A very common AR coating is a single layer of MgF, which reduces the reflection off glass from about 3% to less than 1%. Multiple-layer AR coatings can be made to reduce the reflection per surface to less than 1/4% with a simple stack, to less than 0.1% with more complicated coating combinations. Since the standing wave is a function of the wavelength of light, AR coatings are only effective over a limited region of the spectrum. In general, the more complex the coating (more layers), the narrower the wavelength region is over which it will be effective (they can be designed wide for a price). Outside of the designed wavelength region, the AR coating may reduce the reflection or may enhance the reflection. Using the improved coating, such as one peaked in the visible spectrum used in the infrared, will generally be worse than having no coating at all.

2.7.2 Reflecting

Perhaps the other most popular coatings are the mirror coatings. A standard, inexpensive mirror will often have a coating of aluminum, with an overcoating of SiO. The overcoating is needed because the aluminum will quickly oxidize, producing a hazy reflection (this is actually why second-surface mirrors are made, so the reflecting surface is not exposed to the air, at the expense of the image degradation caused by looking through the glass). A good aluminum coating can reflect up to 90% of the visible light and a little better in the infrared. Many inexpensive aluminum coatings are single layers of aluminum, evaporated onto the substrate. These single-layer coatings tend to have pinholes in them which can add to the scatter of the light (creating a slight haze which gets worse at each surface) and reduce the reflectivity to typically 80% in visible light. A better coating will often be multiple thin layers (like they tell you on the paint can), put down with the hope that the pinholes in successive layers will not line up.

Other types of reflecting coatings include protected silver, which gives reflectivities in excess of 97%, but can oxidize quickly if there are holes in the over coat, gold, which is a moderate reflector in the blue-green region but over 95% in the infrared, and dielectric coatings. The dielectric coatings can be made to have over 99% reflectivity but also have some drawbacks. Dielectric coatings actually work on an interference effect, which means they are sensitive to both wavelength and angle (since a change in angle will change the path length for the light going through the coating). A dielectric mirror made for HeNe laser light can look quite transparent at other wavelengths. If the dielectric mirror is made to work at 45° incidence angle, it will likely have very poor reflection properties at normal incidence.

Many of these same coatings are used when only a partial mirror is desired, a so-called beam splitter which will reflect some of the light and transmit the rest. Beam-splitting coatings can be put on flat glass plates to form a plate beam splitter, be sandwiched between two right angle prisms to form the diagonal of a cube (cube beam splitter), or be put on a thin plastic or "cellophane" membrane, which is called a pellicle beam splitter (see Figure 2.21). The surface figure and quality considerations are the same as already discussed for windows and mirrors.

A thick glass plate is the most common and easiest to fabricate beam splitter. A tilted glass plate will introduce a certain amount of spherical (as discussed before) and astigmatism

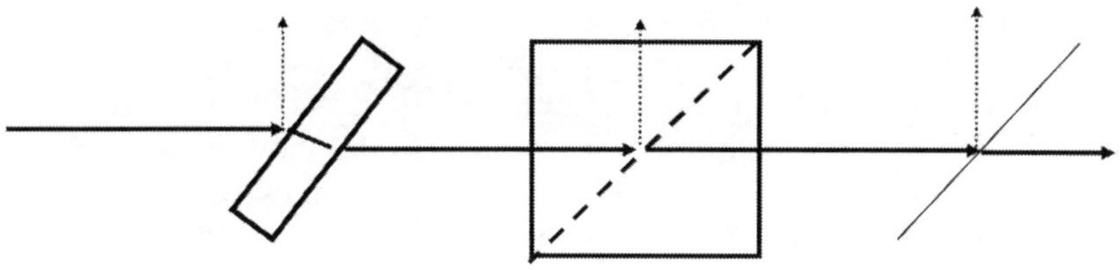

FIGURE 2.21
Plate, cube, and pellicle beam splitters.

aberration into the beam. Being that there are two surfaces to the plate, the second surface which does not have the beam-splitting coating will often have an AR coating to remove the ghost reflection from the unused surface. Cube-type beam splitters can also have ghost reflections if the faces of the cube are not AR coated. A cube beam splitter can introduce even more spherical aberration, but since it is not a tilted glass plate, it will not exhibit the beam displacement or astigmatism evident with the plate beam splitter. Perhaps the primary motive for a cube beam splitter is it permits the beam-splitting coating to be protected from scratches or even just the air (like a second-surface mirror). A pellicle beam splitter introduces effectively no aberrations, simply because it is very thin. The second-surface reflection from a pellicle is not typically a problem either since it is not displaced by a noticeable amount from the reflection off the first surface. The primary drawback of pellicle beam splitters is they are very fragile, therefore easily damaged, and tend to pick up vibrations like a microphone. The vibrations can modulate the reflected light or just blur out the image.

The most common type of beam splitters typically uses a neutral color metal such as inconel. The primary disadvantage of inconel is it will absorb upward of 20% of the light that hits it, meaning the best split you can get gives 40 reflected and 40 transmitted light. The dielectric beam splitters have losses of typically under a percent. There are dielectric beam splitters that cover most of the visible spectrum with only minimal polarization effects (a few to 10% difference for the two polarizations), but they are not typically useful over ranges extending into the infrared and may be angle sensitive. A special case of the dielectric beam splitter coatings is the polarizing beam splitter, which will be covered in more detail later in our discussion.

All of these reflecting coatings can affect both the color balance and the polarization state of the light. A gold coating will make white objects look gold in color (more red and yellow), while a green dielectric coating will make objects look green. This argument also holds for the AR coatings. If the AR coating makes the transmission of green light more efficient than red light by reducing reflection losses in the green, but increasing reflection losses in the red, things will look a bit green. Even if the camera is black and white, if the color is relevant to the inspection, the color balance can be a concern. If most of the light from the source is of a particular color (such as near infrared from an incandescent light), the light efficiency of the system may not be as expected (not to mention unexpected ghost images in living color).

The polarization of the light is especially affected by the dielectric coatings, but the metal coatings change the state of polarization as well. The reflectance of any mirror at nonnormal incidence will be different for vertically or horizontally polarized light. Although typically only a few percent difference for most metals, for the dielectrics, the difference can be 50% or more. This change in polarization is actually used to measure the properties of coatings (complex optical index and thickness) by a method known as ellipsometry. I refer the reader to text on the subject for more information on the method.[19] This effect does mean that if linearly polarized light is being used to reject glints, for example, the analyzing polarizer should be before any mirrors in the viewing system for maximum effect. After a few reflections, linearly polarized light can easily become elliptically or circularly polarized or, in the case of dielectric mirrors, may be lost from the system entirely due to poor reflectance of that particular polarization.

2.7.3 Apertures

Physical blocks and apertures also are actually a kind of filter which needs to be considered in the overall design of the system. The apertures put in most lenses actually serve to reduce glare and unwanted scattered light off the edge of the lens elements and in some cases can be used to reduce aberrations. Apertures can also be used to select certain features of the light. For example, placing an aperture at the focal plane (not to be confused with the image plane), one focal length behind the imaging lens, the range of angles of the light reflected off the part passed by that lens can be restricted. Restricting the range of angles passed by the optical system can be used to isolate glints off a part, to enhance surface irregularities such as scratches on near-specular parts, or to control the viewing direction of the optical system. When used in close proximity to the part, a telecentric view, as discussed previously, can be created such that every point on the part is viewed at the same angle. This perspective normalization can be very valuable when viewing parts with 3D shapes to be measured. With a constant viewing angle to the part, a change in focus will not change the magnification of the part (we've rejected images of other magnifications from entering the image plane), thereby making a measurement system less sensitive to size changes caused by defocus (the image can still get blurred).

The selection of rays of light associated with specific features on the part is greatly enhanced with the use of coherent light. When coherent light such as from a laser is used, a given size feature on the part will diffract the light at an angle which is directly related to the size of the feature. Small features will diffract light at large angles, while large features diffract light at smaller angles. On a diffuse part, the surface can be thought of as a set of small mirrors, each pointing in a slightly different direction. This diffusion of the light limits the ability to select out rays associated with just one size feature. However, if the part is mirrorlike, or optically transparent, then physical blocks placed at the focal plane of the lens can be used to pass or block only light coming from selected sizes or orientations of features. Coherent spatial filtering has been used to find submicron defects in ICs or hairline cracks in glass parts. (This area too is a field in itself.[15])

Physical blocks are a useful part of any optical system for rejecting unwanted light but also affect the final performance of the system. As discussed previously, a small aperture can limit the resolution of the system by diffraction effects. A larger aperture will also create a smaller depth over which small features will stay in focus while leaving larger features to be more easily inspected (see Figure 2.22). Apertures can also vignette wanted light, causing the image to get dark near the edges even if the part is evenly illuminated. Even if the field is not vignetted, a small aperture, such as in the lens, can block much of the light presented to the optical system, making the system light inefficient.

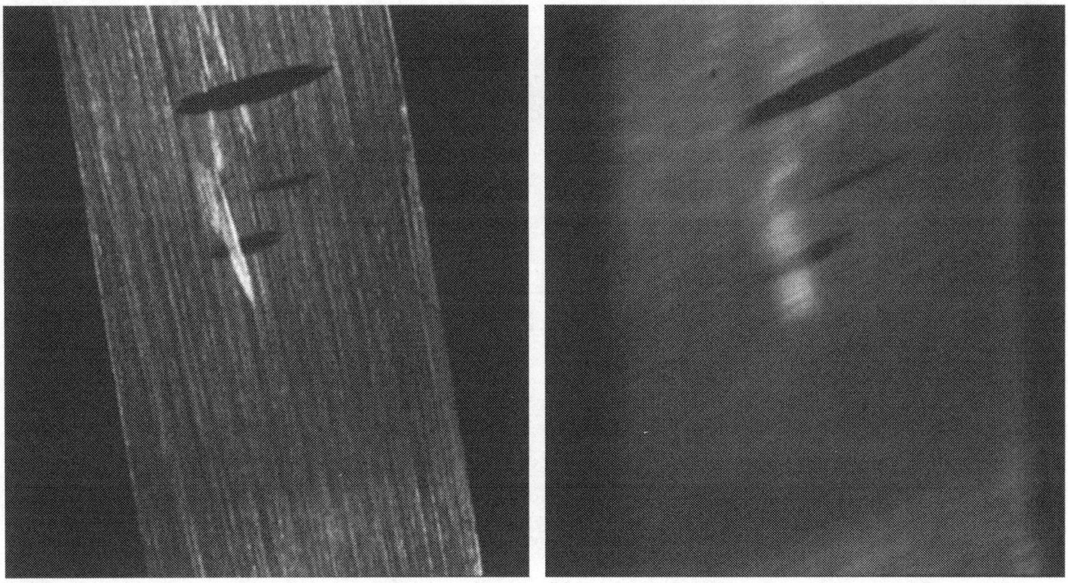

(a) (b)

FIGURE 2.22
Decreasing the focus depth allows an image with a lot of structure (a) to be converted to a simpler one by defocusing the fine structure but leaving larger features (b).

2.8 Prisms

A few types of prisms have already been mentioned, for polarizing or splitting a beam or acting as a 90° reflector. There are also a few specialty types of prisms to consider for use in a machine vision optical system and some special concerns imposed on the performance of the system by the use of prisms.[11] A 90° prism can be used to efficiently reflect a beam at 90° by entering the prism normal to one leg of the prism and reflecting off the hypotenuse. The light will reflect off the hypotenuse by means of total internal reflection (described previously). If the entrance and exit faces of the prism are not AR coated, there will be losses of 3%–4% per surface, plus possibly some ghost images present. The total glass path (the distance traveled through the glass by the light) will contribute to the spherical and other aberrations (often chromatic) of the final system. However, in cases needing robustness, a solid prism is often preferred since the reflecting surface can be protected by putting a solid mount behind it, thereby reducing light losses due to the deterioration of the reflector.

2.8.1 Constant Deviation Prisms

If alignment stability is of concern, then another type of prism known as a constant deviation prism can be useful. Two special cases of these prisms are the pentaprism and the corner cube. A pentaprism, often called an optical square, will deviate a beam of light at a constant angle of 90° independent of any rotation of the prism in the plane of incidence, as shown in Figure 2.23a. As the prism translates, the beam will still translate, but this is only

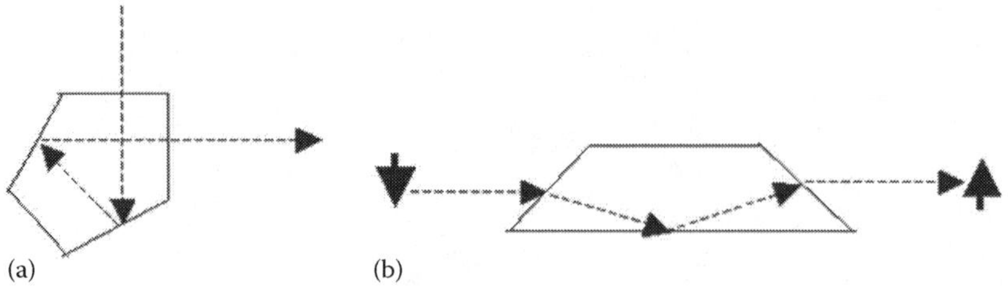

FIGURE 2.23
Constant deviation pentaprisms change the angle of a beam by a constant angle (a), while image rotating prisms will rotate an image (b).

a linear relationship which can be minimized by proper positioning of the prism in the optical train, whereas the tilt of a mirror by even a small amount can cause drastic changes in the image position. For example, if a mirror is placed 1 m (about 3 ft) from a subject, a movement of the edge of a 2.5 cm (1 in.) mirror of 5 μm (0.0002 in.) would displace the apparent object position by about 1 mm. This amount of motion would be enough to begin to blur out details in the image which may be of interest. A translation of a pentaprism by 5 μm would move the apparent object position by only 5 μm, which in this case is below the resolution limit of the optical system described (by diffraction limit).

In a like manner, a corner cube will return a beam back along its original path (180° turn) but also translates the image. Corner cubes have been used to return a beam of light directly back from the moon. Constant deviation is truly constant. Mirrors place on the moon would tend to bounce the return beam all over the heavens so to speak. A corner cube is invariant to variations in angle in both azimuth and elevation. A pentaprism will deviate a beam in elevation (out of the plane of incidence) on a one-to-one basis, with only a minor error introduced to the right angle (actually a quadratic). Two right angle prisms fixed with respect to each other can serve as the equivalent of a corner cube to return a beam along a parallel path. A right angle prism pair, such as described, can often be a more stable option when folding an optical path due to space restrictions.

2.8.2 Image-Rotating Prisms

A more specialized version of prisms useful for a limited class of applications is the image rotators. A dove prism is the simplest version of these, such as shown in Figure 2.23b, but a wide range of others are available as well such as Pechan, Schmidt, and Abbe to name of few. These prisms have the unique ability to rotate an image as the prism is rotated, with the image rotating at twice the speed of the prism. An image rotator permits a scene to be optically rotated, or derotated, to permit viewing the image in different orientations, or to view a rotating part as if it were stationary. A case where image rotation ability may be useful is viewing a large donut-shaped part, creating a continuous image with a linear array camera viewing along part of a radius of the part. A high resolution can be built up quickly by this method, without the need to quickly rotate a large part. Another application example may be a rotating fan which must be inspected face on for asymmetric rotation or even vibrations (much work has been done with optically derotated holographic and moire contour imaging).

In all of these prisms, the effect of the glass path must be considered on limiting the resolution of the system due to introduced aberrations and by producing a "tunnel" which will restrict the working F-number and hence the diffraction-limited performance of the optical system.

2.9 The Part as an Optical Element

We have been considering a variety of specialized optical elements which can serve to either enhance or degrade the performance of the optical system. Perhaps the least controllable but absolutely necessary part of the optical system is the part under inspection. If the part to be inspected was a traditional optical element, such as a prism or a lens, we could plug it into the formulas and predict its performance like any other element. In most cases, the optical performance of the part is not considered in the part design, so it is not well characterized. Despite the lack of knowledge of the optical performance of the part under test, the part has a major impact on the ultimate performance of the optical system we are using (if it did not, we would not be able to inspect it).

The questions we pose about other elements in the system are equally relevant to the part:

- How does the part direct the light which illuminates it?
- What is the effect the part has on the color content of the light?
- What is the effect the part has on the polarization of the light?
- How efficiently does the part transfer the light (losses)?

The last item of light loss is generally the major impact the part has on the optical system, often introducing more losses than the entire rest of the optical system combined. The light efficiency of the part is greatly affected by how the part directs the light and how much it filters the light in color or reflectance (100% of the light does not necessarily reflect off the part, e.g., steel is only 50% reflective), and the effect changes in polarization of the light have downstream in the optical system.

There are a number of direct questions we can ask about the surface of the part:

1. Surface finish
   a. Highly specular
      i. Flat surfaces
      ii. Curved surfaces
      iii. Irregular surfaces
   b. Highly diffuse
      i. Flat surface
      ii. Curved surfaces
   c. Partially diffuse, directional reflections
      i. Directionally sensitive (fine grooved surfaces)
      ii. Directionally uniform reflections
   d. Mixed surfaces of specular, diffuse, and directional
2. Surface geometry
   a. Flat
   b. Gently curving
   c. Sharp radii/prismatic
   d. Mixed surfaces

3. Surface reflectance
   a. Highly reflective (white or gray)
   b. Poor reflectivity (dark or black)
   c. Mixed high reflections and dark areas
   d. Translucent
   e. Transparent

   All of the earlier properties will contribute to how the part directs the light incident upon it.
4. Coloration
   a. Monocolor
      i. Broad color (gray)
      ii. Single color
   b. Color variations of interest
      i. Subtle variations
      ii. Discrete colors
   c. Mixed broad and discrete colors

If the colors reflected by the part are colors where the light source is weak, then the part will appear dim. If the color of the part does not reflect the colors of light which are strong from the source, then light will be lost to absorption by the part. The color of light reflected from the part acts like a filter, to which any coatings or lens designs in the system should be matched for best performance. If the part primarily reflects red light, and the optics are designed for use with a xenon lamp (more blue light), the AR coatings will likely be optimized for blue light but be mediocre in the red, meaning the optimum lens performance is not being used. Understanding these many factors introduced by the part provides a more complete picture that will drive the design and application for any machine vision application.[5,6,20,21]

2.9.1 Specular Surfaces

The surface finish and geometry can have an even more devastating effect on the light efficiency of the optical system. A truly specular, mirrorlike finish can direct all of the light into the optical system, or virtually none, or only all the light from specific areas, such as glints (see Figure 2.24a). If a specular part has a simple geometric shape to its surface, such as flat or spherical, the illumination light can actually be matched to the part so as to use the part as an optical element to relay the light to the viewing optics. This same argument can hold true for transparent objects as well (they act like a lens or window). If the shape is irregular with many twists and turns, it is much more difficult to see light on the entire surface. In some cases, seeing light from only selected points on the surface is all that is needed. For example, in some cases, a turned part shape can be identified, checking for shoulders or diameters, by only looking at the specular reflection from a line-shaped source parallel to the axis of the cylinder.

When specific points on a specular, irregular-shaped object must be seen, then it may be necessary to trace out where a ray of light would need to illuminate the part from in order to be directed by the part into the imaging optics. In the extreme case, it is often possible to provide all directions of illumination light to a part by placing the part in a light tent.

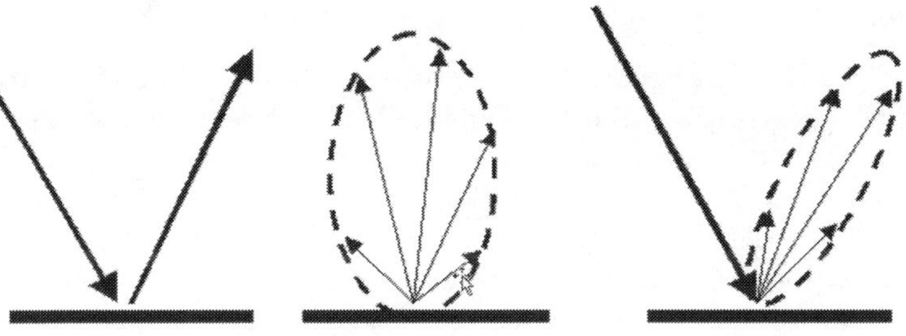

FIGURE 2.24
A part surface can produce (a) a mirror-like reflection that is specular, (b) a diffuse reflection that goes in all directions, or (c) a directional reflection that is brighter in one direction.

A light tent is a diffuse surface which surrounds the part, which in turn is illuminated by a light source whose direct rays are blocked from illuminating the part (see Figure 2.8). The result of a light tent is to make the part under test appear to have a uniformly diffuse-reflecting surface, even though it is specular. At times, this uniform illumination is not desirable since it can make edges on a continuous, specular surface disappear. In cases where the part is partially diffuse and partially specular or a diffuse feature like a scratch is what is of interest on the part, the light tent can be very effective.

2.9.2 Diffuse Surfaces

The light tent discussed previously and a perfectly diffusing part are typically very light inefficient. A diffuse surface, also called a Lambertian surface,[15] is a surface that will deflect any light, incident at any angle, into a wide distribution pattern (described by the cosine to the fourth as viewed of the angle from normal), which is centered about the normal to the surface, such as shown in Figure 2.24b. This means that for a fixed aperture, if the angle of view changes from about 5° from normal (with a 10° collection angle) to about 45°, the light seen can go down by a factor of 10. Looking at the problem from a different view, if you illuminate a flat surface from 5° with a fixed field angle (angle viewed from the lens), then illuminate the surface from 45°, you will be covering about 10 times more area and have less irradiance (watts per area) on the surface by a factor of about 10. The change in the area covered with a collimated beam changes as the cosine of the angle to the surface. In like manner, the area viewed by the sensor increases by the cosine of the angle of view. Treating the surface as a source, the projected area of the source also decreases as the cosine of the angle of emission from normal (this is Lambert's cosine law of intensity). As the distance from the surface increases, the light expands into increasing size hemispheres, thereby decreasing in intensity as the distance squared (the so-called inverse square law), which brings us to cosine to the fourth (see the second reference to understand this).

The actual irradiance on the detector will be determined by the f-number and magnification of the viewing lens. However, one can see from this that the angle of view of a diffuse surface can greatly affect how much of the light is used by the system. Even collecting about an f/5.6 (about 10°) solid angle of light at 45° (remember, this is an f/2.8 lens at 1:1 magnification), only about half a percent of the light is collected, assuming the part reflects all the light that illuminates it, which is unlikely, and does so perfectly diffusely. The rest of the light is simply lost and not collected by the optical system.

2.9.3 Directionally Reflective/Transmitting Surfaces

In many cases, the light reflected by the part's surface will be diffused; therefore, we cannot use it as a mirror, but the light may also be directional, centered about where the specularly reflected beam would go, such as shown in Figure 2.24c. We may want to look along this preferred direction to maximize our light, but the center of the field may exhibit a bright glare. Diffuse reflections do not preserve polarization, meaning that if the light was polarized before it strikes the part, it will be largely unpolarized after reflection (this varies to degrees depending on how good of a diffuser the part actually acts). If the reflected beam is part diffuse and part specular, giving bright "glints" or specular glare in the field, these factors can be removed with polarization techniques we are familiar with using. Glare removal from images of diffuse parts may not work because the glare may be the natural diffuse distribution of light and thereby not effectively maintain polarization.

On some surfaces with machining marks, or partially transparent surfaces which are translucent, there is actually a set of ridges which can act like separate surfaces from the overall geometry of the surface so as to direct the light in a direction set by the slopes of the machining marks or other microstructure of the surface. This grating effect can often be seen just by moving a light source around the normal to the part (at a constant angle down from the normal) while viewing the part from a constant (usually nonnormal) direction. Selective directional reflectance can cause the light level from a moving part, or a part side lighted which can be in various orientations on a plane, to vary like someone is opening and closing a venetian blind.

2.9.4 Cosine Diagrams

Mapping out the directions of specular reflections or preferred directions of light reflection from a part with a variety of surfaces can be difficult. One possible aid in visualizing how the part directs the light is by means of cosine diagrams.[19] I will not attempt a detailed description of cosine diagrams here, but rather attempt to introduce some tools in one variation of their use which are useful in visualizing how light might reflect off or transmit through a part. The direction cosines of the light are simply the range of angles of propagation of the light. Referring to Figure 2.25, the angle of light collected by a lens can be normalized by projecting the solid angle of the light collected up to a unit sphere. In like manner, the light reflected from or transmitted through the part can be projected up to this same unit sphere. For the purpose of graphically seeing where the light is going, the geometric projection of the patterns of the light on this unit sphere dome can be brought to a flat circle, by simply drawing lines from the edges of the projections on the dome down normally (straight down) onto the circle.

A lens viewing a surface at normal incidence (straight down) will have a collection of the light which will appear as a circle in the center of the diagram's circle. If the lens is off center, then the projection of the collection cone angle of the lens on the circle will be a kidney bean shape off center (see Figure 2.25). The reflected illumination can be added in the same manner. A specular reflection will give a well-defined pattern on the circle, whereas a diffuse reflection will cover the entire circle. A diagram of this type provides a means to visualize what light will be collected by the viewing system and how much light is lost. Specular or directional light patterns reflected by a part can be estimated by graphically tracing a few rays, typically of the steepest slopes of the part surface to see what range of

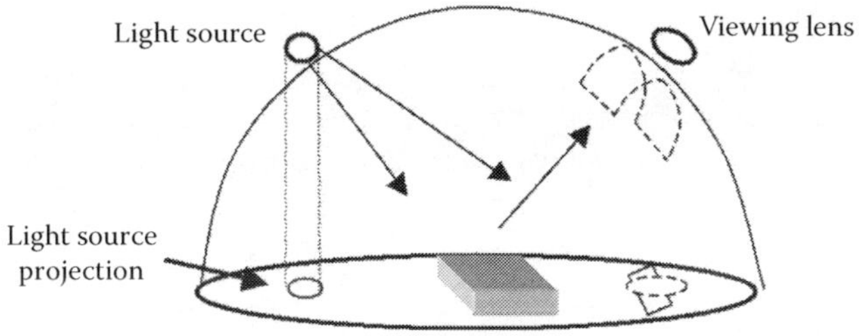

FIGURE 2.25
A cosine diagram allows the projection of the footprint of a light source and reflection on a unit sphere to better visualize how light reflects from a part.

angles exist, or by experimentally looking at the reflected light field. Using these diagrams can provide a visual feel for what the light is doing, by analyzing how the part will direct the light rather than just moving a camera around a field randomly.

The cosine diagram gives us a feel for how large of a lens is needed and where we would need to position the lens in order to see the entire glint. An alternative here is actually to diffuse the illumination source in the direction along the length of the part, which would give us a diffuse line along a 180° arc of our diagram.

Putting quantitative numbers on the effect that the part plays as a component of the optical system is not always easy. However, considering the part as a component in the system in a worst case situation, based upon what is known about the part, can provide some guidelines as to what is needed in the rest of the system and may provide a means of identifying the best approach if not the actual performance of the final optical system.

2.10 Setting Up the Design of the Optical System

Unless you are an optical designer, you probably will not design specific lenses for an application. The many vendors and design houses can do these for you. However, knowing the effect of individual components within the optical system is a first step to guiding an optical design for a machine vision system. As has been shown in some of the examples, the sum total of the effect of a series of components on such parameters as resolution, polarization, or noise can be surprisingly large, despite there only being a small perturbation to the image by each component. So the question at hand is how to start. As with any problem, the definition of the problem itself is a major step toward a viable solution. The problem definition is fed by the information at hand which can be gathered on the constraints imposed on the system. Some of these constraints may include

- The amount of light required by the camera to obtain a good signal-to-noise ratio
- The final resolution required of features on the part
- The FOV and depth of field over which the features on the part must be resolved
- The type of finish or surface character of the part under inspection

- The requirements on polarization where applicable, including the effect of the part on the light polarization
- The requirements on color, including the color range available from the source, as well as the color reflectance properties of the part (how much useful light is available)
- Physical constraints on the positioning of the components of the system
- Environmental constraints such as temperature changes, vibrations, and noise which must be endured (this will affect the resolution in most cases)
- Limitations on the type or power of the lighting (so that you do not "fry" your subject)
- Cost limitations—considered in light that saving money here will usually be more costly elsewhere (it is desirable to find out what the right system is first, then see if there is a real cost consideration)

In addition to these questions which apply to almost any application, there may be special needs imposed by the inspection process, such as the need to maintain color information about the subject, the need to take a picture of a moving part, or just how to access hard to see places on the subject. For example, a moving subject may require a strobe light to freeze its motion. The strobe light still puts out light which looks the same on a schematic of the optical layout, but it actually has a number of unique properties which place constraints on the rest of the system. Some strobe lights are xenon sources, which have a high degree of ultraviolet. If the ultraviolet light is to be used, the coatings on the optics must be appropriate to those wavelengths or the losses and ghost images may be very high. Some glass lenses will absorb the ultraviolet and ultimately break due to thermal stresses, so quartz optics may be required. Quartz optics are not available in the variety of regular glass lenses, which may pose a practical constraint on the size, focal length, or design of lenses used in the system. If the part has organic-based parts (which include many plastics), the strobe may make the part fluoresce. The fluorescence will be at a different wavelength than the light source and may cause ghost images since the AR coatings are not meant to deal with this new wavelength. Fluorescence is also unpolarized, so even if most of the light is specularly reflected (such as off of grease), a polarizer may not be effective at blocking the light reflected from that area of the part. Blocking the glare from the grease may be desirable because the grease may be smudges on the part. However, with ultraviolet illuminations, the fluorescence can leave an apparent uneven lighting where grease smudges are present.

As is evident from the example earlier, very few variables in a system are truly independent. The system as a whole needs to be considered to find the optimum design. Given the types of constraints discussed earlier, a system design will typically follow a fairly standard procedure:

1. Collection of pertinent information on the constraints which must be met by the system
2. A problem definition of what the optical system must do—what constitutes a good image for a particular application (I want to see "that" feature)
3. Review of the possible components which would be required
4. A conceptual design of the system based on the best guess of the possible solutions identified

5. Analysis of the system in accordance with (a) those constraints which must be met, (b) those parameters which are flexible (maybe a bigger image is okay), and (c) the features of the system which would be desirable, but not necessary
6. Redesign of the system with real component parameters (what you can buy)
7. Building, testing, and refinement of the system

These steps are not especially unique to designing an optical system. The unique aspect comes in the understanding of what constraints exist and how the tolerances at each component be in lighting, lenses, filters, or the part itself contributes to the final image.

2.11 Cameras

The next piece in a machine vision system is the camera. In the early days of machine vision, the application engineer was stuck with cameras made for closed circuit video or needed to pay very high prices for higher-quality cameras. The typical analog camera provided 50–80 usable gray levels, with a varying degree of noise that was often different from one camera to the next. Today, the consumer market has helped to push the digital revolution to the old camera industries. Many vision systems today use, or at least have the capability to use, digital cameras that provide better stability, higher dynamic range, and more pixels.

Cameras with over 1000 by 1000 pixels of resolution are commonplace today. A camera offering 10 or 12 bits of pixel depth, that is, a light dynamic range of over 1000–1 (even allowing for a few counts of noise), is within the price range of many machine vision applications. This has made it easier and more reliable to cover larger fields or obtain better resolution without complicated multicamera systems. The advent of new communication options for cameras, including FireWire, high-speed Internet, USB, GigE, and the camera-oriented camera link (which offers very high image transfer speeds), has also made it easier to install cameras in a manufacturing environment, network them together, and ultimately collect more data.

But even with all these great advances in the camera options of digital interfaces, more pixels, and lower noise, the big impact of growing digital capabilities today is in the form of "smart" cameras. A smart camera combines together a video camera with an on board processor and memory to create a vision system that can fit in the palm of your hand. Smart cameras in one form or another have been around for some time, going back to early systems that were very simple in rather clunky boxes in the 1980s. These early systems could typically do only one of a handful of operations at one time. Simple edge detectors that could find the distance between two edges were useful in measuring alignments or gaps in assembled parts such as automotive bodies. Basic blob recognition provided simple optical character verification or verification of any shape pattern. But these systems were so limited, and they were often little more than high-end bar code scanners but at a much higher price.

2.11.1 Smart Cameras: Machine Vision in the Palm of Your Hand

Today, smart cameras have a much larger range of operations (see Figure 2.26), using more memory and processing power (typically Pentium class) than older desktop computers

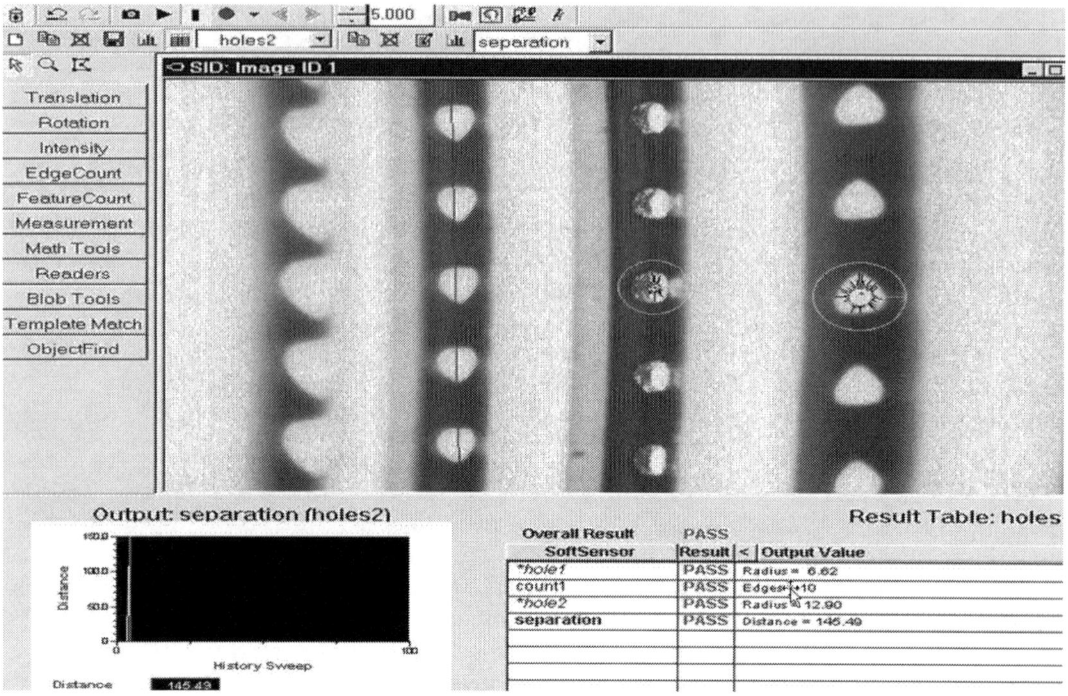

FIGURE 2.26
A smart camera typically can perform a range of operations such as feature positions, size, or counts, accessible through simple menus or icons (left).

had even 10 years ago. The types of operations available on a typical smart camera system can include

- Identifying part position and rotation (allows for part position variation)
- Analysis of multiple edge locations including counts, separations, and angles between edges
- Blob analysis to match complicated patterns, including doing full optical character reading (not just verification)
- Providing a wide range of outputs ranging from simple logic outputs to detailed numerical reports of fits to tolerances, amount of errors, and statistical information

In many cases, these cameras are made to go onto internet connections to allow them to be networked together. Since most of the processing is local, only the results or daily reports need to go over the network, removing the need for separate dedicated computers.

The degree of sophistication and cost of these smart cameras is still fairly wide, ranging from modern versions of the simple pattern matchers, now in the thousand dollar range rather than ten thousand, to full systems costing a few thousand. In general, the software is made user friendly, using pull-down menus and icons to set up applications rather than C-code and low-level communication protocols. However, these systems do not do all the operations that are needed by metrology needs.

Complicated operations like image preprocessing, morphological operations, Fourier analysis, or similar mathematical analysis and correlations are typically beyond what is reasonable to do with smart cameras. However, with the current capability of smart

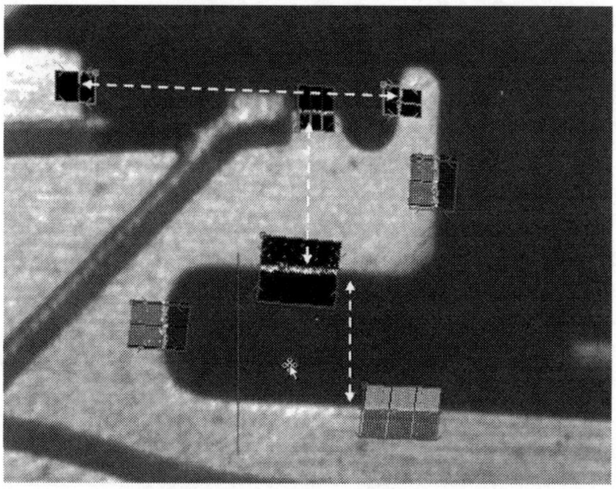

FIGURE 2.27
Finding edges to find the separation and orientation of features with a smart camera.

cameras, there are a vast number of "good" vision applications that they can do well. Applications such as simple inspections, hole finding or counting such as shown in Figure 2.27, as well as basic part measurements are very reasonable to do both in effort and cost using smart cameras. What does this mean to the industry? Many of those applications that just could not be justified because the vision system itself was going to cost 30,000–60,000 dollars can now be done with a smart camera at an equipment cost of a tenth of these amounts. It is still important to do a good application engineering job, but the hardware investment is less costly now.

2.12 Machine Vision Software

This brings us to the third basic building block of machine vision, the software component. Not too many years ago, to set up any vision application required either expertise in programming using C or similar language or learning a dedicated vision language, built around mnemonics, command strings, and controller codes. Basic C-code libraries are available today and in fact have become fairly inclusive, providing a wide range of capabilities for the programmer. For many applications, some dedicated vision system software packages offer a user interface with access to the common functions, filters, communications, and the like without writing a single line of code. In some systems, an operator or user interface can generate an executable file for use in production to provide better speed of operation. Simple additions of setups are often available through small scripts, usually based upon visual basic, visual C++, or C#, with clear examples and instructions for the end user.

A few vision systems use some form or proprietary hardware today. They may use multiple processors, high-speed memory, and fast graphics, but it has become infrequent that the end user needs to perform low-level programming of any type to get the system to work properly. Camera setups, which at one time required the writing of special configuration files, are often plug and play using the standards mentioned earlier. Modern high-speed computers have replaced high-end workstations costing a 100,000 dollars, with entry-level systems selling for less than 500 dollars. The exception to this rule of nondedicated hardware and software includes the high-speed production systems in continuous process applications such as roll-to-roll or primary metals operations. In these applications, the standard PC may not yet be up to the speeds needed.

The user-friendly nature of modern machine vision software has brought the setup, programming, or changes into the capabilities of plant engineers without extensive programming experience, as well as many shop floor maintenance personnel who can now easily maintain systems that at one time could only be serviced by a high-level programmer not typically available in a plant. New part programming rarely needs to be done by the vendor anymore (though most are willing to do so). This puts the control and schedule of new part introduction, tolerance changes, or the ability to add new checks to diagnose manufacturing issues within the hands of the production manager. This ease of use of modern machine vision, along with the lower prices, has made vision systems more attractive for many industrial inspections, alignment, and simple gaging applications that in the past may have been too expensive to purchase, to maintain, and to set up for many applications.

2.12.1 Conventional Algorithms Used in Machine Vision Software

Machine vision software is the implementation of mathematical algorithms representing processes performed on camera-captured images that enable making a decision about the target object being imaged and inspected. For instance, a decision may be whether the object is dimensionally correct, has surface acceptable defects, or is missing some components. Next, we will briefly review basic terms related to digital imaging and image processing algorithms which are implemented in every machine vision software package.

2.12.1.1 Grayscale and Color Image Generation

A grayscale image is an image where each pixel holds a single value, a gray level, corresponding to the amount of light that was irradiating it. The camera producing grayscale images is called a black-and-white camera. In contrast, a camera that produces three different intensity scales corresponding to red, green, and blue colors is called a color camera. An important trade-off in color cameras is most often the use of a single chip with a matrix filter on its front with dividing the pixels into combinations of red, green, and blue (called the Bayer filter), but in doing so the usable resolution of the camera is reduced. The gaps of data in each of the color bands are bridged by interpolation which does not constitute a true increase of resolution (see Figure 2.28).

The exception is the three-chip camera which has separate red, green, and blue chips, which must be accurately aligned to each other. The three-chip color camera design requires more light, since the light must be split three ways, and is more expensive to make.

For the reasons of process time, light level, and resolution and cost stated earlier, the majority of machine vision operations today can be and are performed using grayscale images. Particularly in the case of gaging applications, the operation is typically concerned more with physical features such as edges or holes rather than the color of a part. But there are instances where color may be used to see such an edge. In these cases when seen in manufacturing, it is typically easier to use a colored filter on the camera to make the feature stand out in a grayscale image rather than doing the extra processing and addressing the issues earlier with a color camera.

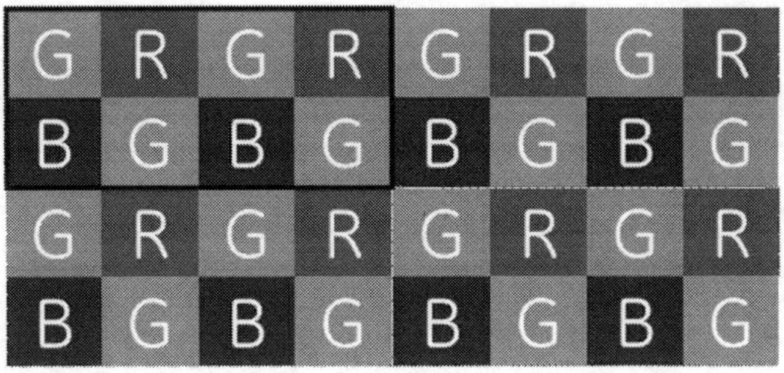

FIGURE 2.28
Beyer filter applied to the camera sensor is a repetition of this pattern.

2.12.1.2 Thresholding

One of the most common tools used in machine vision processing is thresholding. Thresholding is the process of mapping a continuous or a discrete scale into two or more discrete levels. The most common thresholding is called grayscale thresholding. Thresholding is applied to a grayscale image to produce a binary image with two gray levels, which often distinguish between foreground (e.g., an object or a local defect) and background such as shown in Figure 2.29. The selection of the threshold is either manual or automated. One simple case of automated thresholding is where the original image has high object to background contrast. The algorithm may use the histogram of the image pixel values and determine the threshold that would provide the best grouping. When the part is not uniformly illuminated, often the threshold values can be changed automatically to fit each section of the image (see Figure 2.30).

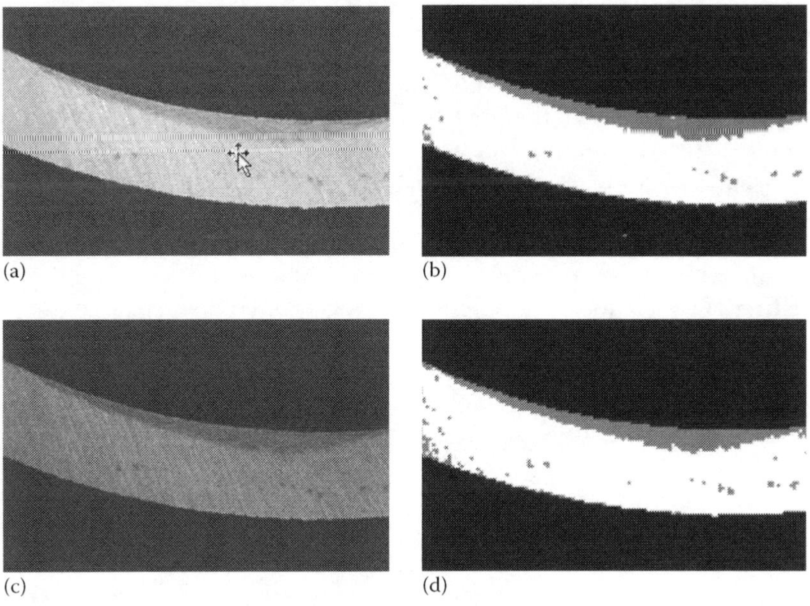

FIGURE 2.29
(a) Simple grayscale thresholding, (b) multiple thresholding, (c,d) location-dependent thresholding.

FIGURE 2.30
Quantization effects on image clarity.

2.12.1.3 Grayscale Quantization

Grayscale quantization is a multithreshold process where the original number of gray levels in the image is mapped to a lower number of gray levels. Along with segmentation of an image, quantization provides the benefit of reducing the computational load during calculation so that it can accelerate the processing. Common machine vision digital cameras produce images with 256–65,536 gray levels (corresponding to 8–16 bit cameras, respectively). When the algorithm requires a small number of gray levels to start with, the camera can often be adjusted to produce a lower bit rate and therefore smaller image files to be moved and stored. However, when optimal, nonlinear mapping into a few of the gray level is required, a high-bit-rate image is captured and the mapping is done via software-based quantization.

2.12.1.4 Edges and Edge Detection

Once segmentation has been done using thresholding or similar methods to reduce the amount of data to be processed, the next step in gaging applications is often to define the boundaries of features as reference points of the measurement. Edges present in an image signify the existence of changes, such as in an object's outer boundaries. Edge detection is a procedure to extract only the edges appearing in the image and is used for applications where the edge describes, for instance, a boundary of an object. A cluster of edges may represent certain texture, depth, geometric features (like a hole), or material properties. An image consisting of edges does not include gray levels or colors. An edge image is usually extracted by applying a certain threshold to the gradients or intensity level changes in the original image. Referring to Figure 2.31, first, a gradient image is calculated from the raw (original) image. This gradient image delineates the regions where the grayscale level has changed by some amount defined by the program. Thresholding then maps the gradient image into an edge image.

Sobel, Robert, and Prewitt are simple gradient edge detectors. Marr's edge detector relies on finding zero crossings of the Laplacian of Gaussian (LOG) filtered image (Figure 2.32 demonstrates the process). An original image is operated on (convolved with) a Gaussian function that serves as a low-pass filter that removes high-frequency changes, such as image noise. The resultant image is on the right in Figure 2.32, compared to simple averaging in the middle image. The larger the Gaussian in terms of number of pixels used, the

FIGURE 2.31
Original image at 256 gray levels, at 16 gray levels, and thresholded image.

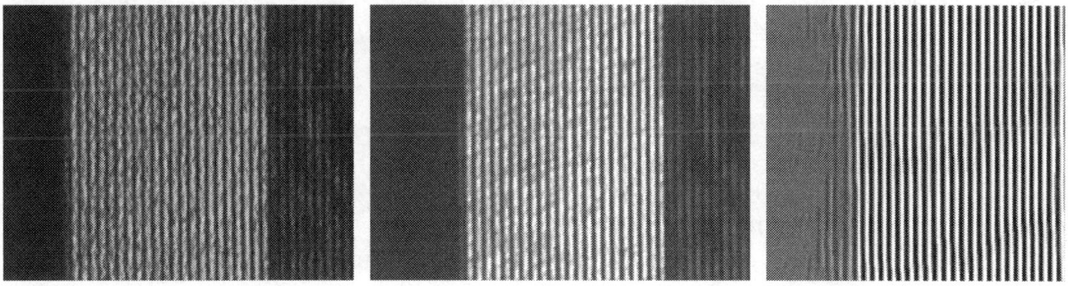

FIGURE 2.32
An original noisy image of lines (a) is simply averaged (b) versus filtered by combining with a Gaussian function (c).

more effective the filter in removing noise. The trade-off is a lower accuracy in preservation of the edge position. Now the Laplacian is applied, and its zero-crossing pixels are represented as the edges. A different and a very common algorithm is the Canny edge detector. Canny edge detector calculates the Difference of Gaussian (DOG), where the edge is considered as a high-pass filter in a noise-reduced low-pass filtered image.

A procedure also used in Canny algorithm is to apply two thresholds to the gradient image, to detect high-gradient edges (which is standard to all the edge detectors), as well as any medium-gradient edges that are connected to the high-gradient edges. It is common in many industrial scenes that the boundary contrast is insufficient in some location, but the human eye is able to connect the weak boundaries to strong ones, the Canny algorithm seeks to mimic this process. First, the edges are detected with a high threshold, then by a low threshold. The result provides a more complete picture of all the edges.

A trade-off consideration in selecting edge detection algorithms is that often the simple gradient algorithms are better suited to real-time inspection applications when a vast number of large images are processed. A low-complexity algorithm typically uses less memory resources, has very short processing time (able to process a full image in a few milliseconds), and is compatible with embedded hardware implementation. However, some algorithms, such as the Canny edge detection, which are typically slow, can be embedded in hardware and thus can be accelerated to comply with production line speed requirements.

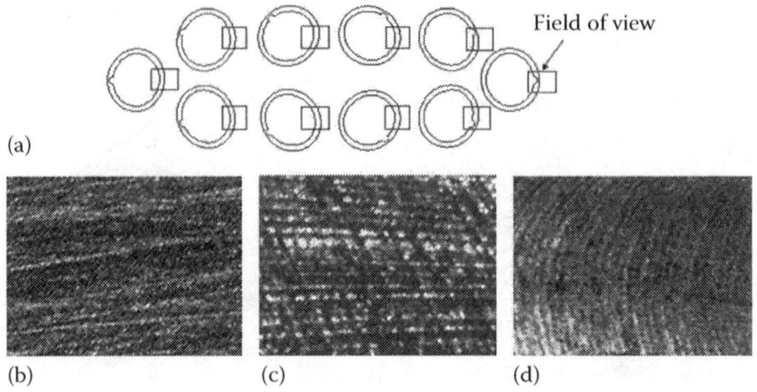

FIGURE 2.33
Different surface texture variations seen on different rings (a) can be used for segmenting regions having close to uniform texture such as straight line texture (b), overlap machining marks (c) or curved machining marks (d).

Edge detection often competes with other methods on the segmentation task. Thresholding is often not effective where either the part or the background have large variations in brightness or color. Edge detection is also the first step in algorithms using contrast in texture-related features between regions, such as edge orientation, texture curvature, and similar features.

In summary, edge detection is useful for the detection of part features or defect detection when the part or the background is not textural (where the part and background are smooth in appearance). Alternatively, edge detection can be useful for segmenting regions based on texture, when each region is characterized by approximately uniform texture such as shown in Figure 2.33.

2.12.1.5 Selection of Region of Interest

Typically a step prior to image analysis is the selection of the ROI in the camera FOV. When possible, one would prefer to focus on the ROI and perform further computation-intensive processes only on the ROI to save memory space and analysis time which even nowadays are typical key bottlenecks in the manufacturing process. The ROI selection can be processed using one of the aforementioned segmentation procedures. An example of a ROI might be the location where a hole exists on a part. Once the hole is located, perhaps by the presence of a curved boundary found by edge detection, further processing can be limited to the local area around the hole which may be a small segment of the entire image. An example of an ROI selection is shown in Figure 2.34.

2.12.1.6 Blob Finding and Analysis

An additional common class of algorithms for segmentation is blob finding and analysis. The assumption of blob finding is that specific objects appear different than the surrounding image. The simplest case is where the object is bright and the surrounding background in the image is black, or vice versa. Such a distinction can often be imposed in machine vision applications using appropriate lighting as discussed previously as well as methods such as background paint and material control. More complex algorithms rely on the relative intensity of each pixel compared to its neighborhood, while others rely on connectivity with similar intensity pixels to form a blob (see Figure 2.35). A typical post-processing

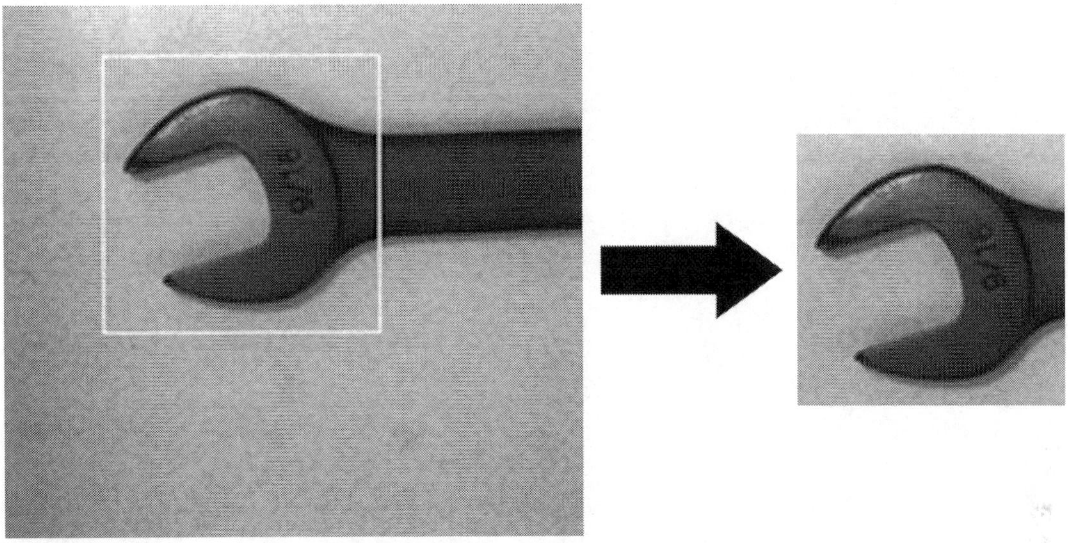

FIGURE 2.34
ROI selection to allow processing of only the information of interest.

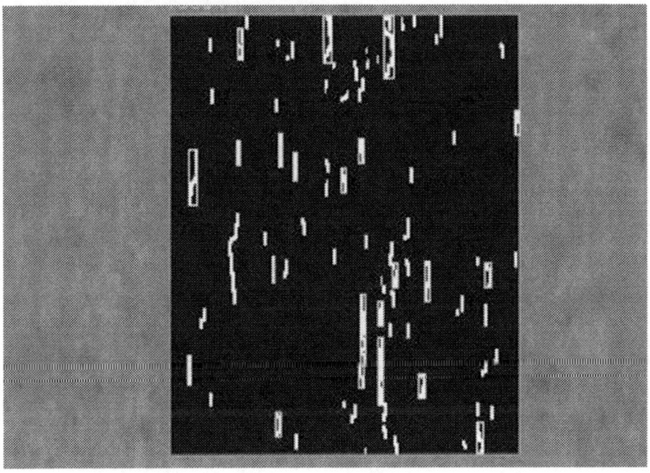

FIGURE 2.35
Blob finding locates segments of a certain shape, then the background is easily filtered.

for blob finding is using prior knowledge to select only "meaningful" blobs. The filters can be related to geometry (area, elongation, aspect ratio, roundness, convexity, etc.), color, intensity, contrast relative to the environment, and so on.

2.12.1.7 Morphological Operations

Another set of simple tools used for segmentation, ROI selection, and image noise removal is morphological operators. These processes manipulate the shape of geometries appearing in the image in an irreversible way. This way, only certain features in an object stand

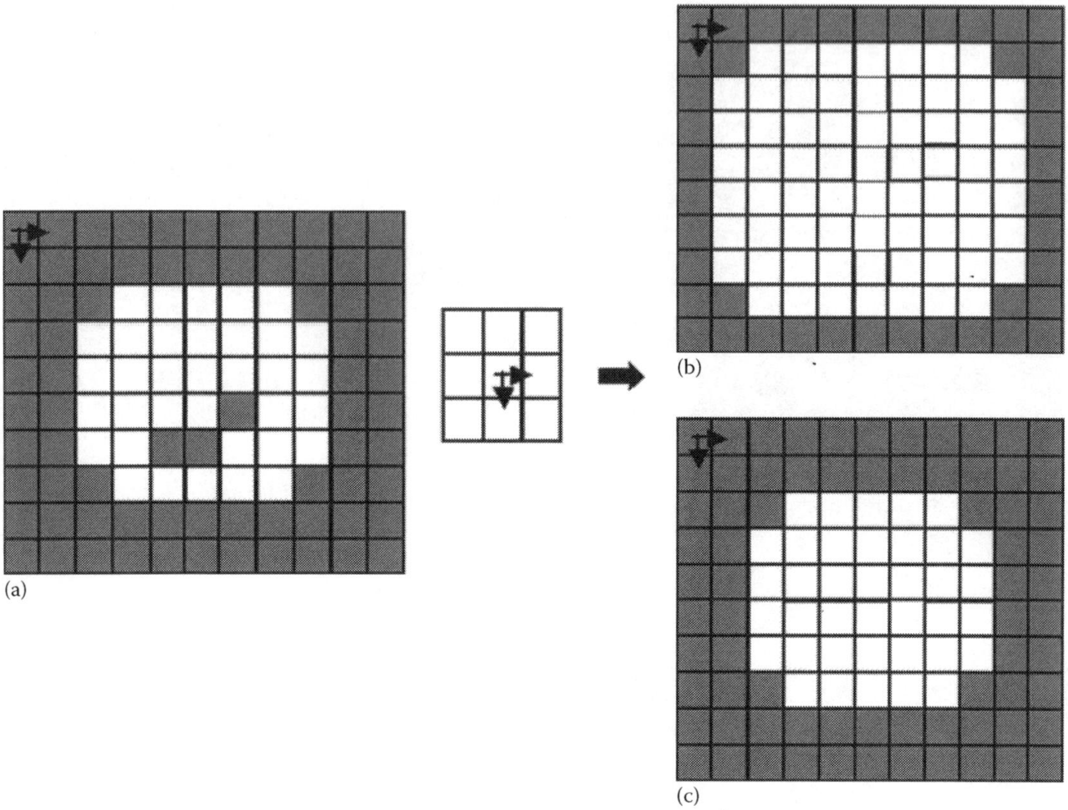

FIGURE 2.36
Morphological operations: original image (a) goes through dilation (b) and then erosion (c) to filter out the noise features in the image.

out and is ready to be measured, noise can be eliminated, and boundaries (e.g., external or internal) can be selected.

While some modern morphological operators are applied on gray-level images, we will demonstrate the common operators applied to a binary image. The concept of morphological operations is to apply a small shape to each of the pixels in the image at a time, changing the value of the pixel according to the way the shape interacts with the pixels in the neighborhood of each pixel. The shape is called a "structuring element," and the way it interacts with the image is determined by the underlying mathematical formula of each operator. A structuring element can be, for instance, a circle, square, or a cross.

Some of the simplest morphological operators are dilation to grow shapes, erosion to reduce the shape, and skeletonizing to define just central threads. As can be seen in Figure 2.36, an application of dilation followed by erosion may eliminate features up to a certain size or noise. Alternatively, comparing the preoperation image and the postoperation image can actually highlight the small features if detecting them is desired.

2.12.1.8 Image Analysis and Recognition

Many algorithms are based upon a method known as machine learning and require training for recognition, for instance, of what is considered an object (or a certain type of object out of a selection of objects) and what is considered background. Generally, these methods have an off-line training process where the software reads images of known types. The methods extract mathematical features such as certain dimensions and proportions of each image, vectors defining the average shape, and distribution of shapes of a certain class. The set of these features that represent an image is called a feature subspace. During online operation, the same process of extracting features is repeated on the inspected images, and a comparison (classification) between the inspected images and the trained images is done in the feature subspace by a classification algorithm to determine if the inspected image looks like the trained image.

Among the machine learning methods, appearance-based methods are used to recognize objects, sections, or defects or scenes based on high-level similarities and differences such as directions of view, shading, overall shape, and size. One simple machine learning method is to extract features using principal component analysis (PCA), where the principal vectors define what the mean or average shapes are in an image. For the purpose of explanation simplicity, let us assume that each image includes only 2 pixels. We will represent all the images used for training in one graph (see Figure 2.37).

If each image is represented by n coordinates (n corresponding to the number of pixels), groups of images may be separated along the directions of maximum distribution. These directions are represented by the principal components. Often, images can be separated into single-class clusters along the principal components. For example, all features on the image

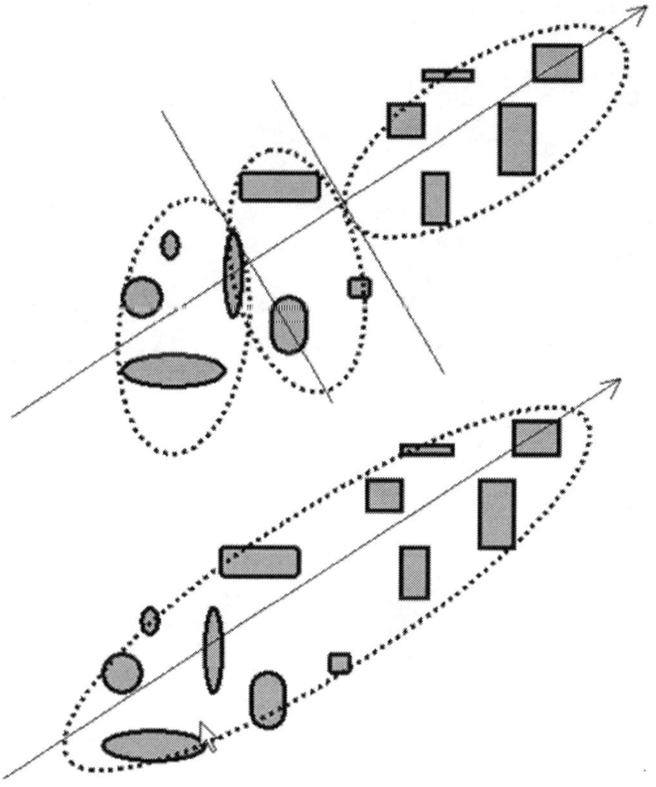

FIGURE 2.37
PCA showing different groupings.

that are tall and narrow might fall into one class, while short and wide features fall into another, regardless of what other characteristics, like roundness, their shapes may possess.

If indeed the images belong to different classes (since the part is different), it is expected that in this depiction, the images would be clustered into a number of distinct clusters corresponding to the number of classes. A class could correspond to the part shape, its shape and orientation (as in Figure 2.38), its surface characteristics, and so on. The first principal vector defines the direction of maximum scatter and thus the direction of where the classes are likely distributed. The second principal vector is orthogonal to the first one

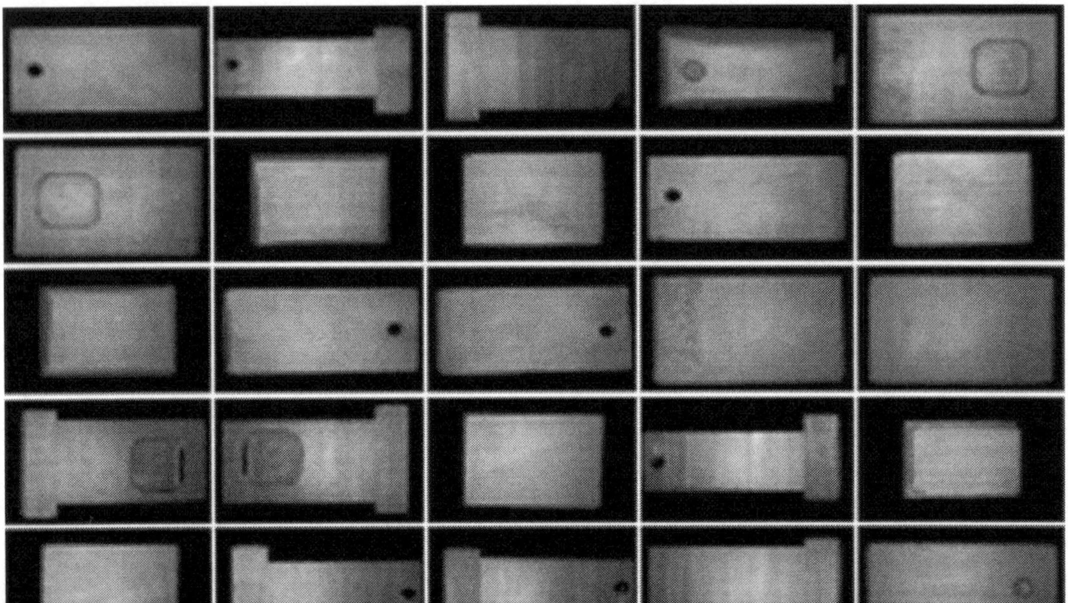

FIGURE 2.38
A set of training images. For this pick and place application, information of both part type and par orientation was needed.

at the direction of maximum lateral scatter. Typically, images have hundreds of thousands or millions of pixels so only the first few principal vectors are selected as features to use in analysis. Vectors associated with the highest principal values (such as the lengths and widths of the ellipses in Figure 2.37) are preserved and serve as features.

Therefore, a feature vector will contain only a few elements rather than the number of pixels of the original images, and the ratio between the number of pixels in each original image and the number of derived features is the dimensionality reduction achieved by feature extraction. When classification is performed in the feature subspace, the time, memory, and CPU consumption is reduced dramatically. In the feature subspace, classification can be performed for instance by a nearest neighbor classifier measuring distance.

Many algorithms, providing smart decisions, are available (Duda, Hart, and Stork). In order to reduce the computation time for metrology and inspection in production applications, the algorithms should

- Be compatible with the application
- Be as simple as possible to implement and maintain, efficient enough for efficient CPU and memory use and fast enough for online inspection
- Provide repeatable results with the same data

There is no one feature extraction and classification method that meets all the requirements for all the applications. So tailoring a combination of methods to a machine vision application is often a customize effort.

Software packages used for machine vision are typically either complete dedicated systems for machine vision or libraries providing functionalities for the specific machine vision system components. These libraries include routines providing camera control, motion control, image acquisition, image processing and analysis, feature extraction and classification, and data graphic display.

The typical programming platform for machine vision is based on the Microsoft Windows operating system, using MS visual C++ or C#. However, hybrids are available for instance where the software framework is coded in C++, while other functions are called from a different language (e.g., visual basic). Common machine vision software packages include the Matrox Imaging Library (MIL),[22] Intel IPP,[23] and Halcon.[24] One cross company and university shared library is the Vision-*something*-Libraries (VXL), which provides a very large set of functions, many of which represent the very last algorithms available. VXL provides an excellent flexible prototyping tool. However, the hierarchical structure of this set of libraries and the minimal attention to efficient programming (a price to pay for incorporating very recent algorithms) prohibits it from being useful in real-time online applications.

In addition to vision and image processing libraries, Intel provides a set of functions—one of the "Integrated Performance Primitives" (IPP) packages—that provides optimization of the code of computer vision algorithms to the Intel CPU, thus vastly increasing the processing speed. When software level computation speed up reaches a limit, the developer is left with a choice to use embedded tools, such as the field-programmable gate array.

2.13 Applications

As with any tool, there are good applications of machine vision-based metrology where it affords good capabilities, and there are poor applications that can best be done by other methods. The specific attributes machine vision has that best distinguish the application domain include

- Noncontact, so low chance of damaging delicate parts.
- Fast data collection, where many points may be needed, for example, SPC applications.
- Line of sight part access is needed; if you can't see it, you can't measure it with vision.
- Interaction is optical, meaning it is more important how a part looks than how it feels.

These attributes are neither inherently better nor worse than the attributes of other types of gages, but they are different. Machine vision has been used in very hostile environments, such as steel foundries to check width and thickness or nuclear reactors to measure defects. But machine vision is not used the same as a mechanical gage. Stray light, reflections, and glints will affect machine vision but not a mechanical gage. Extreme temperatures and mechanical shocks will affect mechanical gages, but not machine vision. This means that the way a vision system is applied is different than how a mechanical technology is used. Optical or mechanical metrology tools may be best for the job, depending on the conditions of the application.

One of the biggest mistakes in applying machine vision as a gaging tool is trying to apply it as though it was some other technology entirely. A screwdriver makes a poor hammer and a worse saw, yet these are perhaps more alike than machine vision and mechanical point probes.

There are some generalizations we can make about what makes for a good potential application of machine vision today and what makes a bad one. There will always be exceptions to these generalizations, but it provides some guidelines. Some typical good parameter of a machine vision metrology application may include

- A large standoff to the part is needed, an assembly with an active robot in the way.
- The environment is mechanically hostile, such as in a hot rolling mill or a forge.
- Part touching is a problem, such as a fragile ceramic part.
- Many small feature must be measured, such as circuit boards.
- The part interacts with light predictably, such as fluorescing grease.

By the same discussion, there are a number of features of an application which may make machine vision a poor choice given the capabilities of machine vision today. Some typical bad application parameters for a machine vision metrology application may include

- The part appearance varies widely, but we don't care about that.
- There is poor access to the features of interest, so they cannot be seen.
- The air is difficult to see through due to smoke or particulates.
- Only a few points are needed, and the current mechanical/electrical gage works fine.

Machine vision need not be a one-to-one replacement of something that is already being done by other means. There are plenty of applications which were not done at all in the past, because other methods were not able to do them or it was impractical, such as measuring the volume of solder paste on a surface mount board or checking the spacing of the

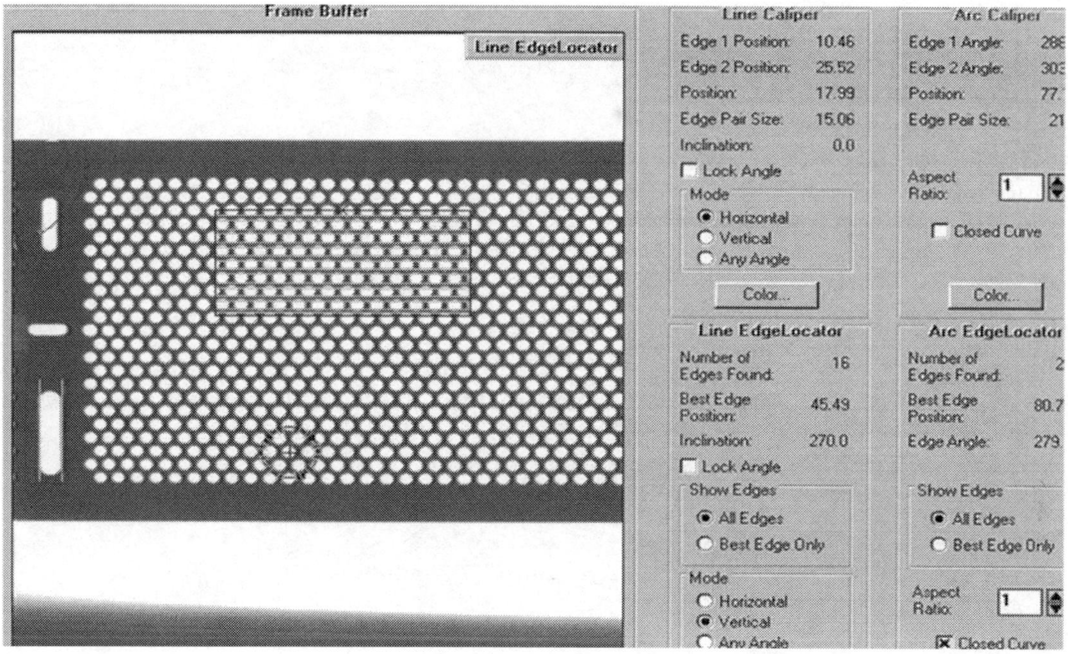

FIGURE 2.39
Checking a large number of holes would be laborious with manual or mechanical means such as pin gages.

leads on an electronics chip. A task such as checking a larger number of holes in stamped parts manually would mean inserting a pin gage into each hole, a very laborious task, but very fast with machine vision (see Figure 2.39). Today, such painstaking tasks are seen as good applications of machine vision. Some examples include inspection of pills for defects such as insufficient material, electronic component and connector measurements for bent or missing leads, and shape verification of gaskets. With technologies like smart cameras, many more of these applications now make economic sense to attempt with machine vision technology.

Machine vision today plays many roles in manufacturing, such as go/no-go inspection of small plastic parts used in circuit breakers, visual defect measurement on paint finishes of consumer product like toys, assembly verification of motors in washing machines or wire lengths in cars, and robot guidance for alignment of parts in welding or assembly of bearings. Although they are closely related, each of these tasks is fundamentally different. Each places different requirements on the system's processor, optics, mechanics, and, in some cases, even the cabling.

A machine vision system today can accommodate a degree of misposition, both in rotation and translation, and still make a reliable measure relative to the part. The critical factor in fixed mechanical gages is the fixturing. The critical factor in machine vision is contrast, presenting an image which contains the dimensions of interest, regardless of where it may be. Highly uniform light sources such as LED arrays and telecentric optics have eased this task for machine vision developers. This is not to say fixturing is not important to machine vision as much processing time can be spent locating a part. Ten or fifteen years ago, this part position finding could take several seconds. Today, a small misposition rarely takes more than a tenth of a second to correct. Flexibility is one of the key features of machine vision as a gaging tool.

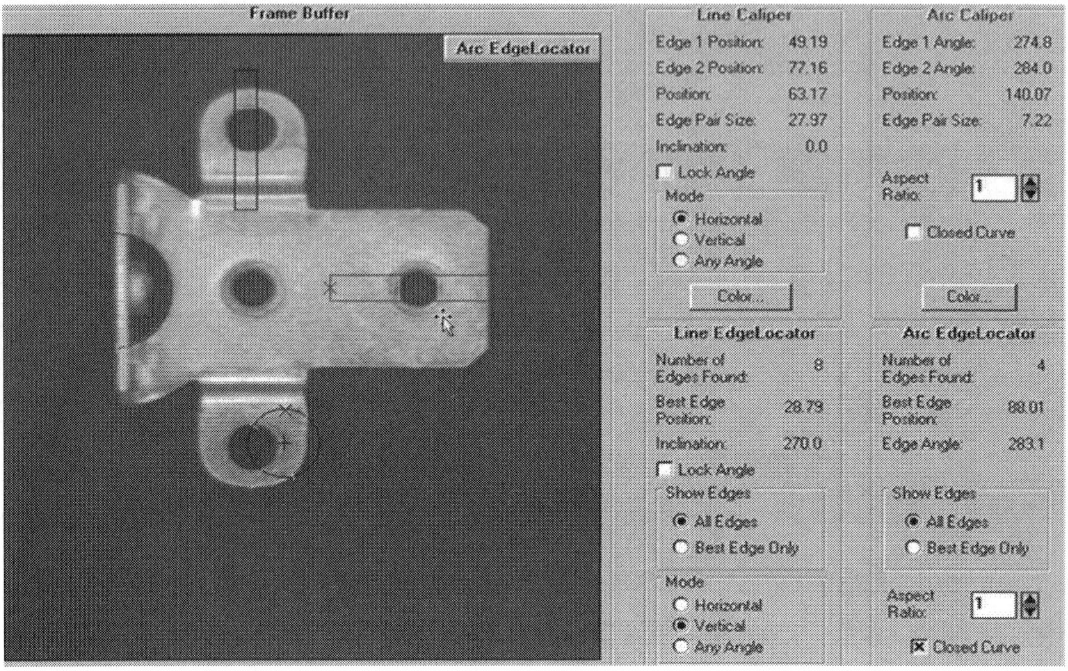

FIGURE 2.40
Finding hole locations, diameters, and arcs using simple edge detection of features.

In general, there are many areas of precision gaging that call for metrology lab precision that is just not viable with most 2D or 3D machine vision today. Submicron measurements of optical components with interferometry, mapping of the microstructure of precision bearing surfaces, or measuring detail in biological cells all require more stability and control than would be practical in a factory environment. However, there are many online applications where machine vision can provide valuable information for control of the manufacturing process.

One of the more popular instances of the use of machine vision has been in the fit and finish of auto bodies. A consortium effort between industry and university workers used machine vision gaging to control the fit of auto assembly to better than 1 mm using vision technology. This was a substantial step toward a consistent level of quality for the manufacturer. Another important area has been the inspection of web products such as paper, plastics, and primary metals for small surface imperfections.

Smart cameras are being used for many applications formerly done by manual optical comparators, such as hole sizing (Figure 2.26), finding the relationships between key edges for mounting (see Figure 2.27), and basic outline tolerance checking of stamped parts (see Figure 2.40). Three-dimensional vision technology is routinely used for electronic component placement guidance. Fast gaging of forgings and stampings is just starting to make headway against hard gages and coordinate measurement machines (CMMs), but with fast computing power is more attractive than ever today as an option. A 3D system capable of measuring a foot size volume at a thousandth of an inch accuracy in a few seconds is in the same price range as a good automated CMM but, for complicated parts, can be over 1000 times faster than even a scanning CMM.

2.13.1 Digital Comparators

For a subclass of applications where the subject may be largely planar with small features, but measurements are needed over larger areas of centimeters to meters, there are many systems today that combine the motion axis of a CMM with the local measurement capabilities of a machine vision camera (see Figure 2.41). Often referred to as digital comparators, these systems serve a similar role to optical comparators or so-called profile projectors. Traditional optical comparators illuminate a part with a highly directional backlight, then use a telecentric optical system to project the outline of the part onto a larger screen for measurements. For manual systems, the measurements are typically done by comparing a line drawing made on transparent mylar to the shadow outline from the part. Tolerance bands on the drawing show the operator and enlarged view of how the part compares to the part drawing.

With a digital comparator, the tools already described are used to make measurements within the local image captured by a camera. The FOV of the camera can be kept small, even down to the millimeter level, in order to provide good local image resolution for measurements. To relate a small feature in one area to another part of the feature or another feature, the camera is moved in X, Y, and Z using the scales on the stages to add to the measurement range within a single image. By defining features like holes and edges using multiple pixels, resolutions on the order of a micron (0.00004 in.) can be provided.

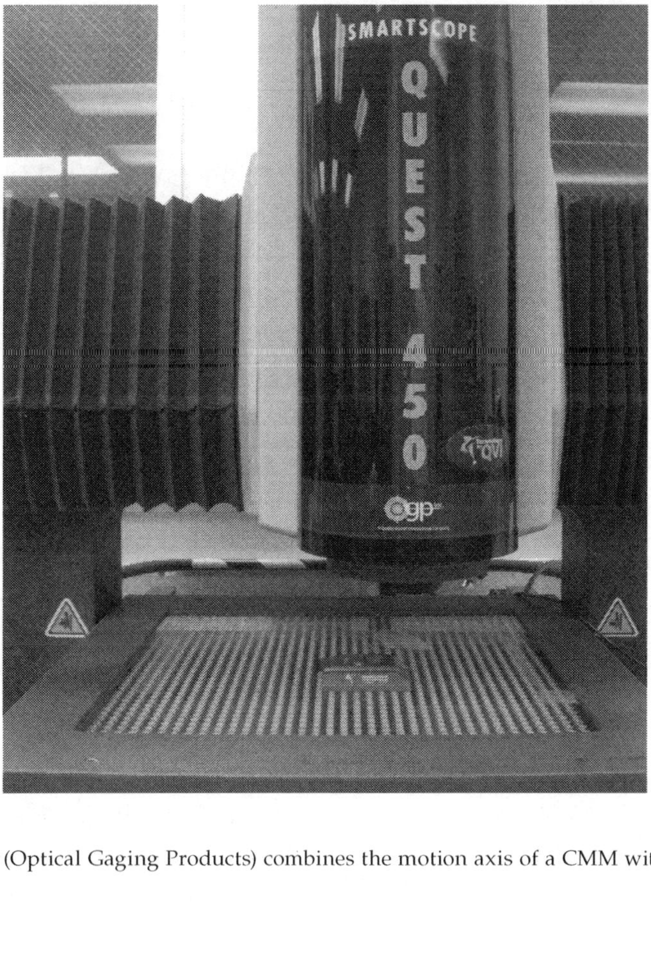

FIGURE 2.41
A digital comparator (Optical Gaging Products) combines the motion axis of a CMM with a video camera for local inspection.

The depth in such systems is typically generated using a focus finding analysis and movement of the camera system. The resolution of the depth measurement is typically going to be limited to area-based measurements rather than very localized points, to take advantage of surface texture of edges to establish the focus location accurately. Some digital comparators incorporate a laser probe (discussed elsewhere) to provide high-resolution Z measurements at localized points or to scan a simple profile, again using the machine stages as a means to extend the measurement range and position the measurement point accurately in the X–Y plane of the part.

An extension of the digital comparator concept using a CMM-like system is to put a vision camera on a robot. Vision cameras used to guide assembly and welding applications are often used on robots for this purpose. Measurement of large parts can be done with a robotic system, but typically higher resolution can be achieved with the fixed, CMM-type system where position scales can provide submicron readouts versus typically 20–30 μm (0.001 in.) level positioning with the best robots.

Examples of applications where digital comparators are in use as a measurement tool include

- Placement of electronics on larger circuit boards
- Measurement of plates with channels in them such as used in automotive transmissions
- Verifying the position of critical hole locations on machined parts
- Measuring a machined radius, slot, or groove on a power transmission part

2.13.2 Summary

A summary of what can typically be expected from 2D machine vision systems, from smart cameras on up, is presented in the following. These numbers are typical, but always improving.

Operation	Part Speed	No. of Features	Resolution
Edge location	30/s	30 per part	1/2000 FOV
Hole center/size	30/s	5 per part	1/10,000 size
Feature location	15/s	10 per part	1/1000 FOV

Economic considerations must always be addressed when designing any production systems. Economic considerations for machine vision applications include

- How important is the inspection function?
- What is the relationship between the costs of capital, service, and labor?
- What are the long-run demand trends?
- What are the costs in a competitive market of not being automated?
- What would be the cost of any retraining?

For example, perhaps you make bearings to go into engines. The inspection you might need to do is look for an oil groove. If you do not inspect for the oil groove and it is missing, the engine may fail. The initial capital for automating the inspection and the retraining cost may be as much as a year of labor for manual inspection. If this part may soon be obsolete, then the automation may not pay for itself. However, if the part is around for the long haul, there is a payback time which some may consider long. However, we must also consider the cost of charges back for failed engines and the competition. If the competition can produce a better part without this flaw (through whatever means), then you stand to lose 100% of your business. So in today's market, you must also ask how much is "the business" worth when deciding on the cost of installing a new tool like machine vision. This is the real potential payoff of machine vision tools today.

2.14 The Future

In the past 30 years, machine vision has grown from a limited tool doing little more than reading bar codes to an accepted process control tool in many areas of manufacturing. The acceptance of machine vision, as with any new technologies, has been slow. Machine vision is different from what manufacturers have used in the past, but it is a tool very in tune with the modern thinking of computer integration in the factory. There is a change occurring in how process control is approached and how the production person thinks about data collection. Industry is no longer satisfied with throwing out bad parts, it doesn't want to make bad parts at all! Most manufacturers have typically not known much about such optical technologies as machine vision, but they are learning.

Just a few years ago, machine vision was viewed with suspicion as some form of black magic, tied into the equally new computers. Just as computers have become accepted in the workplace, and even our everyday lives, so has machine vision made great strides. Even 25 years after the wide spread first uses of machine vision, there are still new applications coming about for machine vision each day. Although machine vision can do much more than it could 25 years ago, it does not do everything. There are some, off-the-shelf applications of 2D vision technology, in standardized systems doing high-volume inspections from bottles to sheet product. Machine vision "standard products" that inspect paper, printed labels, pills, painted surfaces, and plastics can be purchased practically off the shelf today.

Although there are still plenty of new applications being tried every day, not every application is necessarily a new development effort as it was in the past. Much of the development on these early applications has been done, documented, and proven out in practice. Even with this beginning, machine vision systems have not reach commodity status. However, with the foothold of general acceptance growing for machine vision and the many uses of computers in our lives, there is now a new realm of consumer-related products envisioned, ranging from security systems to vision in cars to read street signs to home units to watch the baby. Machine vision will continue to grow as a tool of the future.

References

1. A. Teich and R. Thornton, Eds., *Science, Technology, and the Issues of the Eighties: Policy Outlook*, American Association for the Advancement of Science, p. 7, Westview Press, Boulder, CO, 1982.
2. K. Harding, The promise of machine vision, *Optics and Photonics News*, p. 30, May 1996.

3. N. Zuech, *The Machine Vision Market: 2002 Results and Forecasts to 2007*, Automated Imaging Association, Ann Arbor, MI, 2003.
4. S. P. Parker, *McGraw Hill Dictionary of Scientific and Technical Terms*, 3rd edn., McGraw Hill, New York, 1984.
5. N. Zuech, *Applying Machine Vision*, John Wiley & Sons, New York, 1988.
6. B. Batchelor, D. Hill, D. Hodgson, Eds., *Automated Visual Inspection*, pp. 10–16, IFS Publications Ltd, Bedford, U.K., 1985.
7. K. Harding, The promise of payoff of 2D and 3D machine vision: Where are we today? *Proceedings of SPIE*, 5265, 1–15, 2004.
8. K. Harding, Machine vision—Lighting, in *Encyclopedia of Optical Engineering*, R.G. Driggers (Ed.), Marcel Dekker, New York, pp. 1227–1336, 2003.
9. E. J. Sieczka and K. G. Harding, Light source design for machine vision, in *SPIE Proceedings*, Vol. 1614, *Optics, Illumination and Image Sensing for Machine Vision VI*, Boston, MA, November 1991.
10. K. Harding, Optical considerations for machine vision, in *Machine Vision, Capabilities for Industry*, N. Zuech (Ed.), Society of Manufacturing Engineers, Dearborn, MI, pp. 115–151, 1986.
11. R. Kingslake, *Optical System Design*, Academic Press, New York, 1983.
12. D. C. O'Shea, *Elements of Modern Optical Design*, John Wiley & Sons, New York, 1985.
13. R. E. Fischer, *Optical System Design*, SPIE Press, McGraw–Hill, New York, 2000.
14. J. E. Greivenkamp, *Field Guide to Geometric Optics*, SPIE Field Guides, Vol. FG01, SPIE Press, Bellingham, WA, 2004.
15. E. Hecht and A. Zajac, *Optics*, Addison-Wesley Publishing, Reading, MA, 1974.
16. F. Jenkins and H. White, *Fundamentals of Optics*, McGraw-Hill, New York, 1957.
17. C. S. Williams and O. A. Becklund, *Optics: A Short Course for Engineers and Scientist*, John Wiley & Sons, New York, 1972.
18. D. L. MacAdam, *Color Measurement*, Springer-Verlag, New York, 1985.
19. R. M. A. Azzam and N. M. Bashara, *Ellipsometry and Polarized Light*, North Holland, Amsterdam, the Netherlands.
20. K. Harding, Lighting source models for machine vision, *MVA/SME's Quarterly on Vision Technology*, pp. 112–122, (1989).
21. G. T. Uber and K. G. Harding, Illumination methods for machine vision, *Proceedings of SPIE Opcon 90*, Boston, MA, November 4–9, 728, pp. 93–108, (1987).
22. http://www.matrox.com/imaging/en/products/software/mil/ (last accessed on April 01, 2012).
23. http://software.intel.com/en-us/articles/intel-integrated-performance-primitives-intel-ipp-open-source-computer-vision-library-opencv-faq/ (last accessed on March 30, 2012).
24. http://www.mvtec.com/halcon/ (last accessed on April 03, 2012).

Part II

Optical Metrology of Larger Objects

3

Laser Tracking Systems

Scott Sandwith and Stephen Kyle

CONTENTS

Objective

The objective of this chapter on laser trackers is to offer a concise and practical guide to laser tracker measurement technology and techniques. It should serve as a desk reference resource for those that need a better understanding of laser tracker metrology, its accessories, and targeting requirements in order to make informed decisions and application choices. It should also act as a "how-to" guide for laser tracker use on site and applications in general.

The text aims to offer these key benefits:

- A working understanding of laser tracker capabilities and limitations
- Practical help in judging where laser tracker metrology is best suited for measuring, building, and digitizing larger structures
- A guide to best practice in laser tracker use and practical tips on methods for their effective application

3.1 Introduction

Laser trackers are one of a class of *portable coordinate measuring systems* (*PCMS*) used to construct, align, and inspect large manufactured objects, such as aircraft, ships, submarines, and bridges, and the tooling associated with them. They are also used at even larger scales in the construction of particle accelerators such as CERN's Large Hadron Collider, structures which span kilometers but demand submillimeter accuracy.

Figure 3.1a through e shows applications in aerospace, automotive, marine manufacture and fusion research. Their application significantly improves the precision of such large-volume structures and reduces fabrication and maintenance costs. As portable metrology systems, they can be set up and operational in less than an hour, which has lead to their impact in an increasingly wide variety of applications.[1]

This chapter starts with a brief explanation of how a laser tracker works, followed by methods to deal with line-of-sight restrictions. The key technology components, angles, range (distance) measurement, and target reflector design are then discussed in detail. Discussion continues with environmental effects on tracker measurements and their compensation and a guide to best practice in tracker usage. A review of applications completes the chapter.

3.2 Laser Trackers: Concept of Operation

This section presents an overview of how laser trackers provide three-dimensional (3D) measurements.

3.2.1 From Fixed 3D to Portable 3D

It is important to appreciate the fundamental change in the operation of 3D metrology systems which laser trackers offer. This is mobility—the ability to bring the system to the object being measured rather than bringing the object to a fixed measurement site. Figure 3.2a and b shows examples of small- and large-volume coordinate measuring machines (CMMs) (DEA, Mitutoyo).

Consider the fixed dimensional metrology defined by conventional three-axis CMMs. These were a major technological advance in their time and remain in very extensive use.

CMMs were originally typically designed to measure small items such as engine casings and machined components. They often have a contact probe which slides along three mutually perpendicular axes, with a linear encoder to measure the movement on each. These three encoder readings define the 3D position of a measured object's surface point on contact. In recent years, they have developed to measure large objects such as complete car bodies. However, a common feature is that the CMM is fixed and most probably in an environmentally controlled, indoor location. The object to be measured must be brought to the CMM. For large and heavy objects such as large castings, this may not be possible. Some measurement tasks, by definition, require an on-site measurement,

(a)

(b)

FIGURE 3.1
Aerospace, automotive, and marine manufacture and fusion research.

(c)

(d)

FIGURE 3.1 (continued)
Aerospace, automotive, and marine manufacture and fusion research.

(continued)

for example, to place components in a particular relative alignment at their place of operation. Laser trackers and other large-volume metrology (LVM) systems can provide this on-site capability. Figure 3.3a and b shows a laser tracker application example (FARO Technologies, Inc.). However, CMMs still have a secure place in dimensional metrology, particularly where very high accuracy is required on smaller manufactured parts.

(e)

(f)

FIGURE 3.1 (continued)
Aerospace, automotive, and marine manufacture and fusion research.

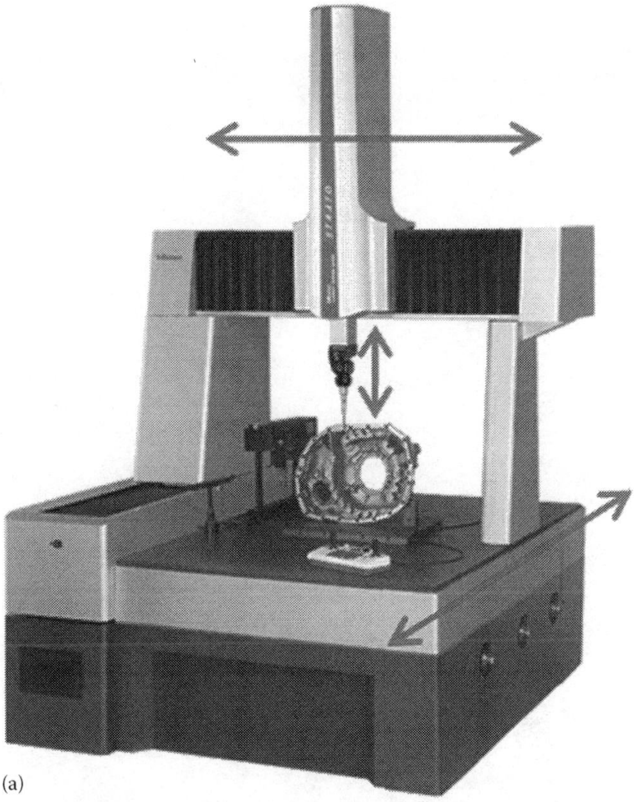

(a)

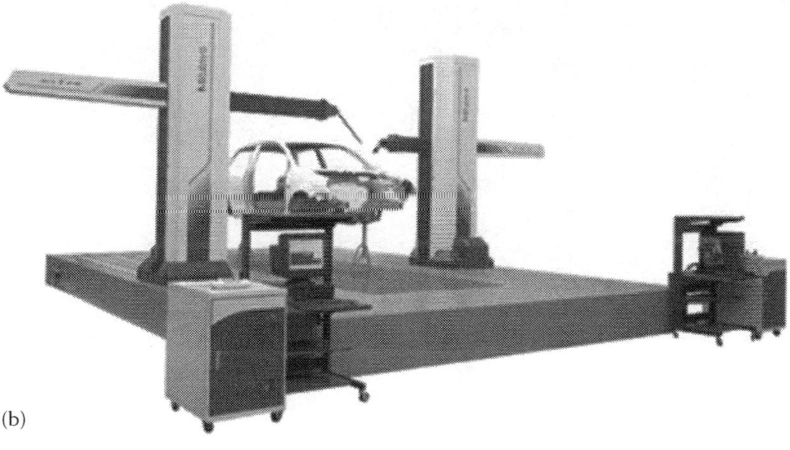

(b)

FIGURE 3.2
(a) DEA CMM. (Courtesy of Hexagon Metrology, London, U.K.) (b) Car-body CMM (2004). (Courtesy of Mitutoyo America Corporation, Aurora, IL.)

So, in contrast to a CMM, a laser tracker can be taken to an object to be measured. It has a fixed base and a rotating head which points a laser beam at a moving reflector. The reflector has the special property that an incoming beam at any angle is reflected back along its own path and returned to the instrument. This is required for tracking and distance measurement. The reflector is mounted in a spherical housing and can be used as a

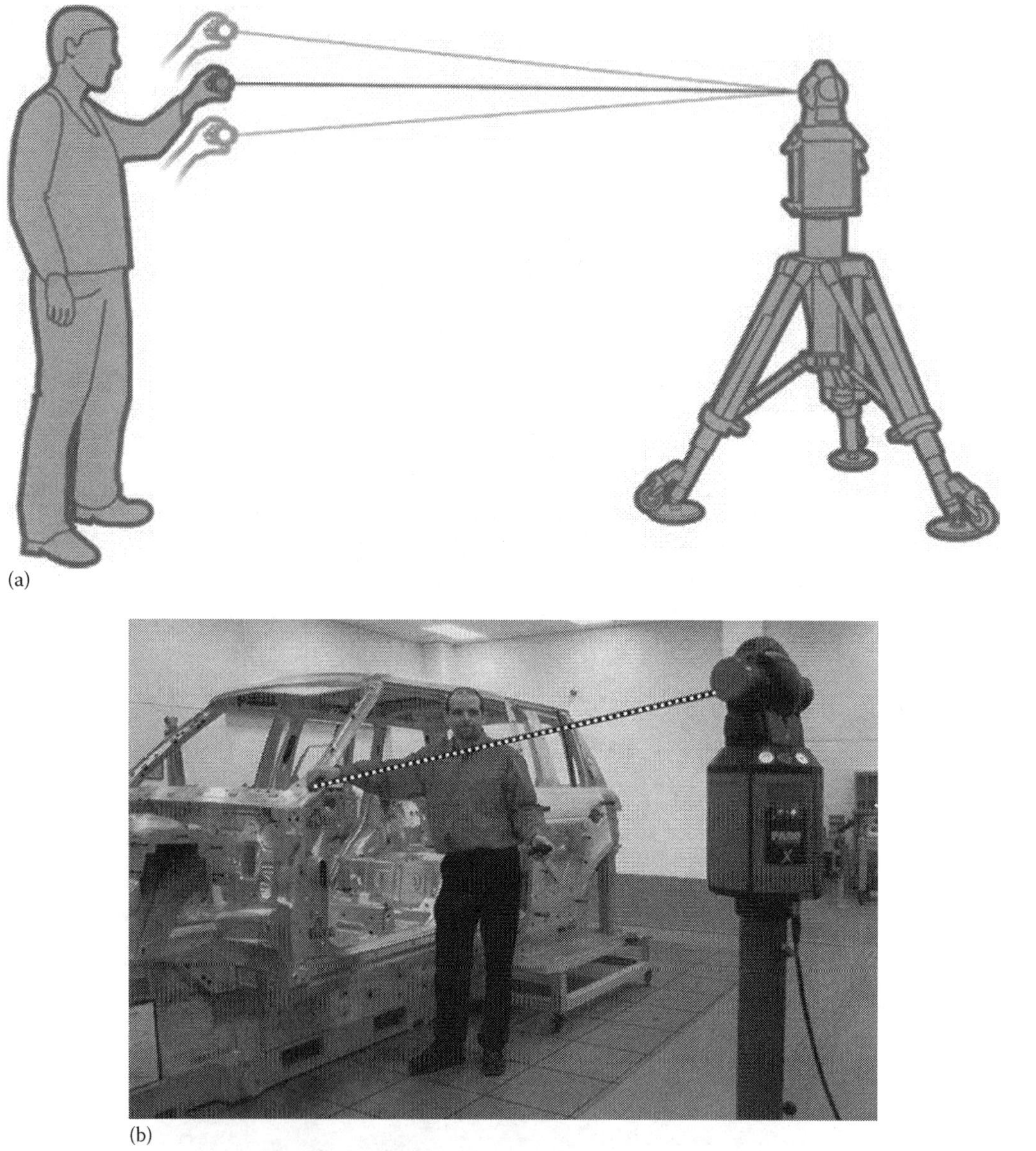

FIGURE 3.3
Laser tracker operation. Handling concept–tracker follows reflector (a), tracking applied to measurement of real object (b). (Courtesy of FARO Technologies, Inc., Lake Mary, FL.)

handheld contact "probe" to determine object shape and position. Figure 3.3a shows the single operator moving the reflector around, and Figure 3.3b shows it in use as a touch probe (dashed line added to indicate laser beam). The laser beam is the beam from either a laser interferometer (IFM) or laser distance meter, and it measures the distance to the reflector. When combined with the beam's horizontal and vertical angles, measured by sub arc second angle encoders, this provides the full 3D position of the reflector in spherical coordinates at a rate typically up to 1000 points per second. Static positional accuracy is of the order of 10–100 μm, depending on environmental factors and range to target.

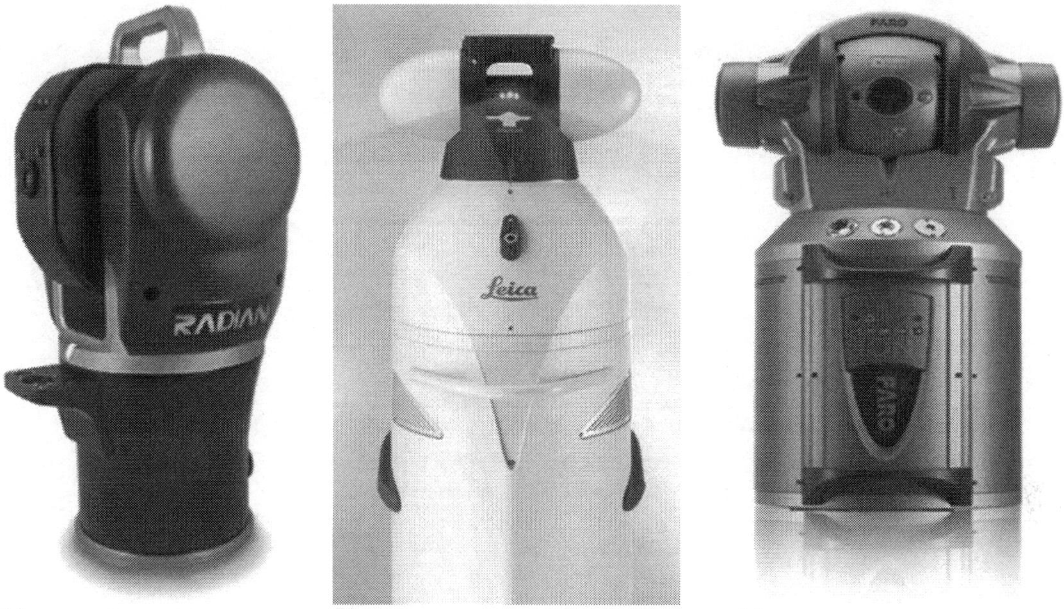

(a) (b) (c)

FIGURE 3.4
(a–c) Current laser tracker models (2012). (a: Courtesy of Automated Precision, Inc., Rockville, MD; b: Leica Geosystems, St. Gallen, Switzerland; c: Faro Technologies, Inc., Lake Mary, FL.)

There are currently three manufacturers of laser trackers:

- Automated Precision, Inc. (API)
- Leica Geosystems (Hexagon)
- FARO Technologies, Inc.

Shown in Figure 3.4a through c are current laser tracker models (2012).

API and FARO Technologies Inc. are American manufacturers. Leica Geosystems is a Swiss manufacturer but part of Hexagon Metrology, a Swedish company with a registered head office in the United Kingdom. The different laser trackers have similar accuracy specifications and similar measuring envelopes, with a usable reach in excess of 30 m.

3.2.2 Tracking Mechanism

Manufacturers incorporate different design features into their instruments, but the basic tracking concept is common to almost all. (Leica's AT401 model introduced in 2010 uses a different method found in surveyor's total stations.) For convenience of illustration, an early model Leica instrument, the LTD 500, will be used to show the mechanism, but note the following principal differences in where the beam is generated and pointed:

- API: The laser beam is generated in an enclosure directly mounted in the rotating head.
- FARO: The laser beam is generated in the fixed base of the instrument and transferred to the rotating head by a fiber optic.

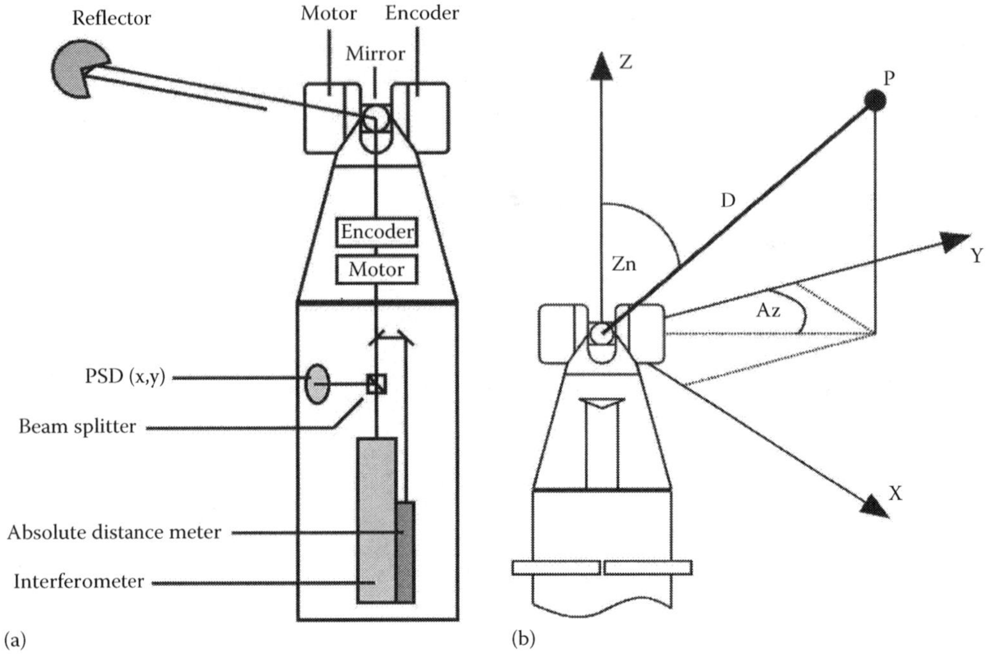

FIGURE 3.5
Schematic diagrams of Leica LTD500: (a) Components and (b) coordinate system (Courtesy of Leica Geosystems, St. Gallen, Switzerland.)

- Leica: The laser beam is generated in the fixed base of the instrument and reflected off a mirror located in the rotating head. (Leica's AT401 model is again an exception to the main range, and its beam is generated in an enclosure mounted in the rotating head.) Figure 3.5a and b shows schematic diagrams of the Leica LTD500 (Leica).

In the Leica LTD 500, an IFM beam in the base reflects off a rotating mirror in the head and onto a target retroreflector. This has the property that it reflects the beam back along the incoming direction so that it re-enters the instrument. Here, a beam splitter diverts part of the return beam for interferometric distance measurement and part onto a position sensing device (PSD). Reflector movement along the beam is detected by the IFM and converted to a distance between tracker and reflector. Any lateral movement of the reflector off the beam causes a corresponding lateral displacement of the return beam which is detected by the PSD as an offset (x,y) value within its flat surface. When the beam is on target, a zero offset is defined. When a non-zero value is present, it is used to calculate corrective values to the beam angles which bring it back on target, reducing the offset back to zero. The angle of the beam is determined by angle encoders on each of the two orthogonal axes of rotation. With an interferometrically measured distance D, a zenith angle Zn and azimuth angle Az, the reflector is located in the tracker's coordinate frame of reference by spherical coordinates which are normally converted to Cartesian XYZ values on output.

3.2.3 Leveled and Non-Leveled Operation

It is not necessary for the tracker to be accurately leveled or otherwise referenced to the vertical. This is an option where it is convenient for X and Y to represent values in a horizontal plane and Z to represent height values. Typically, trackers are normally operated in an only approximately leveled state. In some applications, they have been suspended upside down or even mounted sideways (rotated through 90°) in order, for example, to measure down vertical shafts.

3.2.4 Range Measurement: In Brief

Early laser trackers measured distance, also called range, by laser IFM only. An IFM (Section 3.5.1) measures a change of distance, that is, the distance a retroreflector moves along its beam from a random starting point. For 3D measurement, an absolute distance is required which is relative to the tracker's origin at the intersection of its rotation axes. With a 3D tracking laser IFM, this is achieved by starting measurement with the reflector in a known position, also called home position, which is typically a fixed locating point on the tracker itself. Thereafter, the IFM provides an accurate ± change to the known distance relative to this fixed position.

For interferometry to work, the outgoing and reflected beam must continuously interfere with one another. If the reflection is interrupted, for example, when someone walks through the beam or the reflector's axis is turned too far away from the tracker and no beam is returned (see Section 3.5), then the absolute distance measurement must be re-initialized either back at the fixed point on the tracker or at some temporary nearby location which the operator has setup earlier in the measurement process.

This method of recovery from a beam break was not ideal, so advancements of laser trackers soon moved to the development of absolute distance measuring techniques which have been common in land surveying instruments but which required optimizing for high-accuracy metrology applications. With a coaxial absolute distance meter (ADM), which typically took a couple of seconds to make its measurement, it was possible to initialize an IFM reading to an absolute starting value provided the reflector was in a stable position for the duration of the ADM measurement. This position did not require a known 3D location although the tracker often required a search technique or auxiliary camera view to find the reflector. The use of a coaxial ADM and IFM for this purpose is indicated in the schematic diagram of the LTD 500 in Figure 3.5a and b.

In recent years, ADMs have become capable of operating at very high speed, sufficiently high that they can themselves be used to track a moving target reflector accurately in 3D. IFMs still offer the potentially highest distance accuracy, and so a combination IFM/ADM device may be the one of choice for a particular application. However, many laser trackers are now only offered with an ADM which, if based on a hybrid distance measuring technique involving interferometry, may offer very high distance accuracy.

An ADM considerably improves handling with respect to beam interruptions which are difficult to avoid, particularly in cluttered and busy environments. By adding a coaxial camera view around the beam, a secondary illumination can provide an image in which a reflector appears as a bright spot. If tracking has been lost, but the reflector is still within the camera's field of view, a corrective beam pointing can be calculated using the reflector's image position. This brings the beam back on target, and, with the almost instant ADM facility, distance is reestablished, and 3D measurement can continue. In this way, an interrupted beam can be caught "on the fly."

The ADM is also useful for measuring multiple reflectors, for example, in a monitoring application where a number of locations are being checked for stability. Here, an IFM would not be ideal since it could only work by following a single reflector moved from one location to another. Access difficulties, long cycle times, and physical inconvenience would all argue against this solution and in favor of an ADM.

3.2.5 Reflector

An IFM only works if the outgoing beam is reflected back along itself so that the outward and return beams can be superimposed at the instrument and interfere with one another. This requires a special design of reflector known as a retroreflector. In surveying instruments, ADMs can work without reflectors and utilize the diffuse reflection which results from most surfaces which are not very smooth or polished. However, ADMs only achieve metrology accuracy with the same type of reflector as required by IFMs.

Figure 3.6a illustrates a common type of retroreflector known as an *open-air* or *air-path corner cube reflector*. This has three plane mirrors placed orthogonally to one another, that is, forming the corner of a cube. The target point is defined by the corner point or apex of the pyramid formed by the three planes. The 2D illustrations apply the simple geometry of plane reflection to show how an incoming beam returns along a parallel path (back to the generating instrument). The situation is a little more complicated in 3D, but simple vector geometry can demonstrate that the reflected beam is also parallel to the incoming beam in the 3D case.

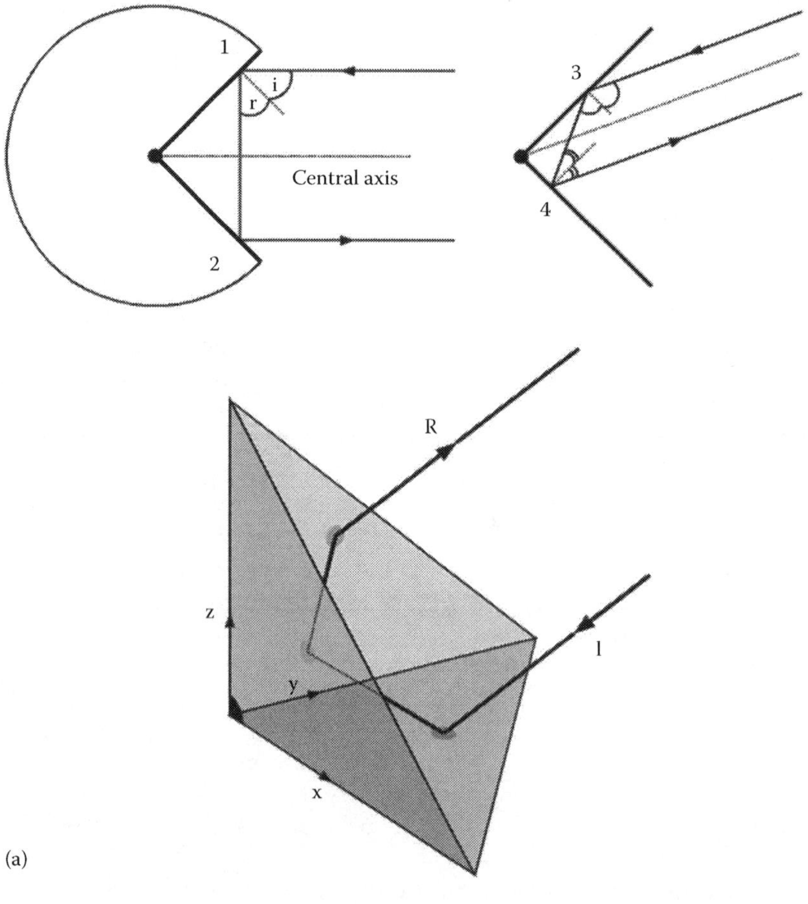

FIGURE 3.6
(a) 2D reflection geometry for corner cube reflector and 3D corner cube axes and beam reflection. (Courtesty of S. Kyle; Leica Geosystems, St. Gallen, Switzerland.)

(continued)

(b)

FIGURE 3.6 (continued)
(b) Corner cube reflectors. (Courtesy of Leica Geosystems, St. Gallen, Switzerland.)

The central axis of the reflector is the line making an equal angle with the perpendiculars to each of the plane surfaces. It will be clear that there is a limit to the angle which an incoming beam can make with this axis, and an incoming beam at a greater angle will not generate a return beam to the instrument. This limit defines the *acceptance angle* of the reflector which may often, and not entirely logically, be stated as a plus/minus value.

It may help to think of the reflector as having a cone with apex at the intersection point of the mirrors, cone axis defining the reflector axis and a half-angle "A" equal to the acceptance angle. The incoming laser beam must lie inside the cone, as shown in Figure 3.7. This in turn means that the reflector must always be pointed approximately at the laser tracker within this angle which is approximately 25° for an open-air corner cube.

Other designs of retroreflector are possible and in current use, such as types where the cube corner is manufactured from solid glass or from concentric glass spheres. These are discussed in more detail in Section 3.6.

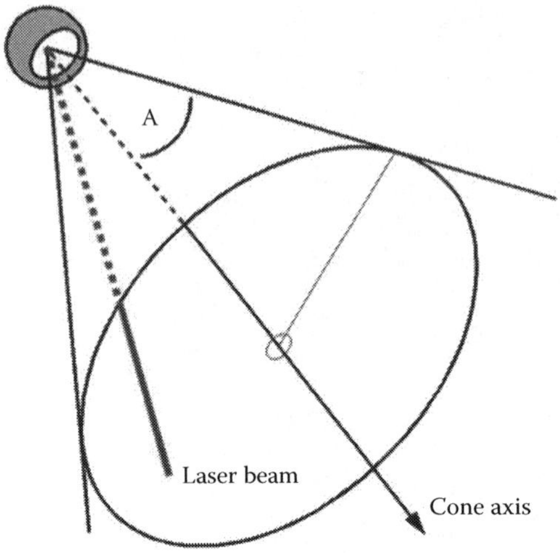

FIGURE 3.7
Acceptance cone. (Courtesy of S. Kyle.)

3.2.5.1 Reflector Offsets

Reflectors are commonly fixed into spherical housings such that the center of the housing and target point of the reflector are effectively at the same location. This *spherically mounted retroreflector* (*SMR*) can then be used directly as a contact probe to measure object points, in much the same way as the ruby ball on a CMM stylus is used to make contact measurement of an object under examination.

However, as Figure 3.8 illustrates, the measured point is the center of the target and not the contact point which is the object point of interest. A correction must therefore be made for the radius of the SMR (as is also the case for the ruby ball on a CMM stylus).

One solution to correcting for reflector radius is shown in Figure 3.9. Here, measured reflector center positions have a curve fitted through them in a plane. This enables calculation of the curve normals in the plane so that corrections can be made along the directions of these normals through the measured points to the true contact points on the surface. Simpler correction methods are based, for example, on knowledge of the local coordinate system of the measured object so that an offset could be corrected along one of the local axes.

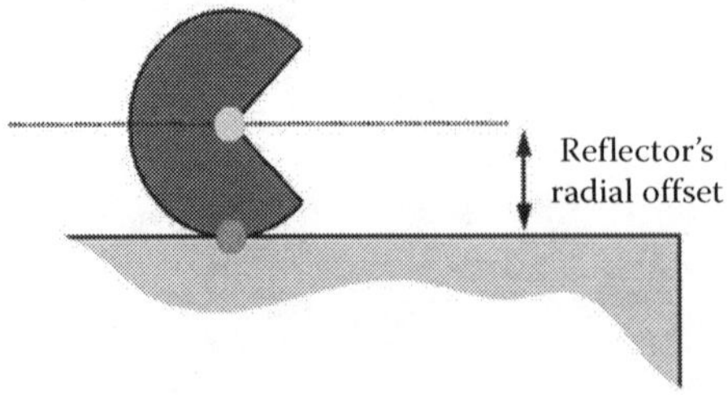

FIGURE 3.8
Radial reflector offsets. (Courtesy of S. Kyle.)

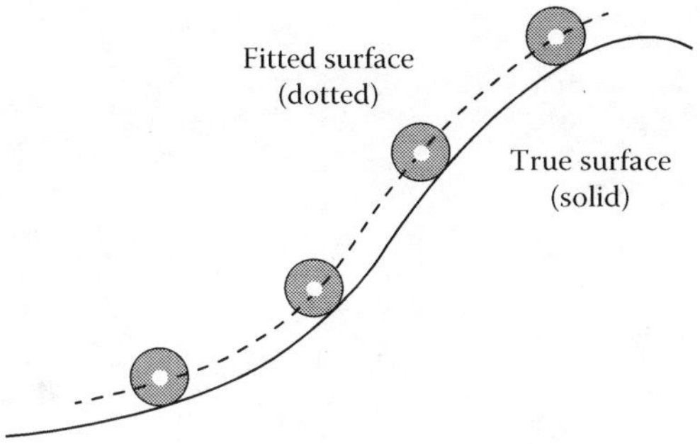

FIGURE 3.9
Correcting for reflector radius. (Courtesy of S. Kyle.)

3.2.5.2 Target Adapters

Sometimes the SMR may be used in conjunction with a *target adapter,* for example, to locate the center of a drilled hole, and again an offset correction must be made as indicated in Figure 3.10. Adapters are useful measurement tools in their own right, and the *pin nest* target adapter will be used to illustrate the adapter's function as a locating and probing device.

The pin nest adapter is shown in Figure 3.11. The adapter has a strong magnet to hold an SMR securely in a kinematically stable three-point mount. It also has a precision shaft or pin whose axis runs through the center of the SMR when it is placed in its mount. The precision pin fits into bushed holes in the object so that the SMR center can be relocated in a consistent and repeatable way. The actual measured point (SMR center) is above the hole at a distance equal to the SMR radius plus the additional adapter thickness. A common pin nest adapter for a 1.5 in. diameter SMR adds an additional 0.25 in. to the radial offset. The net effect is that the measured point has a 1 in. (25.4 mm) offset in the direction of the hole's axis. This value is known and compensated in the tracker software when needed.

The pin nest adapter can also be used to measure machined holes that are larger than the pin diameter, or object spigots (pins), provided the adapter can be held securely against both the top and side of the hole or spigot. The adapter, with its SMR, is placed so that the pin is in contact with the internal face of the hole or side of the spigot and the base of the adapter is in contact with the top of the hole or spigot.

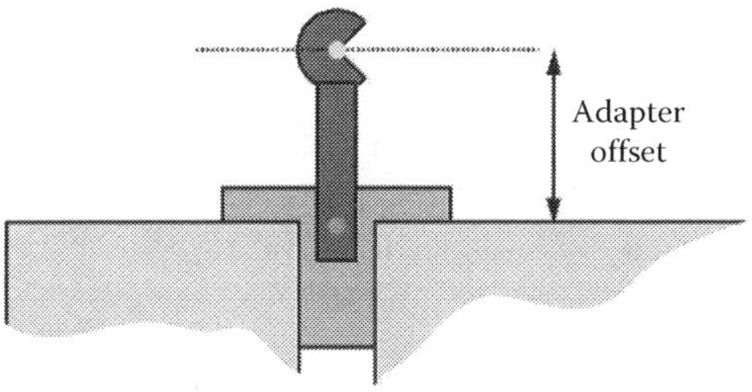

FIGURE 3.10
Adapter offset. (Courtesy of S. Kyle.)

FIGURE 3.11
Pin nest adapter.

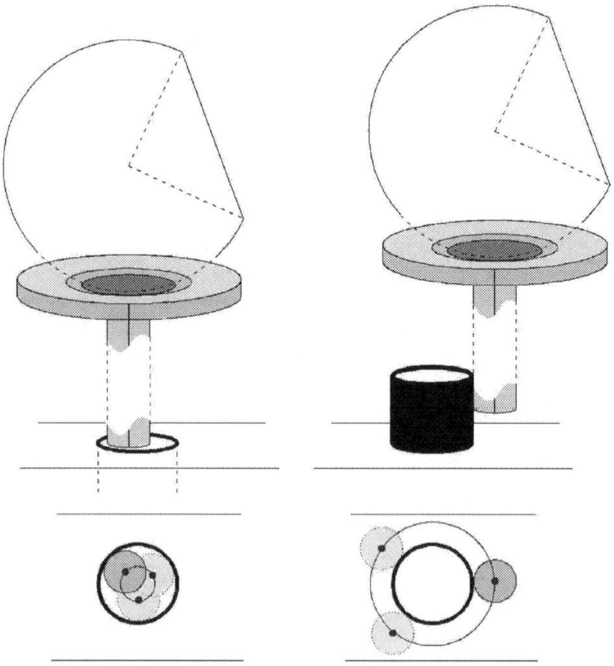

FIGURE 3.12
Pin nest adapter measures hole and spigot. (Courtesy of S. Kyle.)

Figure 3.12 shows an exploded view of the situation (lower surface of SMR mount has been drawn away from the hole and spigot surfaces). In use, the adapter is moved to at least three such positions around the hole or spigot, the SMR's center being recorded at each location. In the case of a hole, the three reflector centers (small circles) generate a circle whose radius is *less* than the measured hole radius by the radial value of the adapter's pin. In the case of a spigot, this circle is *greater* than the spigot radius by the same value. The appropriate correction is made in either case.

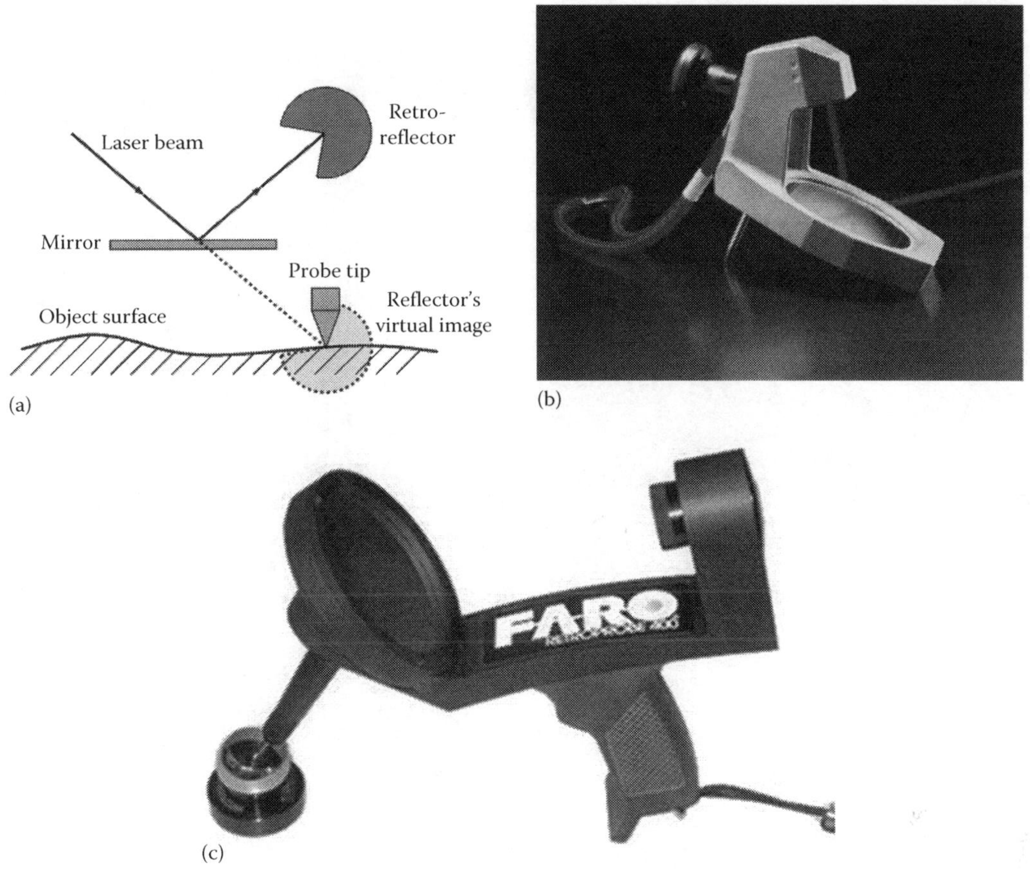

FIGURE 3.13
(a) Virtual reflector. (Courtesy of S. Kyle; Leica Geosystems/FARO Technologies, Inc., Lake Mary, FL.) (b) Surface reflector. (Courtesy of Leica Geosystems, St. Gallen, Switzerland.) (c) Retro-probe. (Courtesy of FARO Technologies, Inc., Lake Mary, FL.)

3.2.5.3 Virtual Reflector (Mirror Probe)

An alternative solution to the problem of correcting for a reflector's radius is to use a virtual reflector, also called a *mirror probe*, as presented in Figure 3.13a, also known as a surface reflector (Figure 3.13b) and a retro-probe (Figure 3.13c). Here, the incoming laser beam is reflected off a plane mirror onto the retroreflector but appears to be measuring to the reflector's virtual image position. By placing a small probe tip at the center of the reflector's virtual image, surface features can be measured directly. However, this does not appear to be a widely used device.

3.2.5.4 Hidden Point Rods

Hidden point rods, also known as *vector bars*, are a common type of target adapter used to measure points and features to which the tracker does not have a direct line of sight. These are known as *hidden points*, and the adapter used to measure them is a rod with two target nests. The target nests are configured so that the measured SMR centers form a straight line that passes through the hidden point.

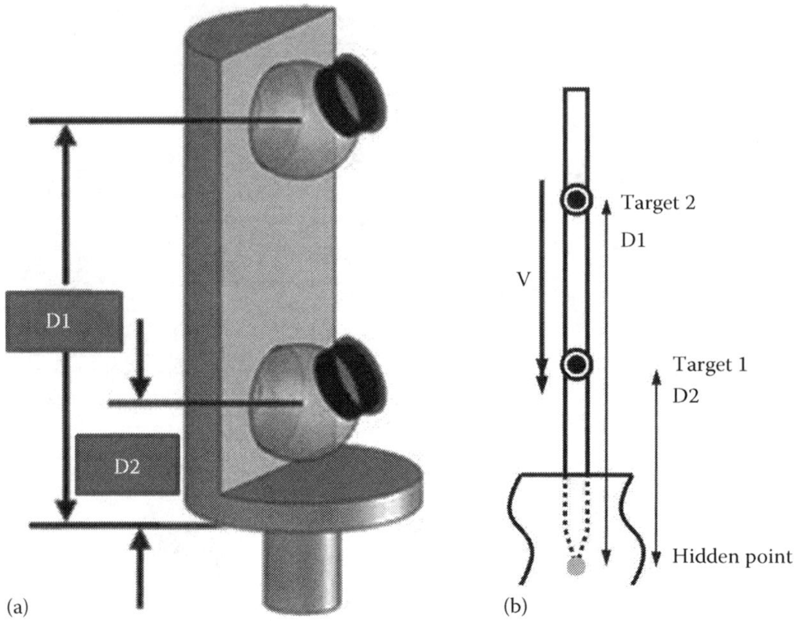

FIGURE 3.14
(a) Hidden point adapter. (Courtesy New River Kinematics, Williamsburg, VA.) (b) Geometry of hidden point adapter. (Courtesy of S. Kyle.)

Figure 3.14 shows a possible adapter design on the left and the elements of the hidden point calculation on the right. By measuring both SMR positions, the vector V between them can be calculated. This can be combined with the known distance from either position to the hidden point (D1 or D2) in order to compute the hidden point's location. Hidden point bars come in many different configurations. Some have both targets on one side of the rod and project to the hidden point. Other hidden point rods have the target nests on both sides and compute a hidden point between them.

3.3 Line of Sight and Measurement Networks

It is an inherent feature of optical dimensional measurement systems that a line of sight is required between the measuring instrument and the target point being measured. However, many objects have complex shapes, and this complexity only increases where multiple objects are to be measured or located relative to one another. Put simply, it is not possible to measure the back side of an opaque object when viewing it from the front. In order to measure all points of interest in a complex object or scene, the essential strategy is to have multiple view points from which all points of interest can be seen. This can be achieved by moving a single instrument from one viewing position to another, having multiple measuring instruments occupying all the viewing positions, or some combination of these arrangements. Regardless of the use of single or multiple instruments, the multiple viewing locations make up a *measurement network*. However, these multiple view points, and the measurements made from them, must be located in a common coordinate system

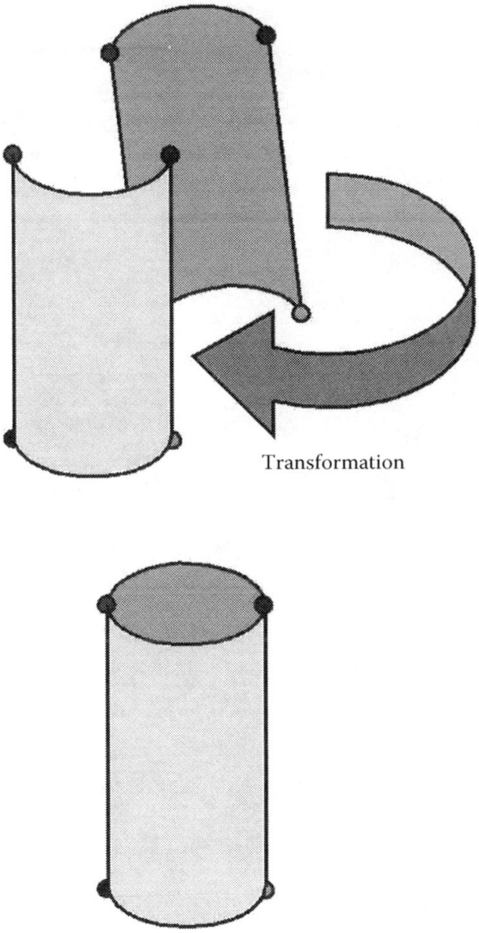

FIGURE 3.15
Measurement in parts. (Courtesy of S. Kyle.)

in order to provide a single picture of the object or scene being measured. The solution is achieved by ensuring there are some common points of measurement between every measurement location (view point) and at least one other measurement location. This enables every set of measurements to be mathematically linked together into a common single framework. There are good and bad ways of doing this, and a separate, detailed discussion is required for a full explanation.

However, a simple strategy illuminates the solution. Consider first the problem of measuring an object in parts in a way which provides a whole object result. In Figure 3.15, a cylinder is shown measured in two parts, light and dark gray, but four points on each part are common to each. A relatively simple 3D mathematical transformation can be applied to the light gray part measurements which make a best fit of the four points on the light gray part to the corresponding four points on the dark gray part, thereby merging all data into a single cylinder. For 3D point measurements, the minimum number of common points is three, and they should form a "good" triangle. Three points on a straight line, for example, would form a hinge and leave the cylinder "open."

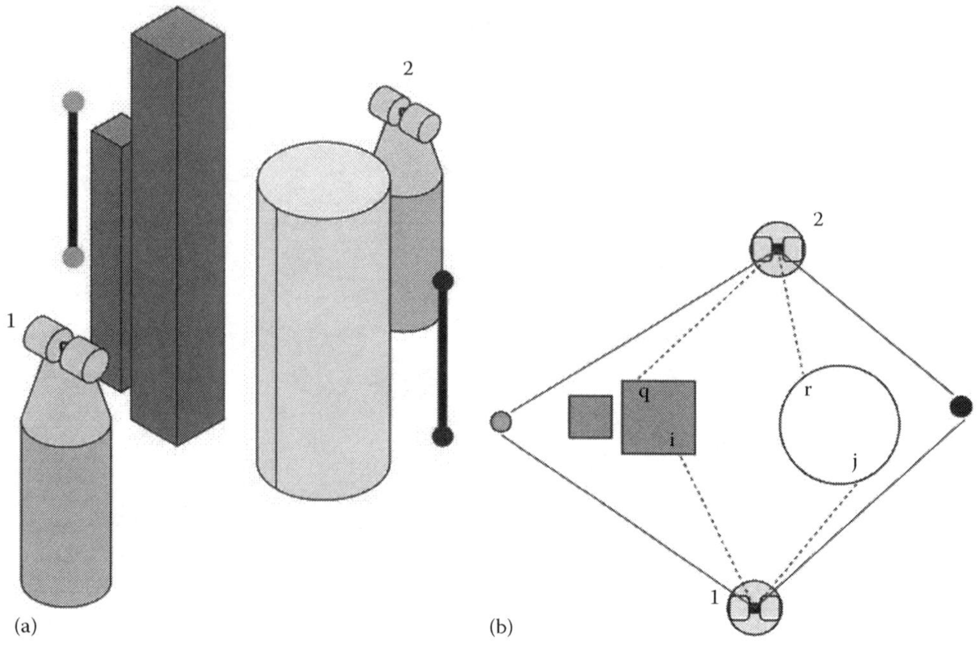

FIGURE 3.16
(a and b) 2-Station network. (Courtesy of S. Kyle; NPL, Middlesex, U.K.)

In Figure 3.16, all points in a multi-object scene, for example, object points at i, j, q, and r, cannot be seen from a single location, so the tracker must be moved from viewing position 1 to viewing position 2 (or two trackers must be used, one at each location). One solution (others are possible) is to place at least three target spheres in positions which can be seen from both tracker locations. In the illustrated example, four are shown, two light gray spheres and two dark gray spheres. In addition to measuring object points of interest at each tracker location, the spheres are also measured as a set of surface points from which their centers can be calculated. This provides four common points in each of the two data sets. Using the same 3D transformation for merging the separately measured cylinder parts mentioned previously, the two separate data sets can be merged here. In fact, by computing the transformation as the first action at viewing position 2, further data generated at position 2 can immediately be presented in the coordinate system defined at position 1. Further viewing positions can clearly be added in a similar fashion. It must, however, be emphasized that simple successive transformations to locate each instrument position will not give the most accurate result over a large volume. A more comprehensive network optimization is able to consider all the observations simultaneously, so yielding the best possible instrument and point positions. The advantages of network optimization are covered in Section 3.8.

3.3.1 6D Probing

Another way to look in and around an object is to use a laser tracker in combination with an articulated arm coordinate measuring machine (AACMM) more commonly known as a CMM arm or simply "arm."

As Figure 3.17a shows, the arm is like a manually operated robot, with rigid links connected by rotating joints, and can optionally have a touch probe as the end effector.

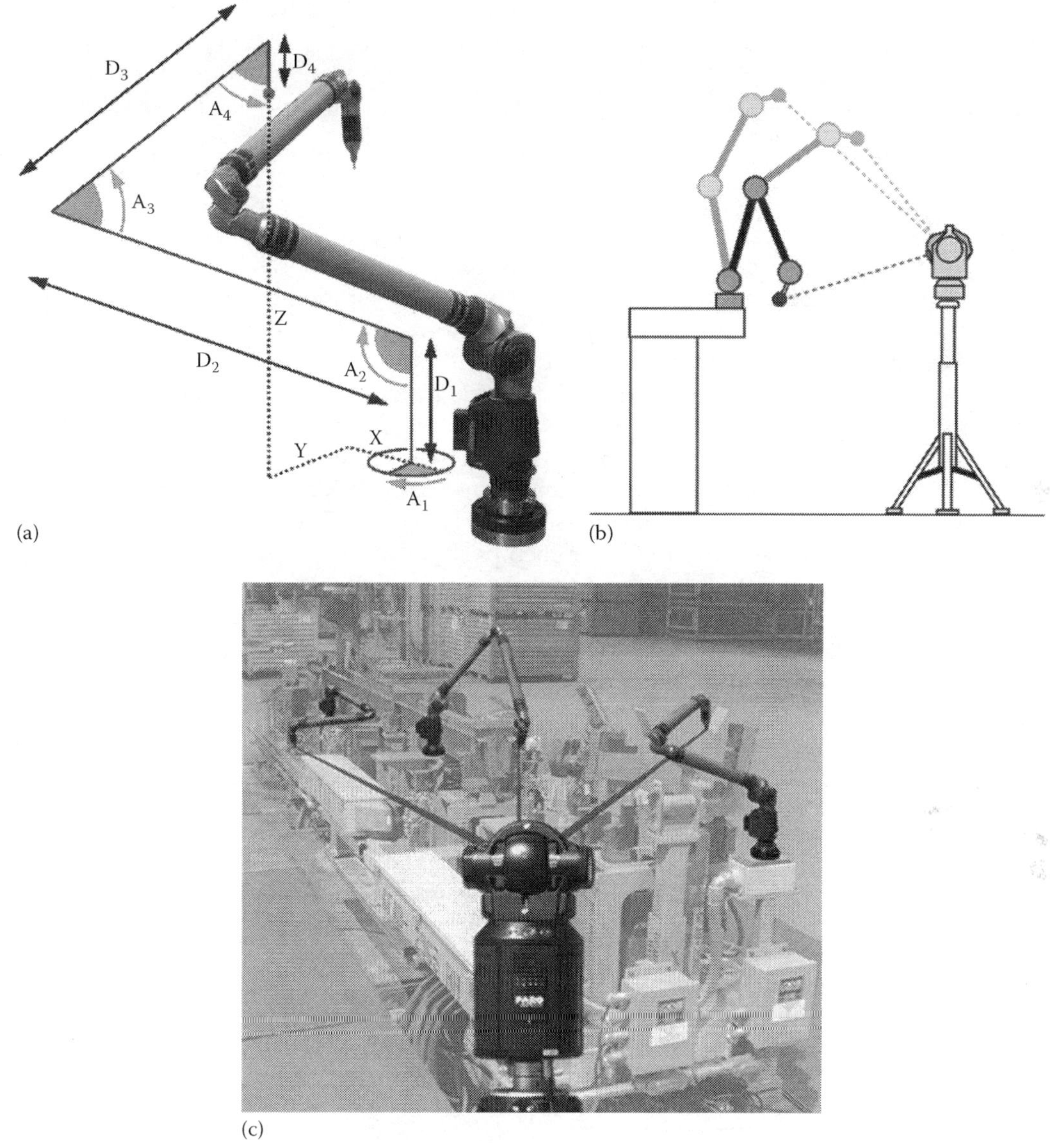

FIGURE 3.17
CMM arm and arm repositioning. (a) Arm concept, (b) arm positioning, and (c) multiple arm locations. (Courtesy of S. Kyle; FARO Technologies, Inc., Lake Mary, FL.)

The operator physically moves the arm into positions where the contact probe can be touched against object points of interest. The 3D position of the probing point in the arm's coordinate system, shown as X, Y, Z, can be calculated from the known link lengths D_n and encoder readings of joint angles A_n.

The arm has a limited reach, typically 1–2 m, but has no line-of-sight restrictions and can reach in and around objects for detailed coverage. However, because the reach is limited, it is sometimes used in combination with a laser tracker. Figure 3.17b shows how a tracker target on the arm is measured in at least 3 arm orientations in order to provide the information which enables its own coordinate values to be transferred into the coordinate system of the tracker. This is the same 3D transformation technique as described earlier for merging multiple data sets. In fact, merging separate data sets is also what is being done here.

Figure 3.17c shows how the arm can be repositioned in multiple locations in order to cover a large object or objects beyond its reach from a single location.

Arm repositioning can be seen as just another variant of a multiple measuring instrument network, but it also helps to introduce the concept of measuring all 6 degrees of freedom (6DOF or simply 6D) of a probing device. Any object such as a manufactured part, probing device, or entire aircraft requires not only the 3D position of one of its points but also its angular orientation in order to describe fully its position in space.

Figure 3.18 uses the example of an aircraft's attitude to show how three angular values, for example, roll, pitch, and yaw, provide the additional angular orientation information which, when combined with a 3D point on the aircraft, gives its full 6D location and orientation.

For CMM arm repositioning, the arm's 6D is effectively determined by the measurement of at least three positions within its own measurement space. This was also the case in the earlier example where a cylinder was measured in two parts. The four common measured points effectively provided the full 6D for each part.

However, multiple point measurement does not permit a laser tracker to track an object's 6DOF in real time as is required if the object is a constantly moving, handheld measuring probe. Consider the problem from a different perspective. Figure 3.19 shows the measurement of a hidden point which is not visible to the tracker. Potentially, the problem is solved with a target probe which is like a miniature CMM arm comprising an SMR, stylus, and offset probing point. To locate the contact probing point from the tracked SMR location requires additional information, namely, the offset length d of the tip and the direction of the offset as defined by the space angle a.

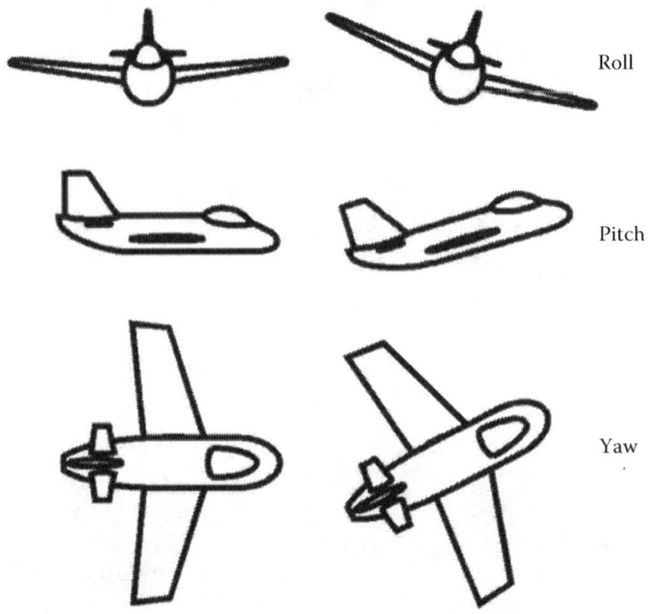

FIGURE 3.18
Aircraft attitude. (Courtesy of S. Kyle.)

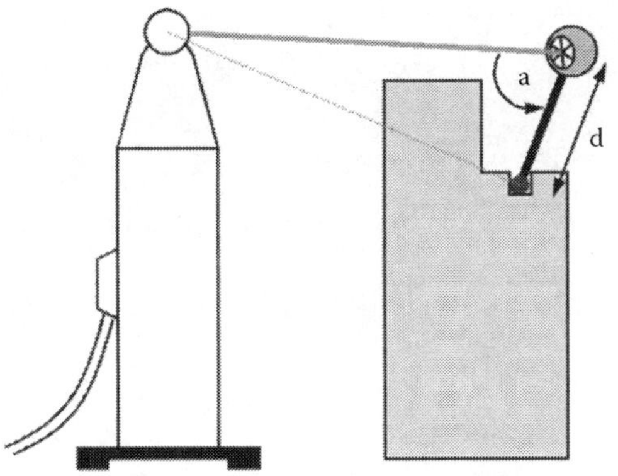

FIGURE 3.19
6D probing. (Courtesy of S. Kyle.)

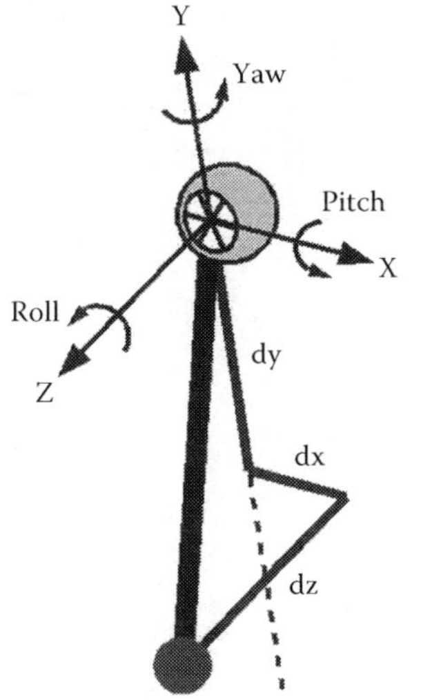

FIGURE 3.20
General 6D probe. (Courtesy of S. Kyle.)

The offset tip is a fixed part of the probe's geometry (and this can be extended to the concept of exchangeable probe tips). Once determined by manufacture and/or calibration, it is known. The real task is to track, in real time, the probe's roll, pitch, and yaw defined relative to some local probe coordinate system (see Figure 3.20).

The tracker itself can only follow one point in 3D—the SMR—so additional measurement capabilities are added which determine the angular orientation values. Currently (2012), Leica and API offer real-time 6D probing, and their techniques for doing this are briefly outlined in the next two sections. FARO currently only offers the two-step solution provided by CMM arm repositioning although it should be remembered that, once in position, the arm also delivers real-time measurement data.

Another point to note about 6D probing is that it is inherently required for surface form measurement using laser line scanners. These devices project a laser fan beam which defines a short profile on an object's surface. The profile is recorded as a dense 2D line of points. By "painting" the fan beam across the surface, a dense 3D point cloud can be built up as a sequence of 2D profiles. While the scanner head can be tracked in 3D, the requirement to locate the offset profile demands a full, real-time 6D solution.

3.3.2 6D Tracking: Leica Geosystems

In the Leica system, a cluster of light-emitting diode (LED) targets surrounds the retroreflector in a 6D probe (Figure 3.21). A vario-zoom camera, which can optionally be mounted onto the rotating head, tracks the LED cluster (Figure 3.22). From the image of

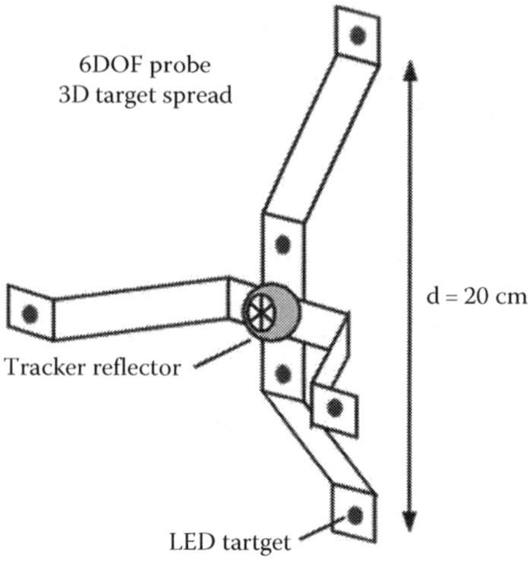

FIGURE 3.21
6D Leica concept probe. (Courtesy of S. Kyle; Leica Geosystems, St. Gallen, Switzerland.)

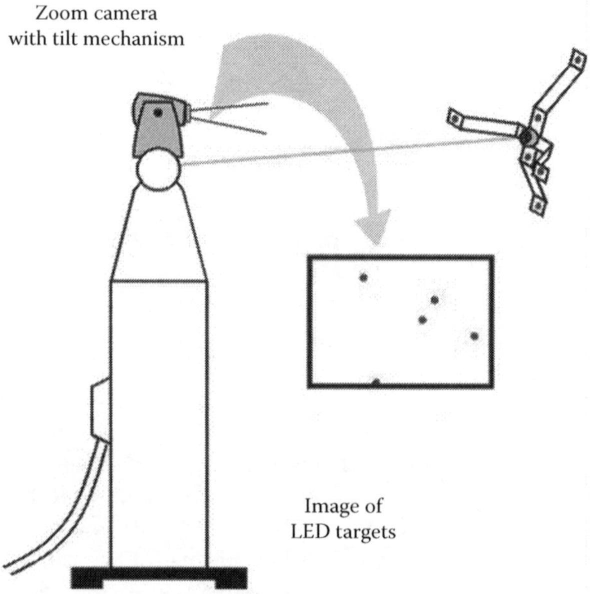

FIGURE 3.22
6D probe location concept. (Courtesy of S. Kyle; Leica Geosystems, St. Gallen, Switzerland.)

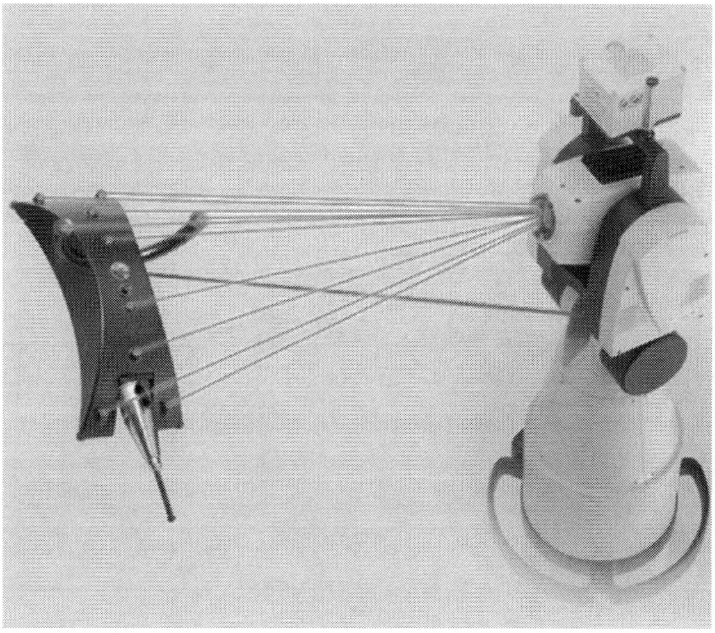

FIGURE 3.23
T-Cam + T-Probe. (Courtesy of Leica Geosystems, St. Gallen, Switzerland.)

the cluster, the relative 6D between camera and cluster can be calculated using a standard mathematical technique known in photogrammetry as a *space resection*. Although the camera provides 6D data, target distance is very inaccurate due to the narrow angle of view. Only the rotational components of the 6D are used since the tracker itself provides very accurate 3D.

For the concept to work accurately, the LED targets must have accurate 3D coordinates in a local probe coordinate system, and they must have a spread in 3D (not just, e.g., in a plane). A calibration process provides this information. Figure 3.23 shows the Leica 6D system. The probe is called the T-Probe and has a tip for touch probing. A T-Scan version is available for surface scanning. On the tracker head is the zoom camera known as the T-Cam. (Note that the additional box over the camera is an optional level sensor.)

3.3.3 6D Tracking: Automated Precision, Inc.

API's first 6D tracking system used geometry very similar to theodolite orientation to deliver 6DOF data. In theodolite orientation (relative positioning) shown in Figure 3.24, each instrument points at the location of the other (the use of substituted targets is possible in surveying). This generates vectors R1, R2 on the baseline between the instruments. By referencing each instrument to gravity (vectors G1, G2), this fixes the roll angle between them. Only distance D must be determined to complete the 6DOF connection (D is determined indirectly when using theodolites). In an intermediate design step, Figure 3.25, one theodolite is replaced by a total station (which provides direct distance measurement) and one by a camera. In practice, the tracker acts in place of the total station and a pinhole prism reflector, Figure 3.26, substitutes for the camera. This is a normal prism reflector (see Section 3.6) for a tracker but with its apex removed to enable part of the tracker's laser beam to pass through and onto an offset sensor such as a PSD. The XY sensor position

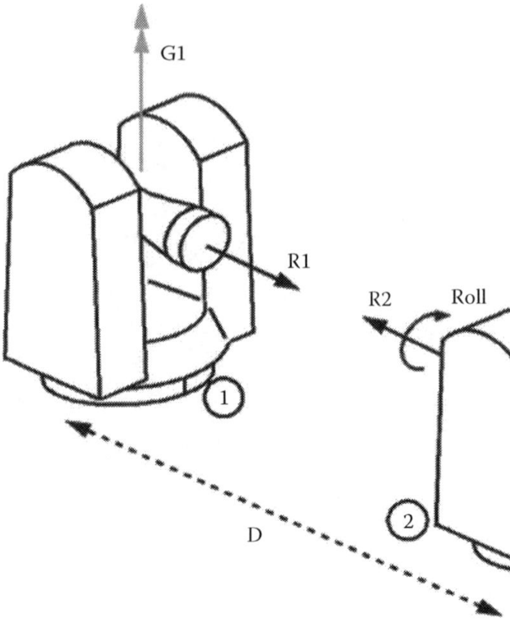

FIGURE 3.24
Leveled theodolites. (Courtesy of S. Kyle.)

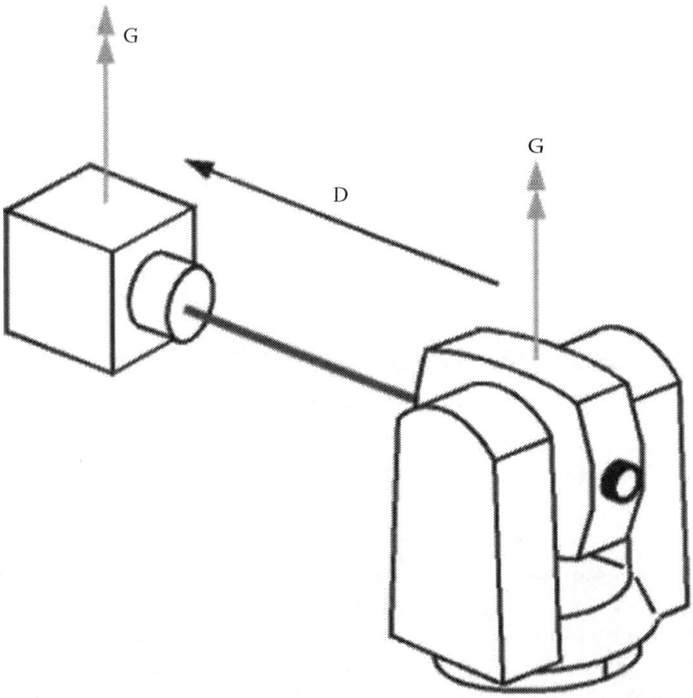

FIGURE 3.25
Leveled theodolite/camera. (Courtesy of S. Kyle.)

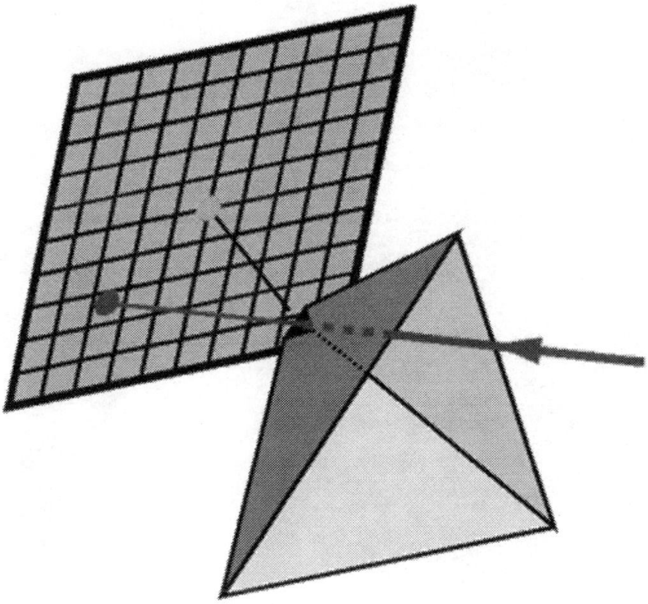

FIGURE 3.26
Pinhole reflector. (Courtesy of S. Kyle.)

of the beam effectively provides the R1 vector in Figure 3.26. Figure 3.27 shows the API instrumentation which uses this concept. API uses this mechanism in their Active Target which uses the offset measured by the PSD as feedback to motors (zenith and azimuth angle) to automatically orient the reflector to face the tracker.

3.4 Laser Tracker: Range Measurement Details

3.4.1 Relative Range Measurement by Interferometer

The technology for range measurement has developed significantly since the first laser trackers. The first laser trackers used a relative IFM, a variation of the Michelson IFM,[2] to measure the distance to the target.

A Michelson IFM splits the laser beam into two separate beams. One part stays in the instrument and provides a reference, and the other goes out to the reflector and back. The returning beam is recombined with the reference beam. Due to the wave property of the laser beams' coherent light, this causes interference of the beams in which the waves either add together to create a *light fringe* (constructive interference) or cancel to create a *dark fringe* (destructive interference), these fringes being observed with a photosensitive detector.

Figure 3.28 shows a conventional arrangement for a Michelson IFM. One mirror is usually fixed, and one moves. In the case of the laser tracker, this is the SMR. As the moving reflector moves toward or away from the coherent light source (laser), the phase of the return beam changes. This causes the interference fringe pattern to alternate repeatedly between constructive and destructive interference.

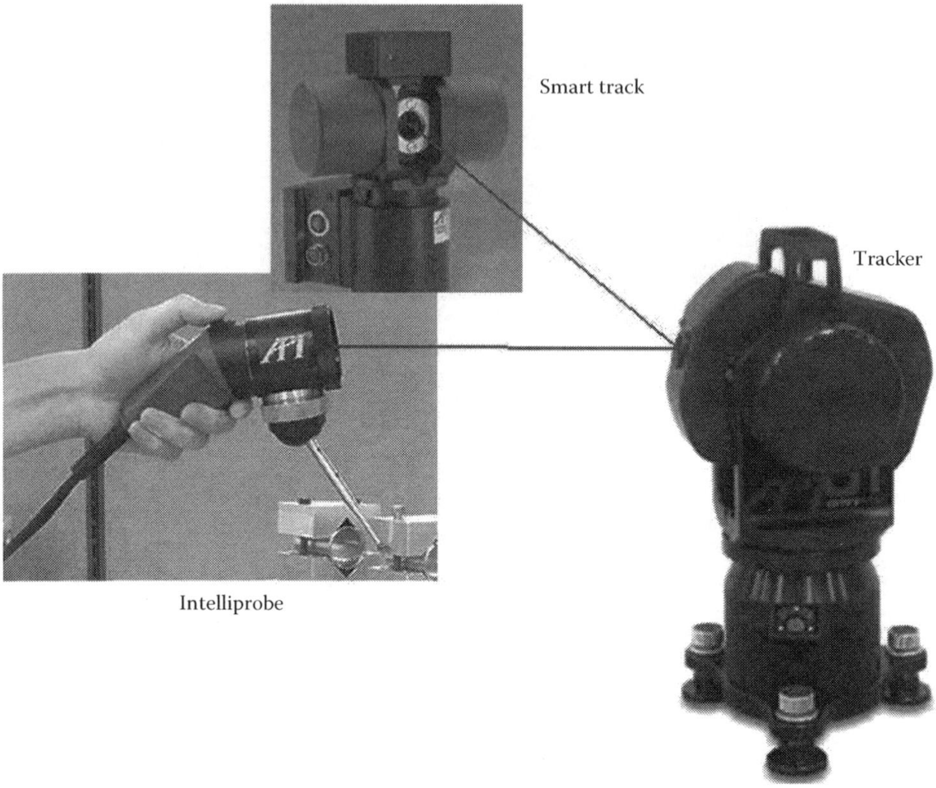

FIGURE 3.27
API 6D system. (Courtesy of S. Kyle; Automated Precision, Inc., Rockville, MD.)

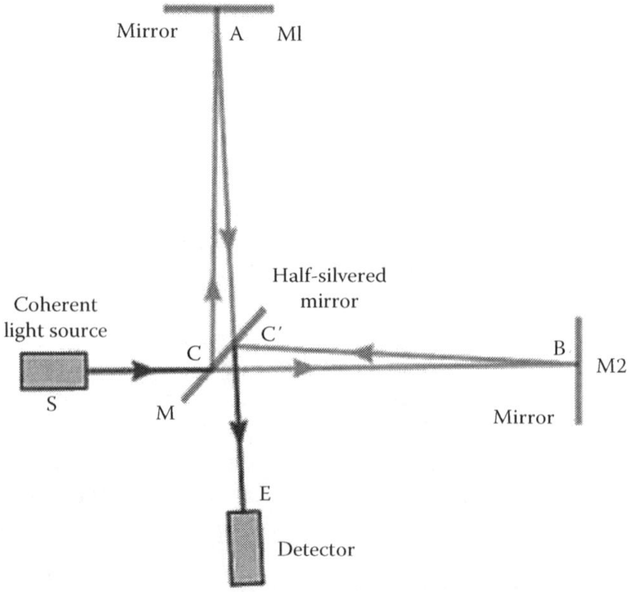

FIGURE 3.28
Michelson interferometer.

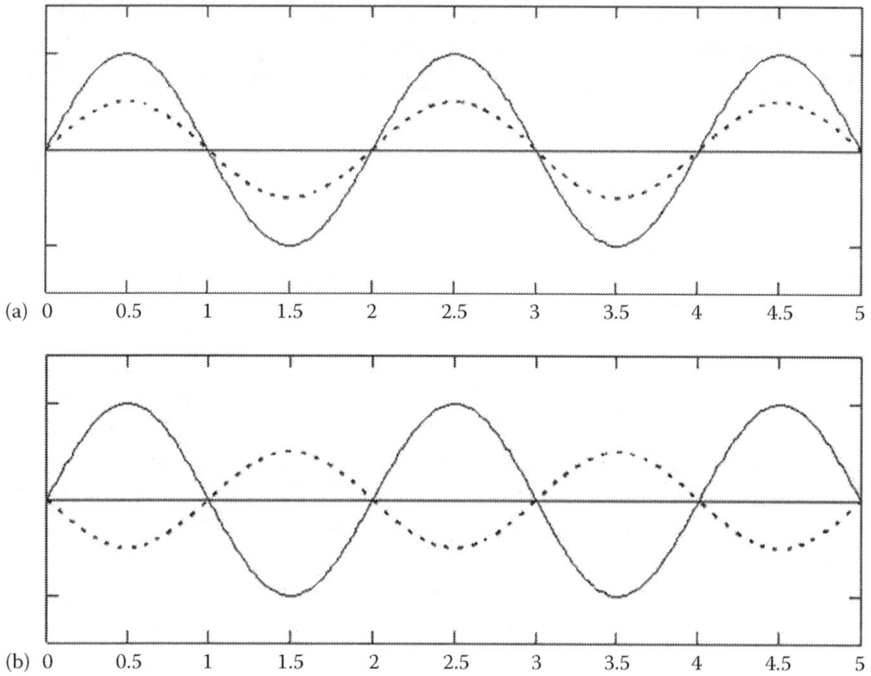

FIGURE 3.29
(a) Constructive and (b) destructive interference of waves. (Courtesy of S. Kyle.)

When the reflector moves a distance equal to ¼ of the wavelength, the round-trip distance traversed by the light out to the reflector and back is a ½ wavelength. This shifts the relative position of the outward and returning light waves by the same amount, corresponding to a change between constructive and destructive interference as shown in Figure 3.29. A further ¼ wavelength movement of the reflector returns the system back to a light (or dark) result. The effect is observed by the sensor system that counts the number of these light or dark fringes. Various techniques are used to determine if a change is added or subtracted, that is, if the reflector is moving away from or toward the instrument.

The wavelength of light is known. For example, a helium–neon (He-Ne) laser has a wavelength of 633 nm in a vacuum. Once corrected for environmental conditions, half this value can be multiplied by the number of fringes counted by the sensor to find the distance moved relative to the initial home position. This is a very fast operation and can be computed continuously in real time.

The accuracy of this system is dependent on the accuracy of a number of parameters. These include the accuracy of the initial home position and how well the system can determine the wavelength of the light in the current environmental conditions which are measured by a *weather station*. In theory, the IFM system is capable of nanometer precision and a correspondingly very high accuracy. As a general guide, the system can produce range measurements between 2 and 10 μm ± 2 parts per million (ppm). These numbers are higher for laser trackers than commonly published values for IFMs due to target manufacturing uncertainties, home position calibration, and the weather station's ability to accurately characterize and compensate for the environmental effects on the laser light along the beam path.

While a relative IFM can provide an accurate and continuous distance measurement, it does require that the laser beam be continuously locked on the reflector. Any break in the beam causes a loss in fringe counting. When the beam is broken, the relative IFM's initial range must be reset to a known value.

3.4.2 Absolute Distance Meter

In a significant percentage of industrial applications, keeping the laser continuously locked on the reflector is not practical. An early development in laser tracking, therefore, was the ADM. Trackers with ADMs are able to set the range to the target directly, meaning they do not require a home or reference point. ADM systems typically use infrared laser sources to measure the range using one of three techniques. Each technique modulates a property of the light source and applies a phase measurement technique, to resolve the range to the target. The three most common light properties that are modulated in laser tracker ADM systems are amplitude (intensity), polarization, and frequency (color).

Amplitude systems modulate the intensity of the light. A primary advantage of this technique is the relatively high speed at which the signal can be controlled and the distance to the reflector can be calculated. In its basic form, the technique requires a relatively clean return beam to function accurately, and this can be affected by differences in target properties.

Figure 3.30 shows a sinusoidal beam modulation, commonly used in many surveying instruments. In the brief discussion of interferometry previously, Figure 3.29a showed constructive interference where the transmitted and return beams add together and destructive interference where they canceled. The waves in that case were the light waves themselves. Rather than detecting fringes where the actual light waves add or cancel, it is possible to superimpose a much longer wavelength on top of the light waves by modulating the light's intensity. Then the *phase difference* (see bars on Figure 3.30) between the modulation on the transmitted beam (solid curve in Figure 3.30) and the modulation on the return beam (dotted curve in Figure 3.30) can be measured. There are well-established electronic techniques to do this. The phase difference is a proportion of one wavelength of the modulation, and this distance value, plus an unknown number of whole wavelengths which the phase measurement cannot detect, gives twice the distance between transmitter and target reflector. Again, there are well-established techniques for determining the unknown whole number of wavelengths, all requiring repeat phase measurements at different modulation wavelengths. The multiple wavelength measurements provide two or more equations for the distance which can be solved to give a unique distance itself based on the number of whole wavelengths.

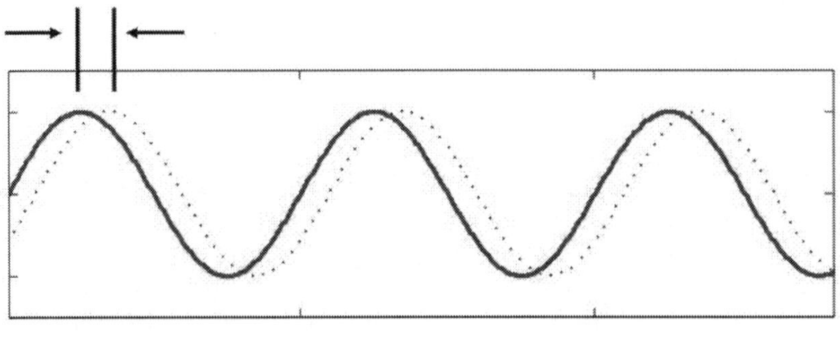

FIGURE 3.30
Phase difference between signals. (Courtesy of S. Kyle.)

Multiple phase measurement by amplitude modulation typically achieves millimeter levels of accuracy in surveying instruments. (This is actually very good as they can measure very long distances, up to kilometers.) API and FARO have optimized the techniques to achieve tens of microns accuracy in laser trackers which have maximum effective ranges up to tens of meters.

A variant phase measurement using amplitude modulation does not actually measure the phase difference between transmitted and return modulations but adjusts the wavelength of the modulation so that the transmitted and return beams either add or cancel. This has some similarities with fringe generation in interferometry. To explain the concept, consider a simplified illustration of the experiment in 1850 by French physicists Fizeau and Foucault to determine the speed of light.

Figure 3.31 shows a transmitted light beam, modulated by passing it through a rotating cogwheel, which continues to a distant fixed mirror where it is reflected back to the transmitter. A semi-reflecting mirror allows an observer at the apparatus to view the reflected beam with a telescope.

As the cogs cut through the beam, they send a string of light pulses to the mirror. When a pulse returns, the wheel will have moved forward slightly, and the next cog will start to obscure the returning pulse. As the wheel is rotated faster, a point is reached where the return pulse is exactly blocked, and the observer no longer sees the light beam. From the measured rotation speed of the wheel in this state, the time taken for a cog to move forward one position can be calculated. This is therefore the time taken for a pulse to travel out to the mirror and back. This distance was known by careful surveying beforehand so that, from known time and distance, the speed of light could be calculated. Once the speed of light is known, the same procedure can be applied to measure an unknown distance, that is, from known time and speed of light, the distance can be calculated.

The Leica laser tracker uses this principle but modulates the angle of polarization of the laser light beam rather than applying an amplitude modulation. Techniques for detecting the polarization angle are more accurate than those for detecting the intensity of amplitude modulation. In this technique, the initial measurement is at a high frequency and contains an unknown number N of whole wavelengths in the measured path. However,

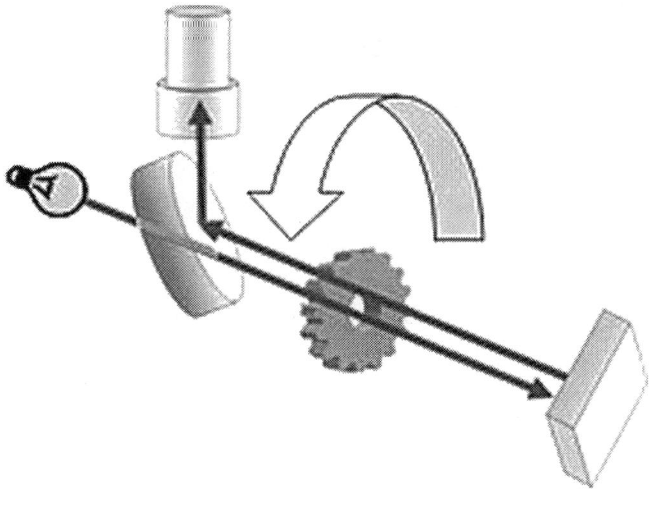

FIGURE 3.31
Amplitude modulation.

by increasing the frequency so that one more wavelength (N + 1) fits into the path, two measurement equations are generated from which it is possible to calculate the two unknowns, that is, the distance D and the whole wavelengths N.

The third measurement technique modulates the frequency of the outgoing light, equivalent to changing its color, and mixes it with the return beam. When signals of different frequency are mixed together, a beat frequency equal to the frequency difference is created. The modulation has the effect that the beat frequency, which is easily measured, is directly related to the time delay which the path length introduces between outgoing and return beams. Again, knowledge of time of travel and speed of light gives the required distance.

A major advantage of this technique is that it does not require a strong return signal, meaning that it can measure directly off of a part's surface or off a cooperative target such as a tooling ball or retroreflective photogrammetric target. A retroreflective mirror target, as used by the other two techniques, is not required.

To understand the differences between the frequency versus amplitude and polarization modulation techniques, consider that a sensor would not need a lot of returning light energy to evaluate changes in its color, while measuring the amplitude of the returning light would require a nice clean beam. Frequency modulation techniques generally require more integration time to determine the range to the object. There are also performance issues with different surfaces and angle of incidence, and the beam is not infinitely small, so the system has to integrate the return over an area on the object's surface. This method is incorporated into the Nikon laser radar system (Figure 3.32), which will not be discussed in further detail here.

3.4.3 Range-Dependent Accuracy

In laser tracker surveys, it is typical for the ranges between target and instrument to vary significantly. Some targets may only be a meter from the instrument, while others are up to 50 m or more from the station. Range (and angle) measurement accuracy decreases with increasing measurement distance. As a result, uncertainties in target positions are different and depend on target distance from the station.

Tracker performance specifications generally define angle measurements to be within 1 arc s and range accuracy with a range-dependent component (e.g., 2.5 ppm) plus a basic ambiguity error (0.0003 in. or 0.0076 mm) due to errors in the target and its mount. It is also important to note that environmental disturbances will add to angle and range errors at increasing distances (see Section 3.7). The working range of trackers is on the order of 50 m, so there is an increasing change in uncertainty of a target's measured location as the distance increases through the working volume of the tracker.

Figure 3.33 shows how the shape and size of the uncertainty field changes relative to the range.[3]

This figure, and two that follow, were computed using a Monte Carlo modeling[4] technique with 1000 samples, assuming the measured components (two angles and range) had Gaussian error distributions.[5]

When the target is close to the tracker (<1 m), the basic range ambiguity is the biggest contributor to the target's uncertainty. This is shown in Figure 3.34[5] where the radial uncertainty is larger than the contribution due to the angle encoders. The spatial effects of angle errors are proportional to distance. For example, a 1 arc s angle error corresponds to 5 μm/m; therefore, at ranges less than 1 m, typical angle errors have a small spatial effect.

FIGURE 3.32
Nikon laser radar. (Courtesy of Nikon Metrology, Leuven, Belgium.)

As the range increases, the uncertainties of the tracker's angle measurements contribute a larger percentage of the target's uncertainty. Figure 3.35 shows the change in shape estimates of uncertainty fields as the range changes from 1 to 20 m. In this example, the instrument is assumed to have angle uncertainties of 1 arc s for both horizontal and vertical angle encoders and a range uncertainty of 7 μm + 2.5 ppm (at 1-sigma coverage). The target uncertainty varies from 12 μm at 1 m up to 156 μm at a range of 20 m (1-sigma).

3.5 Targeting: Retroreflector Design Details

Target reflectors are among the most critical system components for laser tracking. It is a fundamental fact that a tracker cannot measure or track without a target that will return a clean, collimated laser beam back to the instrument. Its opto-mechanical properties also have a direct influence on the quality of the tracker's measurement.[6]

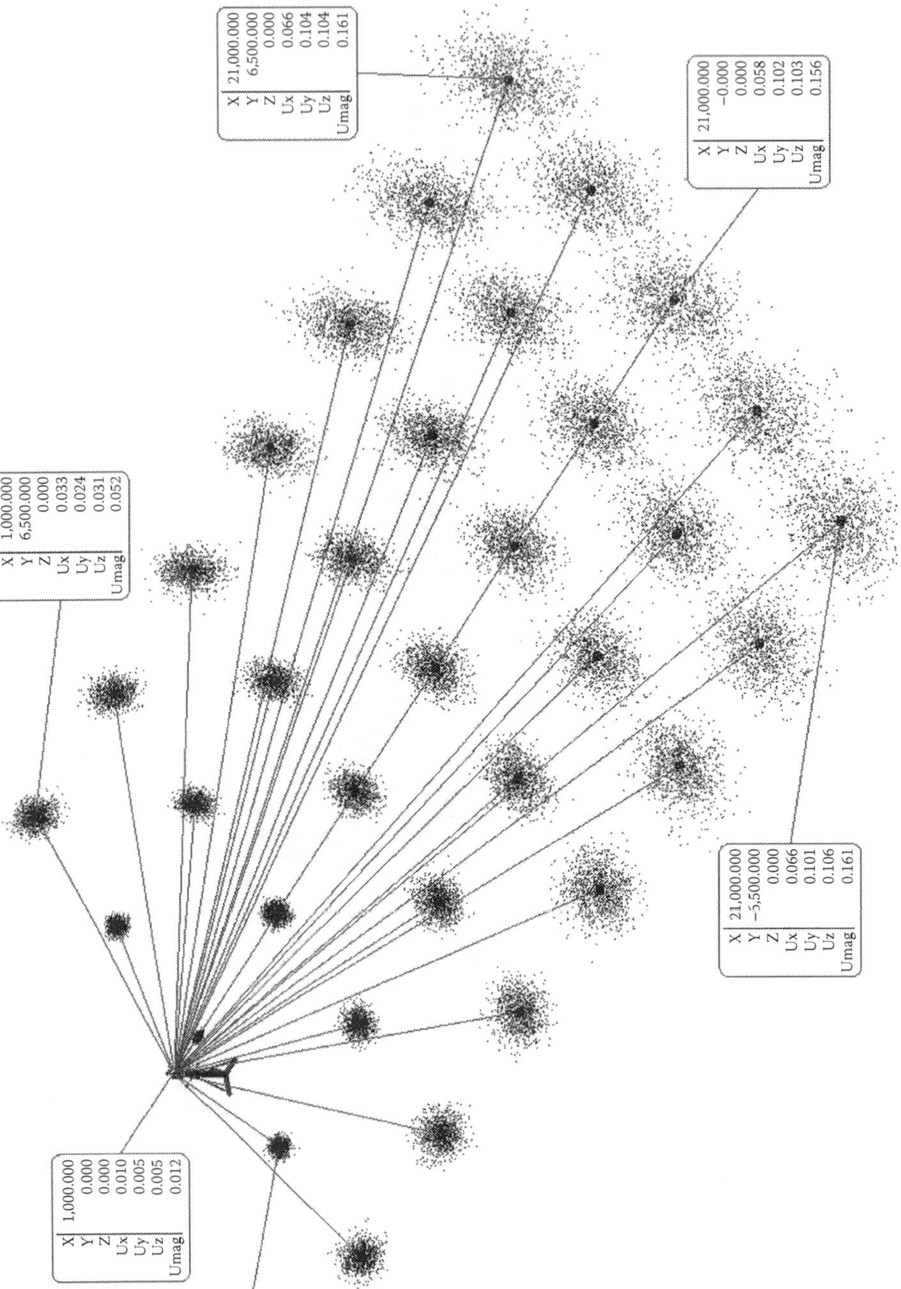

FIGURE 3.33
Range-dependent tracker uncertainty. (Courtesy of New River Kinematics, Williamsburg, VA.)

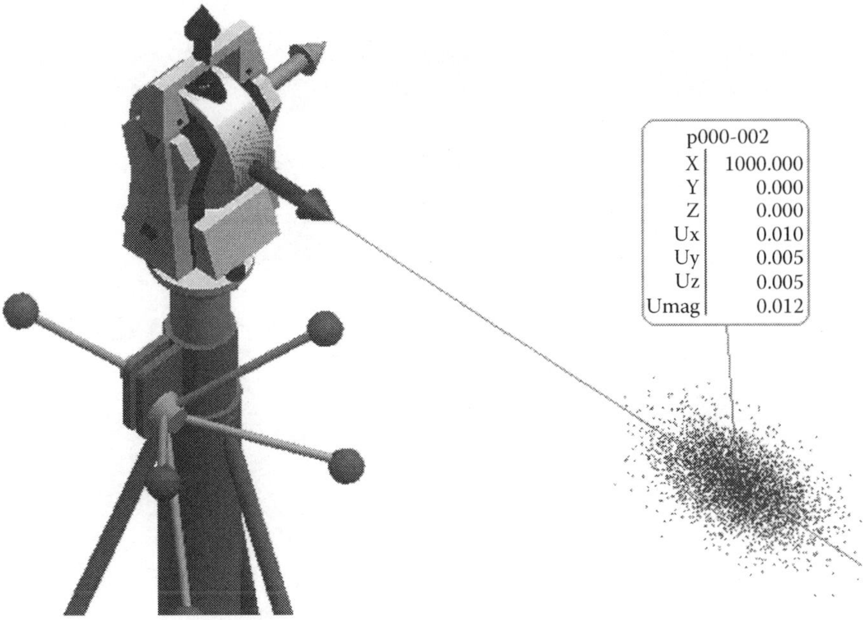

FIGURE 3.34
Close range tracker uncertainty. (Courtesy of New River Kinematics, Williamsburg, VA.)

As explained in Section 3.3, target reflectors are typically mounted in a spherical housing. An accurately mounted reflector (see later comments) is known as a spherically mounted reflector (SMR) and can be used as a contact probe by being held directly against the object being measured. Alternatively, it can be used in conjunction with a target adapter or *target nest* which can be located into a key feature.

Different types and configurations of target reflectors are possible, each with different properties offering advantages and disadvantages. The following three main designs of retroreflectors will be discussed in more detail here:

- Open-air (air-path) corner cube retroreflector, with or without cover glass
- Solid glass prism corner cube retroreflector
- Cat's eye retroreflector

To make proper use of these different reflectors, it is important to understand their optical and mechanical properties, as well as properties particular to their adapters and housings.

3.5.1 Air-Path Corner Cube Retroreflectors

The most common retroreflector design is a configuration of three mirrored glass panels, orthogonally mounted to one another in a spherical housing. It is often called an open-air corner cube (Figure 3.36) or air-path corner cube and has a typical acceptance angle of approximately ±25°. The mirrors are specially coated first-surface mirrors bonded into the mounting sphere with their intersection point (the cube corner) accurately positioned at the sphere center. This technique has a number of key advantages. Specifically, the laser light only travels through air and not through glass. That means there are no additional index offsets or refraction to compensate.

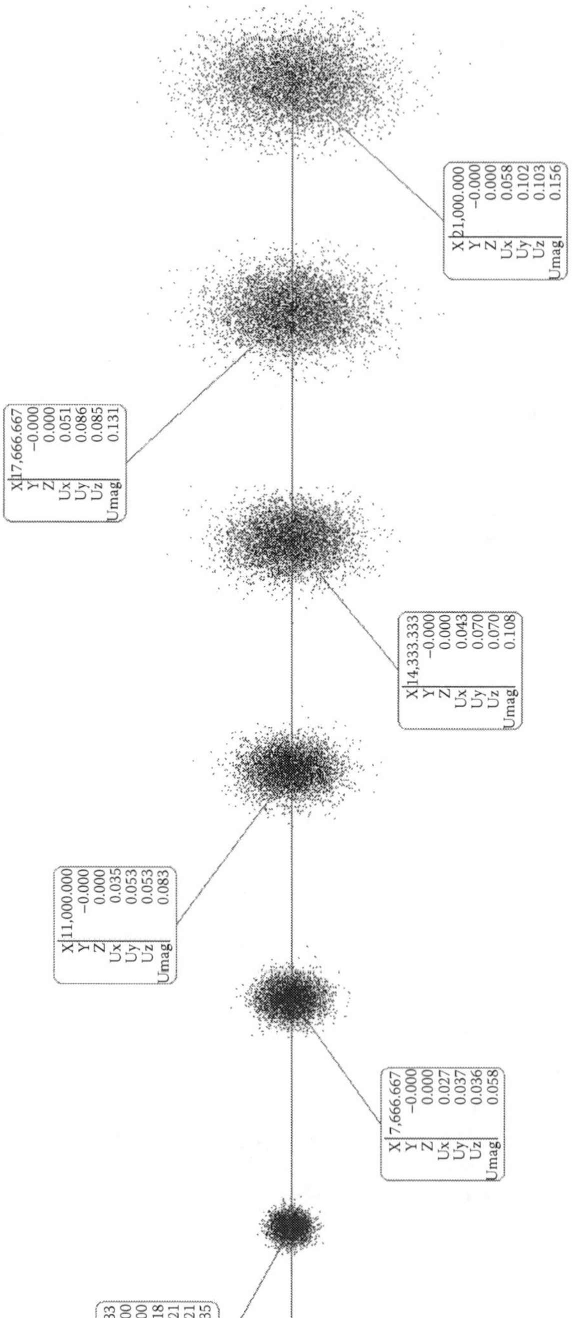

FIGURE 3.35
Range-dependent tracker uncertainty. (Courtesy of New River Kinematics, Williamsburg, VA.)

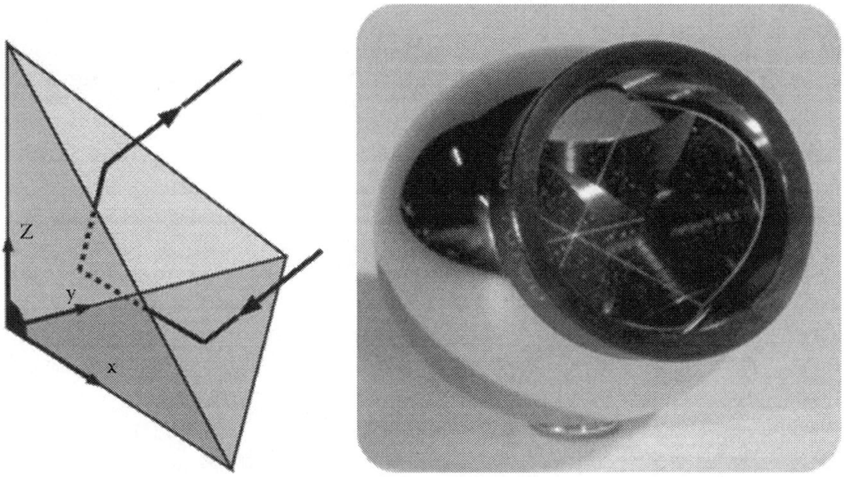

(a) (b)

FIGURE 3.36
(a and b) Air-path corner cube reflector.

These targets are preferred in applications where single points and surface scanning is required to a high accuracy. They typically offer a good balance between cost and precision, but they must be handled with care. For example, they are relatively fragile—SMR mirrors do not often survive being dropped. It is also relatively common to find that the mirror surfaces are damaged either by oxidation or from improper cleaning procedures. Operation in corrosive environments is particularly damaging to the mirror surfaces.

3.5.2 Break-Resistant Reflectors

The break-resistant SMR (Figure 3.37) was developed to deal with the fragility of the standard glass-mirrored SMR. These target's mirror surfaces are directly cut and polished

FIGURE 3.37
Break-resistant SMR.

into the metal sphere. This technique means there is no glass to break, and therefore, it significantly improves these target's geometric stability and durability in shop environments. These targets offer similar performance advantages to the glass SMR but note that their mirror surfaces are also exposed to the environment and will suffer similar effects. They must, therefore, be properly cleaned and maintained.

3.5.3 Solid Glass Prism Reflectors

Instead of manufacturing the retroreflector from three separate mirrors, it is often easier to manufacture the reflector from a solid piece of glass as would be obtained by slicing the corner off a glass cube. This produces a glass prism with four plane surfaces. One is the entry surface for the laser beam and is perpendicular to the prism axis. The other three are orthogonal and back silvered to create the mirror surfaces for the beam as it reflects around the inside of the prism and back out on a parallel return path.

Solid glass targets have some key advantages and disadvantages that should be understood before choosing them. There are two variants of the prism retroreflector—tooling ball reflectors (TBR) and repeatability targets.

These targets are relatively easy to manufacture, so SMRs built to this design represent an economical option. In addition, they are typically more durable and break resistant, when compared to the standard glass-mirrored SMR, and generally easier for the instrument to track.

Solid glass prism targets can have bigger acceptance angles than air-path corner cube SMRs, typically up to ±40° compared with the typical ±25° of an air-path corner cube. This improved acceptance angle is due to refraction when the laser beam travels from air into the glass and is bent in toward the apex of the prism, but refraction introduces two errors:

1. A pointing error due to an apparent shift in apex position (the target point)
2. A range error because refraction means that light travels more slowly in glass, so causing an effective change in path length

The pointing error exists for IFM and ADM systems. This error is dependent on the angle of incidence between the incoming laser beam and the normal direction of the flat entrance surface of the prism. As the incidence angle increases, the pointing error increases. The effect is shown in the following figure and charts.

Figure 3.38 shows the glass prism pointing and radial refraction errors.[7] It is this error which also improves tracking performance since the refraction effectively increases the target's acceptance angle. It is, however, still important to keep the reflector pointed back at the instrument when measuring. When pointed directly at the tracker, the pointing error goes to zero.[8] In general, the measurement error for a 0.5 in. diameter solid corner cube SMR can be up to 0.1 in. laterally and 0.02 in. radially (2.5 mm and 0.5 mm, respectively).

The second error is also caused by the change in index of refraction (I) between air and glass. This is the ratio of the speed of light in air (C) to the speed of light in glass (V) and is given by $I = C/V$. For glass, the index is always greater than one, since light travels more slowly in glass than in air. The longer time of travel causes an apparent increase in range to the target which can be several millimeters.

It is worth noting that accurate correction for the refraction errors described previously is possible in Leica's T-Probe. This uses a glass prism retroreflector to its maximum

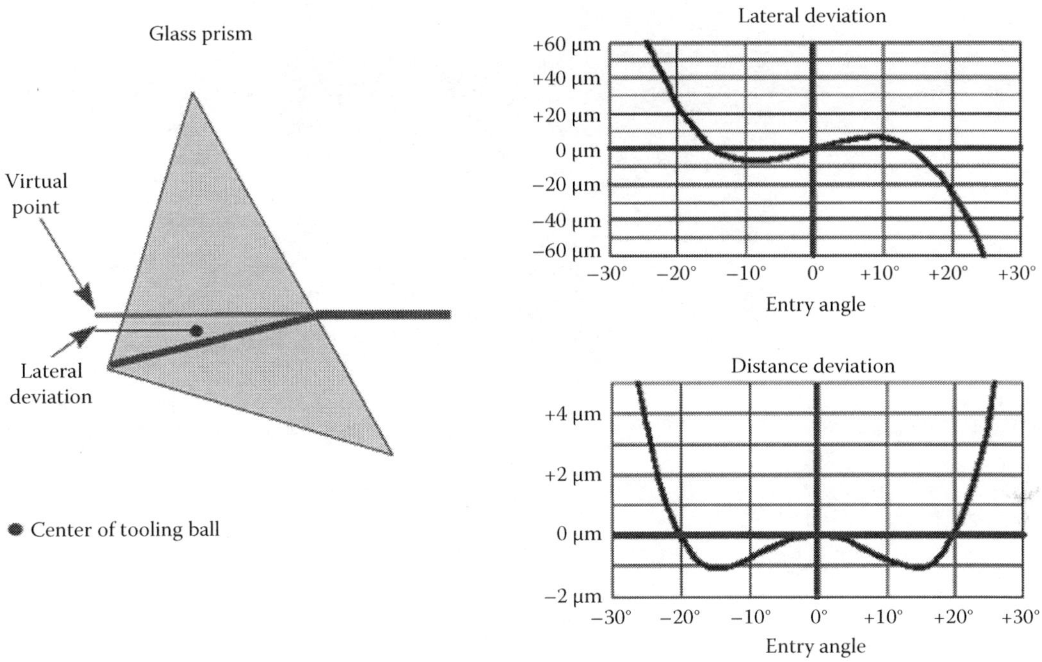

FIGURE 3.38
Glass prism pointing and radial refraction errors.

acceptance angle, which improves its usability to the operator. However, because the probe is being tracked in all 6D, the refraction errors can be corrected since the incident angle of the beam to the prism axis is known. The TBRs and repeatability targets described next have their acceptance angles limited by a mechanical ring in order to limit the errors caused by refraction, but they retain the advantages of lower weight, size, and manufacturing cost.

3.5.4 Tooling Ball Reflectors

TBRs shown in Figure 3.39 are glass prism retroreflectors commonly used in discrete point applications where tracking between points is not required. Their low cost and break resistance make them attractive compared with air-path corner cube targets. They are often fitted with a ring on the front to restrict the acceptance angle to a lower value than theoretically possible in order to limit the refraction errors to acceptable levels. This is partly also achieved by mounting the prism with the apex carefully offset from the center of the sphere, which also helps to offset the errors.

3.5.5 Repeatability Targets

Repeatability targets, shown in Figure 3.40, are typically used to evaluate object drift or deformation over time. They are initially surveyed where reference point locations are required. At regular intervals, or when needed, the same targets are remeasured. The new coordinates are compared against the initial set to determine how much the object or component has deformed or moved relative to a set of fixed points. As repeatability targets,

FIGURE 3.39
Tooling ball reflector.

FIGURE 3.40
Repeatability targets.

their absolute locations are not needed to a high accuracy, so the centering of prism and mounting sphere is not critical. This can effectively make them cheaper to manufacture.

3.5.6 Cat's Eye Retroreflectors

A different design of solid glass retroreflector is the *cat's eye reflector,* often called a *cateye reflector* shown in Figure 3.41. In simple form, this has two glass hemispheres of the same refractive index but different radii, cemented together with a common center. The larger hemisphere

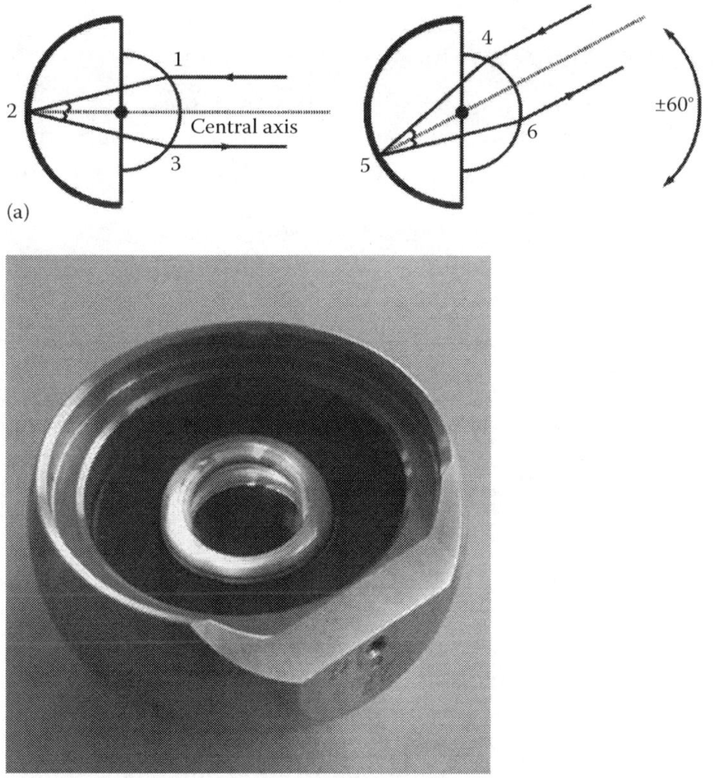

FIGURE 3.41
(a) Cat's eye reflector: concept of operation. (b) Commercial cat's eye reflector. (Courtesy of S. Kyle; Leica Geosystems, St. Gallen, Switzerland.)

is silvered on the back to reflect the incoming beam. Analysis of the ray paths through the device shows that the incoming and reflected beams are parallel and the effective target point is the center of the spheres. In practice, the detailed operation of the reflector is more complex.

Cat's eye reflectors are large, heavy, and relatively expensive in comparison with open-air and prism corner cubes. However, they have a wide acceptance angle of approximately ±60° and effectively a single target point. They are used when the angle of acceptance is critical to the application and are therefore valuable in robotics and machine control applications where the disadvantages of size, weight, and expense are less significant.

3.5.7 Reflector Housing

SMR housings are typically made from chrome-plated steel to maximize life and durability of the target. They are manufactured in several standard diameters including 1.5 in. (38.1 mm), 0.5 in. (12.7 mm), and 0.875 in. (22.225 mm).

It is important in an SMR that the optical target point and the center of the housing sphere are accurately positioned together. Typically, SMR centering tolerances range from ±0.0005 in. down to ±0.0001 in. (0.0127–0.0025 mm).

However, remember that repeatability targets are not designed as SMRs and their spherical housings are a manufacturing convenience.

3.6 Environmental Compensation

There are two environmental compensation issues that frequently concern laser tracker operators in the field. One is atmospheric effects on the laser light, that is, wavelength changes due to changes in air temperature, pressure, and relative humidity. The other is the effect of temperature on the object being measured. Other problems certainly exist, such as mechanical stability, atmospheric refraction, and presence of organic solvents in the air, but there are no straightforward compensations for the errors they can cause. These issues are better handled with proper setup procedures and field checks at the job site.

Wavelength changes to the laser light, and thermal effects on the objects, can be compensated with proper accessories and procedures. Errors made either by not compensating for them, or by making a mistake when compensating, are generally systematic and can be significant.

3.6.1 Wavelength Compensation for Accurate Range

Since the tracker's laser light is traveling through the air, its wavelength depends on the environmental conditions, primarily temperature, pressure, and relative humidity. To achieve accurate results, the wavelength must be accurately known.

The frequency of the laser light on site is calculated by multiplying the frequency of the laser light in a vacuum by the index of refraction for air at the job site. The atmospheric conditions are either read electronically from the laser tracker's weather station, which delivers air temperature, pressure, and humidity, or are manually input by the operator. These readings are required parameters in a formula used to compute an accurate estimate of refractive index on site.

When the transmission medium is denser, its refractive index is higher. The wavelength of light then becomes shorter or, equivalently, increases in frequency. The density of air is higher near sea level than at greater altitudes where pressure is lower and the air is less dense. At higher altitudes, air therefore has a lower refractive index causing the frequency of light to decrease. In the case of temperature, higher temperatures result in lower refractive indexes, and lower temperatures result in higher refractive indexes, so as temperature increases the wavelength increases or correspondingly frequency decreases. The basic relationship is that the laser's wavelength is inversely related to the medium's density.

The wavelength (in a vacuum) of the laser light used in laser trackers generally has a frequency uncertainty not exceeding 10 or 20 parts per billion. The wavelength uncertainty in air is bigger because the index of refraction is affected by atmospheric conditions which are not easily measured with a low uncertainty. Variation of measured distance in ppm due to environmental change is given approximately as follows:

Air temperature: ≈ ±1 ppm/°C

Pressure: ≈ −0.25 ppm/mbar

Relative humidity (RH): ≈ +0.01 ppm/%RH[9]

The tracker's weather station measures the air temperature, pressure, and humidity. These are parameters in formulas, such as the modified Edlén or Ciddor equations, which are used to compute the air's refractive index. This is then used to compute the laser's wavelength in the environment where it is being used. The wavelength in air, λ_{air}, is derived from the known wavelength in vacuum λ_{vac} and the air's computed refractive index n as follows:

$$\lambda_{air} = \frac{\lambda_{vac}}{n}$$

The compensated wavelength of the laser, λ_{air}, is used to determine the range to the target in both IFM and ADM systems.

3.6.2 Bending of Light Due to Temperature Variations

A varying refractive index in the air caused mainly by thermal variations has significant effect on a laser beam such as a change in wavelength. However, one of the most obvious consequences of a change in refractive index is refraction itself, the bending of light as it travels, say, from air to glass or less dense air to more dense air. The effect is routinely seen as shimmer on a hot day. When a laser beam is deviated because it passes through layers of air at different temperatures, the pointing to a target reflector will be in error.

However, although techniques have been evaluated which could compensate for this, robust and economical solutions are not commercially available, and the best compensation technique involves the application of good practice, for example,

1. Avoid sighting near hot or cold objects (e.g., heaters or doors open to cold outside air).
2. Keep lines of sight short where possible.
    a. Bending is less likely at short ranges where environmental effects vary less.
    b. If a pointing is in error, its spatial effect increases linearly with distance.

3.6.3 Weather Station

Laser tracker systems use weather stations to provide local measurements of the air temperature and pressure and often also relative humidity. These are automatically read at frequent intervals to update the laser's wavelength, therefore compensating for variations in the laser's wavelength throughout the measurement job and thereby minimizing systematic error. If the system does not have an automated weather station, the operator needs to input the atmospheric conditions manually.

3.6.4 Scaling Laser Tracker Measurements for Thermal Compensation of Object Shape

Laser trackers normally measure large objects in environments which are not temperature controlled or may even be outside. During measurement, temperature is therefore likely to change due to time of day or season of the year. As a result, objects will expand (as temperature increase) or contract (as temperature fall) during measurement. Compensation for these effects is achieved by scaling the measurements.

Note in the discussion here, that it is assumed objects change uniformly in shape when the temperature changes. This could only be strictly true of objects manufactured exclusively from a single material, but even using a single material, most large objects will be composed of many components joined in different ways (bolting, riveting, welding, etc). The objects might also be mounted rigidly to concrete floors that are, in effect, thermal heat sinks that do not experience significant temperature change. This might introduce different stresses and strains as temperature changes, and a resultant non-uniform change of shape. There is no evidence of research into this, but practical experience suggests it is reasonable to make the assumption of uniform change. However, it is important to be aware of the assumptions and their potential impact.

The goal for the scaling process is to correct all the measurements so that they represent the objects as if they were measured at a consistent reference temperature, typically 20°C (68°F). Once corrected to the reference temperature, object measurements can be compared reliably and consistently either to nominal CAD data, or to other measurements of the object, or to components that have to be assembled together. The key, therefore, is to represent object measurements at a consistent reference temperature.

A simple example can help to illustrate the magnitude of the dimensional change. A 10 m (32.8 ft) long aluminum object changes by approximately 0.24 mm (0.010 in.) per °C due to thermal expansion and contraction. A similar steel object will change by approximately 0.12 mm (0.005 in.) per °C. The amount of change is a significant source of error in most inspection, build, and assembly applications, and it must be corrected with a consistent compensation procedure.

There are a number of techniques which compensate the tracker's measurements for these effects and produce data scaled to a consistent reference temperature. A critical choice for metrology teams is to decide on the best scaling technique for their application and ensure all 3D metrology processes use it and document the scaling result.

The three most common scaling processes for laser tracker measurements are

1. Material coefficient of thermal expansion (CTE)—delta temperature scaling
2. Traceable scale bar of like material
3. Auto-scaling to an established reference network of points

3.6.4.1 CTE Scaling for Thermal Compensation

An object's coefficient of thermal expansion (CTE) is a basic property of its construction material (e.g., aluminum, steel, carbon fiber) and critical for its measurement. The CTE defines how much a particular material expands or contracts due to a change in its temperature.[10] If the object's ambient temperature is known, it can be subtracted from the reference temperature to get its "delta temperature." Multiplying the delta temperature by the object's CTE will yield the amount the object should change per unit length (mm or inches). The length of the object (L_i) at temperatures other than the reference is modeled with the following thermal length compensation function:

$$L_i = L_0(1 + \alpha\Delta T)$$

where
L_i is the length at actual temperature
L_0 is the calibrated length at reference temperature
α is the CTE for object material (ppm/°C)
ΔT is the "delta" temperature difference (actual temperature − reference temperature) °C

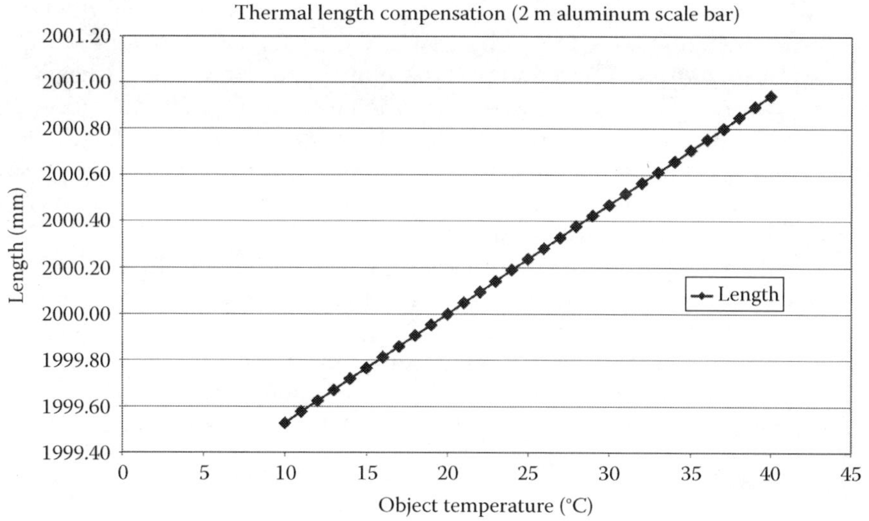

FIGURE 3.42
Object growth versus temperature.

The length change of a 2 m long aluminum object at different temperatures is shown in the graph in Figure 3.42.

At the reference temperature, the object is 2 m long. At 30°C (86°F), the object length is expected to be 2000.47 mm (78.7587 in.) based on the thermal length compensation function. A change in the bar's temperature by 10°C (18°F) changes its length by 0.47 mm (0.0186 in.).

The scale factor is determined using the following function:

$$s = 1 + \alpha\Delta T$$

where
α is the CTE for the object material (ppm/°C)
ΔT is (actual − reference temperature)°C

The scale factor, s, is typically applied as a multiplier of the tracker's range measurements or multiplier of individual coordinate values (relative to the working coordinate frame) before the coordinates are transformed into the part coordinate system.

3.6.4.2 Scaling Using Certified Scale Bars for Thermal Compensation

A certified scale bar can be used to scale laser tracker measurements in a consistent way by comparing a tracker measurement between scale bar points with the certified distance between the points. A scale factor is then applied to tracker measurements to make them fit the certified values which are determined at a reference temperature. Where multiple scale bars are used, a scale factor is determined using an optimizing technique to minimize the differences between measured point-to-point scale bar distances and the corresponding certified point-to-point distances.

The results depend on the relative length of the bar, the uncertainty of the calibrated and measured point-to-point distances, type of material the scale bar is made from, and the number of scale bars used. In general, the scale bar should be made from the same type of material as the object being measured. Like material is preferred so that the object and scale bar have the same (or similar) CTEs and will therefore expand and contract at similar rates. An exception to the use of like material occurs when measuring in temperature-controlled rooms or when trying to evaluate directly the object's change in length due to environmental conditions such as temperature and humidity. In most of these applications, the scale bar is often made of Invar which has a very low CTE and therefore does not expand or contract significantly with temperature.

The length of the bar should ideally be close to the diagonal length of the object. Use of a longer scale bar ensures that measurement errors are not magnified by the ratio of the object's diagonal length to the scale bar length.[11]

When the object length is much greater than the scale bar, increasing the number of measured bar positions can help mitigate the effect of scale bar measurement errors.

There are a number of important practices that help improve scaling by scale bar measurements:

- Use a laboratory-certified and traceable scale bar.
- Use a scale bar of like material to the object being measured.
- "Soak" the bar(s) with the object being measured so that the scale bar and object acquire the same ambient temperature.
- Measure and check the length more than once.
- Ensure that the bar is long enough to show errors in the measurement process.

3.6.4.3 Thermal Compensation Using a Reference Point Network for Scaling

Measurements can also be scaled using a network of reference points integrated with the object. By measuring these at regular intervals during measurement of the object, a seven-parameter best-fit transformation (including scale as the seventh parameter) can be calculated between measured point values and their reference values. This forces the measurements to be consistent with the reference. The transformation is then applied to the object measurements so that these, too, are consistent with the reference data. This technique can produce a very consistent object scale and alignment to a reference coordinate system.

This practice is often called "auto-scaling" or a "7-parameter best-fit transformation." It has the advantage that the scaling will change as the part expands and contracts during and between measurement jobs. The transformation, scale, and reference network of points should be verified and traceably documented using an independent source on a regular or periodic basis.

Critical to this technique is the initial step of establishing a reliable and consistent reference network with a traceable scale. Either one of the two scaling techniques mentioned previously can be used to establish the traceable scale for the reference network.

3.7 Laser Tracker: Best Practice

This section presents 10 of the more common laser tracker setup, measurement, and analysis practices that help ensure the best measurement results.

3.7.1 Stable Instrument Stand and Object Fixturing

The laser tracker is an excellent on-site metrology solution for a wide range of applications. However, a primary criterion for its successful use is that, from first to last measurement, it remains geometrically stable relative to the object or objects that it measures, inspects, or builds. There are three requirements to achieving this stability:

1. Solid and mechanically stable instrument stand
2. Secure and rigidly secured part, tool, or object(s)
3. Drift check at beginning and end of measurements

It is essential that the laser tracker be secured to a solid and mechanically stable instrument stand. Set the stand up in a location that gives the instrument a direct line of sight to the important features on the object. However, care must be taken to choose a location which minimizes the likelihood that the stand may be bumped or disturbed by the operator(s), other nearby personnel, moving vehicles, etc. Ensure the tracker and its stand are not exposed to direct sunlight as it may cause nonuniform thermal expansion in the stand or instrument. Even small changes, particularly to the instrument's angular orientation, can induce significant measurement error. It is also recommended that the tracker and its working volume be properly marked off to avoid disturbance to either the tracker or object. These same issues should also be considered for the part holding fixtures. Taking time to secure the tracker stand and investing in the appropriate part or object fixturing are fundamental to achieving reliable and consistent measurements.

A documented *drift check* (details below) at beginning and end of measurement can confirm that instrument and object have remained stable throughout the measurement process and provide assurance to the laser tracker operator(s) and customer that the instrument and object remained stable.

If a change in relative position of tracker and object is detected, most laser tracker software packages support either the deletion or deactivation of measurements made since the movement occurred. A new station can then be added to the network to allow measurements at this location to be continued.

3.7.2 Ensure Line of Sight

A basic requirement in laser tracker metrology is a clear line of sight between the instrument and target, and it is an important part of the setup procedure to ensure that this is the case. It is not uncommon to find that a single location does not have line of sight to all required target locations or that a single location requires some long-range measurements which would lead to a higher measurement uncertainty that approach or exceed the part or object tolerances. A general rule of thumb is that measurement uncertainty should be four times lower than the object or part tolerances.

Where a single location does not work for all points, it is typical to move the instrument to a series of positions around the object, so creating a measurement network. Section 3.4 explains one technique where common target positions can be established which link together the individual stations into a common coordinate system. A minimum of three targets are required, not all on a straight line, but it is common to use six or more, ideally not all in a plane. If this simple point-to-point connecting method is insufficient to deliver the required accuracy, a more advanced form of network adjustment should be used.

Section 3.4 also discusses an alternative technique to using multiple tracker stations when the tracker does not have a direct line of sight to object features of interest. This either uses special target adapters (e.g., hidden point rods) or a full 6D target probe.

3.7.3 Traceable Environmental Compensation

As Section 3.7 explains, laser tracker measurements are affected by changes in the environment. The two main sources of systematic error which result from this are changes in the refractive index of air which affects the tracker's range measurements and differences in the object's temperature which cause its thermal expansion or contraction.[12] Both can be corrected by using suitable accessory devices and applying proper compensation procedures.

Compensating for the atmospheric effects on the range measurements is relatively straightforward with a weather station that accurately measures the air temperature, pressure, and humidity.[13] All current laser tracker systems have inputs for certified weather stations. Integrating certified weather station updates into the process helps keep the instrument measurements reliable. Recording the actual values with each point contributes to making the measurements traceable.

There are a number of techniques to compensate measurements for changes in object temperature. An important accessory for monitoring changes in the object temperature is a certified part or object temperature probe. All laser tracker systems provide inputs for sensors to monitor and record part temperature. Recording the part or object temperature with a certified temperature probe throughout the measurement process helps to ensure that the measurement results are traceable.

Monitoring processes either automatically by tracker software, or manually at regular intervals by the operator, is common practice in typical industrial applications. When a significant object temperature change is detected (e.g., 4°F or 3°C), the measurement process should be paused to confirm the object temperature change. If a temperature change occurs before all the required measurements are collected then a new tracker station should be added to the measurement network.

3.7.4 Manage Measurement Sampling Strategy

An important factor in achieving reliable measurement results is the number of measurement samples and the length of time over which they are acquired. This sampling strategy is configured in the software and depends on the application and type of acquisition method.

In dynamic data collection where the target is moving over a surface or on a moving machine head, measurement sampling is typically set at one sample per point. However, when trackers measure discrete single points, the software enables the user to configure multiple samples which are averaged to help reduce the effect of local variations caused, for example, by turbulent air or vibration of the instrument, targets, or the part itself. This variation is assumed to be a random error source, so by sampling over a period of time, the system can average out the effects. For discrete single points, it is common to collect between 50 and 1000 samples over 1–2 s. The sample's average angles and ranges are used to compute mean target coordinates. The technique helps to reduce the negative effect of the shop-floor environment. By reporting the sample statistics, overall traceability is improved.

It is common for the tracker software to use sample statistics to provide real-time feedback to the operator(s). The feedback usually takes the form of comparing the sample's standard deviation or root mean square (RMS) value against a threshold set by the operator. If the threshold is exceeded, this indicates a problem during measurement which means the measurement should be repeated.

The appropriate measurement sampling criteria should strike a reasonable balance between a number of factors including the expected part tolerance, variation in the environment (both mechanical and environmental), range from the instrument to the target, time to sample, and a minimum number of samples.

3.7.5 Transforming into Part Coordinates with More Than the Minimum Reference

At initial setup, the tracker's actual location must be determined relative to the part or object coordinate system. One of the possible techniques is to measure a set of reference points that have nominal values in the object's coordinate system. The goal is to determine the spatial transformation for the instrument that optimally aligns the measured target coordinates to their reference values. A mathematical best fit can be created between these data sets by minimizing the differences in the fit using a least-squares solution. Once the instrument is transformed, measured coordinate data can be reported in the part's coordinate system.

The best-fit transformation function requires at least three reference points, but six or more are recommended because they provide a more rigorous alignment, significantly improving repeatability and accuracy and helping to identify errors.

The distribution of the points to encompass the object is important to ensure that the best-fit transformation is not biased and to reduce further the measurement and alignment errors across the volume.

For example, consider the use of three reference points on a 10 × 10 × 1 m object. Further imagine that the target reflector is not set correctly on one of the adaptors, which creates a small systematic error in the transformation. In this example, assume that the target placement causes an error of 0.05 mm (0.002 in.). This can induce an error of 0.25 mm (0.010 in.) in the point on the opposite side of the volume diagonal. Figure 3.43 shows the error induced across the volume when using the minimum number of reference points.

In a contrasting example, consider the use of eight reference points distributed across the volume. Here the least-squares best fit can minimize the errors. In this example, Figure 3.44 shows that the error across the volume diagonal is now 0.01 mm (0.0004 in.), a significant improvement compared with the use of the minimum number of transformation points.

Identifying measurement errors or misidentified points is easier when there is a redundant number of reference points across the measurement volume. In the previous example, the point with the highest residual error was not identifiable using three points but was easy to see when eight reference points encompassed the volume.

3.7.6 Drift Check to Confirm Instrument and Object Orientation

It is critically important when measuring on site to confirm that instrument and object remain rigid and stable during data acquisition. That task can be accomplished using a drift check which employs a set of repeatability targets placed in fixed locations on or around the object prior to measurement. These targets, which are purely for checking purposes, are measured both before and after a set of object measurements are made.

The drift check delivers the differences between the initial and final check point coordinates. If any of the differences exceed the application's measurement tolerance, the measurement team then knows that the drift check targets, the object, or the instrument(s)

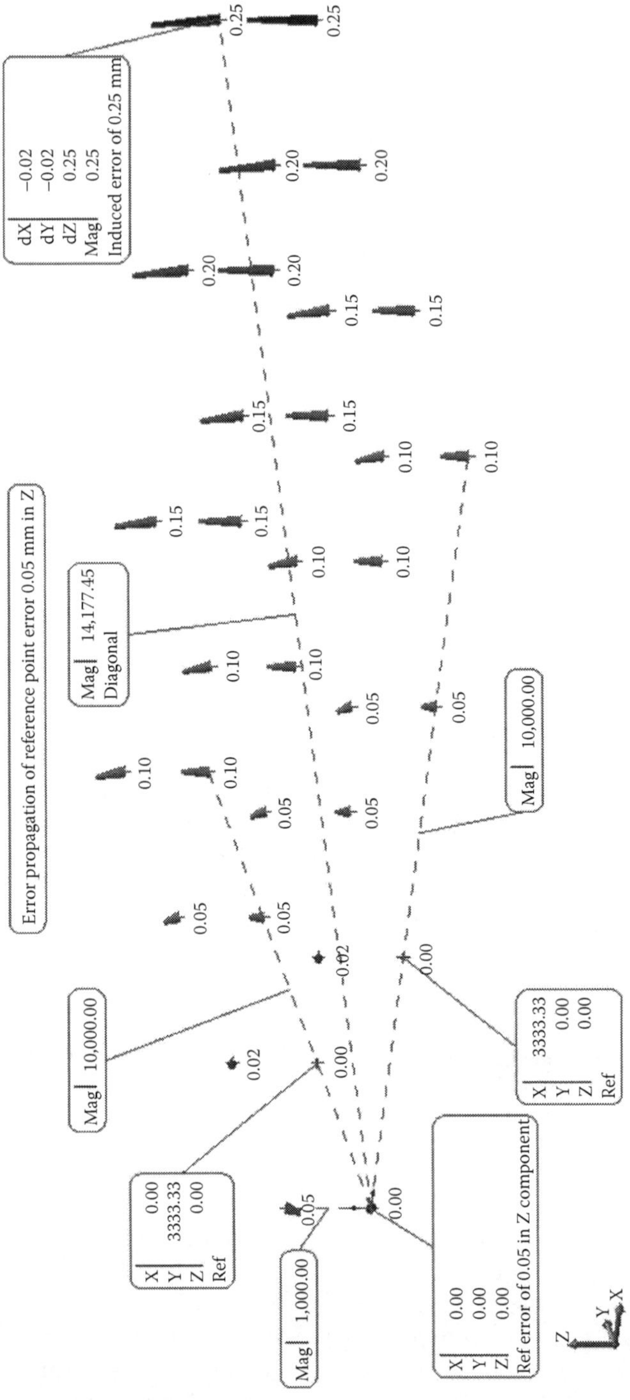

FIGURE 3.43
Alignment errors when using a three point transformation. (Courtesy of New River Kinematics, Williamsburg, VA.)

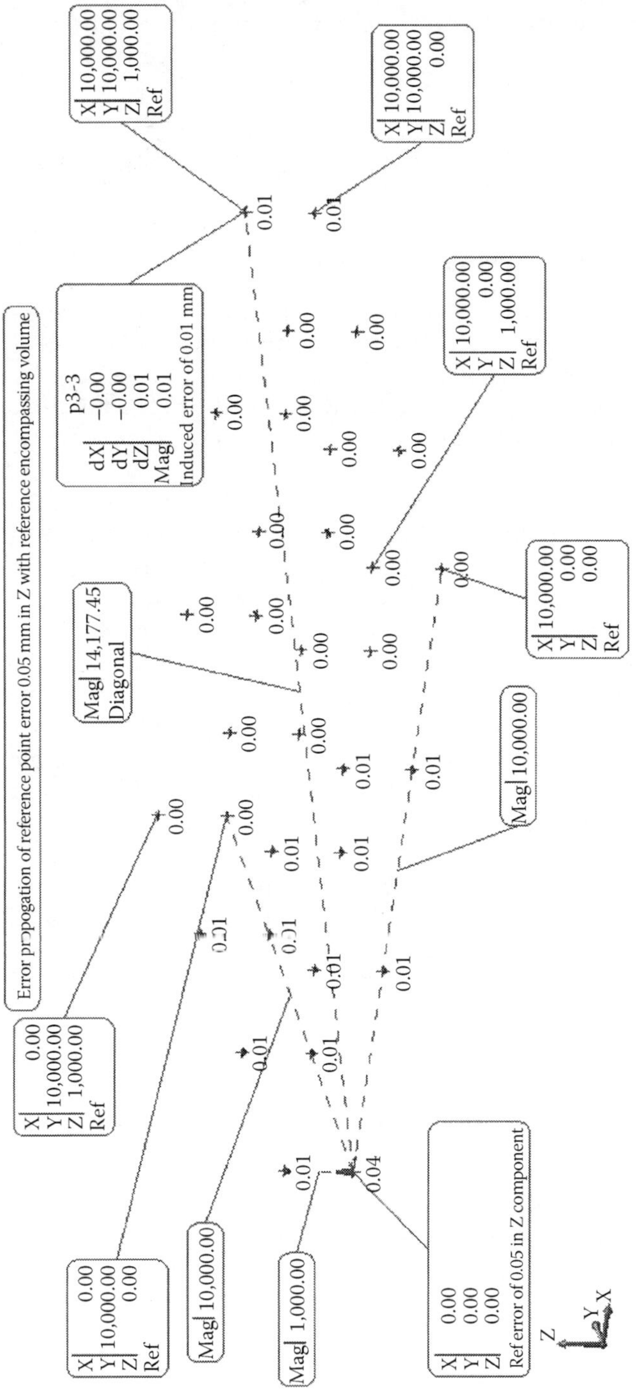

FIGURE 3.44
Improved alignment using eight point transformation. (Courtesy of New River Kinematics, Williamsburg, VA.)

have moved. The object measurements must be evaluated and discarded if shown to be in error. A new instrument station should be created and the measurements repeated to ensure consistent, accurate, and reliable results.

Although three drift check targets are again the minimum required, it is good practice to use more, preferably six. If possible, they should be placed in locations sensitive to movement, but they should also be easy to access.

The drift check report, which confirms the amount the instrument, points, or object moved, scaled, or shifted during data acquisition, is an important document in establishing traceability for the measurement job.

3.7.7 Strong Measurement Network Orientation

Several laser tracker stations are often used in large-volume measurement jobs. In some applications, a single instrument is moved to multiple locations, often called stations, around the object or structure. In other applications, multiple instruments are used either independently or simultaneously. Regardless of the method of implementation, the multiple stations form a measurement network where all the instruments must be located in the same coordinate system.

The process which achieves this typically involves measuring common targets between instruments which effectively link all the stations together. However, the geometry of the configuration affects the relative strength of the station and target network which in turn directly affects the reliability and accuracy of the measurement results. A weak network results in higher measurement uncertainties and is to be avoided. A simple strategy to enhance the strength of a network is to ensure that the common targets and stations are evenly distributed on and around the object.

The software technique used to place the stations and targets in a common coordinate system is often called a *network orientation, network adjustment,* or *bundle adjustment.* It not only derives optimal point coordinates from the measurements but also plays an important role in documenting the measurement uncertainties and delivering traceable results.[13] Software that does a complete network adjustment of all the stations and targets simultaneously is able to produce an optimal alignment. Normally, it also reports the uncertainty of the targets and stations. The uncertainty can be obtained using Monte Carlo[14] or covariance matrix techniques.

Alternative techniques of network adjustment such as a "point-to-point best fit," also sometimes call "leapfrogging," can produce reasonable results but can also hide errors that propagate between stations and result in an error stack up.

3.7.8 Understand Measurement Uncertainty and Ensure Traceable Measurements

Traceability is the property of a measurement result whereby it can be related to appropriate standards, generally national or international standards, through an unbroken chain of comparisons in which all uncertainties are indicated.[15] It helps ensure that the measurements made using different instruments, suppliers, measurement teams, and measurement procedures are equivalent. It is more and more common that components for large systems are manufactured in different parts of the world. Ensuring the equivalence of measurements made by different measurement teams and organizations is established by having measurement processes and reports that are traceable to national and international standards.

Traceability is achieved by calibrating the device and ancillary components at an accredited laboratory and recording the process parameters. Accredited laboratories are linked through national standards to international standards.

Uncertainty in measurement is defined as, "The parameters associated with the result of a measurement that characterizes the dispersion of the values that could reasonably be attributed to the measurand"[16]. A laser tracker's measurement uncertainty can be computed using laser tracker software. Coupling the uncertainty estimate with the measurement results made by certified instruments and accessories is essential in achieving traceability.

3.7.9 Consistent, Rigorous, and Traceable Scale Methodology

Scaling errors are one of the biggest sources of systematic error in laser tracker measurements.[16] An error or variation in scale is magnified throughout the workspace volume. To reduce measurement variation and establish measurement consistency, a scaling technique must be established and consistently followed for the entire project. Several of these techniques were discussed earlier, such as CTE, auto-scaling, and traceable scale bars. Best practice therefore requires the selection of an appropriate scaling process which must then be rigorously followed by all the metrology teams on the job. Scaling and transformation results must be documented at each step in the process to ensure that measurements are consistent and traceable to national or international standards.

3.7.10 Calibration and Regular Checks of the Instrument, Targets, and Accessories

Regular calibration and on-site checks of a laser tracker are another critical aspect to achieving reliable measurement results. This also applies also to scale bars, weather stations, targets, and their adapters used in measurement jobs.

Each laser tracker manufacturer has a procedure and a software package that guides the calibration process from the initial setup and through the specific steps in data acquisition and analysis. Specific calibration criteria must be met and documented to establish that the system meets its performance specifications.

The user should perform the prescribed field check and alignment procedures when setting up the instrument, particularly after shipping to the job site. The field checks are designed as a quick way to identify potential problems with the system calibration.

Best practice further recommends a regular scheduled system calibration at a certified laboratory with processes traceable to a national standards laboratory such as NIST in the United States, NPL in the United Kingdom, and PTB in Germany. Establishing such a calibration program helps to ensure that laser tracker measurement results comply with national and international calibration standards.

3.8 Review of Laser Tracker Applications

The major industries using laser trackers to solve dimensional metrology tasks are aerospace, automotive manufacturing, ship building, energy production, and large-scale research and development. Developing fields such as wind energy have metrology

requirements well suited to laser tracker technology, as do very specialist scientific fields such as particle physics where the accurate alignment of magnets, for example, is critical for these massive machines to reach their research goals. However, many other manufacturing industries, from castings to machine tools to robots, have benefited from the application of laser tracker technology.
The primary application areas can be listed as follows:

1. *Inspection (verification)*: Possibly the most commonly required dimensional metrology task is to check that some manufactured component is the right shape or size and is in the right location/orientation, that is, that it conforms to design within an agreed tolerance. For this reason, it may also be called verification. Often, an inspection or verification procedure will require a comparison of measurements against a CAD model or CAD values. Inspection can involve measuring some specific features or require a detailed surface analysis based on comparing a few or millions of points against the nominal CAD model.
2. *Alignment*: This refers to a class of tasks in which it is critically important to have a set of components arranged in a particular geometrical relationship, for example,
   a. Magnets in a synchrotron (particle accelerator)
   b. Rollers in a paper mill
   c. Sensors in a robotic manufacturing cell.

   Alignment can be a simple checking procedure, but it may also involve physical adjustments to the position and orientation of components, followed by repeat checking.
3. *Tool building*: Many manufacturing processes require a tool or jig to provide a reference set of locations for an object being assembled from parts or to act as fixtures for holding, say, aluminum or composite panels, so that they can be correctly machined using the tool as a template or guide. In constructing the tool itself, critical locations and holding points must be positioned relative to one another. This is a build process for which a tracker's real-time 3D output can provide the necessary feedback loop. There are similarities here with alignment and adjustment procedures.
5. *Reverse engineering:* This is similar to a surface inspection process for which a CAD model does not exist. The purpose of the surface model resulting from the process is not then to be checked against a CAD model, which does not exist, but to generate the CAD model itself. Example cases might include
   a. Reconstructing a replacement component in an old turbine for which drawings no longer exist
   b. Creating a facsimile of a statue for 3D printing at reduced scale
6. *Calibration*: Laser trackers provide a very accurate 3D reference in a large volume, and in a very flexible way. They are therefore excellent reference sources against which to check and calibrate robots and machine tools.

The following selection of applications, taken from equipment manufacturers' publications, illustrates how laser trackers are used and the industries in which they are applied.

3.8.1 Aerospace: Full Aircraft Surface Scan

The one remaining example of a historic Swiss fighter bomber in a museum near Zurich was scanned to create a wind tunnel model for aerodynamic tests as shown in Figure 3.45.

3.8.2 Aerospace: Tooling Jig Inspection

Figure 3.46 shows how Airbus Bremen inspects a tooling jig with a 3D laser tracker.

FIGURE 3.45
Full surface scanning and inspection. (Courtesy of Leica Geosystems, Hexagon Metrology, St. Gallen, Switzerland.)

FIGURE 3.46
Aerospace tool inspection. (Courtesy of Leica Geosystems, Hexagon Metrology, St. Gallen, Switzerland.)

FIGURE 3.47
Inspection of automotive chassis. (Courtesy of Leica Geosystems, Hexagon Metrology, St. Gallen, Switzerland.)

3.8.3 Automotive: Saloon Car Inspection

At the BMW plant in Leipzig, Germany, a 6D-probing system is used to perform a dimensional inspection of a car chassis shown in Figure 3.47.

3.8.4 Motorsports: Formula 1 Legality Checks

Red Bull Racing in the United Kingdom uses a 6D-probing system to check that the dimensions of one of their racing cars conforms to race requirements, for example, the width of the car or position of the pedals, shown in Figure 3.48.

3.8.5 Nuclear Power: Turbine Blade Checks

Siemens in Germany has provided steam turbines for a Finnish nuclear power station. The blades must be a close fit to the housing, checked by 3D laser tracker shown in Figure 3.49.

3.8.6 Wind Turbine Components: Inspection

The Spanish company Labker uses a 6D-probing system to inspect large components of wind turbines as shown in Figure 3.50.

FIGURE 3.48
Racing legality inspection. (Courtesy of Leica Geosystems, Hexagon Metrology, St. Gallen, Switzerland.)

FIGURE 3.49
Power generation turbine blade inspection. (Courtesy of Automated Precision, Inc., Rockville, MD.)

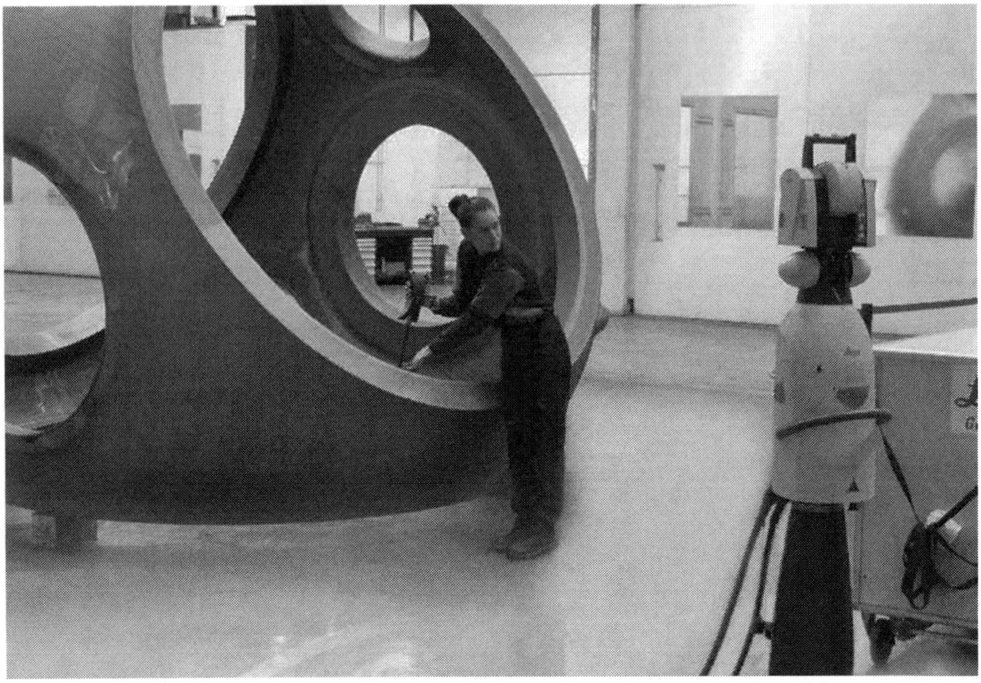

FIGURE 3.50
Wind turbine machine part inspection. (Courtesy of Leica Geosystems, Hexagon Metrology, St. Gallen, Switzerland.)

3.8.7 Machine and Plant Engineering: Large Transmission Case Checks

The German company Ferrostahl uses a 3D laser tracker measurement to inspect a large steel engine transmission case shown in Figure 3.51.

3.8.8 Robot Calibration

At the KUKA robot manufacturing plant in Germany, a laser tracker provides reference 3D coordinates to enable a robot to be calibrated, as shown in Figure 3.52.

3.8.9 Fusion Research: Laser Beam Alignment

At the U.S. National Ignition Facility (NIF), three 3D laser trackers are used to align the 192 power lasers so that their beams intersect the hydrogen fuel capsule and generate a fusion reaction. See Figure 3.53.

3.8.10 Radio Telescope: Assembly

The individual petals of each parabolic dish in the 15-telescope array of the Atacama Large Millimeter Array (ALMA) are aligned and assembled using a 3D laser tracker at the center of the dish shown in Figure 3.54.

FIGURE 3.51
Machine part inspection. (Courtesy of Automated Precision, Inc., Rockville, MD.)

FIGURE 3.52
Robot calibration. (Courtesy of FARO Technologies, Inc., Lake Mary, FL.)

FIGURE 3.53
Research laser beam alignment. (From CMSC, Simultaneous multiple laser tracker alignment at NIF, Archives of CMSC 2010, www.cmsc.org, accessed October, 2012.)

FIGURE 3.54
Antenna assembly and inspection. (Courtesy of FARO Technologies, Inc., Lake Mary, FL.)

3.8.11 Naval Engineering: Gun Positioning Accuracy

On new destroyers for the U.S. Navy, BAE Systems is using a 3D laser tracker to check the positioning accuracy of its advanced gun system (AGS) shown in Figure 3.55.

3.8.12 Marine Heritage: 3D Modeling

An anchor from a historic ship, the USS Monitor, is digitally modeled in 3D by a tracker scanning system to help in its preservation and presentation to the public shown in Figure 3.56.

FIGURE 3.55
Ship building and bore sight alignment. (Courtesy of FARO Technologies, Inc., Lake Mary, FL.)

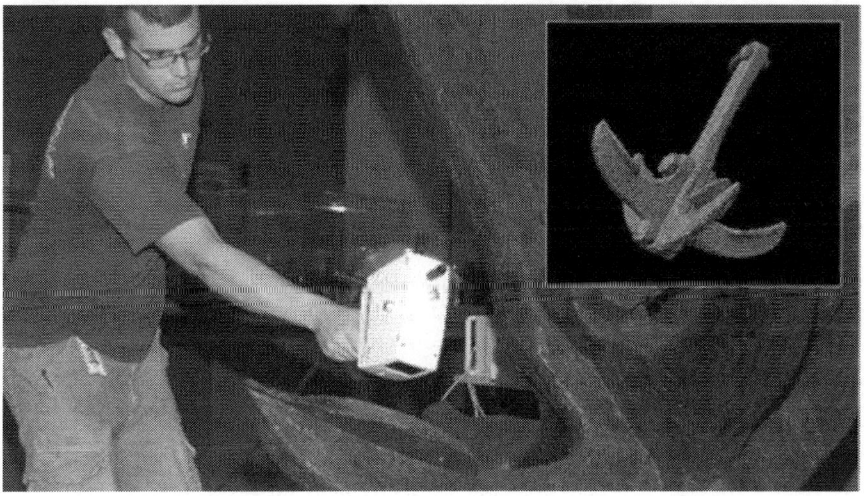

FIGURE 3.56
Historic artifact digitizing and scanning. (Courtesy of Leica Geosystems, Hexagon Metrology, St. Gallen, Switzerland.)

3.9 Summary

For more than two decades, laser trackers have been used on site to deliver high-accuracy 3D coordinate measurements for large-volume and high-value metrology applications. Their initial applications were to build and inspect large aerospace tools and parts in the late 1980s. Their successful application and further development in the form of fast and accurate tracking ADMs, automatic environmental compensation, and real-time 6D acquisition have greatly expanded their range of industrial applications.

Today, laser trackers are a core metrology solution in small-scale, single-operator applications and a key technology integrator for the assembly of components fabricated from around the world. Manufacturers in diverse industries have come to depend upon laser trackers for their on-site operation, portability, accuracy, and reliability.

In future, they will be applied more and more on production lines as embedded sensors in the control processes. Their high accuracy and data acquisition rates are ideal for supporting robotic drilling machines or automating large component assembly such as the wing-to-body join in aircraft manufacture.

Machine guidance and calibration are enabling trackers to deliver accuracy as part of the manufacturing process, as opposed to conventional quality assurance methods.[18] In contrast to checking, and possibly rejecting, a part after it has been produced, the embedded laser tracker enhances the manufacturing machine's capability to deliver better parts. This is also helping manufacturers to reduce costs and build bigger and more flexible machines.

References

1. B. Bridges, D. White, FARO Technical White Paper, Published by *Quality Digest*, February 1998.
2. A. A. Michelson, *Studies in Optics*, University of Chicago Press, Chicago, IL, 1927.
3. J. Calkins, Quantifying coordinate uncertainty fields in coupled spatial measurement systems, Dissertation, 2002, etd-08012002-104658, http://scholar.lib.vt.edu/theses/available/etd-08012002-104658/
4. *Guide to the Expression of Uncertainty in Measurement*, 1st edn., ISO, Geneva, Switzerland 1995.
5. Spatial Analyzer Software, New River Kinematics, Williamsburg, VA, www.kinematics.com
6. J. Palmateer, Those #%&!$ corner cubes, Paper presented at the *Boeing Seminar 1998*; A. Markendorf, The influence of the tooling ball reflector on the accuracy of laser tracker measurements: Theory and practical tests, Paper presented at the *Boeing Seminar 1998*.
7. S. Kyle, R. Loser, D. Warren, Automated part positioning with the laser tracker, Leica Geosystems, St. Gallen, Switzerland.
8. A. Markendorf, The influence of the tooling ball reflector on the accuracy of laser tracker measurements: Theory and practical tests, Leica Geosystems, St. Gallen, Switzerland.
9. Effect of environmental compensation errors on measurement accuracy, Renishaw, New Mills, Wotton-under-Edge, Gloucestershire, U.K.
10. T. D. Doiron, Temperature and dimensional measurement, NIST report on CTE material property, NIST, Gaithersburg, MD, Web Site: http://emtoolbox.nist.gov/Temperature/Slide2.asp
11. S. C. Sandwith, Scale artifact length dependence of videogrammetry system uncertainty, *Proceedings of SPIE* 3204-04, November 1998.
12. S. Sandwith, Thermal stability of laser tracker interferometer calibration, *Proceedings of SPIE* 3835, Vol. 93, November 1999.
13. B. Hughes, W. Sun, A. Forbes, A. Lewis, Determining laser tracker alignment errors using a network measurement, *J. CMSC*, 2010, 5(2), 26–32, National Physical Laboratory, Middlesex, U.K.
14. M. Basil, C. Papadopoulos, D. Sutherland, H. Yeung, Application of probabilistic uncertainty methods (Monte Carlo simulation) in flow measurement uncertainty estimation, *Flow Measurement—International Conference*, Peebles, Scotland, May 2001.

15. VIM, 1993, *International Vocabulary of Basic and General Terms in Metrology*, 2nd edn., VIM.
16. J. Calkins, S. Sandwith, Integrating certified lengths to strengthen metrology network uncertainty, *CMSC Journal* 2007.
17. CMSC, Simultaneous multiple laser tracker alignment at NIF, Archives of CMSC 2010. www.cmsc.org (accessed October 17, 2012).
18. T. Greenwood, Laser tracker interferometers in aerospace applications, SME Paper, 1992.

4

Displacement Measuring Interferometry

Vivek G. Badami and Peter J. de Groot

CONTENTS

4.1 Introduction

The wavelength of light provides an exceedingly precise measure of distance and is the foundation for commercial interferometric measurement tools that monitor object positions with a resolution better than 1 nm for objects traveling at 2 m/s. A wide range of applications include machine tool stage positioning and distance monitoring over length scales from a few millimeters to hundreds of kilometers in space-based systems.

Displacement measuring interferometry or DMI enjoys multiple advantages with respect to other methods of position monitoring. In addition to high resolution, wide measurement range, and fast response, the laser beam for a DMI is a virtual axis of measurement that can pass directly through the measurement point of interest (POI) to eliminate Abbe offset errors. Figure 4.1 illustrates the position of DMI in terms of resolution and dynamic range with respect to capacitive gaging, optical encoders, and linear variable differential transformer (LVDT) methods. The measurement is noncontact and directly traceable to the unit of length. Since the first practical demonstration of automated, submicron stage control using displacement interferometry in the 1950s,[1] DMI has played a dominant role in high-precision positioning systems.

This chapter is intended as an overview of the current state of the art in DMI as represented by the technical and patent literature. The chapter structure is correspondingly encyclopedic and allows for access to specific topics without necessarily reading the chapter linearly from start to finish.

Following this brief introduction, we begin in Section 4.2 with fundamental physical principles of DMI, followed in Section 4.3 by a review of phase detection methods most

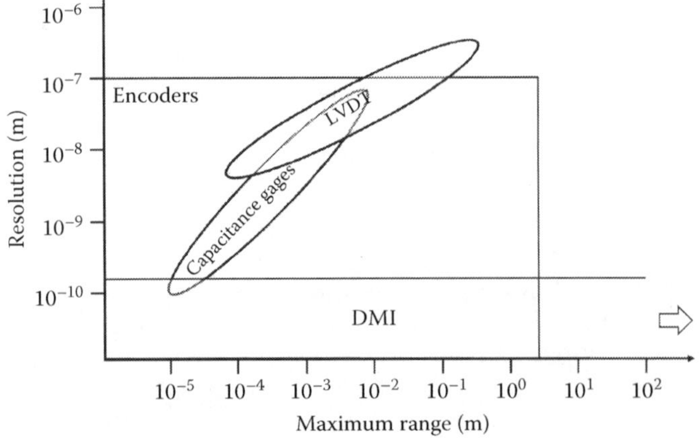

FIGURE 4.1
DMI with respect to comparable devices for distance measurement.

common in practical implementations, introducing the essential concepts of homodyne and heterodyne detection. Section 4.4 considers ways of generating wavelength-stable light having the coherence and modulation characteristics essential for heterodyne DMI. Section 4.5 catalogs some of the more common optical interferometer configurations sensitive to various measurement parameters such as displacement, angle, and refractive index. With these tools in hand, we then examine in Section 4.6 the question of system performance in terms of the measurement uncertainty, encompassing error sources such as wavelength instability, Abbe offset error, cyclic error, air turbulence, and thermal drift. A gallery of practical DMI applications follows in Section 4.7, considering calibration and validation tasks as well as integration of DMI into a complete machine for continuous motion control and/or measurement. The applications section includes examples of unusual technology approaches that may find more common usage going forward.

4.2 Fundamentals

The high precision of DMI leverages the rapid change of phase as light propagates, equivalent to a 2π phase shift for a distance of less than half a micron for visible light. This fine-scale, traceable metric is accessible through interferometric methods of comparing reference and measurement beams generated from a common source light beam and then recombined.

To establish terminology and notation, we provide here a mathematical description of the principles of the technique. The oscillating electric field of a source beam of amplitude E_0 is

$$E(t,z) = E_0 \exp\left[2\pi i\left(ft - \frac{nz}{\lambda}\right)\right] \quad (4.1)$$

where

f is the frequency of oscillation
λ is the vacuum wavelength
z is the physical path length
n is the index of refraction

The frequency at visible wavelengths is very high—approximately 6×10^{14} Hz—making it difficult to detect the phase $2\pi(ft - nz/\lambda)$ of $E(t,z)$ directly. To access the wavelength as a unit of measurement, we need to remove or at least drastically slow down the optical frequency component. This is why we use interferometry.

Referring to Figure 4.2, the nonpolarizing beam splitter (NPBS) splits the source light into two beams, labeled 1 and 2 for the measurement and reference beams, respectively. These beams follow different paths and therefore have different phase offsets related to the propagation term nz/λ:

$$E_1(t,z_1) = r_1 E_0 \exp(2\pi i f t)\exp\left(\frac{-2\pi i\, n z_1}{\lambda}\right) \quad (4.2)$$

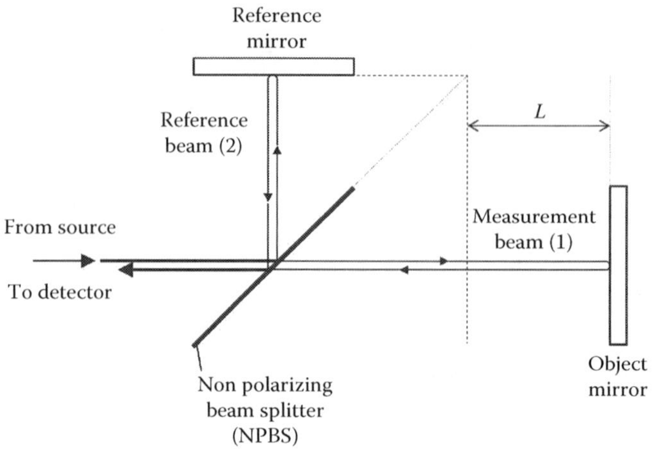

FIGURE 4.2
Michelson-type amplitude division interferometer for monitoring the position of an object mirror.

$$E_2(t, z_2) = r_2 E_0 \exp(2\pi i f t) \exp\left(\frac{-2\pi i\, n z_2}{\lambda}\right) \tag{4.3}$$

where

z_1, z_2 are the path lengths traversed by beams from the point that they are separated by the NPBS to the point that they are recombined

r_1, r_2 are their relative strengths with respect to the original complex amplitude $|E_0|$

When the two beams superimpose coherently on a square-law detector, the time average of the resulting intensity is

$$I(z_1 z_2) = |E_0|^2 \left\langle |\exp(2\pi i f t)|^2 \right\rangle \left| r_1 \exp\left(\frac{-2\pi i\, n z_1}{\lambda}\right) + r_2 \exp\left(\frac{-2\pi i\, n z_2}{\lambda}\right) \right|^2 \tag{4.4}$$

The frequency term *ft* averaged over time becomes a constant:

$$\left\langle |\exp(2\pi i f t)|^2 \right\rangle = 1 \tag{4.5}$$

but the final expression preserves the optical path difference $z_1 - z_2$:

$$I(L) = I_1 + I_2 + \sqrt{I_1 I_2} \cos[\phi(L)] \tag{4.6}$$

where

$$I_1 = |r_1 E_0|^2 \tag{4.7}$$

$$I_2 = |r_2 E_0|^2 \tag{4.8}$$

$$\phi(L) = \left(\frac{4\pi n L}{\lambda}\right) \tag{4.9}$$

$$2L = z_1 - z_2 \tag{4.10}$$

The fundamental principle of DMI is to detect changes in the distance L by evaluation of the interference phase $\phi(L)$ via its effect on the time-averaged intensity $I(L)$.

4.3 Phase Detection

As is the case with all forms of interferometry, there is a heavy emphasis in DMI technology development on methods of determining the phase $\phi(L)$. In the earliest interferometers, dating back to Michelson, phase estimation was visual.[2,3] The visual method relies on the presence of a spatial fringe pattern viewed by the eye, generated, for example, by tilting one of the interferometer mirrors. The experienced observer with the aid of a crosswire or adjustable compensating optical element can estimate to about 1/40 th of a fringe or 15 nm—sufficient for routine gage block measurements.[4,5] Another historical technique especially useful for objects in motion is fringe counting—essentially keeping a tally of the number of times that the output beam goes from light to dark—which is appropriate for low-precision applications for which a displacement estimate with a resolution of $\lambda/2$ is sufficient and for which there is no uncertainty regarding the direction of motion.

Modern DMI systems rely on electronic detection and data processing to allow for an evaluation of the intensity over a range of controlled phase shifts to establish the direction of motion and to interpolate to small fractions of a fringe. In the majority of systems, this means encoding the reference and measurement beams by polarization, so that they can be mixed with three or more phase shifts α between them:[6]

$$I(L,\alpha) = I_1 + I_2 + \sqrt{I_1 I_2}\cos[\phi(L) + \phi_0 + \alpha] \tag{4.11}$$

A sequence of phase shifts allows for signal processing based on quadrature detection, the fitting of sines and cosines, or equivalent methods to solve for $\phi(L)$. The additional phase term ϕ_0 is a constant offset equal to the detected phase for the displacement position defined as $L = 0$.

Figure 4.3 illustrates one arrangement for encoding the reference and measurement beams by polarization. A polarizing beam splitter (PBS) separates source light into a reference and a measurement beam according to s and p linear polarizations, respectively. A suitable source polarization would be linear, at a 45° angle with respect to the plane of the figure. Reference and object corner cubes reverse the beam paths so that they are collinear and aligned parallel when recombined by the PBS.[7] This exit beam now contains the reference and measurement beams encoded by orthogonal linear polarizations.

Two established options for phase detection with polarization-encoded interferometers are the homodyne and heterodyne techniques.

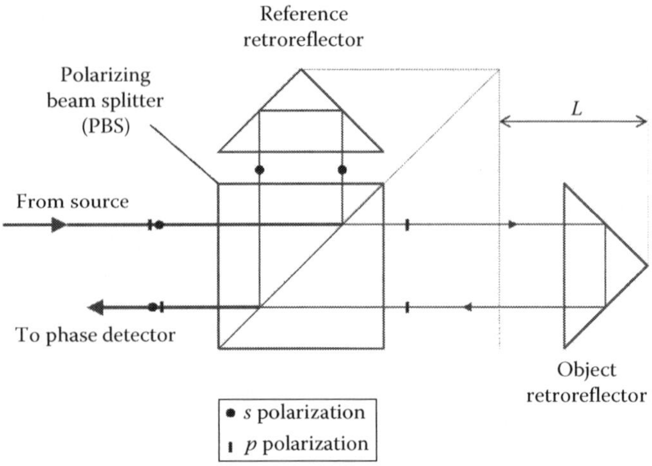

FIGURE 4.3
Linear displacement interferometer using retroreflectors and polarization encoding of the measurement and reference beams.

In the homodyne method, polarizing optics and multiple detectors measure the intensity for several static phase shifts imposed by polarizing optics.[8] Figure 4.4 illustrates an example homodyne detector employing quarter-wave and half-wave retardation plates (QWP and HWP, respectively). The four detectors shown in the figure detect intensity signals with $\pi/2$ relative phase shifts:

$$I_j = I(L, \alpha_j) \tag{4.12}$$

$$\alpha_j = \frac{j\pi}{2} \tag{4.13}$$

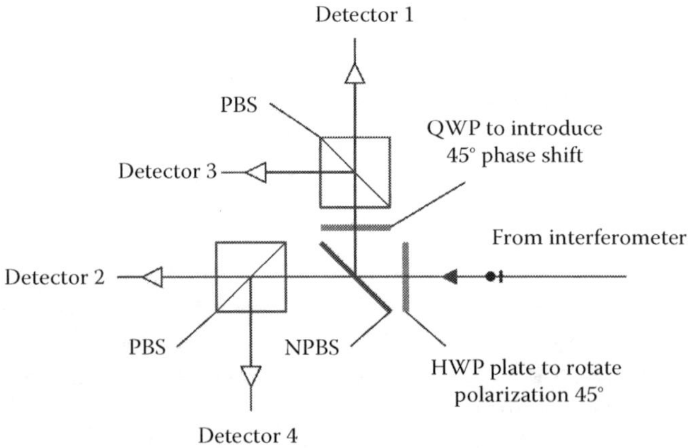

FIGURE 4.4
Optical components of a homodyne phase detector.

The phase follows (to within a constant offset) from the formula

$$\tan(\phi) = \frac{I_1 - I_3}{I_2 - I_4} \tag{4.14}$$

This method is to first order insensitive to fluctuations in the source intensity and fringe contrast and, to the degree that the imperfections in the detection optics can be compensated algorithmically[9] or by design,[10] provides a precise estimate of the instantaneous phase and the displacement L. Homodyne detection is popular for low-cost, single-axis DMI systems[11,12] and for optical encoders.[13] Variations include reversing the roles of the wave plates in Figure 4.4 or creating a spatial pattern of interference fringes by recombining the measurement and reference beams at a small angle, similar to what was done in the earliest days of interferometry, with fixed sampling points at different points on a fringe.[14]

In the heterodyne method, the interference intensity is continuously shifted in phase with time by imparting a frequency difference between the reference and measurement beams.[15–17] Instead of sharing a common optical frequency f, an offset frequency $\Delta f = f_2 - f_1$ is imposed between reference and measurement beams, usually at the source, resulting in a continuous, time-dependent phase shift in Equation 4.11:

$$\alpha(t) = \Delta f t \tag{4.15}$$

The output beam intensity varies sinusoidally at a rate of a few kilohertz to hundreds of megahertz, depending on the system design and the anticipated maximum object velocity. A single detector captures this signal for each measurement direction or axis (Figure 4.5). Several different ways are available to analyze this signal, including lock-in methods, zero crossing, and sliding-window Fourier analysis, with advanced data age compensation for high-speed servo control.[18] Often there is an electronic reference signal, generated either electronically from the drive electronics that generate the frequency shift Δf or optoelectronically, by measuring the phase of the heterodyne signal detected prior to the interferometer optics.

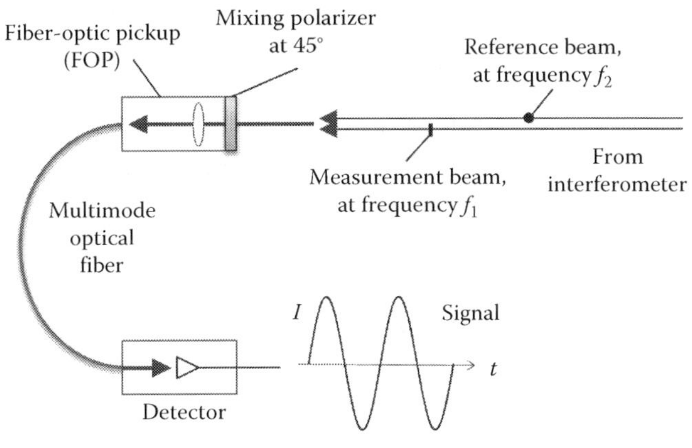

FIGURE 4.5
Heterodyne detection using a fiber-optic pickup (FOP).

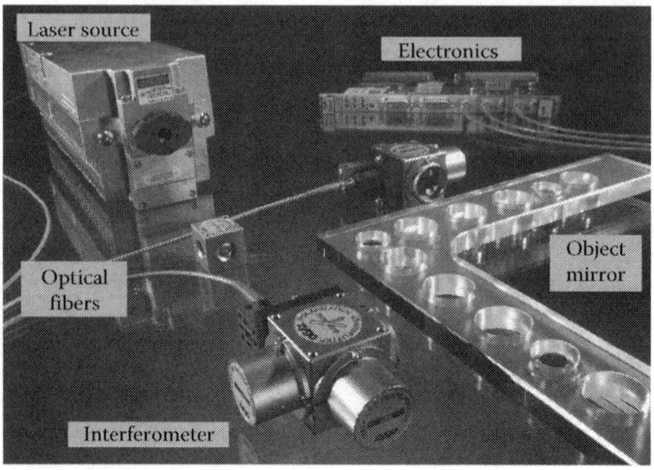

FIGURE 4.6
Components of a heterodyne laser system. (Photo courtesy of Zygo Corporation, Middlefield, CT.)

Of the several advantages of heterodyne systems, we note the simplicity of the detector, which is particularly important in cost-effective multiaxis systems, and the shifting of the signal to a frequency far from DC to avoid thermal and other statistical noise sources that scale inversely with frequency. The various components of a heterodyne interferometer system shown in Figure 4.6 address a wide variety of measurement configurations. In particular, heterodyne interferometry serves high-precision applications such as microlithographic staging (see Section 4.7).

4.4 Light Sources for Heterodyne Displacement Interferometry

Light sources for heterodyne interferometry must satisfy multiple optical and metrological requirements. They generate two linear-polarized mutually orthogonal frequency-shifted beams at a stable and known wavelength. The last two requirements also establish traceability to the unit of length and provide the scaling factor required to convert the change in phase measured at the output of the interferometer into units of length. Common applications also require coherence lengths of the order of meters, although sources with short coherence lengths may be used to advantage in some applications.[19,20] Additional practical demands include compact size, low heat dissipation, high optical output power, high level of stability of the pointing and polarization state orientation relative to a mechanical datum, optical isolation from reflected beams, and high wavefront quality.

Laser sources, especially gas laser sources, satisfy virtually all of the requirements. One of the earliest two-frequency lasers to be applied to displacement interferometry was a modified helium–neon (He-Ne) gas laser[15] shortly after continuous-wave emission at 632.9 nm was demonstrated in 1962.[21] Although specialized applications use other laser types and wavelengths[22,23] and a number of light sources provide a practical realization of the meter,[24] the He-Ne laser operating at 632.9 nm remains the laser of choice for

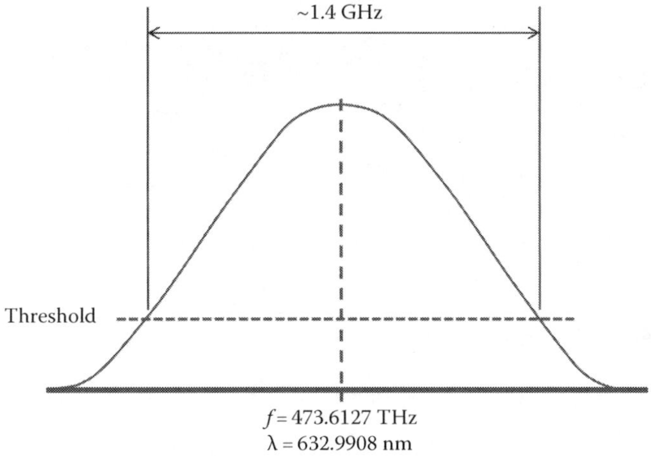

FIGURE 4.7
Gain curve of the He-Ne laser at ~633 nm. Frequency and wavelength are the CIPM recommended values. (From Stone, J.A. et al., *Metrologia*, 46, 11, 2009.)

metrology applications. This preference stems from a number of reasons, including inherent simplicity, ease of alignment due to the visible radiation, and the ready availability of silicon photodetectors.[25] The following discussion confines itself to this type of laser and in particular to the commonly available commercial implementations and the stabilization schemes used therein.

The value of wavelength of the He-Ne laser and the associated uncertainty are key parameters in DMI. The value of the vacuum wavelength (along with the refractive index) scales the measured phase changes into units of length, and any uncertainty in the wavelength propagates directly into the uncertainty in the displacement measurement (see Section 4.6). The basic atomic physics of the He-Ne laser operating at a vacuum wavelength of 632.9 nm ($3s_2 \rightarrow 2p_4$ transition in neon) has interesting implications for the realization of the unit of length in that the atomic transition is a reliable optical standard of wavelength, albeit with an associated uncertainty.[26] As shown in Figure 4.7, the width of the gain curve above the threshold is ~1.4 GHz (or 3×10^{-6} in terms of relative frequency) and is determined by the governing physics of this type of laser. A recent recommendation of the Consultative Committee for Length (CCL) of the International Committee for Weights and Measures (CIPM) has resulted in the inclusion of the unstabilized He-Ne laser operating at 632.9 nm in the list of light sources used to realize the unit of length.[27] The CIPM recommended value of the associated relative standard uncertainty is 1.5×10^{-6}, which with a coverage factor of two exceeds the width of the gain curve above threshold. Although this establishes some degree of traceability for a free-running He-Ne laser, the 1.5×10^{-6} level of uncertainty is insufficient for many DMI applications, and stabilization of the laser wavelength is a necessity.

Figure 4.8 represents a classification of He-Ne laser sources for heterodyne DMI based on the method used to stabilize them. Dual-frequency lasers (hereafter referred to as metrology lasers) are for use in heterodyne DMI systems. Reference lasers in the right-hand branch of Figure 4.8, such as the $^{127}I_2$-stabilized He-Ne laser, realize the definition of the meter to very high levels of accuracy for calibration of metrology lasers when used in accordance with the recommendations of the BIPM.[24] The majority of commercial DMI laser systems follow the left-hand branch of Figure 4.8 and use a single emission mode for

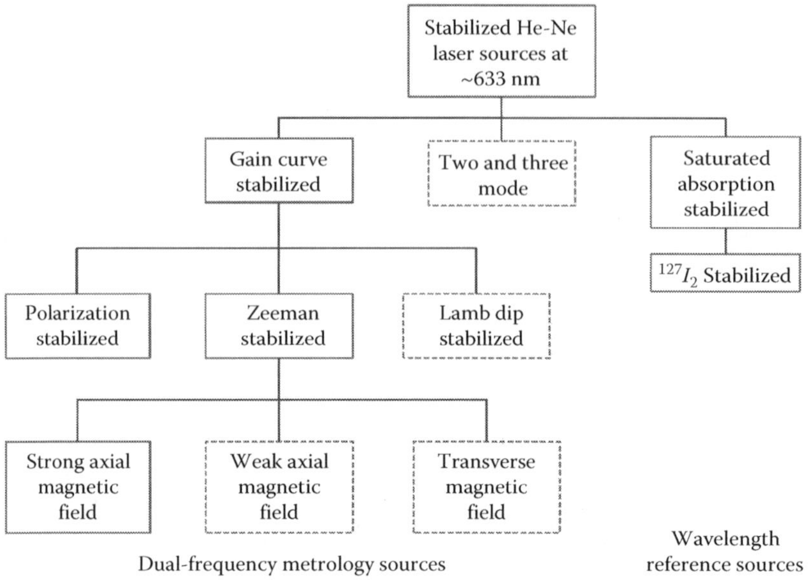

FIGURE 4.8
Classification of laser sources. Dashed lines indicate sources that are restricted to laboratory use or in limited commercial use.

metrology, although they may operate multimode for the purpose of stabilization by reference to the gain curve. A minority of these lasers use two or more lasing modes directly, with the heterodyne frequency established by the difference in optical frequency between these modes.[28–30]

Commercial metrology lasers are stabilized by controlling the position of two orthogonally polarized TEM_{00} modes relative to the gain curve for the He-Ne gain medium. Adjustment of the length of the resonator under closed-loop control drives the difference in intensities of the two modes to zero. Two methods effect gain-curve stabilization: Zeeman splitting and polarization stabilization.

4.4.1 Zeeman-Stabilized Lasers

The Zeeman laser[31–33] traces its roots to the devices described by Tobias et al.[34] and by de Lang et al.[15] The method relies on the Zeeman effect,[35] wherein the application of a relatively strong magnetic field along the axis of the tube results in a "splitting" of the single laser mode into two orthogonally circularly polarized modes as depicted in Figure 4.9, with frequencies slightly above and below the original center frequency f_0, denoted by f_+ and f_-, respectively. Zeeman splitting provides the frequency difference required for heterodyne interferometry as well as a method for stabilization of the laser wavelength. The frequency difference depends on the magnitude of the applied magnetic field and ranges from about 1 to 4 MHz. While commercial laser sources rely on a strong axial magnetic field to produce the two modes, alternate implementations using a transverse magnetic field have also been reported in the literature.[36–38]

Figure 4.10 shows a schematic of a typical Zeeman-split laser source. The internal cavity He-Ne laser operates at a single longitudinal and lateral mode at ~633 nm. The requirement for single-mode operation dictates the maximum length of the resonator cavity,

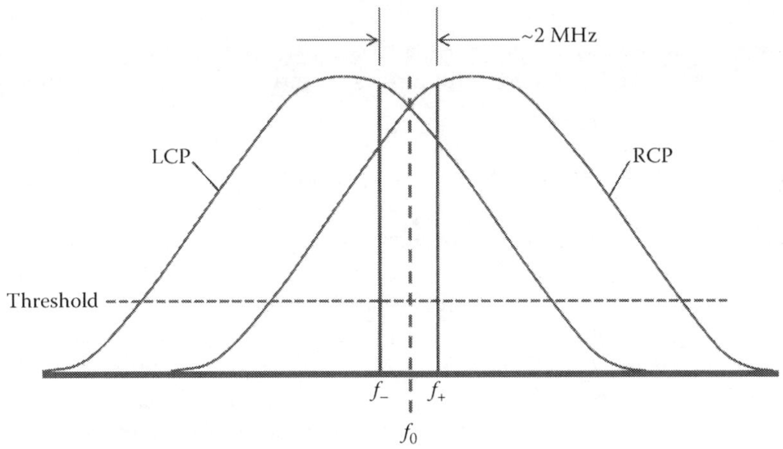

FIGURE 4.9
Splitting of single longitudinal laser mode into two orthogonal polarization states (left circular polarized or LCP and right circular polarized or RCP) due to Zeeman splitting. The split is exaggerated for clarity.

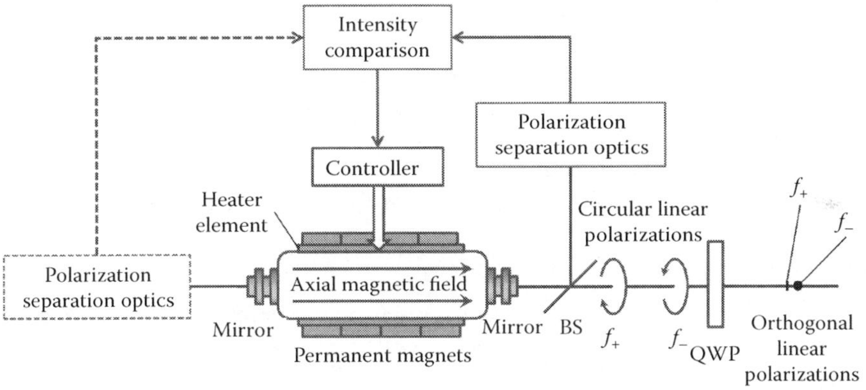

FIGURE 4.10
Schematic representation of a Zeeman-stabilized laser.

which in turn sets an upper bound on the achievable power. Permanent magnets that surround the laser tube apply an axial magnetic field. A beam splitter (BS) samples a small portion of the two orthogonal circularly polarized beams exiting the tube for stabilization. The two circularly polarized states convert to orthogonal linearly polarized states that generate an error signal for the feedback system that controls the length of the laser cavity. Typical commercial implementations separate the two linearly polarized beams and direct them onto two detectors and use the difference in the outputs as the error signal.[31,33] An alternative method that eliminates the need to match the detector gains employs a single detector and a programmable liquid-crystal polarization rotator to alternately sample the intensities of the two polarizations.[33] The intensity difference between the two modes as a function of resonator length exhibits a zero crossing with a large slope at the point corresponding to the unshifted center frequency. This slope makes it possible to stabilize the wavelength by controlling the length of the resonator using either a piezoelectric (PZT) actuator[15,31] or a heating element.[33] An alternate method uses the beat frequency between the two modes,[32] obtained by arranging for the two beams to interfere by means of a linear polarizer prior to directing them onto a single detector. Locating the minimum in the beat frequency as function of resonator length controls the resonator length.

As shown in Figure 4.10 by the dashed lines, the laser output for stabilization can be the weaker beam that exits the rear of the tube, thereby making all of the main output of the laser available to the metrology application. A recent design exploits the elliptically polarized beams that result from the inherent anisotropy in the laser tube to produce the feedback signal without any additional polarization optics.[39,40] The light exiting the source transforms into two mutually orthogonal linearly polarized beams (at two slightly different frequencies) by means of a QWP. In this type of laser, the production of the two frequencies for heterodyne DMI is integral to the stabilization scheme.

While the Zeeman method has some drawbacks from a control standpoint,[41] Hewlett–Packard (now Agilent),[31,33] Zygo Corporation,[42] and other manufacturers produce Zeeman laser DMI systems that have played a significant and widespread role in metrology. A drawback of this method is the low heterodyne frequency and the resulting limitation on the maximum slew rate of the target. A further drawback is the requirement on single-mode operation of the tube,[43,44] which limits the length of the laser tube to less than about 10 cm and in turn reduces the achievable output power compared to the longer tubes used in polarization-stabilized lasers.

The quality of the exiting beams in terms of the orthogonality and ellipticity are key parameters that have an impact on the cyclic errors, as described in Section 4.6. Early characterizations of these parameters report deviations from orthogonality of 4°–7°,[45] while more recent measurements suggest a much smaller value of 0.3° and an ellipticity of 1:170 in the electric field strength.[46]

4.4.2 Polarization-Stabilized Lasers

In contrast to the Zeeman-stabilized system, polarization-stabilized DMI laser source systems separate the stabilization and frequency shifting functions.[17,47–49] The stabilization technique relies on matching the intensities of two adjacent longitudinal modes as shown in Figure 4.11.[50,51] The presence of two orthogonally polarized modes under the gain curve

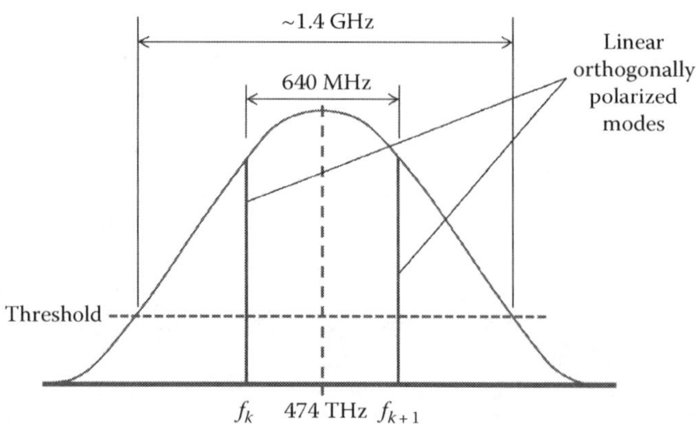

FIGURE 4.11
Two-mode operation of a polarization-stabilized laser.

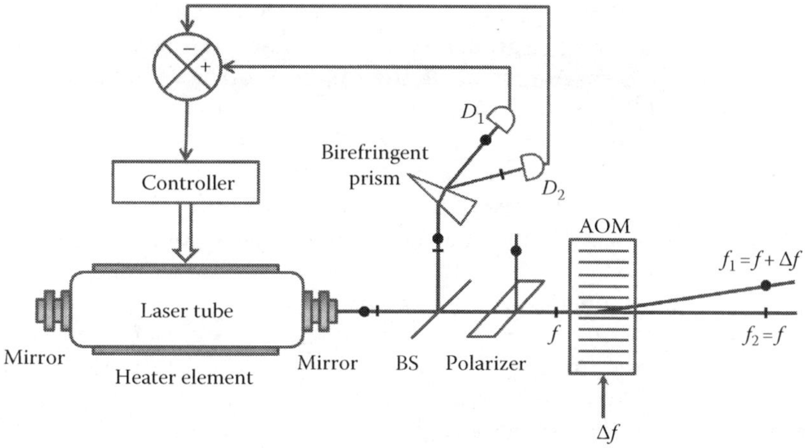

FIGURE 4.12
Schematic of a polarization-stabilized laser.

makes it possible for the tube in a polarization-stabilized laser to be longer than in the Zeeman-stabilized laser (~30 cm).

A BS as shown in Figure 4.12 samples a small portion of the output beam. A Wollaston prism (or other suitable birefringent prism) divides the sampled beam according to the two orthogonal polarizations corresponding to the two lasing modes and directs them on to two detectors D_1 and D_2.[50] The control system strives to minimize the difference in intensity observed at the two detectors D_1 and D_2 by adjusting the cavity length. As shown in Figure 4.11, which depicts the relative intensities of the modes when the laser is in the stabilized condition, the two modes are disposed symmetrically with respect to the wavelength corresponding to the peak of the gain curve.

The polarizer following the BS passes only one of the modes, which then passes through an acousto-optic modulator (AOM) to produce the two mutually orthogonal linearly polarized frequency-shifted beams arranged to overlap and emerge parallel to one another from the laser by using a second birefringent prism (not shown).[47] The frequency difference Δf corresponds to the drive frequency of the AOM. Although one of the modes is rejected, the increase in tube length (when compared to a Zeeman-stabilized laser) for two-mode operation more than makes up for the loss of one of the modes, resulting in a net gain in output power. This method of frequency shifting allows for large frequency differences relative to the Zeeman frequency split, with a 20 MHz frequency difference being common, which supports much higher slew rates of the target. This technique is however less efficient in its use of the available light. One way to recover the lost power is to use the two lasing modes to supply two independent interferometers.[52]

Commercial-stabilized laser sources typically specify a relative "vacuum wavelength accuracy" or "unit-to-unit variability" of ± 0.1–0.8×10^{-6}.[42,52,53] These numbers can be the basis for the estimation of a standard uncertainty in the vacuum wavelength. If a more accurate value of the wavelength is desired, the wavelength should be determined for the source in question by comparison against an iodine-stabilized He-Ne laser[41,54] or against a frequency comb.[55] Stabilities for commercial sources are typically specified over various time scales, with short-term (1 h), medium-term (24 h), and long-term (over the laser lifetime) relative wavelength stabilities of ± 0.5–2×10^{-9}, ± 1–10×10^{-9}, and ± 10–20×10^{-9}, respectively.[53,56] A recent report on the measurement of 28 different laser heads of both types by comparison to an iodine-stabilized He-Ne laser over a period of 20 years confirms that in general the manufacturer's specifications are met or exceeded.[57]

Some special care is necessary to achieve the stated stability of the laser. One cause for degradation in laser stability is optical feedback caused by reflections back into the laser cavity from external sources.[58,59] Even weak reflections of ~0.01% of the output of the laser can cause major instabilities in older systems resulting from the formation of unstable external cavities outside the influence of the stabilization control.[43,58] Traditional methods to combat optical feedback include rotating the polarization of the reflected beam in a nonreciprocal manner by means of a Faraday cell so as to prevent reentry of the reflected beam into the resonator by means of a crossed polarizer.[60]

Using an AOM to generate the two frequencies provides an inherently high degree of feedback isolation.[47,61] Beams reflected back through the laser system are either directed away from their original beam paths or are frequency shifted outside the gain bandwidth of the laser.[41,62] Some modern systems have further enhanced isolation by specialized AOM design, effectively eliminating optical feedback as a source of concern.[52,63]

General considerations common to all types of laser heads include beam size, heat dissipation, pointing stability, and other operational characteristics.

The beam is typically expanded by means of a telescope (not shown) to above 5 mm diameter so as to limit the natural beam divergence that results from Gaussian beam propagation and permit measurements over reasonable displacement ranges.[64] Another consideration that dictates the beam size is tolerance to shear between the measurement and reference beams. Lateral translations and angular misalignments of the measurement beam relative to the reference beam can result from errors in the optics and alignment, and the beam size must provide adequate overlap.[65] Commercial systems most commonly produce a beam with a 6 mm $1/e^2$ diameter, which offers a reasonable trade-off between measurement range and the size of the interferometer optics;[53,56,66,67] although 3 and 9 mm beam diameters are available as options.[53]

Heat dissipation by the laser head is a thermal management issue in demanding applications. Laser heads typically dissipate 20–40 W during operation and must be carefully located to minimize the heat load.[53,56] Specialized heads use liquid cooling to reduce the dissipated power to under 10 W.[66,67] Another option is to locate the head remotely and couple the light into the application via optical fibers.[66]

4.4.3 Stabilization Using Saturated Absorption: Iodine-Stabilized He-Ne Laser

Lasers stabilized by saturated absorption provide the highest level of stability and reproducibility in the field of dimensional metrology and are typically used by National Metrology Institutes (NMIs) as the primary means to realize the unit of length.[68] These single-frequency lasers represented by the right branch of Figure 4.8 calibrate two frequency metrology lasers. The 633 nm iodine (I_2)-stabilized He-Ne developed in the early 1970s remains the most common primary standard laser for dimensional metrology, largely because of the ease of calibration, achieved by mixing the output of the He-Ne metrology laser with the I_2-stabilized laser emission and directly measuring the resulting radio frequency beat signal.

The vastly superior performance of the absorption-stabilized laser stems from the separation of the light generation and stabilization functions, making it possible to use a transition in separate species as a reference. The stabilization subsystem is physically separate from the laser tube and can therefore be placed in a stable environment, uncoupling it from the perturbations that occur during the process of light generation (such as variations in pressure of the discharge tube). The separation results in improvements in the reproducibility of approximately 10^3.[41]

The I_2 He-Ne relies on the saturation of absorption[69] of the iodine cells to eliminate the Doppler broadening and make the hyperfine absorption lines in the iodine spectrum accessible. The absorption saturates because of the bidirectional standing waves within the cavity, resulting in a Lorentzian-shaped dip in the absorption of the iodine cell, the width of which corresponds closely to the natural line width of the transition and occurs at the center of the absorption line. This drop in the absorption of the cell results in sharp Doppler-free peaks with quality factors of $\sim 10^8$ superimposed upon the Doppler-broadened gain curve of the He-Ne laser as shown in Figure 4.13a. These hyperfine lines are highly reproducible and relatively immune to perturbations,[70] in large part because the hyperfine absorption lines are the result of transitions from the ground state, eliminating the perturbing influences of an excitation mechanism[71] and guaranteeing a high level of wavelength reproducibility.

The gain in power at these peaks is typically $\sim 0.1\%$ of the base output power at that wavelength, and any locking technique must be capable of recovering the weak signal from a strong background in order for the peak to serve as a frequency discriminator. The location of the peak is usually determined by the method of third-harmonic locking that effectively determines the third derivative of the signal.[71,72] This produces an

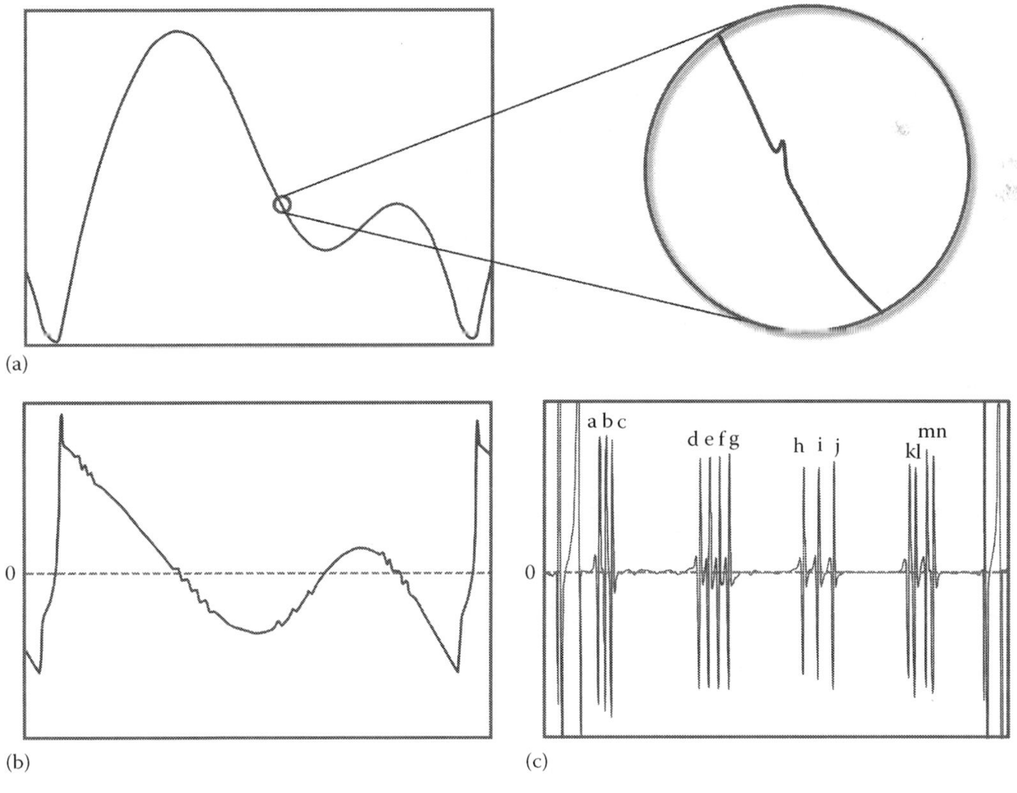

FIGURE 4.13
(a) Gain curve of a cavity with a He-Ne plasma tube and iodine cell. (b) First and (c) third derivatives of the gain bandwidth curve with respect to frequency.

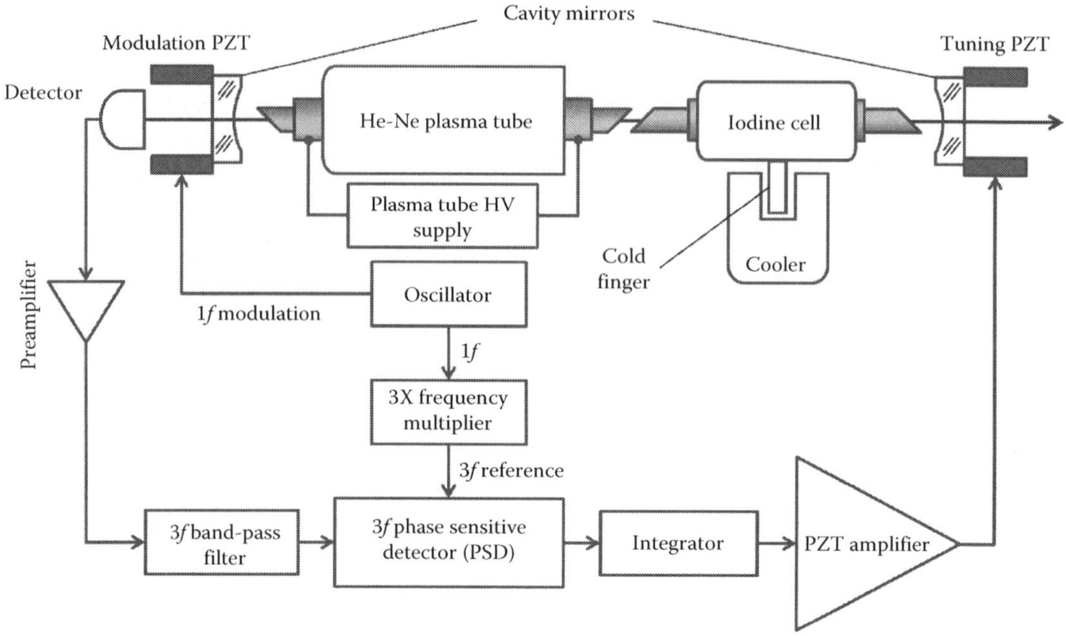

FIGURE 4.14
Schematic of iodine-stabilized He-Ne laser.

antisymmetric zero crossing at the location of the peak as shown in Figure 4.13c in contrast to the first derivative shown in Figure 4.13b, which does not always produce a zero crossing due to a large background slope.

Figure 4.14 shows a representative schematic of an iodine-stabilized He-Ne laser. While several different designs are available,[70,71,73–80] virtually all implementations employ a DC voltage excited plasma tube that contains the He-Ne gas and a separate intracavity I_2 cell, both of which have Brewster windows. The cell contains $^{127}I_2$ at low pressure to reduce contributions of pressure broadening. A temperature-controlled cold-finger integral to the cell controls the pressure of the cell. Two high-reflectance mirrors (typically spherical) form the resonator cavity, and a PZT actuator displaces one or both in a controlled manner. To lock the laser frequency, a PZT actuator imposes a sinusoidal modulation on one of the cavity mirrors, and a photodetector senses the resulting modulation in the output. The third harmonic of the photodetector signal is phase sensitively demodulated against a reference at the third harmonic of the modulation frequency. The resulting third-derivative signal serves as an error signal to a second PZT under closed-loop control that varies the length of the cavity and strives to drive the value of the third derivative to zero.

The CIPM specifies, through periodic publications, the nominal wavelength for a specific transition of the $^{127}I_2$ He-Ne, the associated relative standard uncertainty, and the conditions required to achieve the stated uncertainty. The current recommendations adopted in 2001 specify a slightly revised value for the recommended frequency and a slightly reduced relative standard uncertainty of the wavelength of 2.1×10^{-11} for the a_{16} or f component of the R(127) 11^{-5} transition for $^{127}I_2$ based on measurements using frequency combs.[24,81]

Numerous international intercomparisons between He-Ne lasers and the iodine cells themselves have established the reproducibility of the $^{127}I_2$-stabilized He-Ne operating

TABLE 4.1

Relative Wavelength Accuracies and Stability of Stabilized He-Ne Lasers

			Wavelength Stability (×10^{-9})		
Laser Type	**Frequency**	**Relative Wavelength "Accuracy"**	**1 h**	**24 h**	**Lifetime**
Zeeman stabilized	Dual	Accuracy (3σ) ± 0.1 × 10^{-6}	±2	Not specified	±20
Polarization stabilized		Accuracy or unit-to-unit variability ±0.1 – 0.8 × 10^{-6}	±0.5[a]	±1[a]	±10[a]
Iodine stabilized	Single	$\frac{u_c(\lambda)}{\lambda} = 2.1 \times 10^{-11}$	Allan variance of $10^{-11}/\sqrt{t}$ where t is the measurement time		

[a] 3σ values.

at 633 nm.[82,83] Comparisons over a period of 25 years show that ~85% of the lasers in the comparison agree to within the relative standard uncertainty of 2.5 × 10^{-11} specified by the then current *mise en pratique*,[84] and all the lasers were in agreement to within the relative expanded uncertainty with k = 3.[85] Recent measurements of the frequency of these lasers by direct comparison to the second using femtosecond frequency combs show a reproducibility of 1 × 10^{-12}.[86]

Iodine-stabilized reference lasers typically operate with much lower output powers of approximately 100 μW and produce only a single, user-settable frequency, thereby making them unsuitable for direct use as a metrology laser in a heterodyne system.[87] An additional limitation is that the output intensity of the laser modulates as a natural consequence of the small modulation imposed on one of the cavity mirrors as part of the third-harmonic locking scheme.[88] In high-accuracy applications where the stability of the metrology laser is inadequate and/or uncertainty in the nominal wavelength is too large, embedding these lasers into the machine provides for continuous monitoring of the wavelength of the metrology laser.[89,90] They may also be used as reference lasers for multiple higher-power metrology lasers, which are offset locked to the iodine-stabilized He-Ne, thereby producing a source that combines the high stability of the iodine-stabilized laser with the higher unmodulated output power available from a conventional He-Ne.[88,89,91]

Table 4.1 provides an estimate of the relative accuracies and wavelength stabilities from manufacturer's specification sheets for lasers without any special calibration. Data for the dual-frequency lasers use the manufacturer's nomenclature, because of a lack of consistency in the method of specification. In contrast, at least one commercial iodine-stabilized laser specifies a relative standard uncertainty in the wavelength and specifies the stability in terms of the Allan variance as a function of time.[87,92,93]

4.5 Interferometer Design

Returning to the interferometer optics, the linear interferometer configuration of Figure 4.3 with retroreflectors serves well for applications that involve single-axis motion. The object retroreflector preserves the parallelism of the reference and measurement beams after recombination, even for small tilts of the object.[7]

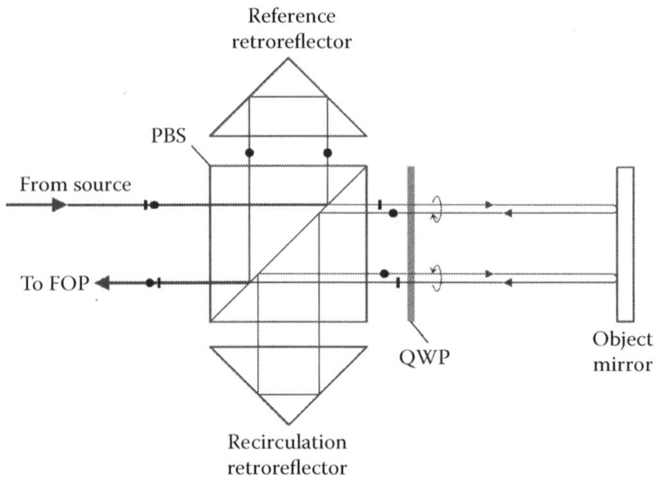

FIGURE 4.15
PMI with recirculating retroreflector.

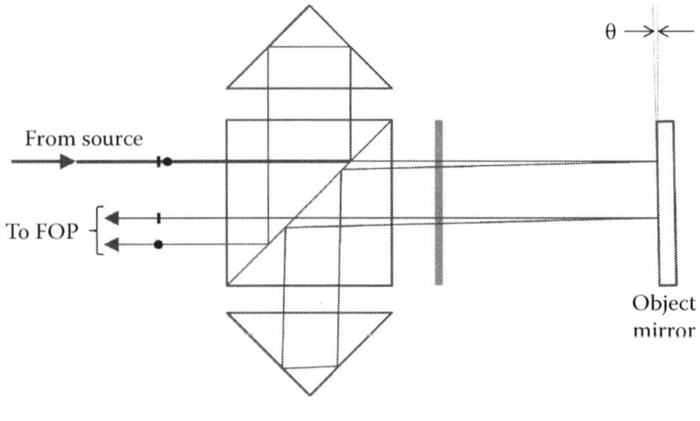

FIGURE 4.16
Function of a PMI when the object mirror tilts.

For dual-axis motion as required by an x–y stage, the preferred object is a plane mirror rather than a retroreflector. Figure 4.15 shows how this may be accomplished using a QWP. The QWP converts linear polarization to circular and then back to an orthogonal linear polarization after reflection from the object mirror, which is now free to move orthogonal to the line of sight without disturbing the beam paths. Figure 4.16 shows how the double pass to the object mirror with a retroreflection compensates for a tilt θ of the mirror. An additional benefit of the double pass is a finer measurement resolution, with one full 2π phase cycle for every quarter wavelength of object motion.

Another popular design for a plane-mirror interferometer (PMI) is the high-stability type or high-stability plane-mirror interferometer (HSPMI), shown in Figure 4.17. This geometry self-compensates for any changes in the interferometer optics—for example, thermal expansion—by configuring the reference and measurement paths symmetrically,

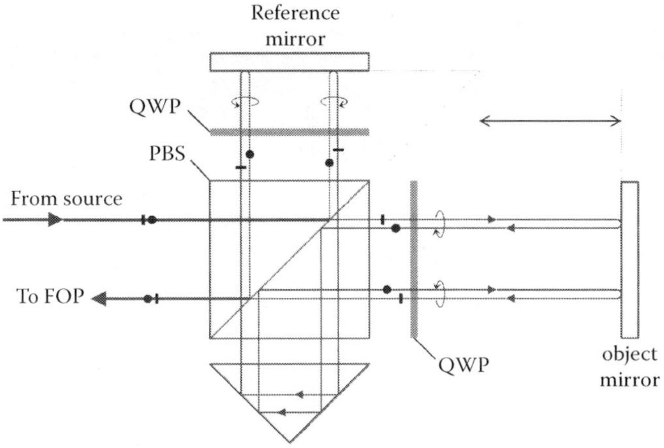

FIGURE 4.17
High-stability plane-mirror interferometer (HSPMI).

with the same amount of glass in each beam.[94] This approach brings thermal sensitivity to below 20 nm/°C in well-designed HSPMI packages.[95]

Although shown as 2D in the figures to clarify their function, actual interferometers are 3D and have more complicated beam paths, as shown in Figure 4.18. Practical designs use optics of BK7 or crystalline quartz, vacuum-grade low-volatility adhesives, and stainless steel housings. Although there are a large number of reflections and transmissions through various optical surfaces, the net light efficiency of a commercial HSPMI is approximately 60%.[95]

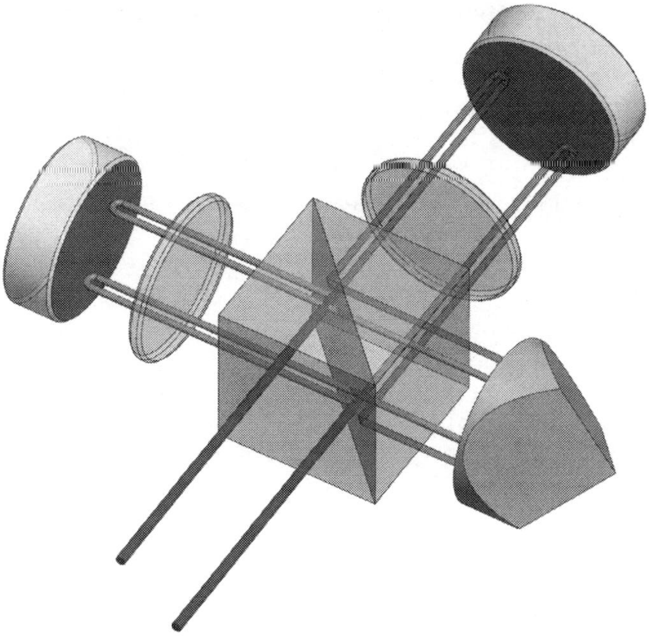

FIGURE 4.18
HSPMI and corresponding beam paths presented in three dimensions.

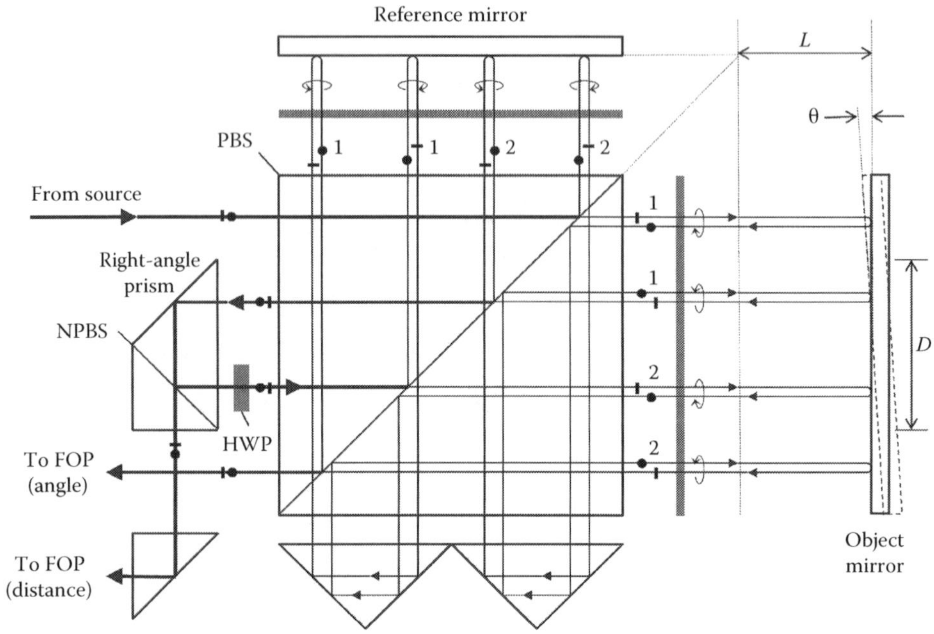

FIGURE 4.19
Dual DMI for measuring both distance and angle.

Stage metrology in particular often requires measurements of multiple degrees of freedom (DOF), including stage pitch and yaw. For this reason, a variety of angle measurement interferometers have been developed, often using the HSPMI as a building block. Figure 4.19 illustrates one way to achieve this by integrating two HSPMI subsystems into a single, partially monolithic system with high thermal and mechanical stability.[96] Here the distance measurement involves only the upper pair of beam paths to the object mirror, whereas the angle measurement involves both upper and lower paths to the object, with the roles of measurement and reference beams (labeled 1 and 2, respectively) reversed between the two pairs. Figure 4.20 shows one implementation using cemented bulk optical components and intended for a 3 mm beam diameter. Multiaxis interferometer systems can be quite complex, including half a dozen distinct motion measurements, often referenced to different parts of a mechanical system or metrology frame, with individualized beam steering to compensate for any imperfections in the optical components.

In addition to the basic metrology function of interferometer optics, it is common to use interferometers to perform the auxiliary function of tracking variations in the effective wavelength λ/n in the ambient medium that surrounds the DMI system. The differential plane-mirror interferometer (DPMI) as shown in Figures 4.21 and 4.22 has the ideal geometry for a wavelength tracker, reporting variations in the measured path $(n - 1)L$ and hence the index and effective wavelength.[17,97]

Figure 4.23 illustrates one way to measure motions orthogonal to the nominal target motion, for example, to monitor the straightness of travel of a stage.[98] In this case, the interferometer is measuring the lateral displacement of the birefringent prism, which is mounted to the moving stage. The straightness deviation Δx is detected as a distance change given by

$$\Delta L = 2\sin(\gamma)\Delta x \tag{4.16}$$

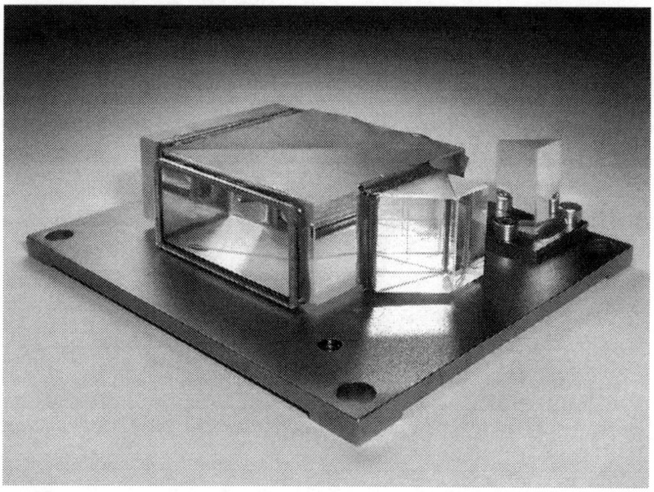

FIGURE 4.20
Compact distance and angle interferometer as in Figure 4.19. The baseplate measures 54 mm. (Photo courtesy of Zygo Corporation, Middlefield, CT.)

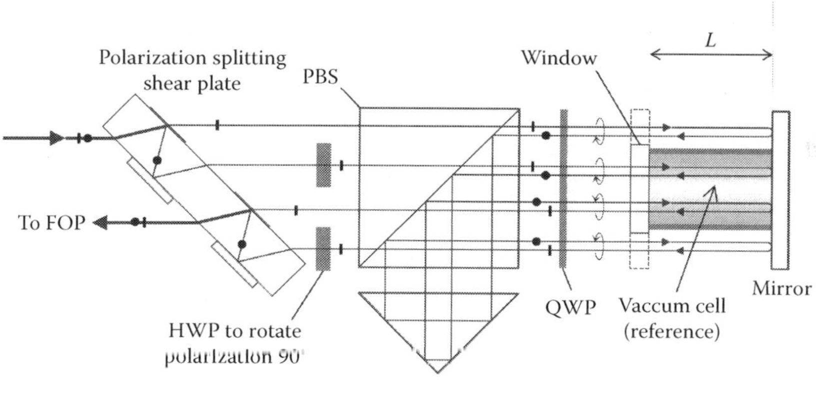

FIGURE 4.21
DPMI configured as a wavelength tracker, using a vacuum cell to detect changes in the ambient index of refraction.

where γ is the angle between the two parts of the dihedral mirror. To first order, this design is insensitive to the tip and tilt of the birefringent prism. In an alternative configuration using the same components, the stage carries the dihedral mirror and the birefringent prism is fixed. This measurement also follows Equation 4.16 but requires the measurement and compensation of the angular motion of the dihedral mirror in the plane of Figure 4.23. With more advanced geometries and considerable care, straightness of travel can be monitored and corrected to within a few nanometer over several hundred millimeter of travel.[99]

In addition to various configurations to measure different types of motion, interferometer designs vary according to strategies for reducing error sources. These include specialized components such as polarization-preserving retroreflectors,[100] split wave plates to reduce ghost reflections,[101] single-pass plane-mirror configurations using dynamically

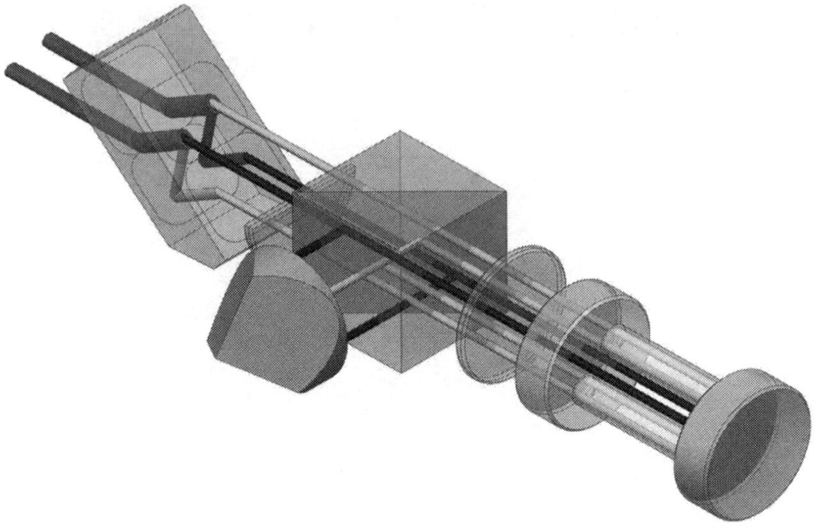

FIGURE 4.22
Three-dimensional representation of a DPMI employed for wavelength tracking.

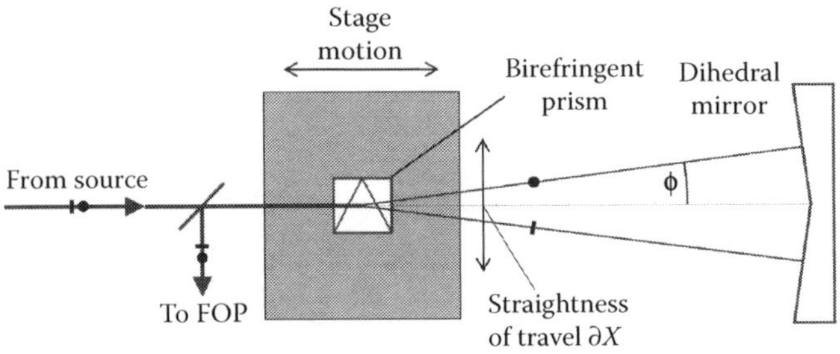

FIGURE 4.23
Straightness interferometer.

steerable components,[102] and roof mirror assemblies in place of plane mirrors for stage metrology.[103] Figure 4.24, for example, shows a design that reduces errors related to beam mixing using birefringent prisms to encode and decode the measurement and reference beams according to a small difference in propagation angle.[104,105] Figure 4.25 illustrates a method of eliminating most sources of cyclic error related to polarization leakage by maintaining the measurement and reference beams fully separated spatially.[106–110]

4.6 Error Sources and Uncertainty in DMI

Displacement interferometry is capable of producing high-accuracy measurements and is often the ultimate reference standard in many applications. Nonetheless, measurements of displacement using interferometry are like any other measurements in that they have

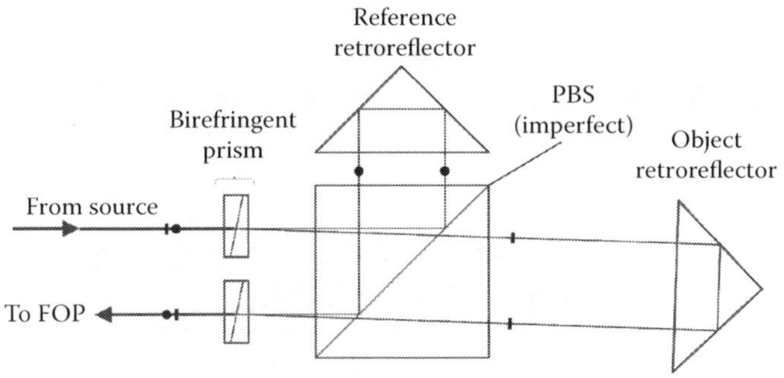

FIGURE 4.24
Interferometer using birefringent elements to overcome imperfect polarization separation of the reference and measurement beams at the beam splitter. (From de Groot, P.J. et al., *US* 6, 778, 280, 2004.)

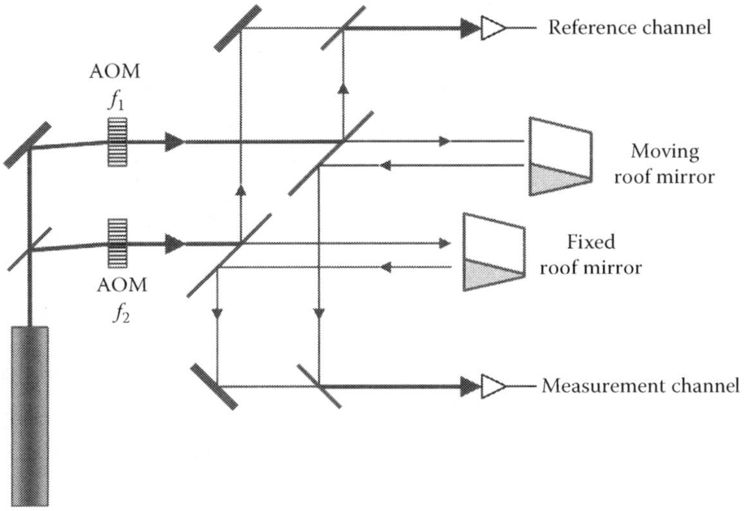

FIGURE 4.25
Unpolarized, separated beam interferometer using AOMs. (From Tanaka, M. et al., *IEEE Trans. Instrum. Meas.*, 38, 552, 1989.)

sources of error and the required measurement uncertainty determines the suitability of DMI for a given application.[111] Any uncertainty estimation requires the identification of the various contributing sources as a first step followed by appropriate combination of these.

This section discusses the most significant contributing sources of uncertainty in the context of a typical measurement—a simple linear interferometer configuration with a corner-cube target for the measurement of the displacement of a linear stage. Figure 4.26 indicates two distinct displacements: the measurand D, which is a displacement of the POI on the stage, and the measured displacement D_m. In general, because of the various sources of error, the measurand D and the measured displacement D_m are not the same.

The relationship between the various quantities defined earlier is

$$D = D_m + \sum_i \delta D_i \tag{4.17}$$

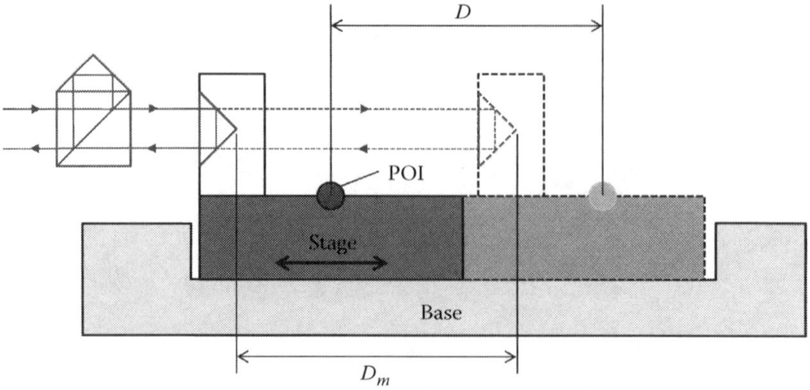

FIGURE 4.26
An arrangement to measure the displacement of a linear stage.

where δD_i represents displacements not directly attributable to the measurand and will henceforth be referred to as spurious displacements, each of which contributes to the uncertainty in the measurand $u(D)$ through associated uncertainties $u_i(D)$. In the absence of any spurious displacements, D would equal the measured displacement D_m, and $u(D)$ would equal the uncertainty in the measured displacement $u(D_m)$. Expressing D_m in Equation 4.17 in terms of the phase change $\Delta\phi$ that corresponds to the measured displacement D_m, we have[112]

$$D = \left(\frac{\lambda'}{N\cos(\beta)}\right)\left(\frac{\Delta\phi}{2\pi}\right) + \sum_i \delta D_i \tag{4.18}$$

where

N is an integer factor that depends on the number of passes (round-trips) of the measurement beam to the target and back

λ' (the effective wavelength) is the ratio of the vacuum wavelength λ and the index of refraction n

$\cos(\beta)$ results from the angular misalignment β between the direction of motion of the stage and the beam

For the particular configuration of Figure 4.26, the number of round-trip passes $N=2$. Equation 4.18 serves as a mathematical model of the measurement and forms the basis of the discussion of uncertainty sources in the following sections. An overall uncertainty for the measurement combines the various contributions according to the procedures of the *Guide to the Expression of Uncertainty in Measurement* (GUM).[113]

4.6.1 Uncertainty Sources

The majority of the contributors fall into two broad categories:

Category 1—uncertainty in the value of a parameter. Uncertainty contributors in this category are relatively few in number and correspond to most of the quantities in the first term of Equation 4.18. These contributors generate uncertainties that are proportional to the measurand, that is, their magnitude scales with the displacement D.

TABLE 4.2

Typical Sources of Uncertainty in Displacement Interferometry

Category 1	Category 2
Uncertainty in vacuum wavelength	Uncertainty in the measurement of phase change
Uncertainty in refractive index arising from either absence of correction or uncertainty in inputs to correction	Effect of deadpath
Cosine error	Cyclic errors
	Optics thermal drift
	Abbe errors
	Target errors
	Beam shear
	Data age uncertainty
	Turbulence
	Target mounting effects

Category 2—uncertainty from sources that affect the measurand but in fact are not directly attributable to it, that is, from spurious displacements. These contributors are represented by $\Delta\phi$ and the second term of Equation 4.18, and unlike the contributors in Category 1, the magnitude of these contributions is not dependent on D.

Table 4.2 is a nonexhaustive list of common contributors in each of the earlier categories. Many of the contributors common to other measurement techniques (e.g., cosine error and Abbe error) are included here clarify their interpretation in the context of interferometric displacement measurement. Some uncertainty sources that are common to any measurement technique, for example, thermal changes in the metrology frame, are not discussed here but would contribute as they would in any measurement.

4.6.2 Vacuum Wavelength

The vacuum wavelength λ establishes the linkage between the unit of length and the change in phase, and any associated uncertainty $u(\lambda)$ contributes directly in the calculated displacement. The uncertainty in the measurand $u_\lambda(D)$ depends on the relative uncertainty in the wavelength $u(\lambda)/\lambda$ and is proportional to the measurand as given by[114,115]

$$u_\lambda(D) = \frac{u(\lambda)}{\lambda} D \tag{4.19}$$

Uncertainty in the vacuum wavelength λ has multiple contributors: the uncertainty in the determination of the absolute value of the wavelength and a contribution arising from short- and long-term drift.[116,117] Displacement interferometers typically use stabilized He-Ne lasers emitting at 633 nm as the light source, and the uncertainty in the wavelength and the associated stability depends on the method of stabilization as summarized in Table 4.1. For applications demanding a lower uncertainty in the vacuum wavelength, a reference laser such as an iodine-stabilized laser may be included within the system and used to monitor the wavelength of the metrology laser used for interferometry.[89,90]

In the case of two-frequency laser sources intended for use in heterodyne systems, a rather subtle error can result from the slightly different nominal vacuum wavelength for the two beams of slightly different frequencies (and different polarizations). The difference depends on the frequency difference (or split frequency). In such systems, it is important to keep track of the wavelength of the beam in the measurement arm and to scale the measured phase by the appropriate wavelength. Failure to do so results in a systematic error in the measured displacement, which can be significant depending on the application.

4.6.3 Refractive Index

Both the refractive index n and the vacuum wavelength λ relate directly to the traceability of displacement measurement.[118] Since $\lambda' = \lambda/n$ scales the measured phase change into units of length, any uncertainties associated with the index $u(n)$ affect λ' and the measurand directly, that is, the uncertainty in the measurand due to index uncertainty $u_n(D)$ is a direct function of the relative uncertainty in the index and is also proportional to the displacement:[115]

$$u_n(D) = \frac{u(n)}{n} D \tag{4.20}$$

DMI measurements are typically in air, and the refractive index of air n_{air} is dependent on several environmental variables such as the pressure P, the temperature T, the moisture content or humidity H, and the exact composition. This can lead to significant variations in n_{air} depending on the prevailing environmental conditions and/or deviations from the assumed composition at the time of the measurement.

A nominal value for n_{air} follows from the so-called weather station approach[119] or from a direct measurement using refractometers of various kinds.[97,120] This first method is more pragmatic and relies on the measurement of P, T, and H, using instrumentation to provide the required inputs for index calculation. Fortunately, the refractive index of air is a well-characterized function of P, T, and H in the form of the empirical equations developed by several investigators.[121–125] An expression commonly used today for dimensional metrology applications under laboratory conditions, that is, at temperatures near 20°C, is a modified form[126,127] of the expression introduced by Edlén.[122] The modified Edlén equation, sometimes referred to as the NPL revision, was developed to account for discrepancies between the experimentally measured and calculated values and to also incorporate changes to the International Temperature Scale (ITS) instituted after the publication of Edlén's original equation. The modified equation agrees with experimental values[128] to within $3 \times 10^{-8} (3\sigma)$ and to a few parts in 10^8 in more recent investigations.[129,130] The complete (and somewhat cumbersome) expression, which is actually a set of three linked equations, appears in the paper by Birch and Downs.[126,127] More recently developed equations by Ciddor apply over a larger wavelength range and more extreme environmental conditions to the stated uncertainty.[131–133]

While the full expression calculates the nominal value of the index corresponding to a set of environmental conditions, the uncertainty associated with this calculated value follows from the partial derivatives with respect to each of the environmental parameters for a vacuum wavelength of ~633 nm. These derivatives are then evaluated at standard conditions ($P = 101325$ Pa, $T = 293.15$ K, and $H = 50\%$) to give sensitivities K_T, K_P, and K_H of the index of air to changes in temperature, pressure, and humidity,[134] numerical values for which are provided in Equation 4.21:

$$\begin{aligned} K_T &= -0.927 \times 10^{-6} / \text{K} \quad \text{or} \quad \sim -1 \times 10^{-6} / \text{K} \\ K_P &= +2.682 \times 10^{-9} / \text{Pa} \quad \text{or} \quad \sim +1 \times 10^{-6} / 3\,\text{mm Hg} \\ K_H &= -0.01 \times 10^{-6} / \%\text{RH} \quad \text{or} \quad \sim -1 \times 10^{-6} / 100\%\text{RH} \end{aligned} \tag{4.21}$$

The equation also gives (in round numbers) the required changes in key environmental parameters to produce a 1×10^{-6} change in the index in more commonly used units. The refractive index is most sensitive to changes in temperature and pressure with relatively large changes in humidity being required to cause a comparable change in the index. Note that the coefficients are signed, and while this usually is not of consequence in the determination of the uncertainty, it is critical in establishing the sign of the error. The associated uncertainty $u(n_{air})$ is a function of the uncertainties in the temperature $u(T)$, pressure $u(P)$, and humidity $u(H)$ and is given by[115,134]

$$u(n_{air}) = \sqrt{K_T^2 u^2(T) + K_P^2 u^2(P) + K_H^2 u^2(H)} \tag{4.22}$$

There is also an "intrinsic" uncertainty associated with this empirical expression that derives from the uncertainty in the data used to derive them.[121,135] In other words, even if the inputs to the equation are known exactly, there is an uncertainty in the calculated index that results from the uncertainty in the coefficients of the modified Edlén equation. The developers of the equation estimate this uncertainty contribution to be $3 \times 10^{-8}(3\sigma)$, or a standard uncertainty of 1×10^{-8} or 10 parts per billion (ppb), corresponding to 10 nm in a displacement over 1 m.[126,127] This intrinsic uncertainty should be included in determinations of $u(n_{air})$.[136] In almost all practical applications, contributions from the uncertainties associated with the determination of the input parameters dwarf this contribution.

The weather station method depends on fixed assumptions regarding the composition, potentially leading to an important uncertainty in the calculated index. The modified Edlén assumes a CO_2 concentration of 450×10^{-6}.[126] The CO_2 concentration can vary—a prime example is a higher than assumed concentrations of CO_2 caused by human respiration.[137] Fortunately, changes in the index of $\sim 2 \times 10^{-8}$ require a change in CO_2 concentration of 150×10^{-6} and only become significant in the most demanding applications.[138] A more significant contribution is due to changes in composition due to the presence of hydrocarbons, for example, acetone, which causes an index change of 10^{-7} for a contamination level of 130×10^{-6}.[59] Solvent vapors are present in many metrology environments, for example, the metrology of optics, where solvents such as acetone, alcohol, and various other volatile solvents are often used for cleaning, resulting in marked deviations from the assumed composition. The composition of the air is not typically monitored in such environments, although in general the presence of a detectable odor is a good indication of the presence of a volatile hydrocarbon at levels that are significant for the highest accuracy measurments.[59] One way this contribution may be incorporated into Equation 4.22 is via an additional term that captures the uncertainty in the concentration of the particular hydrocarbon and the appropriate sensitivity.

For high-accuracy determinations of the index, great care is required in making measurements of temperature, pressure, and humidity, as described in the classic paper by Estler.[139] The measurements of the environmental parameters now also become part of the traceability chain.[118] Recent refractometer comparisons have shown that it is possible with careful measurements to reduce the contribution from uncertainties in the input parameters to the point where the uncertainties in the parameters of the modified Edlén equation are comparable or even dominate.[120,130] This however requires extremely careful measurements of the input parameters, something that is typically not easily achieved in nonlaboratory measurements. Factors such as the location of the sensors, gradients in temperature and pressure, thermal inertia, and self-heating of sensors also complicate the measurement of environmental parameters.[140]

Higher accuracies may warrant an alternative to the weather station approach based on in situ experimental determination of the index using a refractometer.[141] Refractometers are essentially fixed physical path length interferometers. The goal of the measurement is the change in refractive index, any changes in the length of the paths and uncertainty in the absolute value of their lengths being sources of uncertainty in the index measurement.[142,143] Absolute determinations of index are made by measuring the phase change that accompanies a change in pressure in the measurement arm from vacuum to the ambient, while the reference arm is maintained in an evacuated state. Phase changes after the pressure in the reference arm reaches the ambient represent index changes that may also be tracked by the same instrument. The measurand is the refractivity $n - 1$, and relatively modest relative uncertainty in the determination of the refractivity of a few parts in 10^5 results in relative uncertainties $\sim 10^{-8}$ in the refractive index.[143] Many different interferometer configurations can be used to measure the phase change corresponding to the change in index.[97,144] While refractometer- and tracker-based measurements are not immune to gradients between the refractometer and the measurement beam path, unlike the method described earlier, they are sensitive to composition changes. This method also eliminates the careful calibration and maintenance of the various environmental sensors required to make a parametric determination of the index with the requisite uncertainty. One criterion for the selection of the method of determination of the refractive index is the desired measurement uncertainty, with a parametric compensation technique for measurement uncertainties $\geq 10^{-7}$ and a refractometer for measurement uncertainties below this.[43]

A final subtlety concerning the measurement of refractive index by any of the methods described earlier is gravitationally induced variation of pressure with altitude. This gradient is ~ 0.1 mmHg/m resulting in a decrease in the index of $\sim 3.5 \times 10^{-8}$ for each meter of altitude gain,[139] suggesting that the difference in elevations of the measuring beam path and index measurement instrumentation can be a significant contributor to index uncertainty. This also means that the refractive index varies along the beam path for a vertical axis measurement and corrections are necessary for this variation. A similar argument also applies to vertical temperature gradients caused by stratification of air. Horizontal gradients of several tenths of Kelvin are also common in a horizontal plane in laboratories with closely controlled mean temperatures, thus requiring monitoring at multiple points along a horizontal beam path and the use of an appropriate value for the temperature derived from these multiple measurements.[139]

The uncertainty contribution from the refractive index may also be reduced by performing the measurement in a medium other than air. Helium, for example, has a sensitivity of index to changes in the environmental parameters that are smaller by approximately an order of magnitude.[145] Measurements in vacuum produce the highest level of accuracy and have the added advantage of virtually eliminating the uncertainty contribution from turbulence.[89,90,146]

A hybrid approach to measurement of both the nominal refractive index and its change over time is afforded by the use of a wavelength tracker or compensator, examples of which are shown in Figures 4.21 and 4.22.[147] The optical configuration of a tracker is virtually identical to that of a refractometer, and the significant difference is in the construction details and the way it is used.[148] In contrast to a refractometer, a wavelength tracker does not measure the absolute index, but tracks changes in the index from the initial value specified at the time that the measurement begins. The initial value follows from a parametric evaluation based on the measurement of the environmental parameters, and as such, the uncertainty in the index derived from the compensator measurement can be no better than the uncertainty associated with the initial index. The compensator offers the advantage of higher bandwidth in that it responds virtually instantaneously to changes in index when compared to the time delay inherent in the measurement of the environmental parameters.

4.6.4 Air Turbulence

Air turbulence arises from the high airflow rates that are required in high-precision machinery in order to control the temperature, ensure cleanliness, etc., and their interactions with the often cramped confines of such equipment. Turbulence causes time-dependent variations in the index of refraction in the beam paths through localized pressure and temperature fluctuations, and the resulting optical path length changes manifest themselves as relatively high-frequency random fluctuations in the measured displacement.[149] In high-accuracy applications such as microlithography, turbulence can often be the leading source of uncertainty.[150] Turbulence also affects the measurement through direct interactions of the airflow with components within the metrology loop such as mirrors and mounts, resulting in vibrations that contribute to the observed optical path length fluctuations.

While much data are available on the effects of atmospheric turbulence on the propagation of radiation, data on turbulence as they relate to typical interferometric applications (particularly lithographic applications) are scarce, and much of the discussion here draws from Bobroff's now classic paper on the subject.[149] The magnitude of the contribution from turbulence scales with the length of path of the beams and a first level mitigation strategy is to minimize the lengths of air path, a measure that also minimizes the contributions due to deadpath (see Section 4.6.8). The RMS magnitude of the fluctuations can vary from subnano-meter levels for enclosed beam paths to approximately 2.5 nm for flow rates of 100 linear feet per minute (LFM) perpendicular to the beam direction over a 150 mm length of exposed beam path.[149]

Another consequence of turbulence is its effect on correlation in the OPL fluctuations observed in two parallel beams. This has important implications for interferometer configurations of the column-reference type, wherein the measurement and reference beams monitor targets attached to two metrologically significant parts of the machine or structure between which a differential measurement is desired, for example, the projection lens and the wafer in a microlithographic tool. In this configuration, the two beam paths are typically parallel and almost the same length (except for the stage travel in the measurement arm). In the absence of turbulence, index changes are correlated in the common portions of the two arms, thereby rejecting deadpath contributions to a high degree. In the presence of turbulence, this correlation decreases and cancelation is not as complete at high frequencies. A similar situation prevails in situations where two spatially separated interferometers working against a common target mirror measure angular motion of the mirror. In this configuration, the measurement arms of the two interferometers are parallel, and the lack of correlation is a source of uncertainty in the angle measurement.

Time averaging can minimize the effects of turbulence. The averaging period depends on the time scale of the phase fluctuations, and in general this gain comes at the expense of data rate, making it impractical for applications with high servo loop frequencies. Turbulence can also lead to mechanical noise in high-bandwidth systems as the controller attempts to compensate for the high-frequency changes in the measured displacement.

Another approach to minimizing the effects of turbulence is the careful design of flow paths and in general reducing the exposed paths to a minimum. Shielding the paths is effective, although this is not always practical in the measurement beam path. Turbulence can be eliminated by operating the interferometers in evacuated pathways[89,151] or by enclosing the entire device in a vacuum as is often done in extreme ultraviolet lithography (EUVL) machines[152] and electron-beam lithography machines.[153] Another, more speculative approach is dispersion interferometry, described in Section 4.8.3.

4.6.5 Cosine Error

Cosine error results from an angular misalignment between the measurement direction and the average line of motion of the target. Although this error is common to all kinds of displacement measuring devices, the manifestation of this error in the context of displacement interferometry is somewhat more subtle and dependent on the type of target.[59,154] In the treatment that follows, although misalignments are depicted as being confined to a plane represent the various directions, they are in fact in 3D space, necessitating the vector dot product notation, wherein the angles are between the vectors in 3D space. Two distinct cases are discussed: cosine errors for a corner-cube retroreflector target and for a plane-mirror target.

Figure 4.27 depicts the relevant beam path geometry for a corner-cube target. The vector $\vec{D}$ defines the direction of motion, and the unit vector $\hat{i}$ defines the direction of the incident light. The measured displacement D_m is then

$$D_m = \vec{D} \cdot \hat{i} = D\cos(\beta) \quad (4.23)$$

From the earlier equation it is clear that the measured displacement is always smaller than the actual displacement D. Neglecting the uncertainty contribution from D_m, the uncertainty component attributable to just the misalignment $u_\beta(D)$ is[134]

$$u_\beta(D) \approx Du^2(\beta) \quad (4.24)$$

The contribution is proportional to the measurand and is unique in that it also creates a bias proportional to the variance of the misalignment angle.

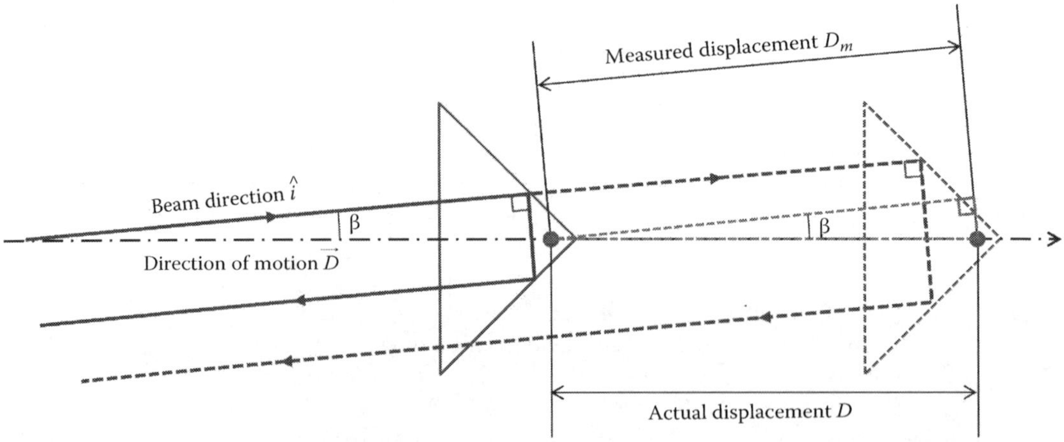

FIGURE 4.27
Cosine error with a cube-corner target.

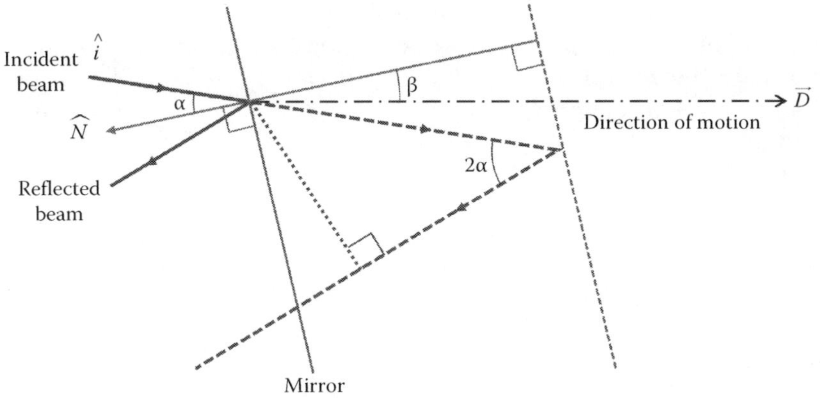

FIGURE 4.28
Misalignment parameters for a plane-mirror target. (Adapted from Bobroff, N., *Precis. Eng.*, 15, 33, 1993.)

The situation for plane-mirror targets is somewhat more complex in that it involves the mutual alignment of three vectors, as shown in Figure 4.28.[154] An additional unit vector $\widehat{N}$ is required to describe the orientation of the mirror normal along with an additional angle α that represents the angle between the unit vector along the direction of the incident beam $\hat{i}$ and $\widehat{N}$. The angle β now represents angle between $\widehat{N}$ and the direction of motion $\vec{D}$. The resulting expression comprises two vector dot products, each corresponding to one of the angles.[154] The measured displacement is then

$$D_m = D\left(\widehat{N}\cdot\hat{i}\right)\left(\vec{D}\cdot\widehat{N}\right) = D\cos(\alpha)\cos(\beta) \tag{4.25}$$

An expression for the associated uncertainty contribution analogous to Equation 4.24 follows from the procedures outlined in the ISO GUM.[113]

In general, careful alignment minimizes the contribution from misalignment. Typically, alignment involves monitoring the translation of the laser beam spot on a target in a direction perpendicular to the beam axis as the target moves through the measurement range. The best alignment is achieved when the target translation is large, which is fortuitous in that the best alignment is often desired in applications where the target travel is large.

Cosine errors are also an insidious source of uncertainty in angle interferometers such as the dual corner-cube configuration that relies on the differential displacement between the parallel beams to measure angle.[155] In the presence of misalignment between the beams or the measurement beams of multiple interferometers, the different cosine error contributions produce differential displacement even as the target executes pure translation along the beams, resulting in an apparent change in angle.[156,157]

4.6.6 Phase Measurement

Uncertainties in the measurement of the change in phase $\Delta\phi$ directly affect the measurand. The contributing uncertainties to the measurement of $\Delta\phi$ include the linearity and the resolution (noise floor) of the phase meters. Oldham et al. discuss phase measurement using the time interval analyzers (TIA) that were just becoming available at the time.[158] The technology of phase measurement has changed a great deal in the intervening period, and a sense of the magnitude of these contributions may be gleaned from the datasheets of commercially available phase measurement electronics[159] and manufacturer's published guides to uncertainty.[111] Accuracy figures are quoted for various measurement velocities of the target, with the accuracy number when the target is stationary representing the "static accuracy" or alternately the noise floor of the electronics. The accuracy numbers at different target velocities are a measure of the "dynamic accuracy" and in general tend to be larger than the noise floor of the instrument as they also include contributions from nonlinearities of the phase meter. The noise floor of the instrument is a strong function of the signal levels, and manufacturer's specifications assume a certain signal level, both in terms of the total amount of light at the detector as well as the depth of modulation of the AC signal.[18] Heterodyne interferometers have a distinct advantage over homodyne interferometers in the achievable noise floor for a given optical power. The shift in the operating point of the heterodyne interferometer away from DC and the large $1/f$ noise background makes it possible to achieve an extremely low noise floor in the presence of very low optical powers. This in turn makes it possible to illuminate a dozen or more measurement channels from a single laser source without sacrificing the noise performance.

In some applications where the length standard is established independently, for example, by the spacing between fiducials on a physical artifact, the uncertainty in the value of scaling factor resulting from uncertainties in the index, vacuum wavelength, alignment, etc., is much less important. This scenario is encountered in applications such as lithography where the measurement between two or more fiducials on the wafer sets the length scale and the interferometer system then simply serves to access locations relative to the coordinate system so established. The stability of the various scaling terms becomes far more important to guarantee the stability of this calibration for the duration of the measurement.

4.6.7 Cyclic Errors

Periodic or cyclic errors are characteristic of all systems that rely on phase estimation. Figure 4.29 shows small signal fluctuations and the typical periodic or cyclic error related to known component and alignment imperfections. Although most generally a cyclic error may have any cause that correlates to interference phase, in DMI, the most common origins relate to insufficient isolation of the measurement and reference beams as they travel through the optical system or the admixture of unwanted reflected or scattered beams.

The estimation of the uncertainty contribution of cyclic errors is somewhat trickier because they are sinusoidal. Monte Carlo simulations graphically demonstrate the

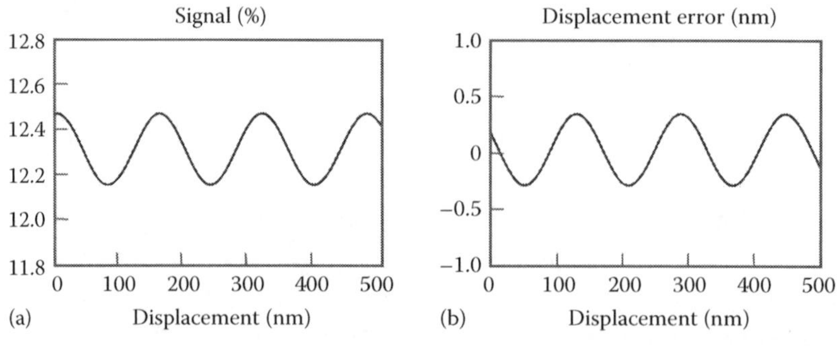

FIGURE 4.29
Simulations of signal fluctuation (a) and cyclic displacement measurement error (b) in a plane-mirror DMI system.

U-shaped distributions that result from these errors.[160] The standard uncertainty for these contributors may be calculated by assuming an underlying sinusoidal probability distribution function and scaling the amplitudes appropriately.[115]

Evaluating cyclic error sources and magnitudes begins with a model of the optical system, starting with a functional block diagram such as is shown in Figure 4.30. A significant amount of attention has been devoted to quantifying and mitigating errors arising from polarization mixing, ghost reflections, and other sources of spurious signals that result in periodic nonlinearities.[161,162] The importance of polarization separation implies that Jones calculus notation is a suitable approach to representing each of the blocks or functional elements mathematically for subsequent analysis.[163,164]

In one such representation of the DMI system, the heterodyne optical signal is given by[165]

$$I = \left|\mathbf{Mix} \cdot \mathbf{Int} \cdot \mathbf{Src} \cdot \mathbf{Frq} \cdot \mathrm{Las}\right|^2 \tag{4.26}$$

where the three-letter variables are mnemonics correspond to the functional blocks in Figure 4.30, in reverse order. A vector Las represents the laser source, emitting equally in the s and p polarization states, represented by the two elements of the vector:

$$\mathrm{Las} = \frac{1}{\sqrt{2}}\begin{pmatrix} 1 \\ 1 \end{pmatrix} \tag{4.27}$$

The next functional matrix **Frq** corresponds to the device that provides a frequency shift between the two polarizations:

$$\mathbf{Frq}(t) = \begin{pmatrix} e^{+i\pi\Delta f t} & 0 \\ 0 & e^{-i\pi\Delta f t} \end{pmatrix} \tag{4.28}$$

The product

$$\mathbf{Frq} \cdot \mathrm{Las} = \frac{1}{\sqrt{2}}\begin{pmatrix} e^{+i\pi\Delta f t} \\ e^{-i\pi\Delta f t} \end{pmatrix} \tag{4.29}$$

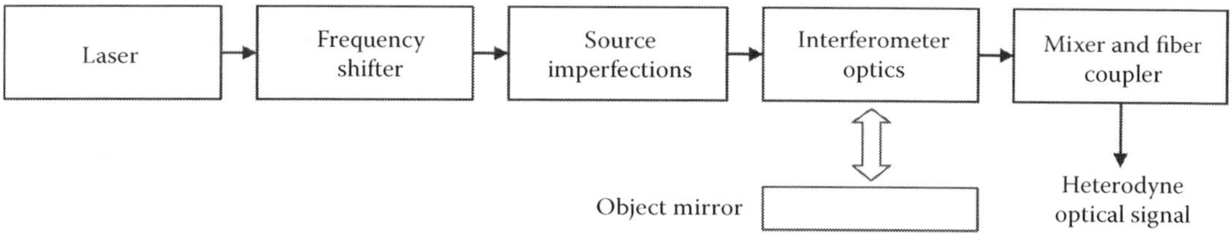

FIGURE 4.30
Functional block diagram of a heterodyne DMI measurement axis.

represents a heterodyne light source with a two-frequency emission encoded by polarization, with a frequency difference Δf.

Continuing with Equation 4.26, the Jones matrix **Src** for source imperfections comprises potential contributions such as alignment errors, polarization ellipticity, and noise. An example contributor is orthogonality of the polarization states carrying the two source frequencies. The corresponding Jones matrix is

$$\mathbf{ort}(\delta\chi) = \begin{pmatrix} \cos(\delta\chi/2) & -\sin(\delta\chi/2) \\ -\sin(\delta\chi/2) & \cos(\delta\chi/2) \end{pmatrix}, \tag{4.30}$$

where typically $\delta\chi$ is on the order of 10 mrad. The **Src** matrix is the product of **ort** and all of the other Jones matrix representations of source errors.[165]

The Jones matrix **Int** corresponds to the interferometer optics and depends of course on the specific optical geometry. Taking as an example a Michelson interferometer as in Figure 4.2 and assuming perfect alignment,

$$\mathbf{Int}(L) = \mathbf{Ref} + \mathbf{Mes}(L), \tag{4.31}$$

where

$$\mathbf{Mes}(L) = \mathbf{pbs}_\mathrm{T} \cdot \mathbf{U}(L) \cdot \mathbf{rtr} \cdot \mathbf{U}(L) \cdot \mathbf{pbs}_\mathrm{T} \tag{4.32}$$

and

$$\mathbf{Ref} = \mathbf{pbs}_\mathrm{R} \cdot \mathbf{U}(L) \cdot \mathbf{rtr} \cdot \mathbf{U}(L) \cdot \mathbf{pbs}_\mathrm{R} \tag{4.33}$$

Table 4.3 lists the component matrices. The final matrix to complete the model is the polarizer to combine the reference and measurement beams:

$$\mathbf{Mix}(\alpha_{mix}) = \mathbf{anl}(\alpha_{mix}) \tag{4.34}$$

There are several additional error sources not discussed here, including surface reflections and errors that vary with stage angle. All of these errors are incorporated into more extensive Jones matrix models. Computer simulation is straightforward using modeling of this type, leading to Figure 4.29.

Of several possible techniques for quantifying cyclic errors experimentally, the most effective is by Fourier or spectral analysis of the interference signal while the object mirror is in motion, resulting in a Doppler shift that separates the errors by frequency.[166–168] Figure 4.31 illustrates such an analysis for a DPMI (as in Figure 4.21), showing the desired heterodyne peak labeled 1 in the figure and a number of secondary peaks, all corresponding to unwanted periodic errors.[169] The displacement-equivalent amplitudes of peaks 2, 3, 4, 5, and 6 in this example are 4, 2, 1.2, 0.6, and 0.1 nm, respectively. Each of these peaks is associated with specific cyclic errors. For example, peaks 4 and 5 are leakage at the PBS (interferometer errors) and polarization mixing in the light source (source imperfections). In many cases, an adjustment of the interferometer optics, for example, by tilting components or adjusting wave plate orientation, can reduce several of these errors interactively while observing the spectral analysis in real time.[170]

TABLE 4.3

Interferometer Component Jones Matrices

Matrix Name	Matrix Form	Typical Values
Beam propagation	$\mathrm{U}(x) = \mathbf{I}\cdot\exp\left(\frac{2\pi i x}{\lambda}\right)$	$\lambda = 632.8$ nm
Antireflection coatings	$\mathbf{arc} = \mathbf{I}\cdot\sqrt{T_A}$	$T_A = 99.5\%$
Polarizer	$\mathbf{pol}(a,b) = \begin{pmatrix}\sqrt{a} & 0\\ 0 & \sqrt{b}\end{pmatrix}$	—
Cube beam splitter reflection	$\mathbf{pbs_R} = \mathbf{arc}\cdot\mathbf{pol}(Rc_s, Rc_p)\cdot\mathbf{arc}$	$Rc_s = 99.9\%$ $Rc_p = 0.1\%$
Cube beam splitter transmission	$\mathbf{pbs_T} = \mathbf{arc}\cdot\mathbf{pol}(Tc_s, Tc_p)\cdot\mathbf{arc}$	$Tc_s = 0.1\%$ $Tc_p = 99.9\%$
Beam propagation	$\mathrm{U}(x) = \mathbf{I}\cdot\exp\left(\frac{2\pi i x}{\lambda}\right)$	$\lambda = 632.8$ nm
Polarization rotation	$\mathbf{rot}(\vartheta) = \begin{pmatrix}\cos(\vartheta) & \sin(\vartheta)\\ -\sin(\vartheta) & \cos(\vartheta)\end{pmatrix}$	—
Corner-cube retroreflector	$\mathbf{rtr} = \mathbf{arc}\cdot\sqrt{R_{rtr}}\ \mathbf{rot}(\zeta)\cdot\mathbf{arc}$	$R_{rtr} = 80\%$ $\zeta = 85$ mrad
Mixing polarizer (dichroic analyzer)	$\mathbf{anl}(\delta\alpha_{mix}) = \mathbf{arc}\cdot\mathbf{rot}(45° + \delta\alpha_{mix})$	$\delta\alpha_{mix} = 10$ mrad
	$\cdot\mathbf{pol}(T_{mix}, 0)\cdot\mathbf{rot}(-45° - \delta\alpha_{mix})\cdot\mathbf{arc}$	$T_{mix} = 80\%$

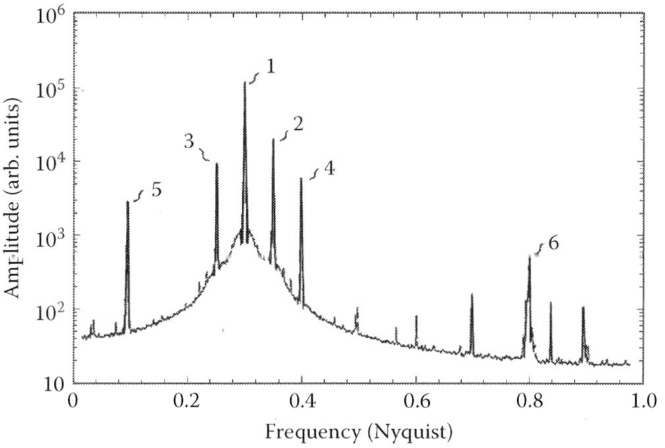

FIGURE 4.31
Spectral analysis of the heterodyne interference signal in a DPMI. (From Hill, H.A., Tilted interferometer, U.S. Patent 6,806,962, 2004.)

4.6.8 Deadpath

Deadpath is the difference in optical path length of the reference and measurement beams when the interferometer electronics are zeroed.[171] This is not necessarily the position of closest approach of the target to the interferometer as emphasized by Figure 4.32.

The imbalance between the two arms results in a sensitivity to both changes in index Δn and source wavelength $\Delta\lambda$, the consequence of which is to shift the point in space at which

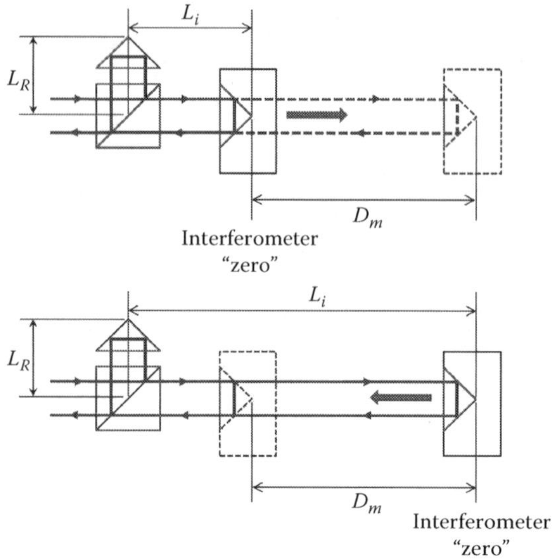

FIGURE 4.32
Initial position determines deadpath L_i. Lower figure has longer deadpath.

the zero location of the interferometer was established (hence the alternate use of the term "zero-shift" error). An equal index change results in an unequal change in the optical path lengths in the two arms, resulting in a spurious displacement.[166] Variation of the source wavelength generates a similar effect. Even though changes in wavelength and index are typically compensated, these corrections usually apply only to the measured displacement, because a displacement interferometer is blind to any initial imbalance between the two arms.

The deadpath error δD_{DP} for a length difference $L_0 = L_i - L_R$ between the initial position of the target and the reference arm is[172]

$$\delta D_{DP,\Delta n} = -L_0 \left(\frac{\Delta n}{n} \right) \tag{4.35}$$

for changes $\Delta n = n_f - n_i$ in refractive index and

$$\delta D_{DP,\Delta\lambda} = L_0 \left(\frac{\Delta\lambda}{\lambda} \right) \tag{4.36}$$

for changes $\Delta\lambda = \lambda_f - \lambda_i$ wavelength. Here the subscripts i and f stand for the value at the beginning and end of the measurement, respectively. This definition of the deadpath assumes that the computation of the displacement uses the final values of wavelength and index in Equation 4.18, a different expression being applicable when a value other than the final value is used.[173] The deadpath contribution is proportional to the length of deadpath, although it is independent of the magnitude of the measurand. The corresponding uncertainty contributions from the deadpath, neglecting the term due to the uncertainty in the deadpath length itself, are[172]

$$u_{DP,\Delta n}(D) = L_0 \frac{u(\Delta n)}{n} \tag{4.37}$$

and

$$u_{DP,\Delta\lambda}(D) = L_0 \frac{u(\Delta\lambda)}{\lambda} \tag{4.38}$$

where $u(\Delta n)$ and $u(\Delta\lambda)$ are the standard deviation in the index and source wavelength variation over the period of the measurement.

There are a number of different ways to address the deadpath contribution. A simple software solution is to preset the phase meter with a value corresponding to L_0 such that the interferometer displays this value at the "zero" position. This makes the interferometer system "aware" of the deadpath length and thus subject to the corrections for index and wavelength. The correction afforded by this technique has a residual uncertainty caused by the uncertainty in the determination of the deadpath length that must be included into the overall measurement uncertainty. Another solution relies on reducing L_0 to the minimum allowed by the constraints of the interferometer layout. This typically involves moving the interferometer as close to the zero position of the target as possible, by a combination of physical arrangement of the optics and by zeroing out the interferometer when the target is at its point of closest approach to the interferometer.[171] Another option may be to increase the length of the reference arm to match the distance to the zero position in the measurement arm, at the expense of the stability of the system. This of course assumes that both arms experience the same changes in index. Another zero-shift estimation applicable to certain applications such as scale measurement requires retaking a "zero" reading after a measurement cycle, any discrepancy being a representation of drift including the zero shift over the measurement cycle.[43]

4.6.9 Abbe Error

Abbe in 1890 first enunciated the Abbe principle, sometimes referred to as the comparator principle.[174] Revised by Bryan,[175] the Abbe principle articulates the notion that the line of measurement should coincide with the line along which the displacement measurement is desired in order to reject the contributions that arise from parasitic angular error motions of the object being monitored. Figure 4.33 is a schematic representation of a measurement scenario wherein the measurement axis of the linear interferometer is offset from the mounting surface of the stage where the POI in the definition of the measurand D is located. The nodal point of the retroreflector target establishes the location of the measurement axis, and the beam direction establishes the orientation of this axis,[59] resulting in an Abbe offset Ω between the line of measurement and the POI. The figure depicts the effect of a parasitic pitch or angular motion of the stage in the plane of the figure. For the sake of clarity, only the retroreflector and its mount are shown rotated by an angle θ, rather than showing the entire stage rotated through this angle. In the presence of an angular error motion θ and an offset Ω, the Abbe error is δD_{Abbe}. The magnitude of the error is

$$\begin{aligned}\delta D_{Abbe} &= \Omega \tan\theta \\ &\approx \Omega\theta \,(\text{for small } \theta)\end{aligned} \tag{4.39}$$

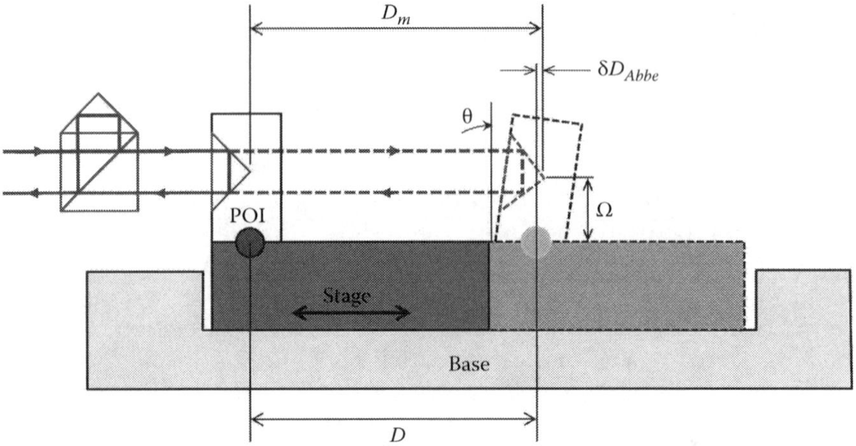

FIGURE 4.33
Abbe offset and the resulting error.

Simplifying the expression given by Schmitz[134] using the small angle approximation gives the associated uncertainty contribution:

$$u_{Abbe}^2(d) = \theta^2 u^2(\Omega) + \Omega^2 u^2(\theta) \tag{4.40}$$

The value of the angle θ and u (θ) typically derives from a priori knowledge of the angular error motions of the stage, for example, specification sheet for the stage or measurements, while Ω depends on the target type and the beam alignment technique. There are typically two Abbe offsets (usually in two mutually orthogonal directions) for any POI and two accompanying angular error motions (about the two mutually orthogonal directions identified earlier). In other words, an expression analogous to Equation 4.40 describes a second contribution from the Abbe offset in the orthogonal direction and angular motions about that direction. In the simplest case where the error motions in the two directions are uncorrelated, the contributions that result from the two offsets combine in quadrature. In a more realistic scenario where the motions are not likely to be totally uncorrelated, an additional term that accounts for the correlation must be included.[113]

The stipulation that the line of measurement passes through the POI requires a way to identify the line of measurement. This in turn means identifying the orientation in space of the line of measurement and a point through which it passes relative to some features in the measurement setup. While this is typically easy to do with many kinds of displacement measuring devices, for example, LVDT and encoder scales, the identification of the line of measurement for an interferometer system is more subtle and depends on the type of interferometer and target. Figure 4.34, for example, shows the line of measurement for a two-pass PMI (as in Figure 4.15) and a single-pass retroreflector interferometer. For the PMI, the line of measurement is parallel to the normal $\hat{N}$.[154] Further, the fact that the phase change reported by the interferometer is the *average* of the phase change in each of the encounters with the mirror also means that the line of measurement is located midway between the two measurement beams. In contrast, the orientation $\hat{\imath}$ of the incident beam sets the orientation of the line of measurement for a retroreflector interferometer parallel to $\hat{\imath}$ in Figure 4.34, while the nodal point or optical center of the corner cube establishes the

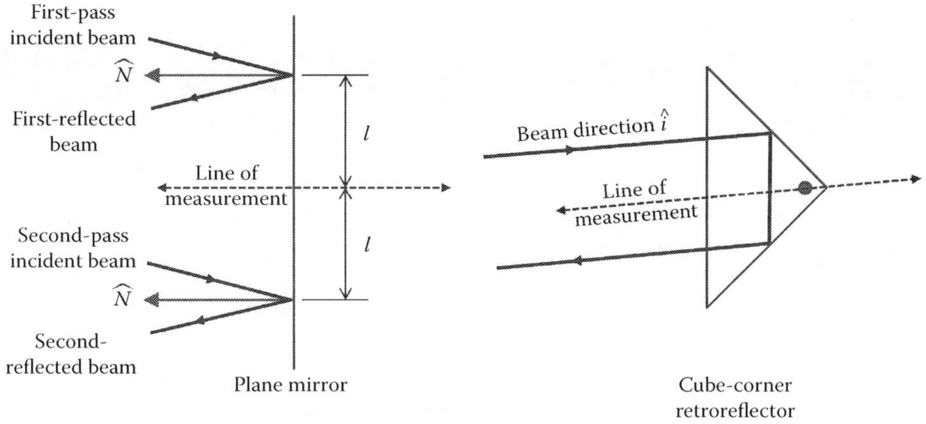

FIGURE 4.34
Line of measurement for plane-mirror and corner-cube targets.

measurement location.[59,176] In the case of the PMI, the location of the line of measurement has to be determined in reference to a virtual datum, that is, the location of the two light beams. This implies a certain level of uncertainty in setting the location of the line relative to the POI. Experience suggests that one can usually minimize the Abbe offsets to approximately 0.5 mm for standard beam diameters without special measures. A smaller value is achievable when using a corner cube because the reference point is precisely manufactured, although it may be inaccessible in some cases due to the retroreflector mounting. The limit to the accuracy of placement is typically the tolerance stack up and/or fixturing used to locate the corner cube.

Interferometers are particularly well suited to measurement configurations designed to minimize the Abbe offset as they establish a virtual axis of measurement that has no mechanical interference with other parts of the system. A family of coordinate measuring machines (CMM) typifies this approach, wherein the interferometer axes are in line with the probe.[177,178] In the relatively rare situations where the POI is not directly accessible or the measurement of multiple DOF is desired, multiple interferometers and appropriately weighted outputs establish a virtual measurement axis.[134,179]

Additional strategies to mitigate the impact of any residual Abbe offset include driving the angular error motion θ to zero or measuring the angular error motions and applying a correction. In a first scenario, eliminating the angular motion minimizes the contribution, usually through active control, so that the angle θ and the first-order Abbe error are nominally zero.[180,181] The uncertainty contribution results from any residual angular motions.

A second approach relies on the measurement of the angular motions.[182] While the angle θ may be measured using a secondary measurement device, interferometers have an advantage in that there are many interferometer configurations that measure displacement and angle simultaneously, making it convenient to implement the desired correction.[183,184] The uncertainty of the correction is now driven by uncertainty in the determination of the angle θ. This type of compensation applies only to rigid body motions, that is, there is the implicit assumption that there are no significant internal bending-type deformations within the monitored body. If these kinds of deformations are anticipated and the desired level of measurement accuracy demands it, a higher level of correction based on the measurement of the deformation is required, or the Abbe offset must be eliminated.

4.6.10 Optics Thermal Drift

Optics thermal drift refers to apparent displacements that stem from optical path length changes caused by thermally induced index and dimension changes within the optics. The optical path length change occurs by virtue of the fact that the measurement and reference beams pass through different amounts and different portions of glass. Two related interferometer configurations exemplify the importance of balanced beam paths through the interferometer: the PMI of Figure 4.15 and the HSPMI configuration shown in Figure 4.17. The first configuration has a large mismatch in the optical path lengths of the measurement and reference beams. In contrast, the HSPMI configuration has matched path lengths resulting in a markedly lower sensitivity to temperature. An interferometer drift coefficient C_I, typically provided by the manufacturer, quantifies this sensitivity to temperature changes. The difference between the two configurations is evident in the drift coefficient C_I values for the unbalanced and balanced configurations, which are ~0.3 μm/°C and 0.02 μm/°C, respectively.[184]

The product of the drift coefficient C_I and the change in temperature gives the magnitude of the apparent displacement related to the interferometer assembly. The corresponding uncertainty (neglecting the uncertainty in the coefficient itself) is simply the product of the drift coefficient and the standard uncertainty in the temperature. Typically, the manufacturer states the specified value of drift coefficient C_I as a positive value, although in practice it can have a positive or negative sign, especially in the case of the nominally balanced configurations where the drift is associated with mechanical tolerances. A similar effect in solid corner-cube targets is quantified by an analogous drift coefficient C_T related to the optical path length, the temperature coefficient of index, and thermal expansion coefficient. The displacement contribution due to the target is as before the product of the drift coefficient C_T and the change in temperature. Again, neglecting the uncertainty in C_T, the corresponding uncertainty contribution is identical to the one derived earlier except C_T replaces C_I.

4.6.11 Beam Shear

Beam shear affects the measurement uncertainty through several mechanisms. In an interferometer with a corner-cube target, beam shear typically occurs due to relative lateral motion of the beam and the corner cube and can originate in either misalignment between the direction of motion and the measurement beam or due to translations of the cube perpendicular to the direction of motion. This motion of the input beam within the aperture of the corner cube causes a proportional motion of the return beam, resulting in a shear between the measurement and reference beams at the detector.[65] In PMIs, a tilt of the target mirror results in beam shear at the detector (see Figure 4.16), with the magnitude of the shear (for a given tilt) being proportional to the distance of the mirror from the interferometer.[59] Beam shear results in a change in the overlap between the two beams with a resulting decrease in the heterodyne component (the AC component) of the beam and an increase in the constant intensity component (the DC component) from the nonoverlapping portions of the two beams. This decrease in the AC-to-DC ratio results in an overall increase in the electronic noise, thereby increasing the measurement uncertainty.

Another consequence of beam shear is an uncertainty contribution attributable to aberrations in the wavefront of the interfering beams. Wavefront aberration is a natural consequence of the fact the detector plane is typically some distance from the beam waist, which is the only point along the beam where the wavefront is a plane wave. A quadratic term dominates the shape of the wavefront at points away from the waist, the magnitude of which is proportional to the distance from the waist. With the exception of balanced interferometer configurations, in most cases, the reference and measurement beams propagate over very different distances, which results in interfering wavefronts with very different curvatures, which in the presence of beam shear contribute a spurious phase term.[59,185] This contribution can become significant in applications where the plane-mirror target undergoes relatively large angular displacements, for example, in X–Y–θ stage applications.[186] Very modest beam shears of a few tenths of a millimeter can contribute ~0.5 nm, even in applications where beam shear results from parasitic motions and misalignments.[185] Bobroff describes a technique for estimating the contribution from sheared aberrated wavefronts based on monitoring the change in the phase meter output as a small aperture is used to profile the interfering beams and suggests using the observed change as a bounding value for the uncertainty contribution.[59]

Specialized "zero shear" interferometer configurations reduce the beam shear to virtually zero in the presence of large angular excursions of the target mirror and may be suitable for applications where beam shear is an issue, for example, applications where long measurement ranges magnify small angular error motions.[187,188]

4.6.12 Target Contributions

Defects in the target contribute to the measurement uncertainty. These defects can range from the obvious such as figure errors in a plane-mirror target to more subtle sources such as refractive index inhomogeneity in solid corner-cube targets. Figure errors typically couple into the measurement when the target translates in a direction orthogonal to the beam. The measured phase change is now a combination of the change due to target displacement along the beam direction as well as contributions due to the deviation of the target from the desired shape (usually a plane). In the case of a plane-mirror target, the contribution is one-for-one for figure errors with spatial wavelengths greater than the beam diameter. The finite size of the beam patch on the target provides an averaging effect over the spot, thereby greatly attenuating the effects of figure variations with spatial wavelengths smaller than the beam size. Figure errors typically result from fabrication, mounting stresses, and gravity loads.[189] Translations of the beam on the surface of the target may be due to application-related translations perpendicular to the beam or due to parasitic motions and misalignments as described in the section on beam shear.

Measuring the figure and generating a corresponding error map can compensate for the effects of figure error. Error mapping usually requires the in situ measurement of the mirrors in the mounted configuration, as much of the mirror deformation can result from mounting stresses. The mirror figure may be measured by comparison to a reference straightedge using a displacement interferometer[190] or by making measurements of the target mirror using multiple spatially separated measuring points.[183,191–193] The latter method does not require an external artifact, and the mirror figure may be recovered in quasi real time. Straightedge reversal techniques separate the straightness error motions of the stage from the out-of-flatness of the target mirror.[189,194] Reversal techniques do not apply to target mirrors with reflecting surfaces oriented perpendicular to the gravity vector, making it necessary to account for the deviation from flatness due to self-weight-induced gravity sag by analytical or finite element methods.[189]

Refractive index inhomogeneity in solid corner-cube targets can produce a contribution in addition to the figure errors of the corner-cube facets. This contribution may be larger where the corner cube rotates as it translates in applications such as angle measurement[195] or servo-track writing.[196] A more subtle contribution results from the change in phase change on reflection (PCOR) at each of the reflecting facets of the corner cube as the internal angles of incidence change with rotation.[59,197] Similar considerations also apply to plane targets coated with metallic coatings[198] or dielectric coatings[199] that undergo significant angular excursions.

4.6.13 Data Age Uncertainty

Data age refers to the time difference between the occurrence of a motion event and the time that the information becomes available to the user control system. Issues related to data age apply to multiple-axis systems that measure multiple DOF nominally simultaneously for the purposes of producing coordinated motion or calculating a parameter based on the readings of two or more axes, for example, angle. In order to achieve full accuracy, all measurements must have the same or at least a known data age. An example of the effect of data age error is an angle measurement derived from interferometric displacement measurement of two points on a stage as the stage executes linear motions. It is easy to see that if the data from the two interferometers do not arrive at the same instant, an error in the difference reading that computes the angle equal to the product of the time lag and the stage velocity results, causing an error in the measured angle.

Data age uncertainty has a root cause in the changes in the time delay in the signal path, including the lengths of optical and electronic cables, and electronic processing delays. Ultimately, data age uncertainty compromises the ability of the controller to produce coordinated motion and results in errors in positioning.

The magnitude of the error δD_{age} due to data age uncertainty is

$$\delta D_{age} = tV \tag{4.41}$$

where

t is the data age uncertainty

V is the velocity of the target[18]

For example, a data age uncertainty of 10 ns will result in an error of 10 nm for a stage moving at 1 m/s. Data age uncertainty has two components: a fixed delay and a variable delay.[200] Modern phase meters provide for onboard adjustment of data age for each channel and allow the user to address the fixed delay component of the data age within 1 ns over 60 channels and over multiple measurement boards.[192,201] The ability to measure the total delay for each channel drives the current limitation on adjusting the data age. Many different techniques exist for the estimation of data age. Some are based on calculation and factory calibration, while others are based on in situ measurements on an assembled system.[192] The variable component of the data age uncertainty arises from the fact that the group delay of the signals through the electronics is a function of frequency and hence velocity due to a nonlinear phase behavior of the electronics. Techniques are also available to make adjustments for the variable component.[200]

4.6.14 Mounting Effects

The mounting of targets to the stage for displacement monitoring has a strong impact on the dynamic performance and stability, as any relative motion between the target and the stage constitutes a source of uncertainty. Slow relative displacements can result from the thermal expansion of the mount and the stage itself because the target cannot be located at the POI for practical reasons. A dynamic contribution also results from vibrations of the target relative to the stage. These vibrations are often excited by the motions of the stage itself and/or by interactions with turbulence in the airstream.[149] In applications where the long-term stability of the measurement is paramount, slow mechanical drift of the optics within the mount can contribute to the measurement uncertainty.

4.7 Applications of Displacement Interferometry

Numerous high-accuracy metrology applications employ DMIs. In addition to measuring displacement, DMIs measure angle, refractive index, and index changes, in application areas as diverse as microlithography[202] and formation flying of satellites.[23]

Displacement-related applications fall broadly into two categories: embedded metrology in a machine or instrument as part of a control system and external reference metrology for performance characterization and correction.

4.7.1 Primary Feedback Applications

Displacement interferometers are the primary source of position feedback in high-accuracy machines where direct traceability to the unit of length,[203,204] high linearity,[205] and large measurement range[206] are critical. Applications in this category range from high-precision machine tools and CMMs to gravitational wave observatories.

4.7.1.1 High-Accuracy Machines

Here we define a machine as any device used to manufacture a part. These parts can range from integrated circuits (ICs) produced by microlithography machines to high-precision components produced by processes such as single-point diamond turning (SPDT) and grinding.

A first machining application of particular historical interest is the manufacture of diffraction gratings. In their original form, specialized machines for ruling gratings (ruling engines) operated under the control of mechanical indexing systems and were later modified to incorporate fringe counting DMI and variants thereof.[1,207–209] Engines such as the MIT "B" engine developed by Harrison and the Michelson engine have since been retrofitted with modern displacement interferometers that monitor and control the positioning of the ruling stylus relative to the grating blank.[210,211] More modern engine designs feature heterodyne DMI as the primary feedback mechanism.[212] Grating ruling represents a particularly challenging application in that a ruling run can last weeks, imposing stringent requirements on the wavelength stability and corrections for index-related effects. More recent techniques for the production of large state-of-the-art gratings by scanning beam interference lithography (SBIL)[213] also rely on displacement interferometers for the synchronization of stage motion to the fringe pattern used to write the grating pattern.

Microlithography (such as wafer exposure and mask inspection) for the manufacture of ICs has driven the development of highly advanced DMIs to accommodate ever-shrinking line widths and overlay requirements (ITRS road map—www.itrs.net). This industry is a consumer of state-of-the-art devices, which are used to measure and synchronize motions

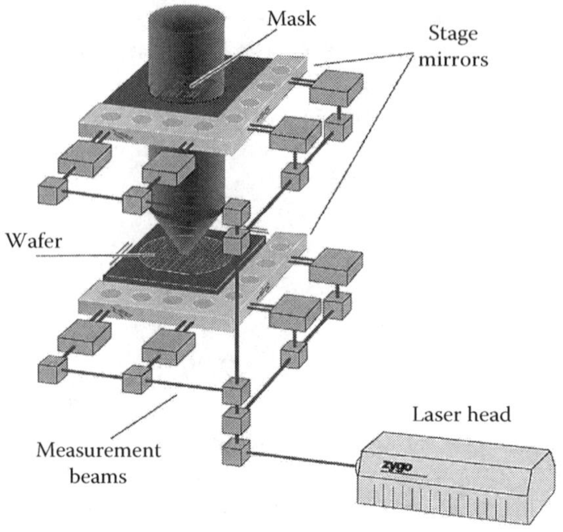

FIGURE 4.35
Schematic of a typical stage control application for microlithography.

of the reticle and wafer stage to nanometer levels to achieve the stringent overlay requirements.[214] Displacement measurements to the required accuracy demand exceptional attention to detail and several advanced features, some of which are discussed in the following.

Although the basic interferometer arrangement in microlithography systems (Figure 4.35) measures displacements and rotations in the plane of the wafer and reticle,[183,215] modern tools measure and control all the DOF of these stages. Additional measurement axes compensate for Abbe offsets and for numerous other measurements and require in excess of 50 channels of displacement metrology per exposure tool. Advances in the signal processing electronics and the low noise floor intrinsic to the heterodyne process make it possible to power multiple measurement channels with subnanometer noise performance from one laser head. These systems also may use specialized interferometers to reject any structural deformations by making a differential measurement between the wafer or mask and the projection or inspection optics.[216,217] Additionally, the deviation from flatness resulting from fabrication and mounting of the target mirrors becomes significant, requiring characterization of mirror shape. Characterization must be carried out in situ to measure the as-mounted mirror figure, by comparison to an external straightedge[190] or by making multiple redundant measurements of the mirror using multiple interferometers.[192,193] Similarly, there is a need to characterize deviations from squareness in the reference mirrors and the motion axes. This is typically performed in situ using reversal techniques.[218,219]

The high velocities and requirements on synchronized motion in microlithography impose stringent requirements in the acceptable variation in the data age, as detailed in Section 4.7. The data age uncertainty results in a positioning error proportional to the velocity, and modern measurement electronics are designed with the ability to adjust the data age in order to minimize or eliminate the data age difference.[18,220]

Modern lithography tools may use a combination of optical encoders (see Section 4.8) and conventional interferometers to overcome air turbulence. These two systems complement each other: The short air path typical of encoders minimizes the effects of air turbulence, thus realizing an improvement in short-term repeatability,[150] while the conventional interferometer provides superior linearity when compared to the nonlinearities encountered in the encoder gratings.[150,205] The limitations imposed by air turbulence disappear in vacuum environments, such as those encountered in reflective electron-beam lithography[153,221] and EUVL [152,222] systems.

Microlithography for flat-panel displays poses its own special challenges. Although the positioning requirements are not as stringent relative to IC manufacture,[223] the ever-increasing size of the substrates necessitates displacement measurements over several meters, an application ideal for interferometers.[206] Long travels may also require long straightedges along multiple mirror segments, which involve an interferometer "hand-off" across the boundaries of the segments.[220] A problem specific to this application is the possibility of large beam shears due to the large lever arm resulting from the long measurement ranges that magnify the effects of angular error motions, an issue that can be addressed by specialized zero beam shear interferometers.[187,188]

Interferometers also enable high-accuracy machines capable of optical quality tolerances and surface finishes such as SPDT machines[145,203,224] and specialized grinding machines.[225,226] The application of interferometers to SPDT originated in diamond turning machines (DTM) built at the Lawrence Livermore National Laboratory (LLNL),[145,203] culminating in the development of the large optics diamond turning machine (LODTM)—arguably the most accurate lathe ever built (Figure 4.36).[203,227] These machines mitigated the dominant uncertainty contribution due to refractive index changes by conditioning the beam paths. The LODTM virtually eliminated this contribution by evacuating the beam paths (Figure 4.36),[89] while another DTM at LLNL, DTM #3, used helium-filled beam paths to reduce the pressure- and temperature-related index changes.[145] Commercial DTMs have also used laser interferometers for feedback, albeit without the conditioning of the beam paths described earlier.[228,229]

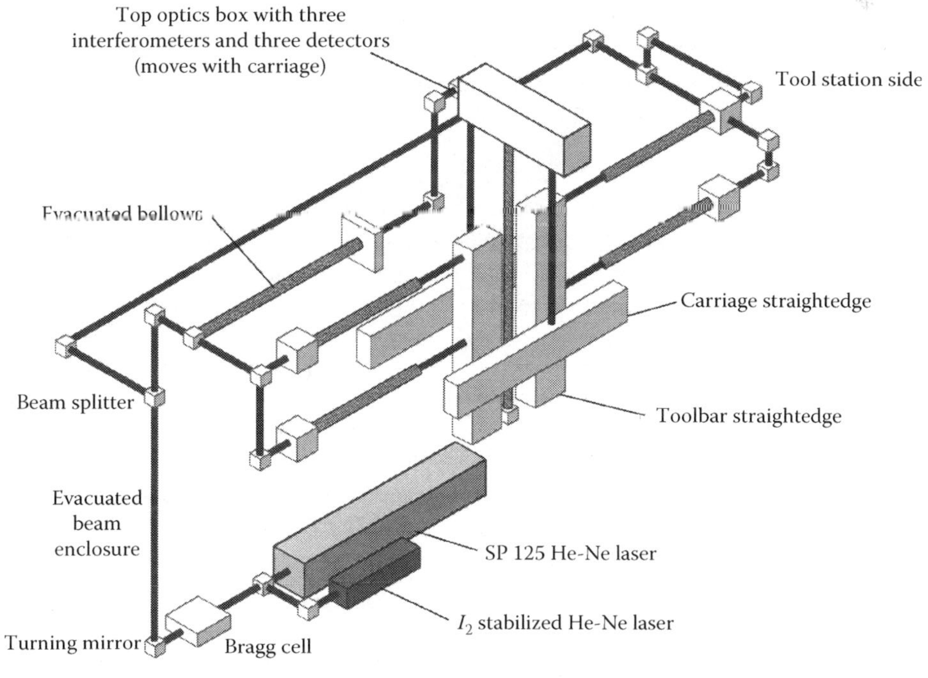

FIGURE 4.36
Interferometer system for the LODTM. (Adapted from Slocum, A.H., *Precision Machine Design*, Prentice Hall, Englewood Cliffs, NJ, 1992.)

A specialized DTM for the manufacture of x-ray optics shares many of the attributes of the machines described earlier. In addition, it uses a combination of glass scales and interferometers for ultraprecision machining. The machine also includes multiple probes with interferometer feedback for the inspection of the profile, roundness, and diameters of the finished components.[224]

Grinding machines for the production of large optical elements for large terrestrial telescopes also employ interferometric feedback.[225,226] These machines incorporate multiple interferometers to reference all the stage motions to a decoupled metrology frame. Multiple refractometers correct for the refractive index. The machine uses a probe adjacent to the grinding spindle for in situ metrology referenced with interferometers to the metrology frame. Another example is an aspheric generator designed for both ductile regime grinding and SPDT.[230,231] A more recent machine for the production of large free-form surfaces uses encoders as feedback for the machining axes but uses an interferometer as feedback on the probe that is used for inspection.[232] Thread grinders operating under interferometric control exploit the direct linkage to the unit of length.[233,234] The interferometer corrects for variations in the pitch of the leadscrews through CNC control of the grinder rather than by hand lapping.

Servo-track writers generate the disk tracks for data storage hard drives at the time of manufacture. The increasing data density requires a finer spacing of the tracks, which are often written using a magnetic head under interferometric control on dedicated machines. The motion of the head may be linear or arcuate. For arcuate motions, an interferometer with a corner-cube retroreflector accommodates the relative large rotations of the target.[196]

4.7.2 Angle Measurement

Angular displacement measuring interferometers (ADMI) do not measure angular displacements directly, but infer them from one of two methods: A first approach relies on a single measurement of the rotation-induced target displacement and knowledge of the perpendicular distance to the axis of rotation R (Figure 4.37a).[235] A second approach relies on the differential displacement of two or more points on a rigid body and knowledge of the separation S between these points. The first configuration has the drawback that parasitic translation of the center of rotation along the measurement direction contaminates the measured displacement. The second method separates the contributions from angular motion and the aforementioned translation and typically measures displacement at two or more spatially separated points on a rigid body (Figure 4.37b).[236,237] Uncertainties in both the measured displacement and the separation drive the uncertainty of the angle measurement, with the uncertainty in the knowledge of the separation being dominant. Measurements at the highest levels of accuracy require an external calibration step to establish the effective separation.[238]

The range of angular motion and the ability to permit translation of the target perpendicular to the measurement direction determines the choice of DMI target. Plane mirrors are sufficient for relatively small angular motions[17,96,183,195] and are used in conjunction with either separate PMIs[195] or specialized interferometers that make two or three displacement measurements within a single compact interferometer package to measure one or two angular motions of the target, respectively (Figure 4.19).[183,184,239] The angular sensitivity is proportional to the perpendicular distance between the beam and the center of rotation for measurements made with a single DMI and to the

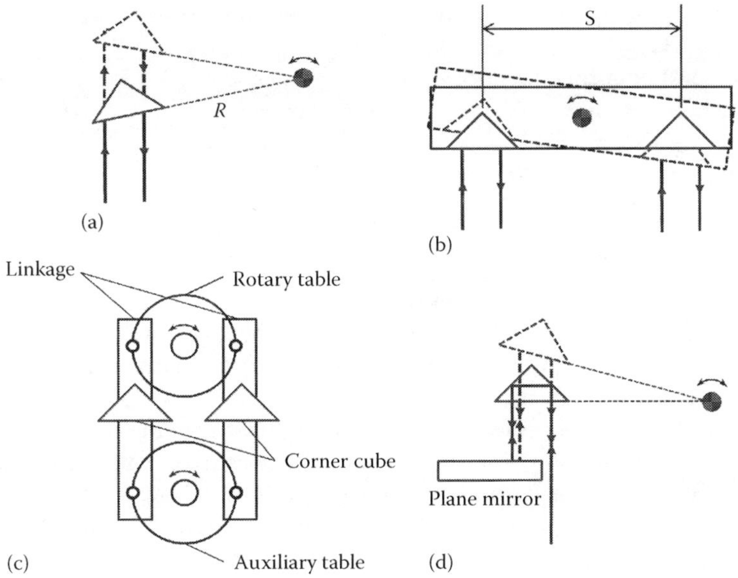

FIGURE 4.37
ADMI configurations. (a) Single reflector. (From Rohlin, J., *Appl. Opt.*, 2, 762, 1963.) (b) Dual reflector. (From Bird, H.M.B., *Rev. Sci. Instrum.*, 42, 1513, 1971.) (c) Linkage mechanism to preserve reflector orientation. (From Shi, P. et al., *Opt. Eng.*, 31, 2394, 1992.). (d) Retroreflector/plane mirror combination for large angular range. (From Murty, M.V.R.K., *J. Opt. Soc. Am.*, 50, 83, 1960.)

perpendicular distance between the measurement axes of the DMIs in the differential configurations. The subnanometer displacement resolution of modern DMI systems makes possible sub-microradian angular resolutions even for relatively modest beam separations of 10 mm.

Corner-cube retroreflectors,[235,236,240] right-angle prisms,[195,238,241–245] and roof reflectors[237] are common for intermediate ranges of motions (~10°). Many of the configurations for large angular displacements use one or more corner cubes as targets[195,235] and trace their origin to the optical configuration implemented by Rohlin.[235] While corner cubes confer immunity to changes in the angular orientation of the target, large angular motions result in shear between the measurement and reference beams, the maximum angular measurement range corresponding to the maximum permissible shear. All corner cube, right-angle prism, and roof prism-based ADMIs display a nonlinear relationship between the measured angle and phase change at the output of the interferometer caused by nonlinear changes in the optical path length through the target.[59]

Some dual corner-cube interferometer configurations exploit the fact that both the measurement and reference beams translate equally, thereby eliminating the beam shear and permitting the measurement of angular displacements of ±10° (Figure 4.37b),[155] although measurements in excess of 40° are possible.[196] This arrangement also results in the cancelation of a majority of the nonlinear behavior by virtue of the fact that the both corner cubes rotate. Many of the limitations outlined earlier may be overcome while extending the measurement range to ±60° by use of a mechanical linkage that prevents the rotation of the corner cube[242] (Figure 4.37c). This approach results in a more complex arrangement wherein the linkages and the rotary joints required for their articulation now form part of the metrology loop.

A combination of plane mirrors and corner cubes have been used for large ranges of motion.[195,235,240] Murty introduces an additional plane mirror into the measurement beam path, which retroreflects the output beam of the corner cube for a second trip through the corner cube before it returns to the interferometer[246] (Figure 4.37d). This arrangement makes the system shear free but requires an additional component whose stability, alignment, and figure now directly influence the measurement.

4.7.3 Measuring Machines

Interferometers serve as the primary metrology in a number of high-performance CMMs. The term CMM is used rather loosely here and encompasses measuring machines of all kinds ranging from conventional multiaxis CMMs[204] and some more unusual multipurpose CMMs[177,178,247,248] to specialized machines such as metrological atomic force microscopes (AFM),[249–252] line scale comparators,[146,253] and devices for the evaluation of encoders.[254,255]

The use of interferometers in CMMs is characteristic of the highest accuracy machines.[177,178,180,204,248,256] One example is the Moore M48 CMM at the National Institute of Standards and Technology (NIST).[204] The configuration of this machine is conventional (see Figure 4.38a), the exceptional performance attributable to the fidelity of the mechanical motions, careful error mapping, and stringent conditioning of the environment. Other machines rely less on the fidelity of the mechanical motions and take a different approach to mitigating the effect of the Abbe offset. Some rely on a combination of an adherence to the Abbe principle and active cancelation of angular error motions.[180] Others locate the

(a)

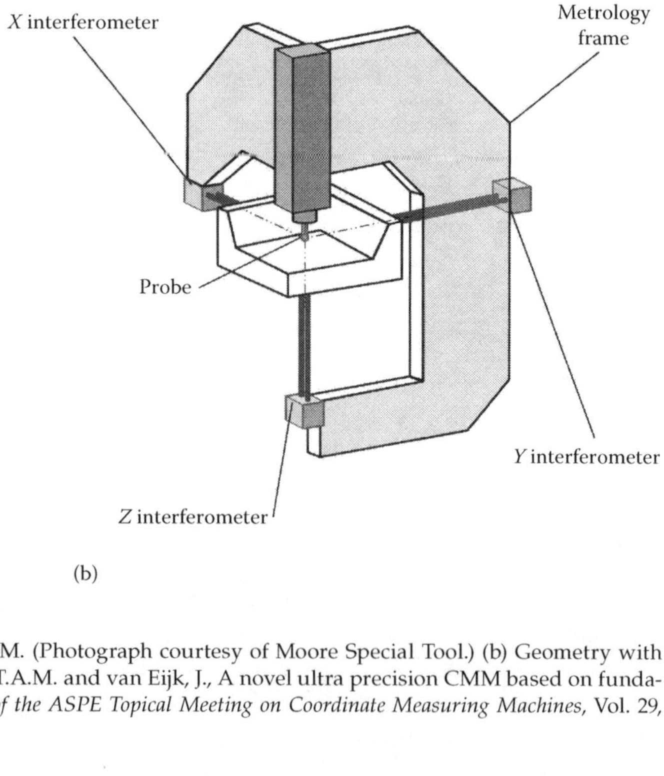

(b)

FIGURE 4.38
CMM configurations. (a) Moore M48 CMM. (Photograph courtesy of Moore Special Tool.) (b) Geometry with nominally zero Abbe offset. (From Ruijl, T.A.M. and van Eijk, J., A novel ultra precision CMM based on fundamental design principles, in *Proceedings of the ASPE Topical Meeting on Coordinate Measuring Machines*, Vol. 29, pp. 33–38, 2003.)

interferometers such that the line of measurement of all the interferometers intersects at the location of the probe, thereby minimizing the Abbe offset to within the levels achievable by alignment (Figure 4.38b). Interferometers have a distinct advantage over encoders in these machines as they provide a means of satisfying the Abbe principle while at the same time allowing motions of the target orthogonal to the beam. This configuration has found application in some ultrahigh-accuracy laboratory machines[248] and in a few commercial machines.[177,178] A hybrid approach relies on retrofitting a conventional CMM with three orthogonal mirrors around the probe as targets for three linear/angular interferometers whose line of measurement passes through the probe and three dual-axis autocollimators mounted in a high-stability auxiliary metrology frame.[247] The DMI measurements are corrected for the residual Abbe errors from any residual offset with the angle information obtained by the angle interferometers/autocollimators.

Line scale comparators measure, calibrate, and certify length artifacts such as line scales, encoder gratings and grids, end standards, and other artifacts that disseminate the standard of length. They are simple devices relative to CMMs in that the measurement is only in one direction, although additional measurement channels may perform differential measurements and compensate for residual Abbe offsets. Measurements take place in carefully controlled environments with additional conditioning of the sample and the interferometer. Comparators consist of a single linear stage with interferometric position monitoring with respect to a suitable probe. A common approach is a differential interferometer that makes a measurement between the probe and the line scale carriage with the beam paths in a carefully controlled environment[253] or with evacuated beam paths.[146,254,257] The Physikalisch-Technische Bundesanstalt (PTB) nanometer comparator (Figure 4.39) goes a step further with an additional axis of interferometric displacement measurement

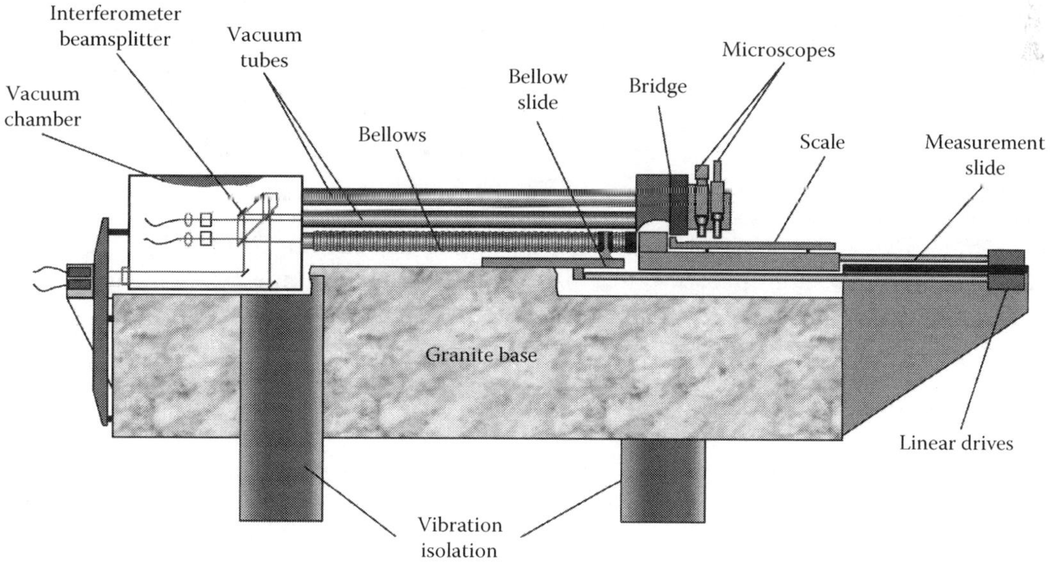

FIGURE 4.39
PTB nanometer comparator. (Adapted from Flügge, J. et al., Interferometry at the PTB nanometer comparator: Design, status and development, in *SPIE Proceedings, Fifth International Symposium on Instrumentation Science and Technology*, Vol. 7133, pp. 713346-1–713346-8, 2009; Köning, R., Characterizing the performance of the PTB line scale interferometer by measuring photoelectric incremental encoders, in *SPIE Proceedings, Recent Developments in Traceable Dimensional Measurements III*, San Diego, CA, Vol. 5879, pp. 587908-1–587908-9, 2005.)

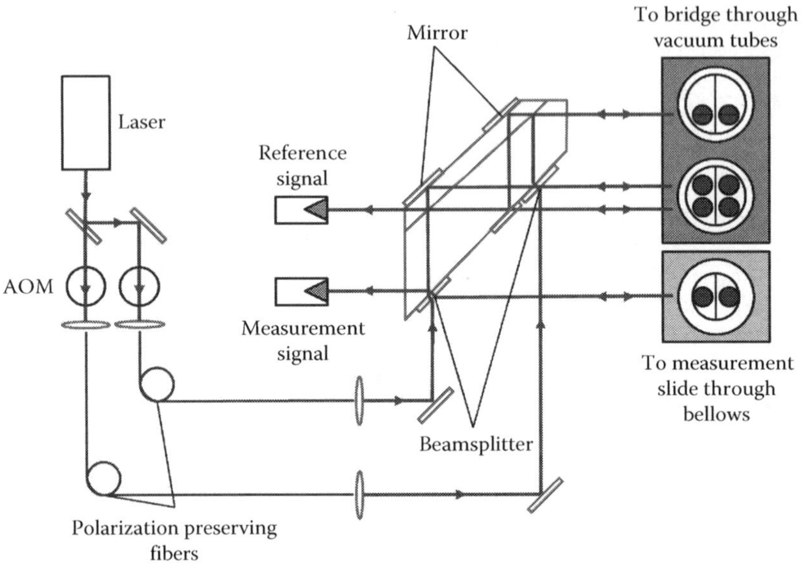

FIGURE 4.40
PTB nanometer comparator interferometer details. (Adapted from Flügge, J. et al., Interferometry at the PTB nanometer comparator: Design, status and development, in *SPIE Proceedings, Fifth International Symposium on Instrumentation Science and Technology*, Vol. 7133, pp. 713346-1–713346-8, 2009; Köning, R., Characterizing the performance of the PTB line scale interferometer by measuring photoelectric incremental encoders, in *SPIE Proceedings, Recent Developments in Traceable Dimensional Measurements III*, San Diego, CA, Vol. 5879, pp. 587908-1–587908-9, 2005.)

to measure the bending of the structure that carries the probe (Figure 4.40) in order to compensate for an Abbe offset between the probing point and the beam axis.[151] Measurement slide angular error motions (pitch and yaw) are also measured by additional interferometers operating through the vacuum bellows.[151] Another instrument for characterizing linear artifacts uses both linear and angular interferometers to compensate for any residual errors from angular motions.[181] Traceability is a key requirement, and all such devices use DMI with well-characterized laser sources as the primary metrology.

Interferometers also find application in a number of specialized CMMs built for the inspection of a particular part or component. An instrument for the measurement of free-form optics, for example, uses a combination of linear and rotary axes and an articulated optical probe.[258] This instrument uses a novel interferometer system that measures the location of the cylindrical rotor of the articulated probe relative to a metrology frame to compensate for error motions of this rotary axis.[259] The interferometer uses a cylindrical lens in one of its arms to measure the displacements of the rotor, which is confocal with the interferometer beam—a configuration also used to monitor a magnetically levitated rotor in a reflective electron-beam lithography tool.[153] The free-form CMM also uses a PMI to track the motions of the probe in conformance with the Abbe principle.[260] Plane-mirror DMIs are also used in a high-accuracy profiler for the measurement of free-form optics.[261] Measuring machines for the inspection of grazing incidence optics use multiple interferometers to simultaneously monitor the displacement of a mechanical probe, the straightness of the slide that carries the probe, and the displacement of the scanning head along the lateral dimension.[262] Another custom machine for measuring inner diameter and circularity of cylindrical x-ray optics works as a combination roundness machine and a comparator.[263] The diameter is determined by comparing the part under test to a length artifact that serves as a reference for the displacement interferometers.

DMIs also find application in high-accuracy instruments for the measurement of ring and plug gages[264,265] and in more specialized instruments for the calibration of pressure standards.[266] DMIs offer a combination of features that are important to such measurements, especially at the NMI level, that is, direct traceability to the unit of length and high resolution over a long measurement range. The last attribute makes it possible to use such instruments in comparator mode between artifacts of widely varying dimensions.

DMIs are common in metrological scanned probe microscopes to characterize standard artifacts for calibration of other instruments such as critical dimension scanning electron microscopes (CD-SEM) and other scanned probe instruments.[249,250] This class of machines is exemplified by the NIST molecular measuring machine (M^3), pictured in Figure 4.41. Figure 4.42 shows an exploded view of one of the interferometers, the x-axis interferometer, which measures the displacement of the probe tip relative to the metrology mirrors that form the box that carries the sample.[249,267] These instruments use DMIs for minimizing Abbe offset errors and for direct traceability to the unit of length. While an ideal arrangement would make a differential measurement between the sample and the probe tip, the small size of the tip and other practical considerations make this extremely difficult. Therefore, virtually all such instruments make a measurement of the displacement of the structure that supports the tip. Interferometer geometries differ, some being designed to make a differential measurement between the sample holder and the tip holder,[112,249,252,268]

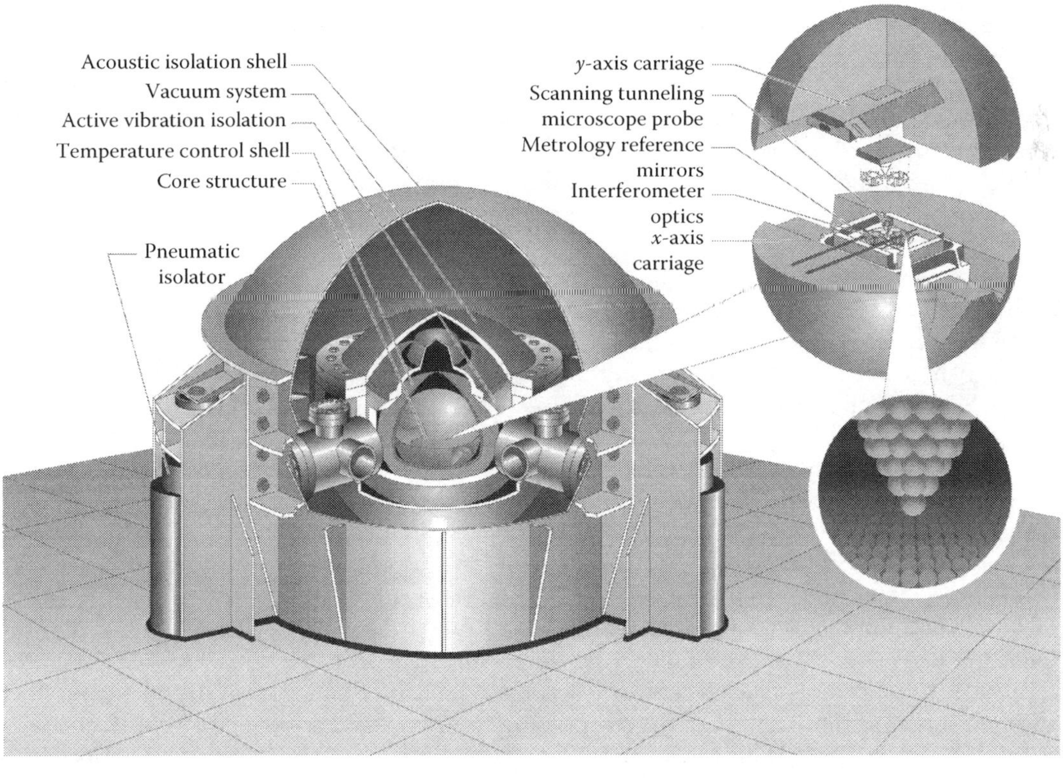

FIGURE 4.41
The NIST M^3. (From Kramar, J.A. et al., *Meas. Sci. Technol.*, 22, 024001, 2011. With permission.)

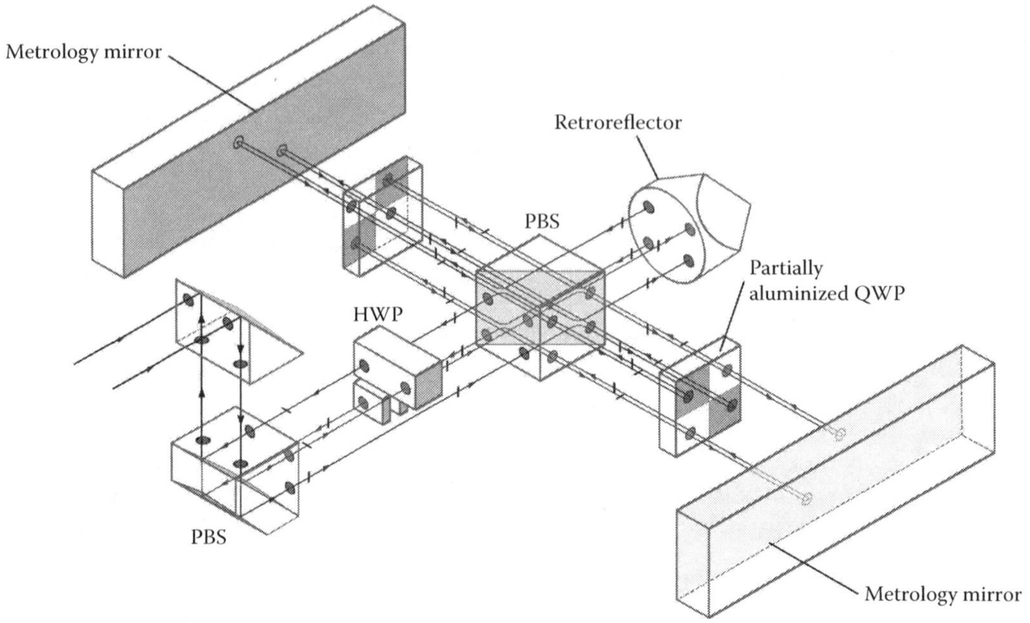

FIGURE 4.42
Exploded view of the NIST M^3 x-axis interferometer. (From Kramar, J.A., *Meas. Sci. Technol.*, 16, 2121, 2005; Kramar, J.A. et al., *Meas. Sci. Technol.*, 22, 024001, 2011.)

while others monitor the motion of the stage in translation and angle. Some designs use interferometers specifically designed to minimize the deadpath,[250] while others exploit the DMI to minimize the Abbe offset.[252] Other examples of metrology stages designed for use with scanned probe microscopes use specially configured PMIs in which the measurement beams makes four round-trips (four passes) to the target to achieve extremely high resolution in a compact package.[269,270]

Dilatometers that perform high-accuracy measurement of the coefficient of thermal expansion (CTE) of artifacts such as gage blocks[271,272] and line scales[273] rely almost exclusively on DMI metrology. A dilatometer for CTE measurement is equipped with a mechanism to vary the temperature of the sample and a means to monitor the temperature change. Dilatometers also measure dimensional stability[274] and the CTE of ultralow expansion materials.[275]

In the highest accuracy DMI-based dilatometer measurements,[276] a modified Michelson interferometer performs a differential measurement in one of two ways—between the end face of the sample and an auxiliary reference surface attached to the other end (Figure 4.43a)[275,277–281] or by using a wraparound configuration that directly interrogates both sides of the sample (Figure 4.43b).[282–284] The first configuration is relatively simple but includes the contributions of the contact between the remote or back face of the sample and the reference optic within the metrology loop and the associated uncertainty. The latter configuration eliminates the uncertainty associated with the behavior of the optical contact and monitors the deformation of the supporting structure and optical path length changes within the beam-directing optics used to wrap the beam around the sample, effectively eliminating the contribution of the structure. The full potential of either measurement technique is realized when the interferometer is operated in a vacuum to eliminate the

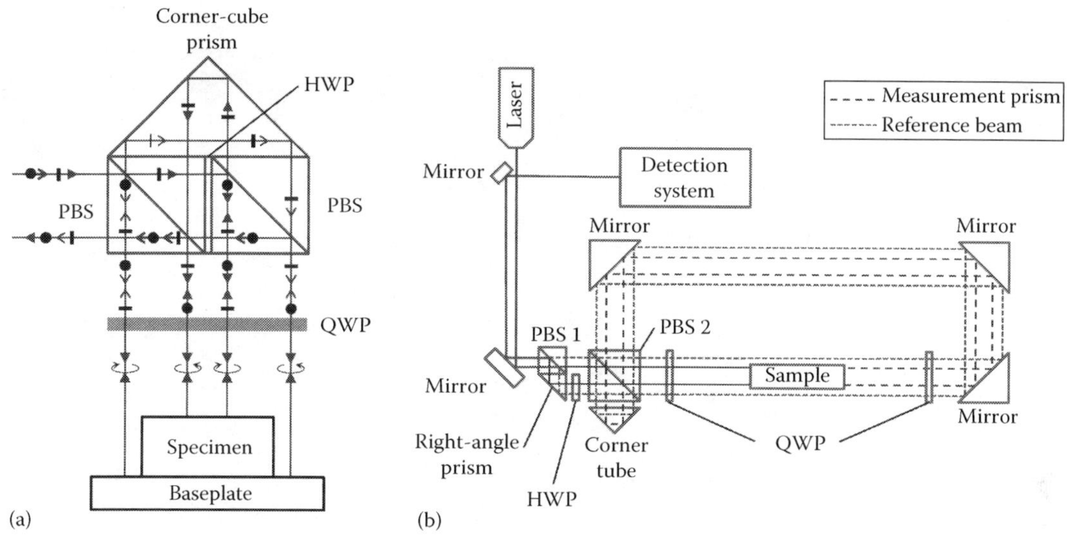

FIGURE 4.43
Two dilatometer configurations. (a) Single-sided. (From Okaji, M. and Imai, H., *Precis. Eng.*, 7, 206, 1985.) (b) Wraparound. (From Ren, D. et al., *Meas. Sci. Technol.*, 19, 025303, 2008.)

effects of the environment-induced optical path length changes or by simultaneous determination of the refractive index, using the equivalent of a wavelength tracker.[172]

Other examples of the measurement of long-term stability include the creep of epoxy joints under shear loading[285] and rotational creep of Elgiloy spiral torsion springs.[286] The latter application is somewhat unusual in that it measures rotational creep through an interferometer setup designed to track the change in angle. Another somewhat unusual approach for the measurement of the stability of line scales also uses DMIs.[273,287]

DMIs are often also embedded within setups that employ other interferometric instruments such as Fizeau interferometers and coherence scanning interferometers (CSI). An example of such a pairing is the measurement of aspheric surfaces, wherein one or more DMIs track the position of the surface under test.[288,289] Such a setup may also measure the radius of curvature of an optical surface, using the DMI to track displacement of the surface from the confocal position to the cat's eye position.[134,290] Many more DOF may be measured using interferometers as is demonstrated in a unique design that is used to determine the position of one part relative to another in six DOF.[99] Nine DMIs determine the six DOF, three of them providing redundant information.

Displacement interferometry combines with CSI (sometimes referred to as scanning white light interferometry or SWLI) for the purposes of performing relational metrology.[291] The fringe localization that results from the low-coherence interferometer locates the surfaces whose relationship needs to be established, while the DMI measures the displacement between the positions of fringe localization. Systems measure step height, as in Figure 4.44, or the complete flatness, thickness, and parallelism of industrial parts using fully integrated DMI metrology.[292] DMIs can also monitor the displacement of the microscope objective in a CSI during its scan.[293] The same general principles extend to measurement systems that establish the relative locations of multiple transparent surfaces, for example, lens surfaces in a lens assembly, in a transmission measurement that uses an extended scan to move the region of coherence using a delay line interferometer. A DMI included in the delay line measures the distance between the surfaces.[294]

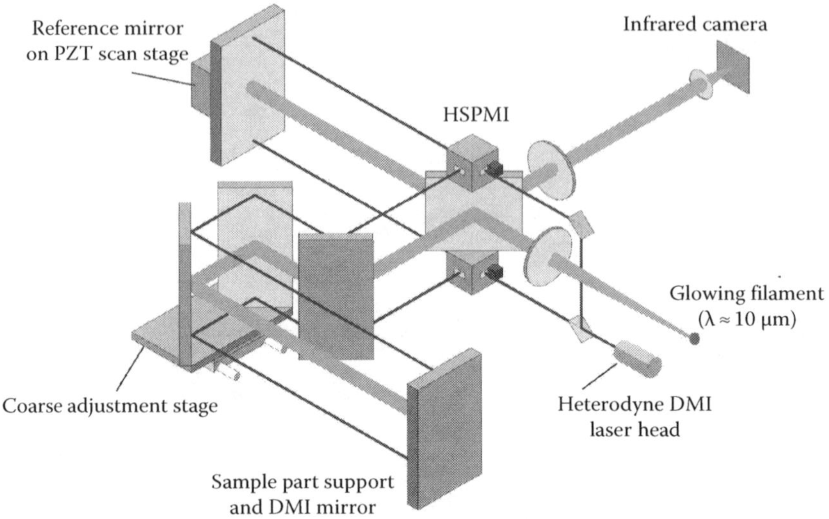

FIGURE 4.44
Example geometry for a step-height measurement system based on infrared CSI combined with a two-axis laser displacement gage and two high-stability plane-mirror interferometers. (From de Groot, P.J. et al., Step height measurements using a combination of a laser displacement gage and a broadband interferometric surface profiler, in *SPIE Proceedings, Interferometry XI: Applications*, Vol. 4778, pp. 127–130, 2002).

4.7.4 Very High Precision

Although commercial interferometer systems routinely measure to 1 nm precision at high speeds, some highly specialized applications demonstrate an entirely different and even more exacting level of performance.

At national standards laboratories, optical interferometry provides a link between laser frequency standards and measurements of mechanical displacements. As part of these developments, systems have been developed to connect optical interferometry to x-ray interferometry.[295] Specialized heterodyne Michelson interferometers have also achieved 0.01 nm positioning uncertainty.[296] Another approach is to use a Fabry–Pérot interferometer, with the object mirror being one of the components of a resonant cavity, together with a tunable source. In this case, the measurement problem is reduced to that of measuring a frequency rather than a phase, which has been shown to provide absolute distance metrology to a relative uncertainty of 4×10^{-10} over a range of 25 mm.[297]

The ongoing effort to detect gravitational waves has involved astonishing requirements for precision in interferometers with path lengths measured in kilometer rather than millimeter. The Laser Interferometer Gravitational-Wave Observatory or LIGO and comparable projects internationally are capable of detecting the tiny force of gravitational waves by measuring the relative positions of free masses, as illustrated in Figure 4.45. The LIGO interferometers combine Michelson interferometry with resonant cavities and beam recycling to reach displacement sensitivities below 10^{-18} m/$\sqrt{\text{Hz}}$ between 70 and 1000 Hz.[298]

4.7.5 Reference or Validation Metrology

Applications in this arena are often referred to as "strap-on" metrology applications where the displacement interferometer is external to the device under test (which has its own

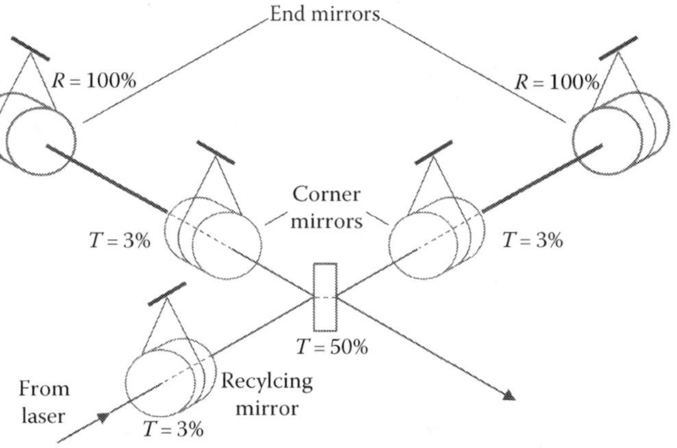

FIGURE 4.45
LIGO optical geometry. (From Spero, R.E. and Whitcomb, S.E., *Opt. Photon. News*, 6, 35, 1995.)

internal feedback mechanism) and is "strapped on" for the duration of the measurement, usually for the purpose of characterizing the performance of or calibrating a machine or sensor.

DMI systems have been used for characterizing the kinematic errors of machine tools almost from the time that the first Zeeman-stabilized laser became available [299] and continue to play a key part in the measurement of errors and their compensation.[300] Commercial DMIs for machine tool characterization consist of a frequency-stabilized laser head, readout, or a dedicated computer to acquire the data and a variety of optical accessories designed to measure specific error motions of a machine tool.[301] These systems measure all of the error motions of a linear axis except for rotation about the axis of the beam, although the same information may be derived from making two straightness measurements along two parallel lines.[157] A simple linear displacement interferometer measures the linear displacement accuracy of the slide (Figure 4.46).[302] A dedicated interferometer setup with a specially designed straightness optic measures deviations from straightness in a direction orthogonal to the beam (Figure 4.23),[98] while an interferometer that measures the differential displacement of a double corner-cube target measures angular error motions.[156,303] This set of optics can also be used in place of an autocollimator for the measurement of the flatness of machine beds and ways and for calibration of surface plates[304,305] according to Moody's method.[306] Straightness errors can also be measured using a straightedge as a target for a PMI.[189] A combination of straightness optics and an optical square establish the out of squareness of nominally orthogonal axes,[301] an alternative being the method of laser diagonals.[157]

While virtually any interferometer system along with the appropriate optics may be adapted to make the required measurement, systems for machine tool metrology are characterized by their ease of use in terms of both the setup of the optics, data acquisition, and dedicated software. Some systems specifically aim at speeding up the machine characterization process, by eliminating the multiple setups that are required for characterization by traditional means. One such instrument is the laser ball bar (LBB),[307] which is based on the principles of trilateration and draws on the telescoping magnetic ball bar.[308] The LBB replaces the short-range sensor in the telescoping magnetic ball bar with a fiber-fed heterodyne displacement interferometer to increase greatly the range of motion while

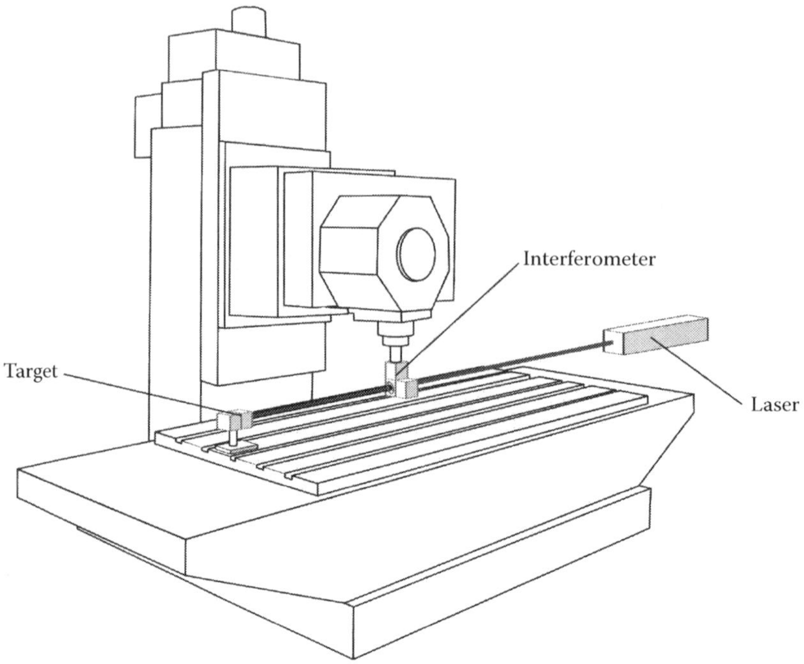

FIGURE 4.46
Calibration of linear displacement for a machine tool. (From Calibration of a Machine Tool, Hewlett-Packard, Application Note 156–4.)

maintaining the high-resolution characteristics. The LBB may also characterize spindle thermal drift in a machine tool by the method of sequential trilateration.[309] Interferometric configurations for the characterization of specific aspects of machine performance such as the measurement of spindle error motions have also been devised and use an interferometric probe against a master ball mounted in the machine spindle.[310] An interferometer offers several advantages over the conventional method that uses capacitance gages such as a much higher bandwidth, higher spatial resolution, greater range of motion, and standoff to accommodate runout without sacrificing resolution and lastly eliminates interactions between the measurement system and the targets.[310] A single-beam configuration with a converging lens typically references a reflective master ball. The ball may be at the focus of the lens in a cat's eye arrangement or in a confocal arrangement, although the latter has an increased sensitivity to lateral motions that can be problematic.[311] The same arrangement may also be used to make high-resolution, noncontact roundness measurements.[312] Interferometric methods are also used indirectly to measure the length of ball bars, which are artifacts of a known fixed length that are used to characterize CMMs by comparing the measured length against the known length at various positions and orientations within the work volume. The length calibration is accomplished in a special instrument based on the principle of self-initialization wherein the displacements measured by a DMI establish the absolute length of an artifact.[313]

The long range afforded by interferometry with laser sources and the fact that the interferometer's high resolution is maintained over this range (see Figure 4.1) make it ideally suited for metrology of large-scale objects such as aircraft structures, ship propellers, layout of large assembly tools, and the measurement of volumetric accuracy on large machine tools,[314,315] CMMs,[316,317] and hexapods. While many different techniques are used in this arena,[318,319] the DMI finds application in a device known as a laser tracker. First developed for robot metrology,[320] this device has numerous applications beyond the original intended purpose. A laser tracker consists of a DMI whose measurement beam is actively steered to follow a target retroreflector. The DMI then measures the radial displacement of a target from a known zero position (usually set when the instrument is initialized), while high-resolution encoders in the beam steering assembly measure the pointing of the beam in horizontal and vertical angular directions. The three measurements in combination establish a spherical coordinate system. The target retroreflector is somewhat different from the corner cube encountered in normal DMI, in that the corner cube is mounted inside a truncated sphere. Many different designs exist,[318] but the most common is the spherically mounted retroreflector (SMR), which consists of a hollow corner cube mounted with its apex mounted at the center of the sphere. Hollow corner cubes are preferred over the solid glass type to minimize refraction errors at large angles of incidence. The validation metrology for trackers also uses DMIs. A reference DMI is used to determine the ranging performance of a tracker by measuring the displacement of a target assembly that can be translated along a rail several meters long whose orientation may be changed to verify the ranging performance in different directions.[321] The target assembly consists of an SMR and the target retroreflector for the reference DMI mounted together in an opposed configuration. Ranging performance over longer distances (restricted to one orientation) is validated in a similar manner on a rail of length 60 m.[321] The measurement of the length of geodetic tapes used in surveying is another application where the long measuring range of a DMI can be used to advantage. The high coherence of the source enables measurement of tapes of 50 m length with an interferometer design that measures the change in separation between the carriages that bear the microscopes that measure the graduations on the scale.[322]

DMIs serve as reference sensors in applications where traceable calibrations[323] are required or for the evaluation and validation of the performance of an actuator[324] or other types of displacement sensor.[325] As shown in Figure 4.47, in both situations, the DMI and the sensor under test (SUT) (or calibration) monitor a common target and the arrangement strives to minimize the Abbe offset between the lines of measurement of the two

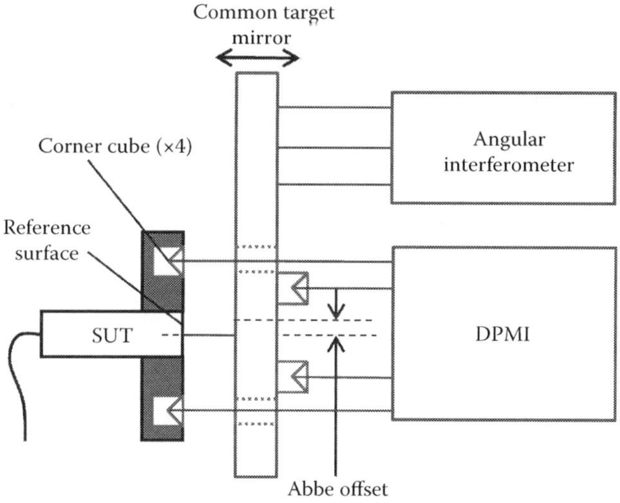

FIGURE 4.47
DMI used in the validation of a displacement sensor.

devices so as to reject uncertainty contributions from angular error motions of the target.[182] Synchronization of the data acquisition from the two sensors is important and is especially critical at high target velocities or in dynamic situations.

Gravimeters use DMIs for the determinations of the absolute value of the acceleration due to gravity. These devices are Michelson interferometers in which the measurement target is part of a free-falling mass within a vacuum chamber and an inertial reference carries the reference mirror. The vacuum mitigates both the effects of atmospheric drag and the uncertainty contributions due to index. Gravimeters require a length standard known to an uncertainty better than one part in 10^9 that is typically provided by an iodine-stabilized laser, although another stabilized laser may be used with periodic recalibration.[326,327]

Primary pressure standards use interferometric measurement of the differential displacement of the liquid columns of manometers, either directly by reflecting the measurement beam from the Hg surface[328,329] or by using a corner cube suspended in the mercury on a suitable float.[330]

4.8 Alternative Technologies

4.8.1 Absolute Distance Interferometry

DMI according to the definition in the introductory part of this chapter refers to measurements of displacements or changes in position. A DMI reports only how object positions change while being measured, not how far away they are from a specific reference point in space. In a conventional DMI, if the measurement beam is blocked and subsequently reestablished, there is no information about any change in position of the object that occurred while the beam was blocked.

For many applications, the distance from the object to a reference position is of importance and needs to be measured at any given moment in time, without relying on a continuous time history of the object motion. Instruments for these applications measure the *absolute* distance L as opposed to a relative displacement $D = L_f - L_i$ from a first position L_i to a second position L_f. There is abundant literature on this topic representing a wide range of solutions.[331] Setting aside ranging systems that operate by pulsed time of flight or microwave intensity modulation,[332,333] the majority of coherent or interferometric systems for absolute distance measurement employ multiple or swept-wavelength sources.

Multiple-wavelength methods have roots in the earliest interferometers for length standards.[3,4] The principle of measurement resides in the dependence of the interferometric phase on wavelength. Recalling Equation 4.9

$$\phi(L) = \left(\frac{4\pi n L}{\lambda}\right) \tag{4.42}$$

it is clear that there is a linear relationship between the phase ϕ and the angular wave number σ, where

$$\sigma = \left(\frac{2\pi}{\lambda}\right) \tag{4.43}$$

The rate of change of phase with wave number σ is proportional to the difference nL in the optical lengths for the measurement and reference paths. The historical method of excess fractions relies on matching up phase values for a sequence of discrete wavelengths to a specific distance using tables or a special slide rule. An alternative methodology involves the concept of an equivalent or *synthetic* wavelength Λ for a pair of wavelengths $\lambda_1 > \lambda_2$ and a synthetic phase $\Phi(L)$ calculated from the difference in the corresponding phase measurements ϕ_1, ϕ_2:[334]

$$\Phi(L) = \frac{4\pi n L}{\Lambda} \tag{4.44}$$

$$\Lambda = \frac{\lambda_1 \lambda_2}{(\lambda_1 - \lambda_2)} \tag{4.45}$$

$$\Phi = \phi_2 - \phi_1 \tag{4.46}$$

For the simplest case of two wavelengths, the absolute measurement is unambiguous over a range defined by at least half the synthetic wavelength. Multiple-wavelength emissions from CO_2 and other gas lasers are a natural choice for this type of absolute distance interferometry.[335] Two or more single-mode lasers locked to a Fabry–Pérot etalon, as illustrated in Figure 4.48, provide high relative wavelength stability for large variations in optical path length.[336,337] A simple and compact configuration for short-path differences is one or more multimode laser diodes.[338] Most recently, work has been carried out to take advantage of

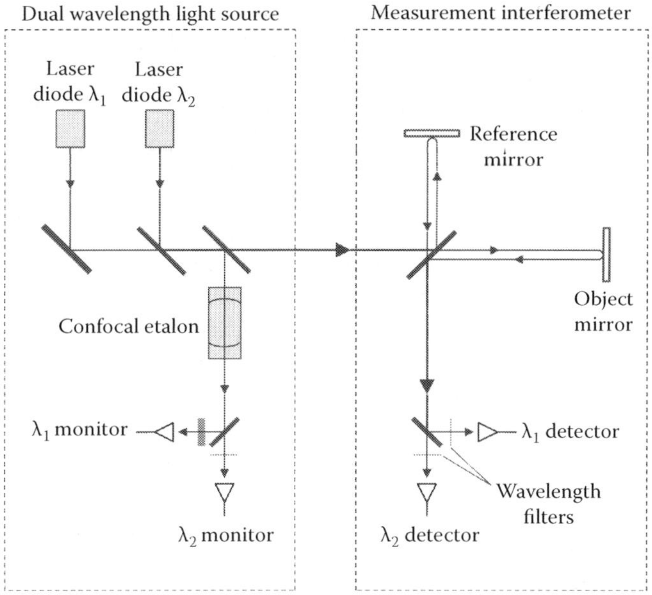

FIGURE 4.48
Simplified illustration of a dual wavelength interferometer based on laser diodes stabilized in wavelength to a common confocal Fabry–Pérot etalon. (From de Groot, P. and Kishner, S., *Appl. Opt.*, 30, 4026, 1991.)

comb spectra from frequency comb lasers[339,340] and to optimize the choice of wavelength for the largest possible unambiguous range.[341]

An alternative approach to multiple wavelengths is a continuously swept wavelength, sometimes referred to as frequency-modulated continuous wave (FMCW) ranging. Just as in multiple-wavelength methods, the principle follows the observation that the interference phase varies linearly with wave number defined in Equation 4.43 at a rate proportional to the distance:

$$\frac{d\phi}{d\sigma} = 2nL \tag{4.47}$$

A linear variation of laser wave number generates an interference signal having a frequency that is also linearly dependent on the distance L, with no limit to the available unambiguous range, provided that the source is sufficiently coherent. Laser diodes have dominated this technique as tunable sources for the past two decades.[342,343] Most often, the available tuning range is insufficient to resolve the absolute distance to within a wavelength, and consequently simple systems for FMCW ranging are usually not capable of the same precision as DMI. High precision on the order of one part in 10^9 of the measured distance is however feasible with advanced sources and sufficient care.[344]

4.8.2 Optical Feedback Interferometry

So far, we have considered interferometer geometries in which splitting the source light into reference and measurement beams followed by recombination at a detector establishes the interference effect. However, it was discovered early in the history of the laser that reflected light directed into the laser cavity would produce wavelength and intensity modulations that could be used directly for distance and velocity measurement.[345] The basic geometry for such systems can be very simple: All that is required is a path for reflected light to enter the laser—a condition that is almost unavoidable in many cases—and a detector for observing the modulations in the laser output in response to the phase of this reflected light.

The optical feedback or *self-mixing* effect is particularly strong in semiconductor laser diodes because of the strong gain medium and weak front-surface reflection of the lasing cavity, which allows even weakly reflected light to influence the behavior of the laser. The amount of feedback need not be high for the effects to be significant—feedback or backscattered light at levels as small as 10^{-9} times the emission intensity is sufficient to measurably alter the power output and frequency of the laser. Figure 4.49 shows a simple system for velocimetry, where the detection is either the oscillation in the laser optical power output or even more simply the variation of the terminal voltage or driving current of the laser itself.[346]

The principles of optical self-mixing in lasers have been extensively analyzed, with most physical models based on the inclusion of the object itself as part of the laser cavity.[347] The additional reflecting surface modulates the threshold gain of the system. A characteristic of self-mixing interferometry is that the signal shape is not sinusoidal but is rather more like a saw tooth, as a result of the laser system phase lock to the external reflection, providing a directional discrimination to the homodyne signal without the need for phase shifting.[348] The phase locking also introduces a frequency shift to the laser output, providing another method of detection that in multimode lasers can be particularly effective.[349,350]

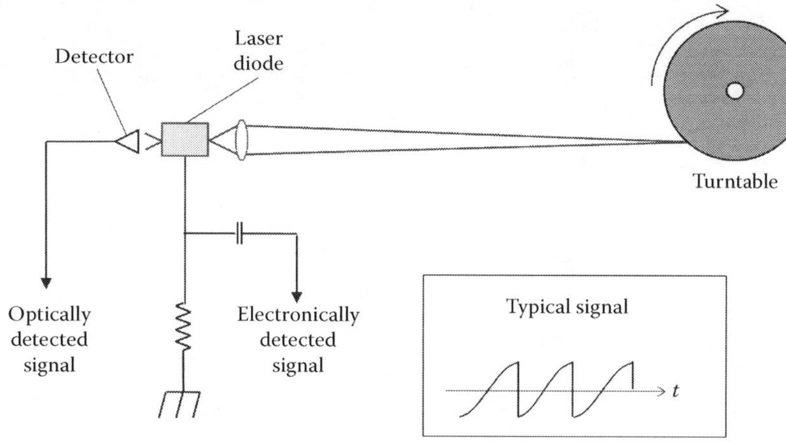

FIGURE 4.49
Self-mixing laser diode interferometer setup to measure the velocity component along the line of sight using the scattered light from the edge of a rotating turntable.

The principle advantage of self-mixing in addition to simplicity may be its very high sensitivity to small amounts of returned light from objects that are not mirrors or other optical elements but may simply be scattering surfaces. Indeed, if the return light intensity is too high, the laser can become unstable and the measurement is no longer useful. Self-mixing has been investigated for a wide range of sensing applications from displacement measurement to absolute distance measurement[351] and has even been proposed for 3D imaging confocal microscopy.[352]

4.8.3 Dispersion Interferometry

DMI systems have achieved a level of performance such that a primary source of uncertainty is the natural variation in the refractive index of the air. Index changes follow environmental fluctuations in temperature and pressure, covering length and time scales ranging from ambient conditions to local turbulence in the beam path. As has been noted in Section 4.6, compensating for ambient conditions involves environmental monitoring and calculation of the refractive index[122,130] or an empirical approach using an interferometer with a fixed measurement path, as in Figure 4.21.[97,353]

Air turbulence within the beam path presents a more challenging problem.[149] A proposed solution is to use two or more wavelengths to detect changes in air density via refractive index dispersion. Dispersion interferometry has the advantage that the compensation can follow even rapid fluctuations in the air path. Figure 4.50 illustrates the basic principle with experimental data acquired by passing a DMI measurement beam comprised of two wavelengths, 633 and 317 nm, through a pressure cell.[354] Pressure is increasing from left to right in the figure as air bleeds into the cell over time. As the air pressure is increased, the optical path nL also increases. The index n_1 for the UV wavelength is higher than the index n_2 for the red wavelength, resulting in different measurements L_1,L_2 for the same physical length. The difference in the measured path length is proportional to the inverse dispersive power:

$$\Gamma = \frac{n_1}{(n_1 - n_2)} \tag{4.48}$$

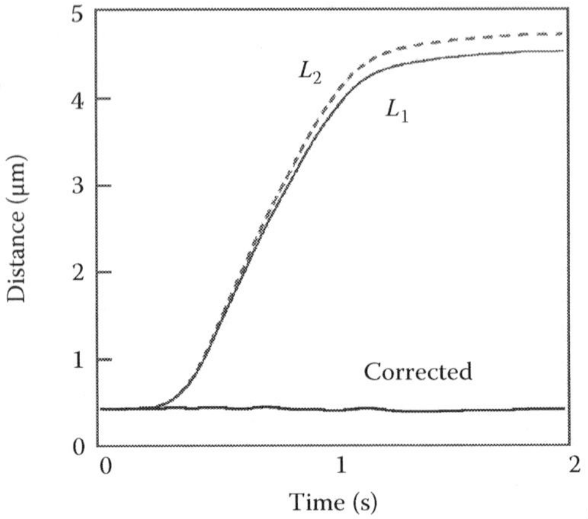

FIGURE 4.50
Atmospheric compensation using optical length measurements at two different wavelengths for an experiment involving a pressure cell. (From Deck, L.L., Dispersion interferometry using a doubled HeNe laser, Zygo Corporation, Middlefield, CT, Unpublished, 1999; de Groot, P. ed., *Optical Metrology*, Encyclopedia of Optics, Wiley-VCH Publishers, Weinheim, Germany, 2004, pp. 2085–2117.)

Note that Γ is an *intrinsic* property of the gas, independent of temperature and pressure. To correct for this effect, one calculates a new length L that is independent of pressure from measurements taken at the two wavelengths, using

$$L = L_1 - (L_1 - L_2)\Gamma \tag{4.49}$$

Dispersion interferometry has most effectively been applied to very long distance geophysical measurements (>1 km) through the atmosphere using interferometry. In an early example from 1972, a He-Ne laser (633 nm) and HeCd laser (440 nm) achieved <1 ppm (parts per million) performance in a portable system.[355]

In the 1990s, considerable effort was invested in adapting two-wavelength dispersion techniques to microlithography stage control, which requires however a much lower uncertainty than a geophysical instrument.[356–358] Example results in Figure 4.51 using a doubled He-Ne laser demonstrate that a standard deviation of 1 nm in a turbulent nitrogen atmosphere is achievable.[354] The cost and complexity of dispersion interferometry has however so far discouraged commercial application of the technique for high-precision applications, with the preferred solution being short-path heterodyne optical encoders wherever practicable.

4.8.4 Optical Fibers and DMI

DMI interferometers are usually physically separated from their light sources and are often in difficult to access areas that complicate beam delivery. There is consequently a strong interest in transmitting the source light through flexible, single-mode optical fibers. Although clearly beneficial, there are several obstacles to a successful fiber-optic delivery system. Firstly, coupling of the light into the fiber demands high precision and optomechanical stability of better than 1 μm for single-mode fibers for visible wavelengths. Next, if the system employs a polarized heterodyne source, the polarization of the two

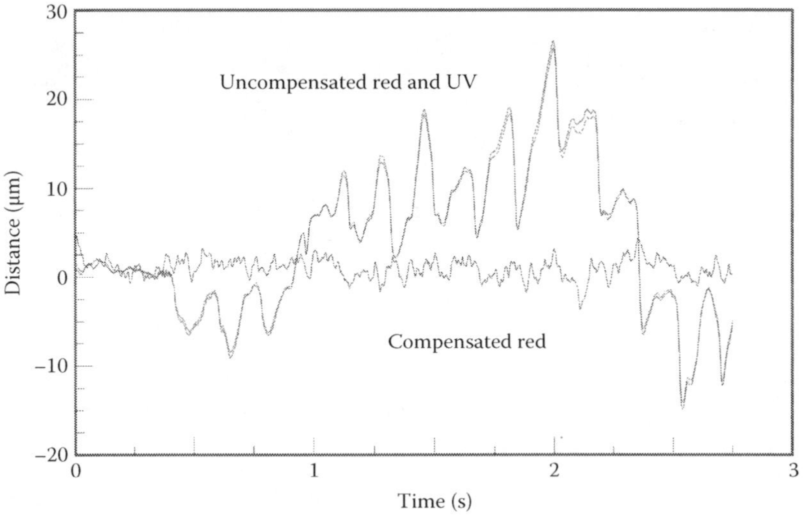

FIGURE 4.51
Dispersion interferometry for correction of air turbulence to the nm level. (From Deck, L.L., Dispersion interferometry using a doubled HeNe laser, Zygo Corporation, Middlefield, CT, Unpublished, 1999.)

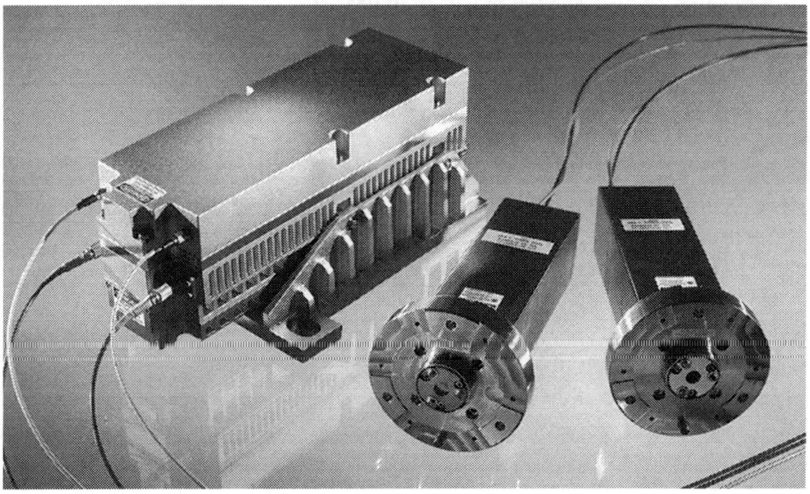

FIGURE 4.52
Fiber-coupled light source with remote frequency shifters. (Photo courtesy of Zygo Corporation, Middlefield, CT.)

frequencies must be carefully preserved,[359] which may require using two fibers in place of one in a separated beam delivery or providing a frequency shifter local to the interferometer itself (Figure 4.52). Finally, an additional detector must be placed at each interferometer of a heterodyne system to monitor the relative phase of the two-frequency source beams, which otherwise would become indeterminate while traveling through fibers.[360] In some proposed systems, this last requirement is satisfied by a push–pull interferometer design such as the one shown in Figure 4.53 in which both of the source beams pass through the measurement and reference paths together.[15,361,362] An advantage of this push–pull approach is that it effectively doubles the resolution of the measurement.

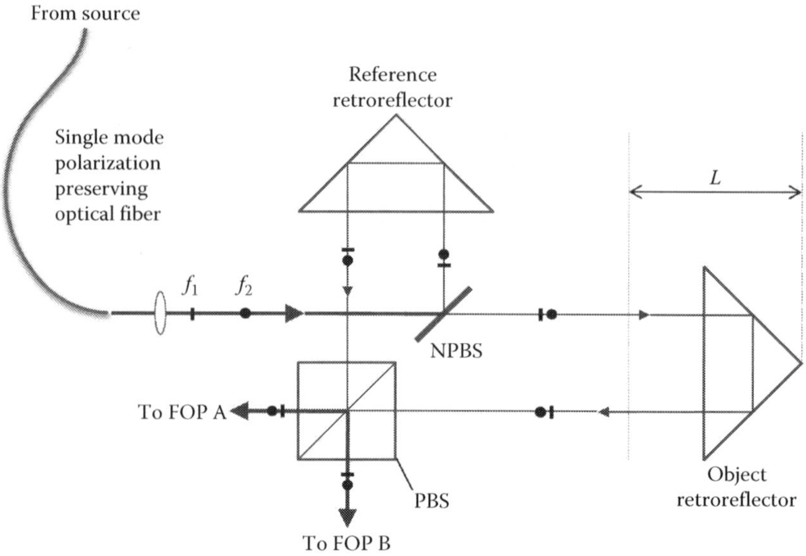

FIGURE 4.53
Fiber-fed interferometer in which beams at both heterodyne frequencies propagate to the object reflector and to the reference retroreflector. (From de Lang, H. and Bouwhuis, G., *Philips Tech. Rev.*, 30, 160, 1969.) A displacement is calculated from the difference in the signals from the two FOPs A and B. (From Bell, J.A., Fiber coupled interferometric displacement sensor, EP 0 793 079, 2003.)

An entirely different use of optical fibers in displacement interferometry relies on precision measurements of the fiber length itself, in an area broadly categorized as fiber sensing. In these instruments, the distance traveled by the light is proportional to a physical or environmental parameter of interest such as strain, temperature, or pressure, now accessible to interferometric measurement by means of a sensing transducer accessed remotely through optical fibers.[363] Common today are optical fiber sensors that use specialized fiber structures such as Bragg gratings.[364]

Fiber-based DMI sensors may be multiplexed through coherence or other mechanisms.[363] This provides the opportunity for a single, perhaps highly complex source and detection system to leverage multiple sensing points cost-effectively. The remote sensors may be entirely passive, that is, without electrical power, and may have multiaxis and absolute positioning capability.

4.8.5 Optical Encoders

Linear and angular displacement can also be measured by optically detecting the lateral motion of a grid or grating pattern. This is the principle behind optical encoders, which have long been a compact and inexpensive alternative to laser DMI systems, particularly in precision engineering and the machine tool industry.[365] As noted in Section 4.7, in recent years, they have become strong candidates for overcoming air turbulence in the most advanced and demanding stage control systems, often relying on 2D XY grids.[366]

Homodyne and heterodyne DMI systems can serve as the basis for optical encoder sensors, as briefly described here with Figure 4.54. Using a modified Michelson interferometer, the reference and measurement beams both diffract from a grating at the Littrow angle.[367,368] The system detects the lateral motion of the grating, with one complete 2π phase cycle for

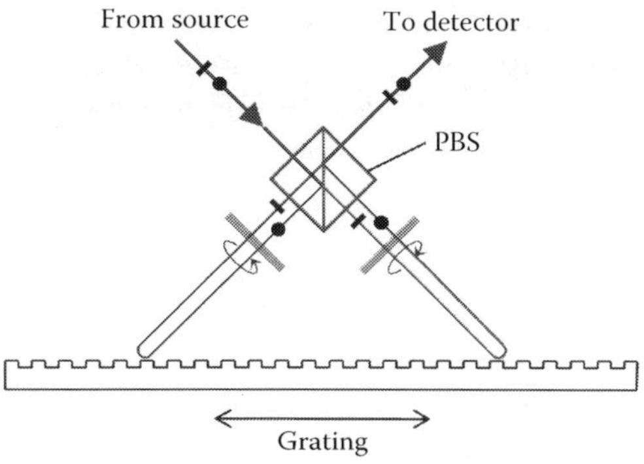

FIGURE 4.54
Optical encoder for detecting lateral motions of a grating using a Michelson interferometer. (From Akiyama, K. and Iwaoka, H., High resolution digital diffraction grating scale encoder, U.S. Patent 4,629,886, 1986.)

each lateral displacement equal to half the grating period. Modern systems target sub-nanometer resolution and can have the benefit of reduced sensitivity to air turbulence.[369] The configuration of Figure 4.54 is also sensitive to grating tilt, because of the wide separation of the points of optical contact for the grating reference and measurement beam, and is useful for monitoring this degree of freedom. Designs have been developed specifically for lateral motions only, with tolerance for tip and tilt, using the same basic double-pass principles as an HSPMI.[370–372]

4.9 Summary: Using Displacement Interferometry

Throughout this survey, consistent themes characterize laser displacement interferometry as an option for precision position monitoring. Chief among this is high resolution (to <1 nm) over wide displacement ranges (>1 m typically, >1 km for geophysical and space applications), low Abbe error, and high-speed data acquisition (Sections 4.1 and 4.2). Applications that rely on these favorable characteristics include microlithography stage position control, machine tool validation, and the calibration of secondary position sensors (Section 4.7). These benefits are offset by the relative high cost of laser interferometry compared to other options, sensitivity to air turbulence, and absence of absolute distance information for the most common commercial systems.

Assuming that the application calls for a displacement interferometer, there are additional choices regarding the configuration and the necessary enabling elements to provide the desired position data. Configuration choices include homodyne or heterodyne method detection (Section 4.3), Zeeman or externally modulated light source (Section 4.4), and the geometry for the interferometer optics (Section 4.5). The choice of interferometer components and system combined with the environmental conditions and the measurement strategy will determine the level of uncertainty (Section 4.6) and dictate modifications or upgrades based on the magnitudes of the various uncertainty contributions.

The advancing requirements for applications continue to drive the advancement of new displacement interferometer systems. Developing solutions (Section 4.8) seek to overcome traditional limitations as well as to push the technology to new performance targets. These developments provide opportunities for innovation as well as new applications for what is arguably the most fundamental and historical application of optical interference in precision engineering.

References

1. G. R. Harrison and G. W. Stroke, Interferometric control of grating ruling with continuous carriage advance, *Journal of the Optical Society of America* **45**, 112–121 (1955).
2. A. A. Michelson, Comparison of the international metre with the wavelength of the light of cadmium, *Astronomy and Astro-Physics* **12**, 556–560 (1893).
3. R. Benoît, Application des phénomènes d'interférence a des déterminations métrologique, *Journal de Physique Théorique et Appliquée* **7**, 57–68 (1898).
4. S. P. Poole and J. H. Dowell, Application of interferometry to the routine measurement of block gauges, in *Optics and Metrology*, P. Mollet, ed. (Pergamon Press, New York, 405–419, 1960).
5. D. C. Barnes and M. J. Puttock, National physics laboratory interferometer, *The Engineer* **196**, 763–766 (1953).
6. D. Malacara, M. Servín, and Z. Malacara, Signal phase detection, in *Interferogram Analysis for Optical Testing* (Marcel Dekker, Inc., New York, pp. 113–168, 1998).
7. E. R. Peck, A new principle in interferometer design, *Journal of the Optical Society of America* **38**, 66 (1948).
8. R. Smythe and R. Moore, Instantaneous phase measuring interferometry, *Optical Engineering* **23**, 361–364 (1984).
9. P. L. M. Heydemann, Determination and correction of quadrature fringe measurement errors in interferometers, *Applied Optics* **20**, 3382–3384 (1981).
10. P. J. de Groot, Homodyne interferometric receiver and calibration method having improved accuracy and functionality, U.S. Patent 5,663,793 (1997).
11. M. J. Downs and K. W. Raine, An unmodulated bi-directional fringe-counting interferometer system for measuring displacement, *Precision Engineering* **1**, 85–88 (1979).
12. A. Dorsey, R. J. Hocken, and M. Horowitz, A low cost laser interferometer system for machine tool applications, *Precision Engineering* **5**, 29–31 (1983).
13. K. Taniguchi, H. Tsuchiya, and M. Toyama, Optical instrument for measuring displacement, U.S. Patent 4,676,645 (1987).
14. S. E. Jones, Unique advanced homodyne laser interferometer system provides a cost-effective and simple position feedback solution for precision motion control applications, in *SPIE Proceedings, Photomask and Next-Generation Lithography Mask Technology XI*, Yokohama, Japan, Vol. 5446, pp. 689–697 (2004).
15. H. de Lang and G. Bouwhuis, Displacement measurements with a laser interferometer, *Philips Technical Review* **30**, 160–165 (1969).
16. J. N. Dukes and G. B. Gordon, A two-hundred-foot yardstick with graduations every micro-inch, *Hewlett-Packard Journal* **21**, 2–8 (1970).
17. G. E. Sommargren, A new laser measurement system for precision metrology, *Precision Engineering* **9**, 179–184 (1987).
18. F. C. Demarest, High-resolution, high-speed, low data age uncertainty, heterodyne displacement measuring interferometer electronics, *Measurement Science and Technology* **9**, 1024–1030 (1998).
19. L. L. Deck, High-performance multi-channel fiber-based absolute distance measuring interferometer system, in *SPIE Proceedings, Instrumentation, Metrology, and Standards for Nanomanufacturing III*, San Diego, CA, Vol. 7405, pp. 74050E-1–74050E-9 (2009).

20. P. de Groot, L. L. Deck, and C. Zanoni, Interferometer system for monitoring an object, U.S. Patent 7,826,064 (2010).
21. A. D. White and J. D. Rigden, Continuous gas maser operation in the visible, *Proceedings IRE* **50**, 1697 (1962).
22. D. A. Shaddock, Space-based gravitational wave detection with LISA, *Classical and Quantum Gravity* **25**, 114012 (2008).
23. D. A. Shaddock, An overview of the laser interferometer space antenna, *Publications of the Astronomical Society of Australia* **26**, 128–132 (2009).
24. T. J. Quinn, Practical realization of the definition of the metre, including recommended radiations of other optical frequency standards (2001), *Metrologia* **40**, 103–133 (2003).
25. S. J. Bennett, Length and displacement measurement by laser interferometry, in *Optical Transducers and Techniques in Engineering Measurement A.R. Lunmoore*, ed. (Applied Science Publishers, London, U.K., pp. 135–159, 1983).
26. The He-Ne Laser, in *Springer Handbook of Lasers and Optics*, F. Träger, ed. (Springer Verlag, Berlin, Germany, pp. 756–757, 2007).
27. J. A. Stone, J. E. Decker, P. Gill, P. Juncar, A. Lewis, G. D. Rovera, and M. Viliesid, Advice from the CCL on the use of unstabilized lasers as standards of wavelength: The helium-neon laser at 633 nm, *Metrologia* **46**, 11–18 (2009).
28. H. Boersch, H. Eichler, and W. Wiesemann, Measurement of length shifts down to 10^{-3} Å with a two-mode laser, *Applied Optics* **9**, 645–648 (1970).
29. M.-S. Kim and S.-W. Kim, Two-longitudinal-mode He-Ne laser for heterodyne interferometers to measure displacement, *Applied Optics* **41**, 5938–5942 (2002).
30. S. Yokoyama, T. Yokoyama, and T. Araki, High-speed subnanometre interferometry using an improved three-mode heterodyne interferometer, *Measurement Science and Technology* **16**, 1841–1847 (2005).
31. G. M. Burgwald and W. P. Kruger, An instant-on laser for length measurement, *Hewlett-Packard Journal* 21, 14–16 (1970).
32. T. Baer, F. V. Kowalski, and J. L. Hall, Frequency stabilization of a 0.633-μm He-Ne longitudinal Zeeman laser, *Applied Optics* **19**, 3173–3177 (1980).
33. R. C. Quenelle and L. J. Wuerz, A new microcomputer-controlled laser dimensional measurement and analysis system, *Hewlett-Packard Journal* **34**, 3–13 (1983).
34. I. Tobias, M. L. Skolnick, R. A. Wallace, and T. G. Polanyi, Derivation of a frequency-sensitive signal from a gas laser in an axial magnetic field, *Applied Physics Letters* **6**, 198–200 (1965).
35. P. Zeeman, On the influence of magnetism on the nature of the light emitted by a substance, *The Astrophysical Journal* 5, 332–347 (1897).
36. R. H. Morris, J. B. Ferguson, and J. S. Warniak, Frequency stabilization of internal mirror He-Ne lasers in a transverse magnetic field, *Applied Optics* **14**, 2808 (1975).
37. H. Takasaki, N. Umeda, and M. Tsukiji, Stabilized transverse Zeeman laser as a new light source for optical measurement, *Applied Optics* **19**, 435–441 (1980).
38. N. Umeda, M. Tsukiji, and H. Takasaki, Stabilized ^{3}He-^{20}Ne transverse Zeeman laser, *Applied Optics* **19**, 442–450 (1980).
39. L. L. Deck and M. L. Holmes, Optical frequency stabilization of a compact heterodyne source for an industrial distance measuring interferometer system, in *Proceedings of the 1999 Annual Meeting of the American Society for Precision Engineering*, Raleigh, NC, Vol. 20, pp. 477–480 (1999).
40. L. L. Deck, Frequency stabilized laser system, U.S. Patent 6,434,176 (2002).
41. R. J. Hocken and H. P. Layer, Lasers for dimensional measurement, *Annals of the CIRP* **28**, 303–306 (1979).
42. *ZMI™ 7705 Laser Head, Specification Sheet* SS-0044 (Zygo Corporation, Middlefield, CT, 2009).
43. M. J. Downs, Optical metrology: The precision measurement of displacement using optical interferometry, in *From instrumentation to nanotechnology*, 1st edn., J. W. Gardner and H. T. Hingle, eds. (CRC Press, Boca Raton, FL, pp. 213–226, 1992).

44. A. D. White and L. Tsufura, Helium-neon lasers, in *Handbook of Laser Technology and Applications Volume II: Laser Design and Laser Systems*, C. E. Webb and J. D. C. Jones, eds. (Institute of Physics, London, U.K., pp. 1399, 2004).
45. Y. Xie and Y.-Z. Wu, Elliptical polarization and nonorthogonality of stabilized Zeeman laser output, *Applied Optics* **28**, 2043–2046 (1989).
46. D. J. Lorier, B. A. W. H. Knarren, S. J. A. G. Cosijns, H. Haitjema, and P. H. J. Schellekens, Laser polarization state measurement in heterodyne interferometry, *CIRP Annals—Manufacturing Technology* **52**, 439–442 (2003).
47. G. E. Sommargren, Apparatus to transform a single frequency, linearly polarized laser beam into a beam with two, orthogonally polarized frequencies, U.S. Patent 4,684,828 (1987).
48. G. E. Sommargren and M. Schaham, Heterodyne interferometer system, U.S. Patent 4,688,940 (1987).
49. P. Dirksen, J. v. d. Werf, and W. Bardoel, Novel two-frequency laser, *Precision Engineering* **17**, 114–116 (1995).
50. R. Balhorn, H. Kunzmann, and F. Lebowsky, Frequency stabilization of internal-mirror Helium-Neon lasers, *Applied Optics* **11**, 742–744 (1972).
51. S. J. Bennett, R. E. Ward, and D. C. Wilson, Comments on: Frequency stabilization of internal mirror He-Ne lasers, *Applied Optics* **12**, 1406–1406 (1973).
52. *ZMI™ 7724 Laser head Specification Sheet*, SS-0081 (Zygo Corporation, Middlefield, CT, 2009).
53. *Optics and Laser Heads for Laser-Interferometer Positioning Systems: Product Overview*, 5964–6190E (Agilent Technologies, Santa Clara, CA, 2009).
54. J. Koning and P. H. J. Schellekens, Wavelength stability of He-Ne lasers used in interferometry: Limitations and traceability, *Annals of the CIRP* **28**, 307–310 (1979).
55. S. N. Lea, W. R. C. Rowley, H. S. Margolis, G. P. Barwood, G. Huang, P. Gill, J.-M. Chartier, and R. S. Windeler, Absolute frequency measurements of 633 nm iodine-stabilized helium–neon lasers, *Metrologia* **40**, 84–88 (2003).
56. *ZMI 7702 Laser Head (P/N 8070–0102-XX) Operating Manual*, OMP-0402H (Zygo Corporation, Middlefield, CT, 2007), p. 9.
57. W.-K. Lee, H. S. Suh, and C.-S. Kang, Vacuum wavelength calibration of frequency-stabilized He-Ne lasers used in commercial laser interferometers, *Optical Engineering* **50**, 054301–054304 (2011).
58. N. Brown, Frequency stabilized lasers: Optical feedback effects, *Applied Optics* **20**, 3711–3714 (1981).
59. N. Bobroff, Recent advances in displacement measuring interferometry, *Measurement Science and Technology* **4**, 907–926 (1993).
60. L. J. Aplet and J. W. Carson, A Faraday effect optical isolator, *Applied Optics* **3**, 544–545 (1964).
61. H. A. Hill and P. de Groot, Apparatus to transform two nonparallel propagating optical beam components into two orthogonally polarized beam components, U.S. Patent 6,236,507 (2001).
62. H. P. Layer, Acoustooptic modulator intensity servo, *Applied Optics* **18**, 2947–2949 (1979).
63. H. A. Hill, Apparatus for generating linearly-orthogonally polarized light beams, U.S. Patent 6,157,660 (2000).
64. P. Gill, Laser interferometry for precision engineering metrology, in *Optical Methods in Engineering Metrology*, D. C. Williams, ed. (Chapman & Hall, New York, pp. 179–211 1993).
65. W. R. C. Rowley, Signal strength in two-beam interferometers with laser illumination, *Journal of Modern Optics* **16**, 159–168 (1969).
66. *ZMI™ 7722/7724 Laser Manual P/N's*: 8070–0257-xx, 8070–0277-xx, OMP-0540C (Zygo Corporation, Middlefield, CT, 2010), pp. 12–14.
67. *ZMI™ 7714 Laser Head Manual P/N's*: 8070–0278-xx, 8070–0279-xx, OMP-0541F (Zygo Corporation, Middlefield, CT, 2011), p. 11.
68. J. Helmcke, Realization of the metre by frequency-stabilized lasers, *Measurement Science and Technology* **14**, 1187–1199 (2003).
69. W. Demtröder, *Laser Spectroscopy: Experimental Techniques,* 4th edn. Vol. 2 (Springer Verlag, Berlin, Germany 2008).

70. G. Hanes and C. Dahlstrom, Iodine hyperfine structure observed in saturated absorption at 633 nm, *Applied Physics Letters* **14**, 362–364 (1969).
71. A. Wallard, Frequency stabilization of the helium-neon laser by saturated absorption in iodine vapour, *Journal of Physics E: Scientific Instruments* **5**, 926–930 (1972).
72. G. Wilson, Modulation broadening of NMR and ESR line shapes, *Journal of Applied Physics* **34**, 3276–3285 (1963).
73. W. Schweitzer Jr, E. Kessler Jr, R. Deslattes, H. Layer, and J. Whetstone, Description, performance, and wavelengths of iodine stabilized lasers, *Applied Optics* **12**, 2927–2938 (1973).
74. W. Tuma and C. van der Hoeven, Helium-neon laser stabilized on iodine: Design and performance, *Applied Optics* **14**, 1896–1897 (1975).
75. V. Dandawate, Frequency stability and reproducibility of iodine stabilised He-Ne laser at 633 nm, *Pramana* **22**, 573–578 (1984).
76. H. P. Layer, A portable iodine stabilized Helium-Neon laser, *IEEE Transactions on Instrumentation and Measurement* **29**, 358–361 (1980).
77. J. Chartier, A. Chartier, J. Labot, and M. Winters, Absolute gravimeters: Status report on the use of iodine-stabilized He-Ne lasers at 633 nm, *Metrologia* **32**, 181–184 (1995).
78. F. Petru, B. Popela, and Z. Vesela, Iodine-stabilized He-Ne Lasers at = 633 nm of a compact construction, *Metrologia* **29**, 301–307 (1992).
79. G. Popescu, J. M. Chartier, and A. Chartier, Iodine stabilized He Ne laser at = 633 nm: Design and international comparison, *Optical Engineering* **35**, 1348–1352 (1996).
80. J. Ishikawa, Portable national length standards designed and constructed using commercially available parts, *Synthesiology-English edition* **2**, 246–257 (2010).
81. T. Yoon, J. Ye, J. Hall, and J. M. Chartier, Absolute frequency measurement of the iodine-stabilized He-Ne laser at 633 nm, *Applied Physics B: Lasers and Optics* **72**, 221–226 (2001).
82. J. Chartier, H. Darnedde, M. Frennberg, J. Henningsen, U. Kärn, L. Pendrill, J. Hu, J. Petersen, O. Poulsen, and P. Ramanujam, Intercomparison of Northern European 127I2-Stabilized He-Ne Lasers at = 633 nm, *Metrologia* **29**, 331–339 (1992).
83. J. Chartier, S. Picard-Fredin, and A. Chartier, International comparison of iodine cells, *Metrologia* **29**, 361–367 (1992).
84. T. J. Quinn, Mise en Pratique of the definition of the Metre (1992), *Metrologia* **30**, 523–541 (1994).
85. J. Chartier and A. Chartier, I2-stabilized 633-nm He-Ne lasers: 25 years of international comparisons, in *SPIE Proceedings, Laser Frequency Stabilization, Standards, Measurement, and Applications*, San Jose, CA, Vol. 4269, pp. 123–133 (2001).
86. A. A. Madej et al., Long-term absolute frequency measurements of 633 nm iodine-stabilized laser standards at NRC and demonstration of high reproducibility of such devices in international frequency measurements, *Metrologia* **41**, 152–160 (2004).
87. *Model 100 Iodine-stabilized He-Ne laser Specification Sheet* (Winters Electro-Optics Inc., Longmont, CO).
88. J. Lawall, J. M. Pedulla, and Y. Le Coq, Ultrastable laser array at 633 nm for real-time dimensional metrology, *Review of Scientific Instruments* **72**, 2879–2888 (2001).
89. E. D. Baird, R. R. Donaldson, and S. R. Patterson, The laser interferometer system for the large optics diamond turning machine, UCRL-ID-134693 (1990).
90. J. Flügge and R. G. Köning, Status of the nanometer comparator at PTB, in *SPIE Proceedings, Recent Developments in Traceable Dimensional Measurements*, Munich, Germany, Vol. 4401, pp. 275–283 (2001).
91. *Model 200 Iodine-stabilized He-Ne laser Specification Sheet* (Winters Electro-Optics Inc., Longmont, CO).
92. D. W. Allan, Statistics of atomic frequency standards, *Proceedings of the IEEE* **54**, 221–230 (1966).
93. D. W. Allan, Time and frequency (time-domain) characterization, estimation, and prediction of precision clocks and oscillators, *IEEE Transactions on Ultrasonics, Ferroelectrics and Frequency Control* **34**, 647–654 (1987).
94. S. J. Bennett, A double-passed Michelson interferometer, *Optics Communications* **4**, 428–430 (1972).

95. *ZMI High Stability Plane Mirror Interferometer (HSPMI) Specification Sheet*, SS-0050 (Zygo Corporation, Middlefield, CT, 2009).
96. C. Zanoni, Differential interferometer arrangements for distance and angle measurements: Principles, advantages and applications, *VDI-Berichte* **749**, 93–106 (1989).
97. P. Schellekens, G. Wilkening, F. Reinboth, M. J. Downs, K. P. Birch, and J. Spronck, Measurements of the refractive index of air using interference refractometers, *Metrologia* **22**, 279–287 (1986).
98. R. R. Baldwin, B. E. Grote, and D. A. Harland, A laser interferometer that measures straightness of travel, *Hewlett-Packard Journal* **25**, 10–20 (1974).
99. C. Evans, M. Holmes, F. Demarest, D. Newton, and A. Stein, Metrology and calibration of a long travel stage, *CIRP Annals—Manufacturing Technology* **54**, 495–498 (2005).
100. H. A. Hill, Polarization preserving optical systems, U.S. Patent 6,198,574 (2001).
101. P. J. de Groot, Interferometer with tilted waveplates for reducing ghost reflections, U.S. Patent 6,163,379 (2000).
102. H. A. Hill and P. de Groot, Single-pass and multi-pass interferometry systems having a dynamic beam-steering assembly for measuring distance, angle, and dispersion, U.S. Patent 6,313,918 (2001).
103. P. J. de Groot, Interferometric apparatus and method for measuring motion along multiple axes, U.S. Patent 6,208,424 (2001).
104. P. J. de Groot and H. A. Hill, Interferometry system and method employing an angular difference in propagation between orthogonally polarized input beam components, U.S. Patent 6,778,280 (2004).
105. Y. Bitou, Polarization mixing error reduction in a two-beam interferometer, *Optical Review* **9**, 227–229 (2002).
106. M. Tanaka, T. Yamagami, and K. Nakayama, Linear interpolation of periodic error in a heterodyne laser interferometer at subnanometer levels (dimension measurement), *IEEE Transactions on Instrumentation and Measurement* **38**, 552–554 (1989).
107. C. M. Wu, S. T. Lin, and J. Fu, Heterodyne interferometer with two spatial-separated polarization beams for nanometrology, *Optical and Quantum Electronics* **34**, 1267–1276 (2002).
108. H. A. Hill, Separated beam multiple degree of freedom interferometer, U.S. Patent 7,057,739 (2006).
109. K.-N. Joo, J. D. Ellis, J. W. Spronck, P. J. M. van Kan, and R. H. M. Schmidt, Simple heterodyne laser interferometer with subnanometer periodic errors, *Optics Letters* **34**, 386–388 (2009).
110. M. Gohlke, T. Schuldt, D. Weise, U. Johann, A. Peters, and C. Braxmaier, A high sensitivity heterodyne interferometer as a possible optical readout for the LISA gravitational reference sensor and its application to technology verification, *Journal of Physics: Conference Series* **154**, 012030 (2009).
111. N. Bennet, *Error Sources* (Zygo Corporation, Middlefield, CT, 2008).
112. J. Haycocks and K. Jackson, Traceable calibration of transfer standards for scanning probe microscopy, *Precision Engineering* **29**, 168–175 (2005).
113. JCGM 100:2008, *Evaluation of Measurement Data—Guide to the Expression of Uncertainty in Measurement*, (JCGM, 2008).
114. H. F. F. Castro and M. Burdekin, Evaluation of the measurement uncertainty of a positional error calibrator based on a laser interferometer, *International Journal of Machine Tools and Manufacture* **45**, 285–291 (2005).
115. H. F. F. Castro, Uncertainty analysis of a laser calibration system for evaluating the positioning accuracy of a numerically controlled axis of coordinate measuring machines and machine tools, *Precision Engineering* **32**, 106–113 (2008).
116. D. Ren, Optical measurements of dimensional instability, PhD dissertation, University of North Carolina, Charlotte, NC (2007).
117. T. Hausotte, B. Percle, E. Manske, R. Füßl, and G. Jäger, Measuring value correction and uncertainty analysis for homodyne interferometers, *Measurement Science and Technology* **22**, 094028 (2011).

118. H. P. Layer and W. T. Estler, Traceability of laser interferometric length measurements, National Bureau of Standards Technical Note 1248, National Bureau of Standards, Washington, DC, 1988.
119. M. L. Eickhoff and J. L. Hall, Real-time precision refractometry: New approaches, *Applied Optics* **36**, 1223–1234 (1997).
120. H. Fang, A. Picard, and P. Juncar, A heterodyne refractometer for air index of refraction and air density measurements, *Review of Scientific Instruments* **73**, 1934–1938 (2002).
121. H. Barrell and J. E. Sears, The refraction and dispersion of air for the visible spectrum, *Philosophical Transactions of the Royal Society A: Mathematical, Physical and Engineering Sciences* **238**, 1–64 (1939).
122. B. Edlén, The refractive index of air, *Metrologia* **2**, 71–80 (1966).
123. J. C. Owens, Optical refractive index of air: Dependence on pressure, temperature and composition, *Applied Optics* **6**, 51–59 (1967).
124. F. E. Jones, Simplified equation for calculating the refractivity of air, *Applied Optics* **19**, 4129–4130 (1980).
125. F. E. Jones, The refractivity of air, *Journal of Research of the National Bureau of Standards* **86**, 27–30 (1981).
126. K. P. Birch and M. J. Downs, An updated edlén equation for the refractive index of air, *Metrologia* **30**, 155–162 (1993).
127. K. P. Birch and M. J. Downs, Correction to the updated edlén equation for the refractive index of air, *Metrologia* **31**, 315–316 (1994).
128. K. P. Birch and M. J. Downs, The results of a comparison between calculated and measured values of the refractive index of air, *Journal of Physics E: Scientific Instruments* **21**, 694–695 (1988).
129. K. P. Birch, F. Reinboth, R. E. Ward, and G. Wilkening, The effect of variations in the refractive index of industrial air upon the uncertainty of precision length measurement, *Metrologia* **30**, 7–14 (1993).
130. G. Bönsch and E. Potulski, Measurement of the refractive index of air and comparison with modified Edlén's formulae, *Metrologia* **35**, 133–139 (1998).
131. P. E. Ciddor, Refractive index of air: New equations for the visible and near infrared, *Applied Optics* **35**, 1566–1573 (1996).
132. P. E. Ciddor, Refractive index of air: 3. The roles of CO_2, H_2O, and refractivity virials, *Applied Optics* **41**, 2292–2298 (2002).
133. P. E. Ciddor, Refractive index of air: 3. The roles of CO_2, H_2O, and refractivity virials: Erratum, *Applied Optics* **41**, 7036 (2002)
134. T. L. Schmitz, C. J. Evans, A. Davies, and W. T. Estler, Displacement uncertainty in interferometric radius measurements, *CIRP Annals—Manufacturing Technology* **51**, 451–454 (2002).
135. J. Hilsenrath, *Tables of Thermal Properties of Gases: Comprising Tables of Thermodynamic and Transport Properties of Air, Argon, Carbon dioxide, Carbon monoxide, Hydrogen, Nitrogen, Oxygen, and Steam*, United States Government Printing Office, Washington, DC, 1955.
136. R. K. Leach, *Fundamental Principles of Engineering Nanometrology* (Elsevier, London, U.K., p. 81, 2010).
137. K. P. Birch and M. J. Downs, Error sources in the determination of the refractive index of air, *Applied Optics* **28**, 825–826 (1989).
138. K. P. Birch, Precise determination of refractometric parameters for atmospheric gases, *Journal of the Optical Society of America A* **8**, 647–651 (1991).
139. W. T. Estler, High-accuracy displacement interferometry in air, *Applied Optics* **24**, 808–815 (1985).
140. M. J. Downs, D. H. Ferriss, and R. E. Ward, Improving the accuracy of the temperature measurement of gases by correction for the response delays in the thermal sensors, *Measurement Science and Technology* **1**, 717–719 (1990).
141. M. J. Downs and K. P. Birch, Bi-directional fringe counting interference refractometer, *Precision Engineering* **5**, 105–110 (1983).

142. K. Birch, M. Downs, and D. Ferriss, Optical path length changes induced in cell windows and solid etalons by evacuation, *Journal of Physics E: Scientific Instruments* **21**, 690–692 (1988).
143. T. Li, Design principles for laser interference refractometers, *Measurement* **16**, 171–176 (1995).
144. M. J. Renkens and P. H. Schellekens, An accurate interference refractometer based on a permanent vacuum chamber—Development and results, *CIRP Annals—Manufacturing Technology* **42**, 581–583 (1993).
145. J. Bryan, Design and construction of an ultraprecision 84 inch diamond turning machine, *Precision Engineering* **1**, 13–17 (1979).
146. M. Sawabe, F. Maeda, Y. Yamaryo, T. Simomura, Y. Saruki, T. Kubo, H. Sakai, and S. Aoyagi, A new vacuum interferometric comparator for calibrating the fine linear encoders and scales, *Precision Engineering* **28**, 320–328 (2004).
147. *ZMI™ Compact Wavelength Compensator Accessory Manual*, OMP-0415F (Zygo Corporation, Middlefield, CT, 2009).
148. G. E. Sommargren, Apparatus for the measurement of the refractive index of a gas, U.S. Patent 4,733,967 (1988).
149. N. Bobroff, Residual errors in laser interferometry from air turbulence and nonlinearity, *Applied Optics* **26**, 2676–2682 (1987).
150. Y. Shibazaki, H. Kohno, and M. Hamatani, An innovative platform for high-throughput high-accuracy lithography using a single wafer stage, in *SPIE Proceedings, Optical Microlithography XXII*, Vol. 7274, pp. 72741I-1–72741I-12 (2009).
151. J. Flügge, C. Weichert, H. Hu, R. Köning, H. Bosse, A. Wiegmann, M. Schulz, C. Elster, and R. D. Geckeler, Interferometry at the PTB nanometer comparator: Design, status and development, in *SPIE Proceedings, Fifth International Symposium on Instrumentation Science and Technology*, Shenyang, China, Vol. 7133, pp. 713346-1–713346-8 (2009).
152. J. B. Wronosky, T. G. Smith, M. J. Craig, B. R. Sturgis, J. R. Darnold, D. K. Werling, M. A. Kincy, D. A. Tichenor, M. E. Williams, and P. M. Bischoff, Wafer and reticle positioning system for the extreme ultraviolet lithography engineering test stand, in *SPIE Proceedings, Emerging Lithographic Technologies IV*, Santa Clara, CA, Vol. 3997, pp. 829–839 (2000).
153. J. di Regolo and U. Ummethala, A novel distance measuring interferometer for rotary-linear stage metrology with shape error removal, in *Proceedings of the 2011 Annual Meeting of the American Society for Precision Engineering*, Denver, CO, Vol. 52, pp. 15–18 (2011).
154. N. Bobroff, Critical alignments in plane mirror interferometry, *Precision Engineering* **15**, 33–38 (1993).
155. *Laser and Optics User's Manual*, 05517–90045 (Agilent Technologies, Santa Clara, CA, pp. 7V1–7V10, 2001).
156. R. R. Baldwin, L. E. Truhe, and D. C. Woodruff, Laser optical components for machine tool and other calibrations, *Hewlett-Packard Journal* **34**, 14–22 (1983).
157. G. Zhang and R. Hocken, Improving the accuracy of angle measurement in machine calibration, *CIRP Annals—Manufacturing Technology* **35**, 369–372 (1986).
158. N. M. Oldham, J. A. Kramar, P. S. Hetrick, and E. C. Teague, Electronic limitations in phase meters for heterodyne interferometry, *Precision Engineering* **15**, 173–179 (1993).
159. *ZMI 4100™ Series Measurement Board Operating Manual*, OMP-0508L (Zygo Corporation, Middlefield, CT, pp. 1–8, 2010).
160. T. Schmitz and H. Kim, Monte Carlo evaluation of periodic error uncertainty, *Precision Engineering* **31**, 251–259 (2007).
161. R. C. Quenelle, Nonlinearity in interferometer measurements, *Hewlett-Packard Journal* **34**, 10 (1983).
162. S. Cosijns, Modeling and verifying non-linearities in heterodyne displacement interferometry, *Precision Engineering* **26**, 448–455 (2002).
163. J. Stone and L. Howard, A simple technique for observing periodic nonlinearities in Michelson interferometers, *Precision Engineering* **22**, 220–232 (1998).
164. S. Olyaee, T. H. Yoon, and S. Hamedi, Jones matrix analysis of frequency mixing error in three-longitudinal-mode laser heterodyne interferometer, *IET Optoelectronics* **3**, 215–224 (2009).

165. P. de Groot, Jones matrix analysis of high-precision displacement measuring interferometers, in *Proceedings of 2nd Topical Meeting on Optoelectronic Distance Measurement and Applications*, ODIMAP II, Pavia, Italy, pp. 9–14 (1999).
166. C. M. Sutton, Non-linearity in length measurement using heterodyne laser Michelson Interferometry, *Journal of Physics E: Scientific Instruments* **20**, 1290–1292 (1987).
167. S. Patterson and J. Beckwith, Reduction of systematic errors in heterodyne interferometric displacement measurement, in *Proceedings of the 8th International Precision Engineering Seminar*, IPES, Compiegne, France, pp. 101–104 (1995).
168. V. G. Badami and S. R. Patterson, A frequency domain method for the measurement of nonlinearity in heterodyne interferometry, *Precision Engineering* **24**, 41–49 (2000).
169. H. A. Hill, Tilted interferometer, U.S. Patent 6,806,962 (2004).
170. H. A. Hill, Systems and methods for quantifying nonlinearities in interferometry systems, U.S. Patent 6,252,668 (2001).
171. C. R. Steinmetz, Sub-micron position measurement and control on precision machine tools with laser interferometry, *Precision Engineering* **12**, 12–24 (1990).
172. J. D. Ellis, Optical metrology techniques for dimensional stability measurements, PhD dissertation, Technische Universiteit Delft, Delft, The Netherlands (2010).
173. J. Stone, S. D. Phillips, and G. A. Mandolfo, Corrections for wavelength variations in precision interferometric displacement measurements, *Journal of Research of the National Institute of Standards and Technology* **101**, 671–674 (1996).
174. E. Abbe, Messapparate für Physiker, *Zeitschrift für Instrumentenkunde* **10**, 446–448 (1890).
175. J. B. Bryan, The Abbé principle revisited: An updated interpretation, *Precision Engineering* **1**, 129–132 (1979).
176. C. D. Craig and J. C. Rose, Simplified derivation of the properties of the optical center of a corner cube, *Applied Optics* **9**, 974–975 (1970).
177. T. A. M. Ruijl and J. van Eijk, A novel ultra precision CMM based on fundamental design principles, in *Proceedings of the ASPE Topical Meeting on Coordinate Measuring Machines*, Vol. 29, pp. 33–38 (2003).
178. I. Widdershoven, R. L. Donker, and H. A. M. Spaan, Realization and calibration of the Isara 400 ultra-precision CMM, *Journal of Physics: Conference Series* **311**, 012002 (2011).
179. A. Davies, C. J. Evans, R. Kestner, and M. Bremer, The NIST X-ray optics CALIBration InteRferometer (XCALIBIR), in *Optical Fabrication and Testing, OSA Technical Digest*, Quebec, Canada, paper OWA5 (2000).
180. J. B. Bryan and D. L. Carter, Design of a new error-corrected co-ordinate measuring machine, *Precision Engineering* **1**, 125–128 (1979).
181. J.-A. Kim, J. W. Kim, C.-S. Kang, J. Jin, and T. B. Eom, An interferometric calibration system for various linear artefacts using active compensation of angular motion errors, *Measurement Science and Technology* **22**, 075304 (2011).
182. V. G. Badami and C. D. Fletcher, Validation of the performance of a high-accuracy compact interferometric sensor, in *Proceedings of the 2009 Annual Meeting of the American Society for Precision Engineering*, Monterey, CA, Vol. 47, pp. 112–115 (2009).
183. G. E. Sommargren, Linear/angular displacement interferometer for wafer stage metrology, in *SPIE Proceedings, Optical/Laser Microlithography II*, Santa Clara, CA, Vol. 1088, pp. 268–272 (1989).
184. *ZMI™ Optics Guide*, OMP-0326W (Zygo Corporation, Middlefield, CT, pp. 2–12, 2010).
185. M. L. Holmes and C. J. Evans, Displacement measuring interferometry measurement uncertainty, in *ASPE Topical Meeting on Uncertainty Analysis in Measurement and Design*, State College, PA, Vol. 33, pp. 89–94 (2004).
186. R. Kendall, A servo guided X–Y–theta stage for electron beam lithography, *Journal of Vacuum Science and Technology B* **9**, 3019–3023 (1991).
187. H. A. Hill, Beam shear reduction in interferometry systems, U.S. Patent 7,495,770 (2009).
188. H. A. Hill, Apparatus and methods for reducing non-cyclic non-linear errors in interferometry, U.S. Patent 7,528,962 (2009).

189. W. T. Estler, Calibration and use of optical straightedges in the metrology of precision machines, *Optical Engineering* **24**, 372–379 (1985).
190. E. W. Ebert, Flatness measurement of mounted stage mirrors, in *SPIE Proceedings, Integrated Circuit Metrology, Inspection, and Process Control III*, Los Angels, CA, Vol. 1087, pp. 415–424 (1989).
191. H. A. Hill and G. Womack, Multi-axis interferometer with procedure and data processing for mirror mapping, U.S. Patent 7,433,049 (2008).
192. S. L. Mielke and F. C. Demarest, Displacement measurement interferometer error correction techniques, in *Proceedings of the ASPE Topical Meeting on Precision Mechanical Design and Mechatronics for Sub-50nm Semiconductor Equipment*, Berkley, CA, Vol. 43, pp. 113–116 (2008).
193. S. Woo, D. Ahn, D. Gweon, S. Lee, and J. Park, Measurement and compensation of bar-mirror flatness and squareness using high precision stage, in *Proceedings of the 2011 Annual Meeting of the American Society for Precision Engineering*, Denver, CO, Vol. 52, pp. 566–569 (2011).
194. C. J. Evans, R. J. Hocken, and W. T. Estler, Self-calibration: Reversal, redundancy, error separation, and 'absolute testing', *CIRP Annals—Manufacturing Technology* **45**, 617–634 (1996).
195. J.-h. Zhang and C.-H. Menq, A linear/angular interferometer capable of measuring large angular motion, *Measurement Science and Technology* **10**, 1247–1253 (1999).
196. Disk drive servo-track writing, Hewlett-Packard, Application Note 325–11,1991.
197. W. Zhou and L. Cai, An angular displacement interferometer based on total internal reflection, *Measurement Science and Technology* **9**, 1647–1652 (1998).
198. M. H. Chiu, J. Y. Lee, and D. C. Su, Complex refractive-index measurement based on Fresnel's equations and the uses of heterodyne interferometry, *Applied Optics* **38**, 4047–4052 (1999).
199. M. H. Chiu, J. Y. Lee, and D. C. Su, Refractive-index measurement based on the effects of total internal reflection and the uses of heterodyne interferometry, *Applied Optics* **36**, 2936–2939 (1997).
200. F. C. Demarest, Data age adjustments, U.S. Patent 6,597,459 (2003).
201. F. C. Demarest, Method and apparatus for providing data age compensation in an interferometer, U.S. Patent 5,767,972 (1998).
202. H. J. Levinson, *Principles of Lithography*, Vol. 146 (SPIE Press, Bellingham, WA, 2005).
203. R. R. Donaldson and S. R. Patterson, Design and construction of a large vertical axis diamond turning machine (LODTM), in *SPIE Proceedings, SPIE 27th Annual Technical Symposium and International Instrument Display*, San Deigo, CA, Vol. 433, pp. 62–68 (1983).
204. J. R. Stoup and T. D. Doiron, Accuracy and versatility of the NIST M48 coordinate measuring machine, in *SPIE Proceedings, Recent Developments in Traceable Dimensional Measurements* Munich, Germany, Vol. 4401, pp. 136–146 (2001).
205. H. Kohno, Y. Shibazaki, J. Ishikawa, J. Kosugi, Y. Iriuchijima, and M. Hamatani, Latest performance of immersion scanner S620D with the Streamlign platform for the double patterning generation, in *SPIE Proceedings, Optical Microlithography XXIII*, San Jose, CA, Vol. 7640, pp. 76401O-1–76401O-12 (2010).
206. T. Sandstrom and P. Ekberg, Mask lithography for display manufacturing, in *SPIE Proceedings, 26th European Mask and Lithography Conference*, Grenoble, France, Vol. 7545, pp. 75450K-1–75450K-18 (2010).
207. G. R. Harrison and J. E. Archer, Interferometric calibration of precision screws and control of ruling engines, *Journal of the Optical Society of America* **41**, 495–503 (1951).
208. H. W. Babcock, Control of a ruling engine by a modulated interferometer, *Applied Optics* **1**, 415–420 (1962).
209. I. R. Bartlett and P. C. Wildy, Diffraction grating ruling engine with piezoelectric drive, *Applied Optics* **14**, 1–3 (1975).
210. R. Wiley, S. Zheleznyak, J. Olson, E. Loewen, and J. Hoose, A nanometer digital interferometric control for a stop start grating ruling engine, in *Proceedings of the 1990 Annual Meeting of the American Society for Precision Engineering*, Rochester, NY, USA, Vol. 2, pp. 131–134 (1990).
211. C. Palmer and E. Loewen, *Diffraction Grating Handbook*, 5th edn. (Newport Corporation, Irvine, CA, 2005).

212. T. Kita and T. Harada, Ruling engine using a piezoelectric device for large and high-groove density gratings, *Applied Optics* **31**, 1399–1406 (1992).
213. P. T. Konkola, Design and analysis of a scanning beam interference lithography system for patterning gratings with nanometer-level distortions, PhD dissertation, Massachusetts Institute of Technology, Cambridge, MA (2003).
214. M. Lercel, Controlling lithography variability: It's no longer a question of nanometers (they are too big), in *Proceedings of the ASPE Topical Meeting on Precision Mechanical Design and Mechatronics for Sub-50nm Semiconductor Equipment*, Berkley, CA, Vol. 43, pp. 3–6 (2008).
215. H. Schwenke, U. Neuschaefer-Rube, T. Pfeifer, and H. Kunzmann, Optical methods for dimensional metrology in production engineering, *CIRP Annals—Manufacturing Technology* **51**, 685–699 (2002).
216. *ZMI™ Column Reference Interferometer (CRI), Vacuum Compatible Specification Sheet*, SS-0067 (Zygo Corporation, Middlefield, CT, 2009).
217. J. Flügge, F. Riehle, and H. Kunzmann, Fundamental length metrology, Colin E. Webb and Julian D. C. Jones, in *Handbook of Laser Technology and Applications: Applications*, Institute of Physics Publishing, Philadelphia, p. 1723 (2004).
218. J. Hocken and B. R. Borchardt, On characterizing measuring machine technology, NBSIR 79–1752 (National Bureau of Standards, Washington, DC, 1979).
219. J. Ye, M. Takac, C. Berglund, G. Owen, and R. Pease, An exact algorithm for self-calibration of two-dimensional precision metrology stages, *Precision Engineering* **20**, 16–32 (1997).
220. D. Musinski, Displacement-measuring interferometers provide precise metrology, *Laser Focus World*, 39, 80–83 (2003).
221. P. Petric, C. Bevis, M. McCord, A. Carroll, A. Brodie, U. Ummethala, L. Grella, A. Cheung, and R. Freed, Reflective electron beam lithography: A maskless ebeam direct write lithography approach using the reflective electron beam lithography concept, *Journal of Vacuum Science and Technology B* **28**, C6C6–C6C13 (2010).
222. K. Suzuki and B. W. Smith, *Microlithography: Science and Technology*, Vol. 126 (CRC, Boca Raton, FL, pp. 383–464, 2007).
223. P. Ekberg, Ultra precision metrology: The key for mask lithography and manufacturing of high definition displays, Licentiate thesis, KTH Royal Institute of Technology, Stockholm, Sweden (2011).
224. W. Wills-Moren, H. Modjarrad, R. Read, and P. McKeown, Some aspects of the design and development of a large high precision CNC diamond turning machine, *CIRP Annals—Manufacturing Technology* **31**, 409–414 (1982).
225. P. B. Leadbeater, M. Clarke, W. J. Wills-Moren, and T. J. Wilson, A unique machine for grinding large, off-axis optical components: the OAGM 2500, *Precision Engineering* **11**, 191–196 (1989).
226. W. J. Wills-Moren and T. Wilson, The design and manufacture of a large CNC grinding machine for off-axis mirror segments, *CIRP Annals—Manufacturing Technology* **38**, 529–532 (1989).
227. J. Klingman, The world's most accurate lathe, *Science and Technology Review* 12–14 (2001).
228. D. H. Youden, Diamond turning achieves nanometer smoothness, *Laser Focus World* **26**, 105–108 (1990).
229. Anon, More diamond turning machines, *Precision Engineering* **2**, 225–227 (1980).
230. P. A. McKeown, K. Carlisle, P. Shore, and R. F. Read, Ultraprecision, high stiffness CNC grinding machines for ductile mode grinding of brittle materials, in *SPIE Proceedings, Infrared Technology and Applications*, San Diego, CA, Vol. 1320, pp. 301–313 (1990).
231. W. J. Wills-Moren, K. Carlisle, P. A. McKeown, and P. Shore, Ductile regime grinding of glass and other brittle materials by the use of ultrastiff machine tools, in *SPIE Proceedings, Advanced Optical Manufacturing and Testing*, San Diego, CA, Vol. 1333, pp. 126–135 (1990).
232. P. Shore, P. Morantz, X. Luo, X. Tonnellier, R. Collins, A. Roberts, R. May-Miller, and R. Read, Big OptiX ultra precision grinding/measuring system, in *Optical Fabrication, Testing, and Metrology II*, SPIE Proceedings, Jena, Germany, Vol. 5965, pp. 59650Q-1–59650Q-8 (2005).
233. M. J. Liao, H. Z. Dai, P. Z. Zhang, and E. Salje, A laser interferometric auto-correcting system of high precision thread grinder, *CIRP Annals—Manufacturing Technology* **29**, 309–312 (1980).

234. J. Otsuka, Precision thread grinding using a laser feedback system, *Precision Engineering* **11**, 89–93 (1989).
235. J. Rohlin, An interferometer for precision angle measurements, *Applied Optics* **2**, 762–763 (1963).
236. H. M. B. Bird, A computer controlled interferometer system for precision relative angle measurements, *Review of Scientific Instruments* **42**, 1513–1520 (1971).
237. G. D. Chapman, Interferometric angular measurement, *Applied Optics* **13**, 1646–1651 (1974).
238. J. R. Pekelsky and L. E. Munro, Bootstrap calibration of an autocollimator, index table and sine bar ensemble for angle metrology, in *SPIE Proceedings, Recent Developments in Traceable Dimensional Measurements III*, San Diego, CA, Vol. 5879, pp. 58790D-1–58790D-17 (2005).
239. *ZMI™ DPMI Accessory Manual*, OMP-0223E (Zygo Corporation, Middlefield, CT, 2002).
240. J. G. Marzolf, Angle measuring interferometer, *Review of Scientific Instruments* **35**, 1212–1215 (1964).
241. P. Shi and E. Stijns, New optical method for measuring small-angle rotations, *Applied Optics* **27**, 4342–4344 (1988).
242. P. Shi, Y. Shi, and E. W. Stijns, New optical method for accurate measurement of large-angle rotations, *Optical Engineering* **31**, 2394–2400 (1992).
243. M. Ikram and G. Hussain, Michelson interferometer for precision angle measurement, *Applied Optics* **38**, 113–120 (1999).
244. T. Eom, D. Chung, and J. Kim, A small angle generator based on a laser angle interferometer, *International Journal of Precision Engineering and Manufacturing* **8**, 20–23 (2007).
245. H.-C. Liou, C.-M. Lin, C.-J. Chen, and L.-C. Chang, Cross calibration for primary angle standards by a precision goniometer with a small angle interferometer, in *Proceedings of the 2006 Annual Meeting of the American Society for Precision Engineering*, Monterey, CA, Vol. 39, pp. 615–618 (2006).
246. M. V. R. K. Murty, Modification of michelson interferometer using only one cube-corner prism, *Journal of the Optical Society of America* **50**, 83–84 (1960).
247. G. Peggs, A. Lewis, and S. Oldfield, Design for a compact high-accuracy CMM, *CIRP Annals—Manufacturing Technology* **48**, 417–420 (1999).
248. I. Schmidt, T. Hausotte, U. Gerhardt, E. Manske, and G. Jäger, Investigations and calculations into decreasing the uncertainty of a nanopositioning and nanomeasuring machine (NPM-Machine), *Measurement Science and Technology* **18**, 482–486 (2007).
249. J. A. Kramar, Nanometre resolution metrology with the molecular measuring machine, *Measurement Science and Technology* **16**, 2121–2128 (2005).
250. I. Misumi, S. Gonda, Q. Huang, T. Keem, T. Kurosawa, A. Fujii, N. Hisata et al., Sub-hundred nanometre pitch measurements using an AFM with differential laser interferometers for designing usable lateral scales, *Measurement Science and Technology* **16**, 2080–2090 (2005).
251. V. Korpelainen, J. Seppä, and A. Lassila, Design and characterization of MIKES metrological atomic force microscope, *Precision Engineering* **34**, 735–744 (2010).
252. C. Werner, P. Rosielle, and M. Steinbuch, Design of a long stroke translation stage for AFM, *International Journal of Machine Tools and Manufacture* **50**, 183–190 (2010).
253. J. S. Beers and W. B. Penzes, The NIST length scale interferometer, *Journal of Research—National Institutes of Standards and Technology* **104**, 225–252 (1999).
254. R. Köning, Characterizing the performance of the PTB line scale interferometer by measuring photoelectric incremental encoders, in *SPIE Proceedings, Recent Developments in Traceable Dimensional Measurements III*, San Diego, CA, Vol. 5879, pp. 587908-1–587908-9 (2005).
255. M. Kajima and K. Minoshima, Picometer calibrator for precision linear encoder using a laser interferometer, in *Quantum Electronics and Laser Science Conference, OSA Technical Digest* (CD), paper JThB128 (2011).
256. D. Thompson and P. McKeown, The design of an ultra-precision CNC measuring machine, *CIRP Annals—Manufacturing Technology* **38**, 501–504 (1989).
257. J. Flügge, Recent activities at PTB nanometer comparator, in *SPIE Proceedings, Recent Developments in Traceable Dimensional Measurements II*, San Diego, CA, Vol. 5190, pp. 391–399 (2003).

258. R. Henselmans, Non-contact measurement machine for freeform optics, PhD dissertation, Eindhoven University of Technology, Eindhoven, the Netherlands (2009).
259. R. E. Henselmans and P. C. J. N. V. Rosielle, Free-form optical surface measuring apparatus and method, U.S. Patent 7,492,468 (2009).
260. L. A. Cacace, An optical distance sensor: Tilt robust differential confocal measurement with mm range and nm uncertainty, PhD dissertation, Eindhoven University of Technology, Eindhoven, the Netherlands (2009).
261. H. Takeuchi, K. Yosizumi, and H. Tsutsumi, Ultrahigh accurate 3-D profilometer using atomic force probe of measuring nanometer, in *Proceedings of the ASPE Topical Meeting Freeform Optics: Design, Fabrication, Metrology, Assembly*, Chapel Hill, NC, Vol. 29, pp. 102–107 (2004).
262. J. R. Cerino, K. L. Lewotsky, R. P. Bourgeois, and T. E. Gordon, High-precision mechanical profilometer for grazing incidence optics, in *SPIE Proceedings, Current Developments in Optical Design and Optical Engineering IV*, San Diego, CA, Vol. 2263, pp. 253–262 (1994).
263. T. E. Gordon, Circumferential and inner diameter metrology for the advanced X-ray astrophysics facility optics, in *SPIE Proceedings, Advanced Optical Manufacturing and Testing*, San Diego, CA, Vol. 1333, pp. 239–247 (1990).
264. M. Neugebauer, F. Lüdicke, D. Bastam, H. Bosse, H. Reimann, and C. Töpperwien, A new comparator for measurement of diameter and form of cylinders, spheres and cubes under clean-room conditions, *Measurement Science and Technology* **8**, 849–856 (1997).
265. J.-A. Kim, J. W. Kim, C.-S. Kang, and T. B. Eom, An interferometric Abbe-type comparator for the calibration of internal and external diameter standards, *Measurement Science and Technology* **21**, 075109 (2010).
266. J. R. Miles, L. E. Munro, and J. R. Pekelsky, A new instrument for the dimensional characterization of piston–cylinder units, *Metrologia* **42**, S220–S223 (2005).
267. J. A. Kramar, R. Dixson, and N. G. Orji, Scanning probe microscope dimensional metrology at NIST, *Measurement Science and Technology* **22**, 024001 (2011).
268. S. Ducourtieux and B. Poyet, Development of a metrological atomic force microscope with minimized Abbe error and differential interferometer-based real-time position control, *Measurement Science and Technology* **22**, 094010 (2011).
269. M. L. Holmes, Analysis and design of a long range scanning stage, PhD dissertation, University of North Carolina, Charlotte, NC (1998).
270. M. Holmes, R. Hocken, and D. Trumper, The long-range scanning stage: A novel platform for scanned-probe microscopy, *Precision Engineering* **24**, 191–209 (2000).
271. S. J. Bennett, Thermal expansion of tungsten carbide gauge blocks, *Metrology and Inspection* 35–37, 1978.
272. M. Okaji, N. Yamada, and H. Moriyama, Ultra-precise thermal expansion measurements of ceramic and steel gauge blocks with an interferometric dilatometer, *Metrologia* **37**, 165–171 (2000).
273. A. Takahashi, Measurement of long-term dimensional stability of glass ceramics using a high-precision line scale calibration system, *International Journal of Automation Technology* **5**, 120–125 (2011).
274. S. R. Patterson, Dimensional stability of superinvar, in *SPIE Proceedings, Dimensional Stability*, San Diego, CA, Vol. 1335, pp. 53–59 (1990).
275. M. J. Dudik, P. G. Halverson, M. B. Levine, M. Marcin, R. D. Peters, and S. Shaklan, Precision cryogenic dilatometer for James Webb space telescope materials testing, in *SPIE Proceedings, Optical Materials and Structures Technologies*, San Diego, CA, Vol. 5179, pp. 155–164 (2003).
276. V. G. Badami and M. Linder, Ultra-high accuracy measurement of the coefficient of thermal expansion for ultra-low expansion materials, in *SPIE Proceedings, Emerging Lithographic Technologies VI*, San Jose, CA, Vol. 4688, pp. 469–480 (2002).
277. M. Okaji and H. Imai, High-resolution multifold path interferometers for dilatometric measurements, *Journal of Physics E: Scientific Instruments* **16**, 1208–1213 (1983).
278. Y. Takeichi, I. Nishiyama, and N. Yamada, High-precision optical heterodyne interferometric dilatometer for determining absolute CTE of EUVL materials, in *SPIE Proceedings, Emerging Lithographic Technologies IX*, San Jose, CA, Vol. 5751, pp. 1069–1076 (2005).

279. Y. Takeichi, I. Nishiyama, and N. Yamada, High-precision (< 1ppb / ° C) optical heterodyne interferometric dilatometer for determining absolute CTE of EUVL materials, in *SPIE Proceedings, Emerging Lithographic Technologies X*, San Jose, CA, Vol. 6151, pp. 61511Z-1–61511Z-8 (2006).
280. S. J. Bennett, An absolute interferometric dilatometer, *Journal of Physics E: Scientific Instruments* **10**, 525–530 (1977).
281. M. Okaji and N. Yamada, Precise, versatile interferometric dilatometer for room-temperature operation: Measurements on some standard reference materials, *High Temperatures—High Pressures* **29**, 89–95 (1997).
282. V. G. Badami and S. R. Patterson, Device for high-accuracy measurement of dimensional changes, U.S. Patent 7,239,397 (2007).
283. V. G. Badami and S. R. Patterson, Optically balanced instrument for high accuracy measurement of dimensional change, U.S. Patent 7,426,039 (2008).
284. D. Ren, K. M. Lawton, and J. A. Miller, A double-pass interferometer for measurement of dimensional changes, *Measurement Science and Technology* **19**, 025303 (2008).
285. S. Patterson, V. Badami, K. Lawton, and H. Tajbakhsh, The dimensional stability of lightly-loaded epoxy joints, in *Proceedings of the 1998 Annual Meeting of the American Society for Precision Engineering*, St.Louis, MO, Vol. 18, pp. 384–386 (1998).
286. K. Lawton, K. Lynn, and D. Ren, The measurement of creep of elgiloy springs with a balanced interferometer, *Precision Engineering* **31**, 325–329 (2007).
287. A. Takahashi, Long-term dimensional stability and longitudinal uniformity of line scales made of glass ceramics, *Measurement Science and Technology* **21**, 105301 (2010).
288. R. Smythe, Asphere interferometry powers precision lens manufacturing, *Laser Focus World* **42**, 93–97 (2006).
289. M. F. Küchel, Interferometric measurement of rotationally symmetric aspheric surfaces, in *SPIE Proceedings, Optical Measurement Systems for Industrial Inspection VI*, Munich, Germany, pp. 738916-1–738916-34 (2009).
290. L. A. Selberg, Radius measurement by interferometry, *Optical Engineering* **31**, 1961–1966 (1992).
291. X. C. de Lega, P. de Groot, and D. Grigg, Dimensional measurement of engineered parts by combining surface profiling with displacement measuring interferometry, in *Fringe 2001: Proceedings of the 4th International Workshop on Automatic Processing of Fringe Patterns*, Berman, Germany, pp. 47–55 (2001).
292. P. de Groot, J. Biegen, J. Clark, X. Colonna de Lega, and D. Grigg, Optical interferometry for measurement of the geometric dimensions of industrial parts, *Applied Optics* **41**, 3853–3860 (2002).
293. A. Olszak and J. Schmit, High-stability white-light interferometry with reference signal for real-time correction of scanning errors, *Optical Engineering* **42**, 54–59 (2003).
294. A. Courteville, R. Wilhelm, M. Delaveau, F. Garcia, and F. de Vecchi, Contact-free on-axis metrology for the fabrication and testing of complex optical systems, in *SPIE Proceedings, Optical Fabrication, Testing, and Metrology II*, Jena, Germany, Vol. 5965, pp. 5965-10–596510-12 (2005).
295. R. D. Deslattes and A. Henins, X-ray to visible wavelength ratios, *Physical Review Letters* **31**, 972–975 (1973).
296. J. Lawall and E. Kessler, Michelson interferometry with 10 pm accuracy, *Review of Scientific Instruments* **71**, 2669–2676 (2000).
297. J. R. Lawall, Fabry-Perot metrology for displacements up to 50 mm, *Journal of the Optical Society of America* **22**, 2786–2798 (2005).
298. M. A. Arain and G. Mueller, Design of the advanced LIGO recycling cavities, *Optics Express* **16**, 10018–10032 (2008).
299. R. R. Baldwin, Machine tool evaluation by laser interferometer, *Hewlett-Packard Journal* **21**, 12–13 (1970).
300. H. Schwenke, W. Knapp, H. Haitjema, A. Weckenmann, R. Schmitt, and F. Delbressine, Geometric error measurement and compensation of machines—An update, *CIRP Annals—Manufacturing Technology* **57**, 660–675 (2008).
301. Calibration of a machine tool, Hewlett-Packard, Application Note 156-4.

302. A. C. Okafor and Y. M. Ertekin, Vertical machining center accuracy characterization using laser interferometer: Part 1. Linear positional errors, *Journal of Materials Processing Technology* **105**, 394–406 (2000).
303. A. C. Okafor and Y. M. Ertekin, Vertical machining center accuracy characterization using laser interferometer: Part 2. Angular errors, *Journal of Materials Processing Technology* **105**, 407–420 (2000).
304. L. J. Wuerz and C. Burns, Dimensional metrology software eases calibration, *Hewlett-Packard Journal* **34**, 4–5 (1983).
305. Calibration of a surface plate, Hewlett-Packard, Application Note 156–2.
306. J. C. Moody, How to calibrate surface plates in the plant, *The Tool Engineer*, 1955, 85–91 (1955).
307. J. C. Ziegert and C. D. Mize, The laser ball bar: A new instrument for machine tool metrology, *Precision Engineering* **16**, 259–267 (1994).
308. J. B. Bryan, A simple method for testing measuring machines and machine tools Part 1: Principles and applications, *Precision Engineering* **4**, 61–69 (1982).
309. N. Srinivasa, J. C. Ziegert, and C. D. Mize, Spindle thermal drift measurement using the laser ball bar, *Precision Engineering* **18**, 118–128 (1996).
310. H. F. F. Castro, A method for evaluating spindle rotation errors of machine tools using a laser interferometer, *Measurement* **41**, 526–537 (2008).
311. P. E. Klingsporn, Use of a laser interferometric displacement-measuring system for noncontact positioning of a sphere on a rotation axis through its center and for measuring the spherical contour, *Applied Optics* **18**, 2881–2890 (1979).
312. W. Barkman, A non-contact laser interferometer sweep gauge, *Precision Engineering* **2**, 9–12 (1980).
313. V. Lee, Uncertainty of dimensional measurements obtained from self-initialized instruments, PhD dissertation, Clemson University, Clemson, SC (2010).
314. P. Freeman, Complete, practical, and rapid calibration of multi-axis machine tools using a laser tracker, *Journal of the CMSC* **2**, 18–24 (2007).
315. H. Schwenke, R. Schmitt, P. Jatzkowski, and C. Warmann, On-the-fly calibration of linear and rotary axes of machine tools and CMMs using a tracking interferometer, *CIRP Annals—Manufacturing Technology* **58**, 477–480 (2009).
316. K. Wendt, H. Schwenke, W. Bosemann, and M. Dauke, Inspection of large CMMs by sequential multilateration using a single laser tracker, in *Sixth International Conference and Exhibition on Laser Metrology, CMM and Machine Tool Performance (LAMDAMAP 2003)*, Huddersfield, UK, pp. 121–130 (2003).
317. H. Schwenke, M. Franke, J. Hannaford, and H. Kunzmann, Error mapping of CMMs and machine tools by a single tracking interferometer, *CIRP Annals—Manufacturing Technology* **54**, 475–478 (2005).
318. W. T. Estler, K. L. Edmundson, G. N. Peggs, and D. H. Parker, Large-scale metrology—An update, *CIRP Annals—Manufacturing Technology* **51**, 587–609 (2002).
319. G. Peggs, P. G. Maropoulos, E. Hughes, A. Forbes, S. Robson, M. Ziebart, and B. Muralikrishnan, Recent developments in large-scale dimensional metrology, *Proceedings of the Institution of Mechanical Engineers, Part B: Journal of Engineering Manufacture* **223**, 571–595 (2009).
320. K. Lau, R. Hocken, and W. Haight, Automatic laser tracking interferometer system for robot metrology, *Precision Engineering* **8**, 3–8 (1986).
321. B. Muralikrishnan, D. Sawyer, C. Blackburn, S. Phillips, B. Borchardt, and W. Estler, Performance evaluation of laser trackers, in *Performance Metrics for Intelligent Systems Workshop, PerMIS'08*, Gaithersburg, MD, pp. 149–155 (2008).
322. S. J. Bennett, The NPL 50-meter laser interferometer for the verification of geodetic tapes, *Survey Review* **22**, 270–275 (1974).
323. T. Eom and J. Kim, Displacement measuring sensor calibration using nonlinearity free laser interferometer, in *Proceedings, XVII IMEKO World Congress*, Dubrovnik, Croatia, pp. 1911–1914 (2003).

324. H. Haitjema, Dynamic probe calibration in the µm region with nanometric accuracy, *Precision Engineering* **19**, 98–104 (1996).
325. H. Haitjema and G. J. Kotte, Dynamic probe calibration up to 10 kHz using laser interferometry, *Measurement* **21**, 107–111 (1997).
326. T. M. Niebauer, J. E. Faller, H. M. Godwin, J. L. Hall, and R. L. Barger, Frequency stability measurements on polarization-stabilized He-Ne lasers, *Applied Optics* **27**, 1285–1289 (1988).
327. J. M. Chartier, J. Labot, G. Sasagawa, T. M. Niebauer, and W. Hollander, A portable iodine stabilized He-Ne laser and its use in an absolute gravimeter, *IEEE Transactions on Instrumentation and Measurement*, **42**, 420–422 (1993).
328. C. R. Tilford, A fringe counting laser interferometer manometer, *Review of Scientific Instruments* **44**, 180–182 (1973).
329. E. R. Harrison, D. J. Hatt, D. B. Prowse, and J. Wilbur-Ham, A new interferometric manometer, *Metrologia* **12**, 115–122 (1976).
330. F. Alasia, A. Capelli, G. Cignolo, and M. Sardi, Performance of reflector-carrying floats in mercury manometers, *Vacuum* **46**, 753–756 (1995).
331. T. Bosch and M. Lescure, *Selected Papers on Laser Distance Measurement*, SPIE Milestone Series, Vol. MS 115 (SPIE Press, Bellingham, WA, 1995).
332. W. R. Babbitt, J. A. Bell, B. A. Capron, P. J. de Groot, R. L. Hagman, J. A. McGarvey, W. D. Sherman, and P. F. Sjoholm, Method and apparatus for measuring distance to a target, U.S. Patent 5,589,928 (1996).
333. A. Biernat and G. Kompa, Powerful picosecond laser pulses enabling high-resolution pulsed laser radar, *Journal of Optics* **29**, 225–228 (1998).
334. C. R. Tilford, Analytical procedure for determining lengths from fractional fringes, *Applied Optics* **16**, 1857–1860 (1977).
335. G. L. Bourdet and A. G. Orszag, Absolute distance measurements by CO_2 laser multiwavelength interferometry, *Applied Optics* **18**, 225–227 (1979).
336. P. de Groot and S. Kishner, Synthetic wavelength stabilization for two-color laser-diode interferometry, *Applied Optics* **30**, 4026–4033 (1991).
337. K.-H. Bechstein and W. Fuchs, Absolute interferometric distance measurements applying a variable synthetic wavelength, *Journal of Optics* **29**, 179–182 (1998).
338. P. de Groot, Three-color laser-diode interferometer, *Applied Optics* **30**, 3612–3616 (1991).
339. P. Balling, P. Křen, P. Mašika, and S. A. van den Berg, Femtosecond frequency comb based distance measurement in air, *Optics Express* **17**, 9300–9313 (2009).
340. J. Jin, J. W. Kim, C.-S. Kang, J.-A. Kim, and T. B. Eom, Thickness and refractive index measurement of a silicon wafer based on an optical comb, *Optics Express* **18**, 18339–18346 (2010).
341. K. Falaggis, D. P. Towers, and C. E. Towers, A hybrid technique for ultra-high dynamic range interferometry, in *SPIE Proceedings, Interferometry XIV: Techniques and Analysis*, San Diego, CA, Vol. 7063, pp. 70630X-1–70630X-8 (2008).
342. H. Kikuta, K. Iwata, and R. Nagata, Distance measurement by the wavelength shift of laser diode light, *Applied Optics* **25**, 2976–2980 (1986).
343. A. J. d. Boef, Interferometric laser rangefinder using a frequency modulated diode laser, *Applied Optics* **26**, 4545–4550 (1987).
344. Z. W. Barber, W. R. Babbitt, B. Kaylor, R. R. Reibel, and P. A. Roos, Accuracy of active chirp linearization for broadband frequency modulated continuous wave radar, *Applied Optics* **49**, 213–219 (2010).
345. M. J. Rudd, A laser doppler velocimeter employing the laser as a mixer-oscillator, *Journal of Physics E: Scientific Instruments* **1**, 723–726 (1968).
346. S. Shinohara, A. Mochizuki, H. Yoshida, and M. Sumi, Laser doppler velocimeter using the self-mixing effect of a semiconductor laser diode, *Applied Optics* **25**, 1417–1419 (1986).
347. P. J. de Groot, G. M. Gallatin, and S. H. Macomber, Ranging and velocimetry signal generation in a backscatter-modulated laser diode, *Applied Optics* **27**, 4475–4480 (1988).
348. E. T. Shimizu, Directional discrimination in the self-mixing type laser Doppler velocimeter, *Applied Optics* **26**, 4541–4544 (1987).

349. D. E. T. F. Ashby and D. F. Jephcott, Measurement of plasma density using a gas laser as an infrared interferometer, *Applied Physics Letters* **3**, 13–16 (1963).
350. P. de Groot, Range-dependent optical feedback effects on the multimode spectrum of laser diodes, *Journal of Modern Optics* **37**, 1199–1214 (1990).
351. T. Bosch, N. l. Servagent, and S. Donati, Optical feedback interferometry for sensing application, *Optical Engineering* **40**, 20–27 (2001).
352. A. Bearden, M. P. O'Neill, L. C. Osborne, and T. L. Wong, Imaging and vibrational analysis with laser-feedback interferometry, *Optics Letters* **18**, 238–240 (1993).
353. H. Haitjema, Achieving traceability and sub-nanometer uncertainty using interferometric techniques, *Measurement Science and Technology* **19**, 084002 4 (2008).
354. L. L. Deck, Dispersion interferometry using a doubled HeNe laser (Zygo Corporation, Middlefield, CT, Unpublished, 1999).
355. K. B. Earnshaw and E. N. Hernandez, Two-laser optical distance-measuring instrument that corrects for the atmospheric index of refraction, *Applied Optics* **11**, 749–754 (1972).
356. A. Ishida, Two wavelength displacement-measuring interferometer using second-harmonic light to eliminate air-turbulence-induced errors, *Japanese Journal of Applied Physics* **28**, L473–L475 (1989).
357. P. de Groot and H. A. Hill, Superheterodyne interferometer and method for compensating the refractive index of air using electronic frequency multiplication, U.S. Patent 5,838,485 (1998).
358. H. A. Hill, Apparatus and methods for measuring intrinsic optical properties of a gas, U.S. Patent 6,124,931 (2000).
359. B. Knarren, S. Cosijns, H. Haitjema, and P. Schellekens, Validation of a single fibre-fed heterodyne laser interferometer with nanometre uncertainty, *Precision Engineering* **29**, 229–236 (2005).
360. R. J. Chaney, Laser interferometer for measuring distance using a frequency difference between two laser beams, U.S. Patent 5,274,436 (1993).
361. J. A. Bell, B. A. Capron, C. R. Pond, T. S. Breidenbach, and D. A. Leep, Fiber coupled interferometric displacement sensor, EP 0 793 079 (2003).
362. K.-N. Joo, J. D. Ellis, E. S. Buice, J. W. Spronck, and R. H. M. Schmidt, High resolution heterodyne interferometer without detectable periodic nonlinearity, *Optics Express* **18**, 1159–1165 (2010).
363. A. D. Kersey, Interferometric optical fiber sensors for absolute measurement of displacement and strain, in *SPIE Proceedings, Fiber Optic Sensors: Engineering and Applications*, the Hague, Province of South Holland, the Netherlands, Vol. 1511, pp. 40–50 (1991).
364. A. Othonos and K. Kyriacos, *Fiber Bragg Gratings* (Artech House, Norwood, MA, 1999).
365. A. H. Slocum, *Precision Machine Design* (Prentice Hall, Englewood Cliffs, NJ, 1992).
366. J. Gargas, R. Dorval, D. Mansur, H. Tran, D. Carlson, and M. Hercher, A versatile XY stage with a flexural six-degree-of-freedom fine positioner, in *Proceedings of the 1995 Annual Meeting of the American Society for Precision Engineering*, Austin, TX, Vol. 12, pp. 203–206 (1995).
367. K. Akiyama and H. Iwaoka, High resolution digital diffraction grating scale encoder, U.S. Patent 4,629,886 (1986).
368. C.-F. Kao, S.-H. Lu, H.-M. Shen, and K.-C. Fan, Diffractive laser encoder with a grating in littrow configuration, *Japanese Journal of Applied Physics* **47**, 1833–1837 (2008).
369. C.-C. Wu, C.-C. Hsu, J.-Y. Lee, H.-Y. Chen, and C.-L. Dai, Optical heterodyne laser encoder with sub-nanometer resolution, *Measurement Science and Technology* **19**, 045305 (2008).
370. W.-W. Chiang and C.-K. Lee, Wavefront reconstruction optics for use in a disk drive position measurement system, U.S. Patent 5,442,172 (1995).
371. N. Nishioki and T. Itabashi, Grating interference type displacement meter apparatus, U.S. Patent 5,035,507 (1991).
372. J. William R Trutna, G. Owen, A. B. Ray, J. Prince, E. S. Johnstone, M. Zhu, and L. S. Cutler, Littrow interferometer, U.S. Patent 7,440,113 (2008).
373. M. Okaji and H. Imai, Precise and versatile systems for dilatometric measurement of solid materials, *Precision Engineering* **7**, 206–210 (1985).

374. P. J. de Groot, X. C. de Lega, and D. A. Grigg, Step height measurements using a combination of a laser displacement gage and a broadband interferometric surface profiler, in *SPIE Proceedings, Interferometry XI: Applications*, Seattle, WA, Vol. 4778, pp. 127–130 (2002).
375. R. E. Spero and S. E. Whitcomb, The laser interferometer gravitational-wave observatory (LIGO), *Optics and Photonics News* **6**, 35–39 (1995).
376. P. de Groot, ed., *Optical Metrology*, Encyclopedia of Optics (Wiley-VCH Publishers, Weinheim, Germany, 2004), pp. 2085–2117.

5

Metrology of Large Parts

H. Philip Stahl

CONTENTS

5.1 Introduction

As discussed in Chapter 1 of this book, there are many different methods to measure a part using optical technology. Chapter 2 discussed the use of machine vision to measure macroscopic features such as length and position, which was extended to the use of interferometry as a linear measurement tool in Chapter 3, and laser or other trackers to find the relation of key points on large parts in Chapter 4. This chapter looks at measuring large parts to optical tolerances in the submicron range using interferometry, ranging, and optical tools discussed in the previous chapters. The purpose of this chapter is not to discuss specific metrology tools (such as interferometers or gages) but to describe a systems engineering approach to testing large parts. Issues such as material warpage and temperature drifts that may be insignificant when measuring a part to micron levels under a microscope, as will be discussed in later chapters, can prove to be very important when making the same measurement over a larger part.

In this chapter, we will define a set of guiding principles for successfully overcoming these challenges and illustrate the application of these principles with real-world examples. While these examples are drawn from specific large optical testing applications, they inform the problems associated with testing any large part to optical tolerances. Manufacturing today relies on micrometer-level part performance. Fields such as energy and transportation are demanding higher tolerances to provide increased efficiencies and fuel savings. By looking at how the optics industry approaches submicrometer metrology, one can gain a better understanding of the metrology challenges for any larger part specified to micrometer tolerances.

Testing large parts, whether optical components or precision structures, to optical tolerances is just like testing small parts, only harder. Identical with what one does for small parts, a metrologist tests large parts and optics in particular to quantify their mechanical properties (e.g., dimensions, mass), their optical prescription or design (i.e., radius of curvature, conic constant, vertex location, size), and their full part shape. Just as with small parts, a metrologist accomplishes these tests using distance measuring instruments such as tape measures, inside micrometers, CMMs, and distance measuring interferometers; angle measuring instruments such as theodolites and autocollimators; and surface measuring instruments including interferometers, stylus profilers, interference microscopes, photogrammetric cameras, or other tools. However, while the methodology may be similar, it is more difficult to test a large object for the simple reason that most metrologists do not have the necessary intuition. The skills used to test small parts or optics in a laboratory do not extrapolate to testing large parts in an industrial setting any more than a backyard gardener might successfully operate a farm.

But first, what is a large part? A simple definition might be the part's size or diameter. For optics and diffuse surface parts alike, the driving constraint is ability to illuminate the part's surface. For reflective convex mirrors, large is typically anything greater than 1 m. But, for refractive optics, flats, or convex mirrors, large is typically greater than 0.5 m. While a size definition is simple, it may be less than universal. A more nuanced definition might be that a large part is any component which cannot be easily tested in a standard laboratory environment, on a standard vibration isolated table using standard laboratory infrastructure. A microswitch or a precision lens might be easily measured to nanometer levels under a microscope in a lab, but a power turbine spline or a larger telescope mirror will not fit under that microscope and may not even fit on the table.

5.2 Metrology of Large Parts

The challenges of testing large parts are multiple, and they typically involve one or more of the following: infrastructure, gravity sag, stability (mechanical/thermal) and vibration, atmospheric turbulence or stratification, measurement precision, and spatial sampling. But these challenges can be overcome by good engineering practice and by following a structured systems engineering approach. No matter how small or how large your testing or metrology task is, the following simple guiding principles will insure success:

1. Fully understand the task
2. Develop an error budget
3. Continuous metrology coverage
4. Know where you are
5. "Test like you fly"
6. Independent cross-checks
7. Understand all anomalies

These rules have been derived from over 30 years of lessons learned from both failures and successes. As a validation of these rules, they have been applied with great success to

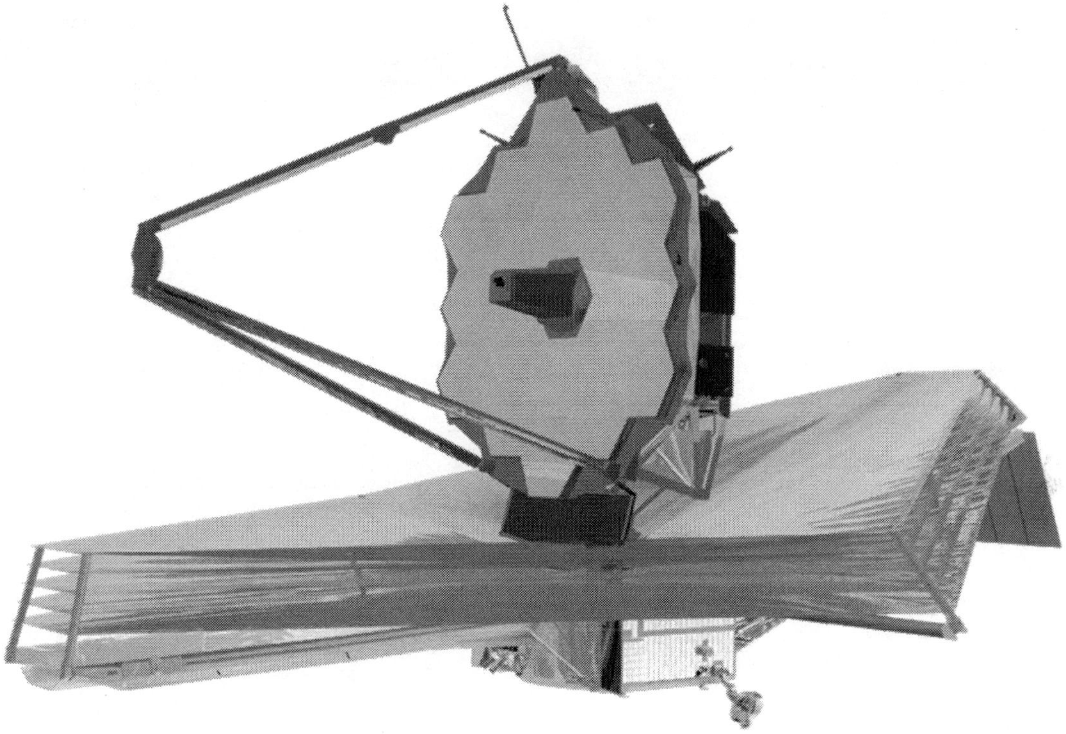

FIGURE 5.1
JWSTs 6.5 m primary mirror consists of eighteen 1.5 m segments.

the in-process optical testing and final specification compliance testing of the James Webb Space Telescope (JWST) Optical Telescope Element (OTE) mirrors (Figure 5.1) [1,2].

5.2.1 Fully Understand the Task

The first step to insure success is to make sure that you fully understand your task. Who is your customer? What parameters do you need to quantify and to what level of uncertainty must you know their value? Do you have the tools and infrastructure to perform the task? And, who is your manufacturing interface?

Before accepting any testing task, study your customer's requirements and understand how they relate to the final system application. Then develop a preliminary metrology plan for how you will quantify each required parameter. This metrology plan should identify the test method to quantify each parameter, the tools and infrastructure required to execute the test, and a preliminary estimate of the test uncertainty. We will explore test uncertainty further in the next section. Summarize all requirements and how they will be quantified into a simple table which can be shared with your customer and your manufacturing methods engineer. Make sure that your customer agrees that what you will quantify satisfies their requirements and the manufacturing methods engineer agrees that they can make the part based upon the data you will be providing. Figure 5.2 shows the meterology plan comphance table which summarizes the final cryogenic temperature requirements for each JWST primary mirror segment assembly (PMSA).

Parameter	Spec	Tol	Units	Verification	Validation
Clear aperture (edge specification)	1.4776 (5)	Min (Max)	mm^2 (mm)	Measure edges at ambient using Tinsley HS Interferometer	Measure area at cryo using XRCF CoC Interferometer
Scratch-dig	80–50	Max		Ambient Visual Inspection	Independent Visual
Conic constant	−0.99666	±0.0005		Measured at cryo and defined by null geometry for XRCF CGH CoC test	Ambient test at Tinsley, compare CGH CoC test with auto-collimation test
Radius of curvature	*	±0.15	mm	Set at XRCF using ADM	ROCO Comparison
Prescription alignment error					
Decenter	*	≤0.35	mm	Cryogenic test at XRCF, defined by residual wavefront error relative to CGH CoC test and fiducial alignment	Ambient test at Tinsley, compare CGH CoC test with auto-collimation test
Clocking	0	≤0.35	mrad		
Piston	N/A			Ambient CMM measurement at AXSYS	Ambient CMM measurement at Tinsley
Tilt	N/A				
Total surface figure error:					
Low/mid frequency	20	Max	nm rms	Cryo-Test at XRCF	Cryo-Test at JSC
High frequency	7	Max	nm rms		
Surface roughness	4	Max	nm rms	Ambient Chapman measurement at Tinsley	NONE

FIGURE 5.2
The JWST PMSA Compliance Table lists the final cryogenic optical performance requirements, the test used to verify that each requirement is met, and the validation test used to cross-check each requirement. (From Stahl, H.P. et al., *SPIE Proc.*, 7790, 779002, 2010, DOI: 10.1117/12.862234.)

Developing a metrology plan for large parts is complicated by the scale of the required infrastructure. For example, while one can easily transport an 8 cm mirror, an 8 m class mirror with a 16,000–20,000 kg mass requires special transport, lifting and handling fixtures, as well as metrology mounts. But, in practice, any part which cannot be safely lifted by two persons also requires special fixtures and should be considered a large part. Safety applies both to the technicians doing the lifting and to the part being lifted. Sometimes, the value of a part is such that it requires special lifting and handling equipment regardless of its size. Furthermore, infrastructure is more than just lifting and handling fixtures. It includes industrial scale work spaces with appropriate temperature, humidity, and cleanliness controls; computer CMMs and test towers; and grinding and polishing machines.

Figure 5.3 shows an illustration of the Itek Autocollimation Test Facility used for the Keck Telescope's 1.8 m mirror segments. The mirror segments radius of curvature was 24 m for a total air path of 48 m. The distance from the mirror under test to the fold flat was approximately 12 m [3]. Figure 5.4 shows the Steward Observatory Mirror Lab (SOML) test

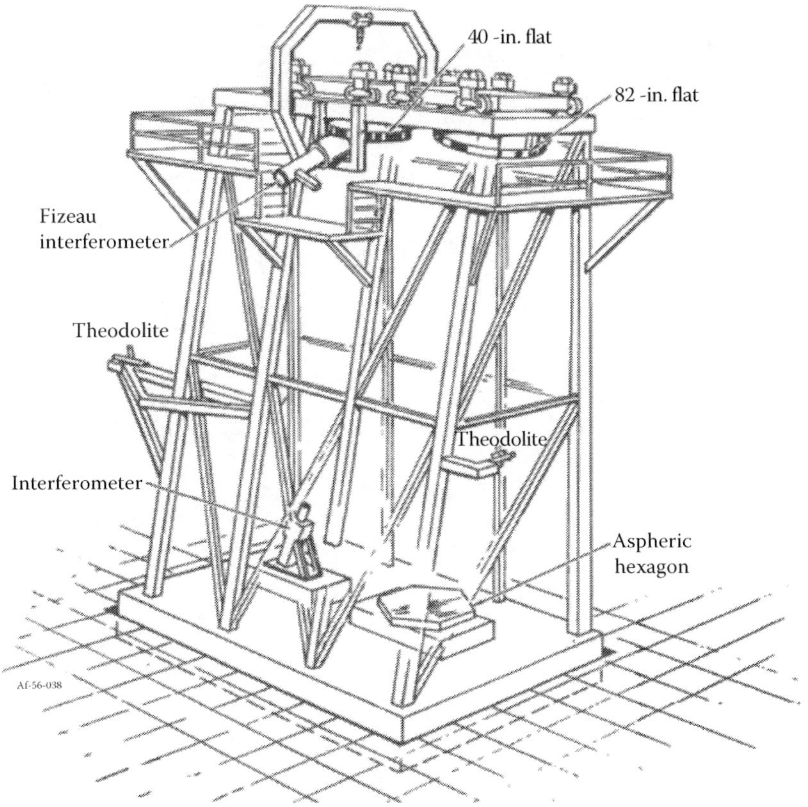

FIGURE 5.3
Itek Autocollimation Test Facility. Each Keck segment was tested in over a 48 m air path. (Figure courtesy of Itek Optical Systems, Lexington, MA; From Stahl, H.P., *Photonics Spectra*, 12, 105, 1989.)

tower which stands 24 m tall and has a mass of 400 tons [4]. Finally, grinding and polishing equipment is important because their capabilities drive metrology requirements such as spatial sampling, test wavelength, and measurement precision.

5.2.2 Develop an Error Budget

The second and most important step is to develop an error budget for every specification and its tolerance. An error budget has multiple functions. It is necessary to convince your customer that you can actually measure the required parameters to the required tolerances. It defines which test conditions have the greatest impact on test uncertainty, and it provides a tool for monitoring the test process. An error budget predicts test accuracy and reproducibility (not repeatability) of the metrology tools. If the variability in the test data of any element of the error budget exceeds its prediction, then you must stop and understand why. Finally, all elements of the error budget must be certified by absolute calibration and verified by independent test. Figure 5.5 shows the JWST PMSA high-level error budget for each of its major requirements.

Mathematically, one constructs an error budget by performing a propagation of error analysis. First, write down the equation which calculates the specification value. Then take

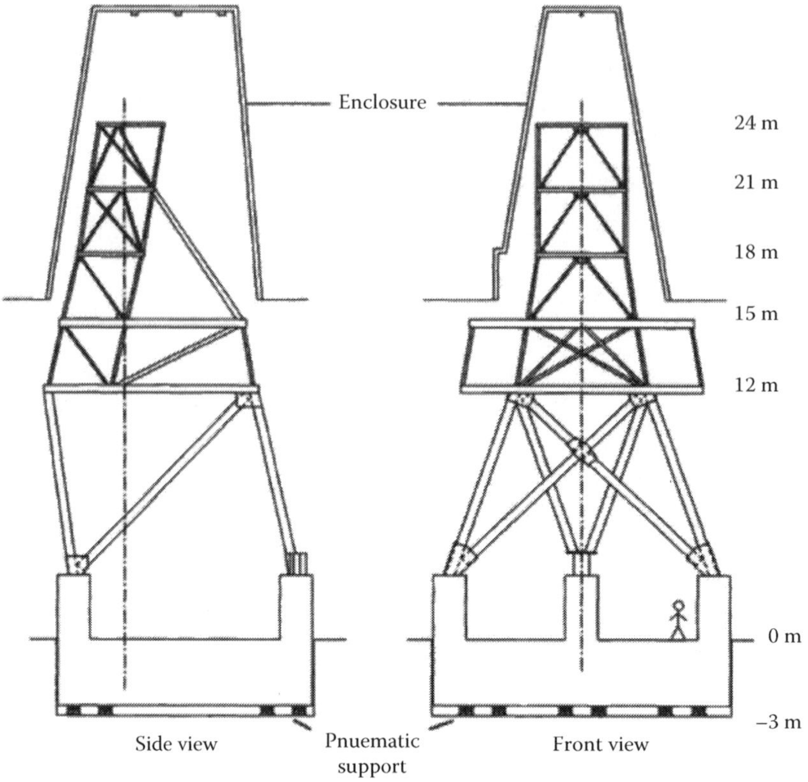

FIGURE 5.4
SOML test tower. Entire 400 ton concrete and steel structure is supported by 40 air-filled isolators. (Drawing by E. Anderson; From Burge, J.H. et al., *SPIE Proc.*, 2199, 658, 1994.)

the partial derivative of that equation as a function of each variable. Square each result and multiple times the knowledge uncertainty (i.e., variance in data) for the measurement of each variable. Then take the square root of the sum. For example, assume that a requirement R is a function of variables (a,b,c), that is, R = f(a,b,c). The uncertainty of the knowledge of the requirement R is given by

$$\sigma_R = \sqrt{\left(\frac{\delta f(a,b,c)}{\delta a}\right)^2 \sigma_a^2 + \left(\frac{\delta f(a,b,c)}{\delta b}\right)^2 \sigma_b^2 + \left(\frac{\delta f(a,b,c)}{\delta c}\right)^2 \sigma_c^2} \quad (5.1)$$

If the defining equation is a linear sum, then the result is a simple root mean square of the individual standard deviations. But if the equation is not linear, then there will be cross terms and scaling factors.

When building an error budget, use the standard deviation of measurement reproducibility not of repeatability. Repeatability will give an "optimistic" result. Reproducibility gives a realistic result. Repeatability is the ability to get the same answer twice if nothing in the test setup is changed. Reproducibility is the ability to obtain the same answer between two completely independent measurements [5,6]. If one is measuring the reproducibility of the ability to align a part in a test setup, then to obtain two independent measurements, one must physically remove the part from the test setup and reinstall it between

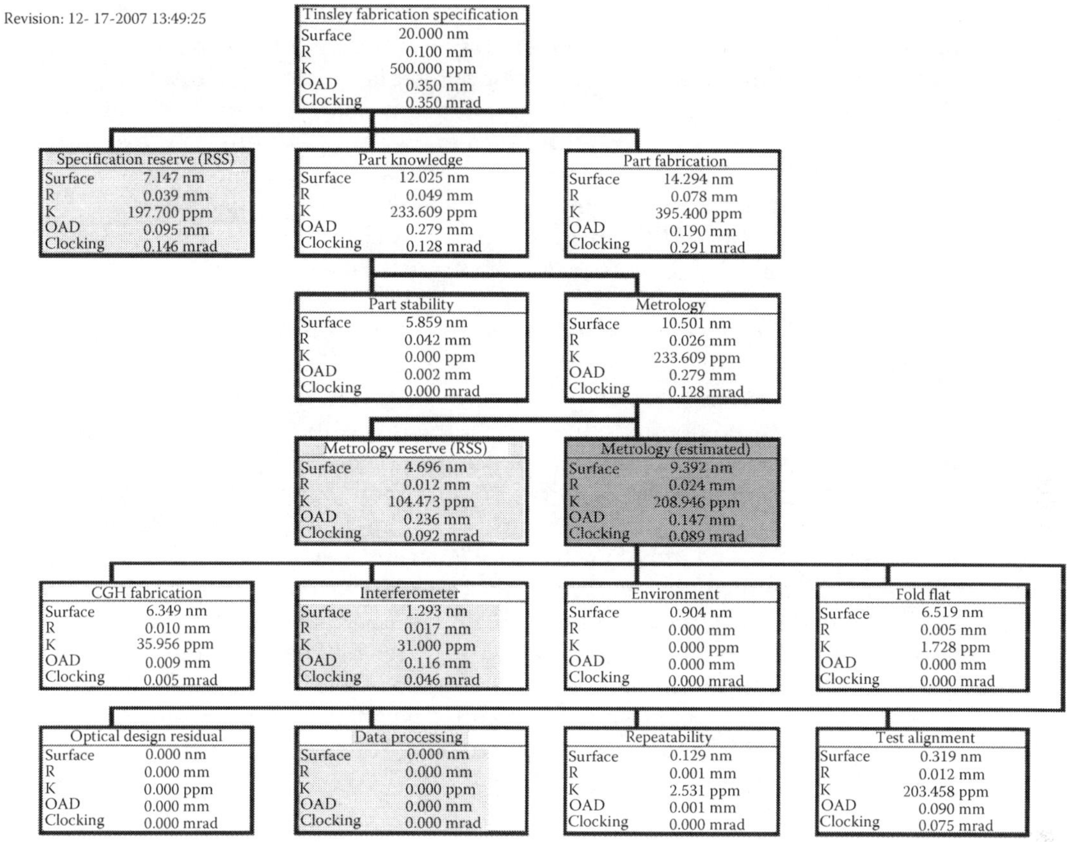

FIGURE 5.5
Each JWST PMSA specification had a separate error budget, that is, surface figure, radius of curvature, conic constant, decenter, and clocking of the prescription on the substrate. For every item in this figure, there was a highly detailed error budget. (From Stahl, H.P. et al., *SPIE Proc.*, 7790, 779002, 2010, DOI: 10.1117/12.862234.)

measurements. If one is measuring the reproducibility of atmospheric turbulence, then all that is required is to make sure sufficient time has passed since the last measurement to insure that the two measurements are not correlated.

From a real-world perspective, reproducibility is much more important than repeatability. The reason is that a part is never tested just once. The components are tested multiple times during fabrication. This is commonly called "in-process" testing. Therefore, the error budget must quantify the knowledge uncertainty of how well the test results can be reproduced from test to test from day to day and even month to month. For example, on JWST, PMSAs were moved not only back and forth between manufacturing and test at Tinsley but also from Tinsley to Ball Aerospace & Technologies Corporation (BATC) and to the Marshall Space Flight Center (MSFC) x-ray and cryogenic test facility (XRCF). On JWST, a complete understanding of each metrology tool's test uncertainty was critical. Data from Tinsley, BATC, and the MSFC XRCF were required to reproduce each other within the test uncertainty. Certified cryo-data must be traceable from the XRCF where they were tested on their flight mount at 30 K to BATC where they were changed from the flight mount to the fabrication mount at 300 K to Tinsley where they were polished on their fabrication mount at 300 K. Accuracy is the ability to get the true answer. The only way to get an accurate measurement is to perform an absolute calibration to quantify any systematic errors which must be subtracted from the data.

Finally, the most important element of an error budget is contingency reserve. All error budgets must have contingence reserve. No matter how much one thinks about every potential risk, one cannot think of everything. No matter how carefully one executes the test plan, something will go wrong. Based on many years of experience, a 33% reserve is recommended. Also, don't wait too long to validate the error budget. On the Infrared Technology Testbed Telescope (ITTT) program (which became Spitzer), this author was responsible for the secondary mirror. A complete error budget was developed, but some elements were allocations. The secondary mirror was manufactured to a Hindle sphere test (Figure 5.7), and the optician achieved an excellent result. Unfortunately, the Hindle sphere was not absolutely calibrated until it was time to perform the final certification, and to my horror, it had a trefoil gravity sag mount distortion. Furthermore, because the secondary mirror had a three-point mount, every time it was inserted into the test, it was aligned to the Hindle sphere's trefoil error. As a result, the optician polished in three bumps which exactly matched the holes in the Hindle sphere. Fortunately, there was sufficient reserve in the error budget such that the mirror still met its figure specification; it just was no longer spectacular. The moral of the story is to not only validate the error budget early but also, as much as possible, randomize the alignment from test to test. Sometimes, bad things happen from being too meticulous. (This could almost be an eighth rule.)

In constructing an error budget for large parts, the three biggest potential error sources are gravity sag, mechanical stability, and atmospheric effects. Of these, gravity sag may be the most important because it can be significant and a metrology engineer's intuition often fails to fully account for its effect. The intuition challenge arises from the fact that gravity sag is nonlinear. To first order

$$\text{Gravity sag} \propto \frac{mg}{K} \propto mg\left[\frac{1}{E}\left(\frac{D^2}{T^3}\right)\right] \tag{5.2}$$

where
- m is the mass
- g is the gravitational acceleration
- K is the stiffness
- E is the Young's elastic modulus
- D is the diameter
- T is the thickness

Therefore, for constant thickness, a 2 m part is four times less stiff than a 1 m part. If they both have the same mass, then the 2 m part will have about 4 times more gravity sag, and if they both have the same area density, then the 2 m part will have about 16 times the gravity sag. Thus, for most small parts, their intrinsic stiffness is such that any bending or shape change caused by gravity is negligible relative to the surface figure specification and thus can be ignored. But, for large parts, gravity sag can be orders of magnitude greater than the surface figure error being measured. For example, an 8 m diameter, 300 mm thick, solid glass mirror (which must be fabricated to a surface figure requirement of less than 10 nm rms) has an edge-supported gravity sag of approximately 2 mm. Now, one would never make or test such a mirror using edge support, but if they did, this amount of sag would not be a problem only if the mirror will be used in the same gravity orientation as it is made and tested. However, it is a problem if during operation the mirror is to be tilted with respect to gravity or if it is going to be used in space. In these cases, the gravity sag must be quantified and if necessary removed from the data.

The key to testing large parts is that the metrology mount must simulate the part's "as-use" gravity orientation or operational support system. The problem is that metrology mounts are not perfectly reproducible. And the less stiff the part under test, the more its gravity sag might vary from test to test. When testing large parts, it is desirable to design a metrology mount with sufficient stiffness to hold the part under test such that the uncertainty in its gravity sag knowledge is 10× smaller than the surface figure specification. For example, if the mirror surface figure requirement is 10 nm rms, then the metrology mount should support the mirror in a known orientation with respect to gravity with an uncertainty of less than 1 nm rms. To accomplish this task requires a support structure which is both mechanically (and thermally) stable and introduces predictable stress/strain and force loads into the part under test. As the part size increases, metrology mounts and handling fixtures become more complicated.

Mechanical stability and vibration errors must be included in any error budget. Small parts are typically tested on a small vibration isolated table with sufficient stiffness to maintain micrometer-level test alignment for arbitrary periods of time. But, large test setups require large structures. And for structures sometimes tens of meters in size, it can be difficult to achieve micrometer (and/or micro-radian) alignment stability between components. Furthermore, at such sizes, the structural material's coefficient of thermal expansion (CTE) can cause the test setup to "breath" as a function of room temperature. When operating at large scale, test uncertainty is impacted by static and dynamic stability.

Static stability is the ability of the structure to maintain the alignment of the test elements relative to each other for long periods of time. Insufficient static stability manifests itself in systematic or even unpredictable drifting of the test alignment during the measurement period. Static stability is also the ability to reproducibly position the test elements in the aligned state from test to test. Static instability primarily occurs when strain, which is introduced via mechanical preload or misalignment or thermal gradients, is released via stick/slip motion. As a rule of thumb, a test setup should be designed such that the ability to reproducibly position the part under test is sufficiently precise that the uncertainty is 10× smaller than the parameter to be measured. Similarly, any error introduced by drift in the test setup should be 10× smaller than the parameter to be measured.

Dynamic stability is vibration, and it can be driven by either seismic or acoustic sources. Small test structures tend to be very stiff and have first mode frequencies which are much higher than the measurement period. If the vibration is at least 10× higher than the data acquisition rate, then their effect will average to zero—with a small reduction in data "contrast" due to blurring [7]. But large structures can have first mode frequencies which are on the order of tens to tenths of Hertz. For example, the SOML test tower moves as a rigid body with a resonance of about 1.2 Hz and an internal first mode of 9.5 Hz [4]. Motions in these frequency bands can introduce significant measurement errors. To minimize these errors, it is necessary to minimize the amplitudes of their motions. This is done by vibration isolating the test structure from the ambient environment. One way (as shown in Figure 5.3) is to bury in a sand pit a very thick concrete slab on which the test structure is setup. The sand dampens vibrations from being propagated from the building into the test structure. As shown in Figure 5.4, the sand can be replaced by pneumatic supports. A third approach is to build large support legs which are physically attached to the building with pneumatic supports at the top from which the test structure hangs.

Regardless of the approach used, it is virtually impossible to eliminate all vibrations. Therefore, additional means are needed to minimize their impact. The Hubble Space

Telescope program mitigated vibration errors by acquiring and averaging many short-exposure measurements [8]. Short-exposure measurements "freezes" the vibration error. And averaging reduces the error contribution to zero because vibration is Gaussian normal (i.e., has a mean value of zero), but it only works if enough measurements are acquired over a long enough time (i.e., over several periods of the vibration) to yield a statistically significant zero mean average.

Another approach is to optically or structurally connect the test components such that the vibrations are synchronized. The Keck segments were tested in the presence of significant vibration by employing a common path technique. The Twyman–Green reference beam was transmitted alongside of the test beam, reflecting from the autocollimating fold flat three times and twice from a small flat physically attached to the segment under test (Figure 5.6) [3]. Another trick is to synchronize vibration between test components by structurally connecting them. Figure 5.7 shows the Hindle sphere test setup used to test the Spitzer secondary mirror. A 2 × 4 board is connecting the Hindle sphere with the Fizeau interferometer phase modulator. If every test element sees the same vibration such that there is no relative motion, then there are no measurement errors. On JWST, it was

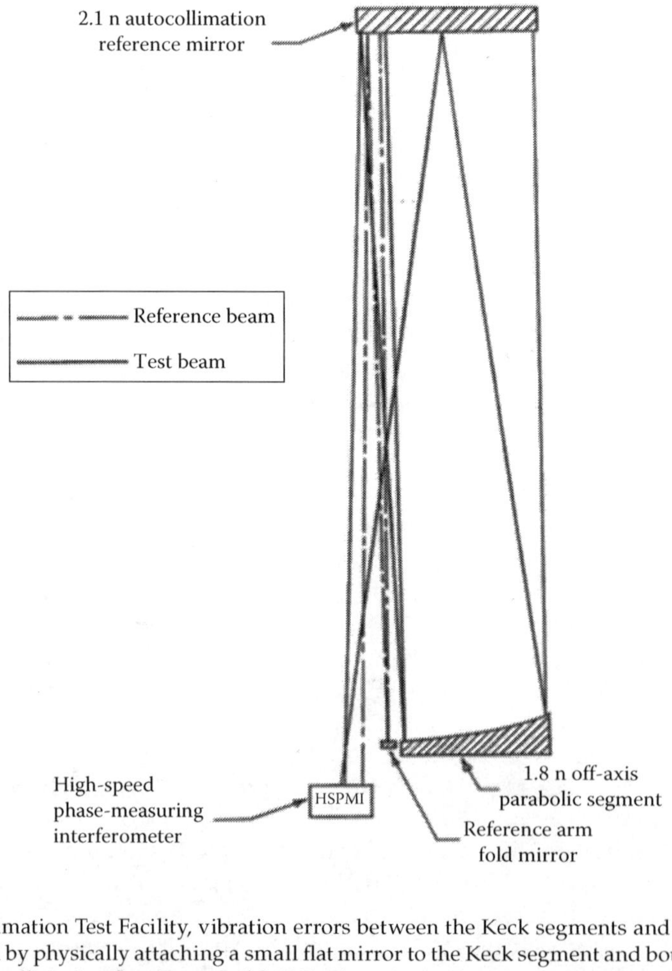

FIGURE 5.6
In the Itek Autocollimation Test Facility, vibration errors between the Keck segments and the autocollimating flat were minimized by physically attaching a small flat mirror to the Keck segment and bouncing the reference beam off of the autocollimating flat. (From Stahl, H.P., *Photonics Spectra*, 12, 105, 1989.)

FIGURE 5.7
Hindle sphere test setup to measure the Spitzer Telescope secondary mirror. A 2 × 4 is used to structurally connect the interferometer phase modulation head and Hindle sphere to minimize vibration-induced relative motion. (Photo courtesy of Goodrich Corporation, Charlotte, NC.)

necessary to characterize the PMSAs at 30 K. This was done by testing them horizontally inside the XRCF with the optical test equipment located outside the chamber at the 16 m center of curvature (CoC) (Figure 5.8). But, because the test equipment was on a vibration isolation slab which was unconnected to the slab on which the XRCF sat, there was a low frequency structuring bending mode which introduced a 0.5 mm pistion motion between the PMSAs and the test equipment. Because of this piston error, State-of-the-art commercial temporal phase-measuring interferometers could not measure the mirrors to the required precision because low-frequency structural bending introduced 0.5 mm of piston motion

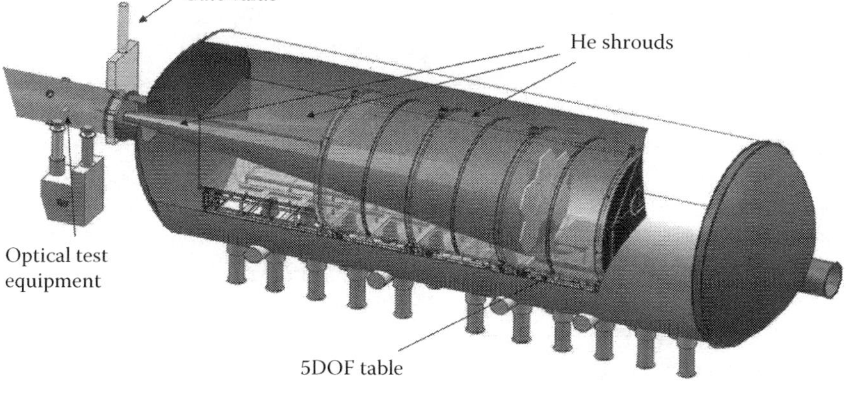

FIGURE 5.8
JWST primary mirror segments were tested at center of curvature. Because the PMSA radius of curvature was 16 meters, the optical test equipment was outside the chamber on an isolated concrete slab different from the isolated slab on which the chamber was supported.

FIGURE 5.9
4D Vision Systems PhaseCAM instantaneous phase-measuring interferometer. (From Stahl, H.P. et al., *SPIE Proc.*, 7790, 779002, 2010, DOI: 10.1117/12.862234.)

between the PMSAs and the test equipment. To solve this problem, MSFC funded the development of the 4D Vision Systems PhaseCAM instantaneous phase-measuring interferometer (Figure 5.9) [9,10].

Atmospheric turbulence and atmospheric stratification are also important error budget elements. These effects may be easier to understand because they can be seen. Anyone who has ever driven down a hot highway and observed the shimmering thermal boundary layer has an intuitive understanding of its affect. Or anyone who has stuck a hand into an optical test beam has seen how rising heat distorts the fringes. Thermal variation causes measurement errors because the refractive index of air varies as a function of temperature. A simple illustration of how this can be a problem is if a small pocket of cooler air (which is more dense and with a higher index) moves across an optical surface, it appears as a "hole" in the surface figure. A more accurate explanation is that optical rays traveling through different parts of the atmosphere with different temperatures experience a differential optical path length error. But turbulence flow is difficult to model and is another area where an optical metrologist's intuition is frequently inadequate. The challenge for large optics is that for a constant F/# component, air path volume increases as the cubed power of aperture diameter. Also while mechanical vibrations are typically periodic, turbulence is chaotic.

Stratification occurs when air forms layers of different temperature, typically cold on the bottom and hot on the top, but temperature inversions are also possible. Normally, one sees this effect in air that is still or not moving, but it can also occur in laminar flow (which is defined as parallel flow with no lateral mixing). Because refractive index varies as a function of temperature, light going through the colder layers has a longer optical path length than light going through the warmer layers. Thus, based on the geometry of how the light traverses the layers, wave front errors can be introduced by the atmospheric stratification. If linear stratification occurs in a parallel optical beam, it introduces a tilt error which can be ignored. But if linear stratification occurs laterally (perpendicular to the optical axis) in a diverging/converging beam, it acts like a tilted plate and introduces an astigmatic wave front error. If linear stratification occurs axially along a diverging/converging beam, it acts like a gradient index lens and introduces power (or focal length change) and a small amount of spherical wave front (or conic constant) error.

An analysis of the Gemini 6.5 m F/11.25 primary mirror predicted that a 0.5 C top to bottom gradient would produce a 2 ppm conic constant error and a 0.3 ppm radius error [4]. In general, it is best to avoid stratification. An interesting exercise for the reader is to set up a CoC test in the laboratory. Take and save a measurement, then "tent" the test and wait for stratification to occur. Then take another measurement and subtract the first. For best results, use a mirror that is larger than 0.5 m.

Turbulence is caused by the convective flow of warmer/cooler air pockets moving through ambient air (or lateral mixing and eddy current mixing at air temperature boundaries). Because refractive index varies as a function of temperature, pocket-to-pocket (or across boundaries) temperature differences manifest themselves as measurement errors (caused by differential optical path length variations). These fluctuations can be distributed laterally as well as axially along the test beam. These pockets can be large and moving slowly, or (with increased mixing) they can be small and moving rapidly. This size and rate of motion is described by diffusion, the greater the mixing or the more turbulent the flow, the shorter the diffusion length.

Ideally, the best test environment is an atmosphere with no temperature variation. In such a case, even if there was significant air flow, there would be no optical turbulence. But such an environment is difficult to achieve. Typical air handling systems are good to 1 C. The Hubble program solved the atmospheric turbulence problem by testing the primary mirror at CoC in a vertical vacuum chamber (Figure 5.10a) [8].

(a) (b)

FIGURE 5.10
(a) The Hubble primary mirror was inside a vacuum chamber to eliminate atmospheric turbulence as a source of measurement error. (From Montagnino, L.A., *SPIE Proc.*, 571, 182, 1985.) (b) Hubble primary mirror being loaded into the vertical test chamber. (Photo courtesy of Goodrich Corporation, Charlotte, NC.)

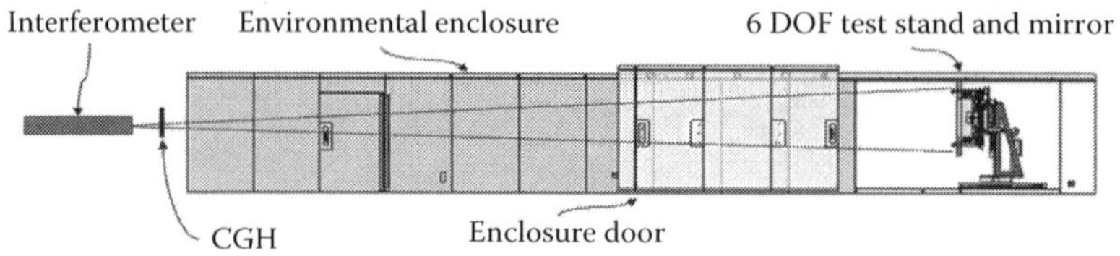

FIGURE 5.11
Ball Optical Test Station (BOTS) for ambient environment testing of JWST PMSAs. (From Stahl, H.P. et al., *SPIE Proc.*, 7790, 779002, 2010, DOI: 10.1117/12.862234.)

When it comes to optical testing in air, there are many different opinions. Some think that the best approach is to maximize turbulent mixing to minimize the size of pockets (diffusion length). Others believe that you should stop the mixing and test as soon as the air becomes quiet but before it becomes stratified. Some believe that the air should flow along the optical axis, while others believe that it should flow perpendicular to the optical axis. This author recommends perpendicular flow with maximum turbulent mixing. The problem with axial flow is that pressure gradients can form in front of the mirror and eddy current vortices can be produced around the edge. The best test environment that this author has ever experienced was a 10-m-by-20-m room whose air flowed from one end to the other, was exchanged approximately every 5 min, and was controlled to 0.01 C [11]. The next best test environment was the BATC Optical Test Station (OTS) for the JWST PMSAs (Figure 5.11) [12]. Each PMSA was tested at CoC in a thermally insulated test tunnel. Thermally controlled air was flowed down the tunnel with fans producing vertical mixing.

An important fact to understand about testing in ambient atmosphere is that turbulence is not statistically random. Turbulence does not average to zero. Rather, atmospheric turbulence is chaotic with a diffusion length. Thermal pockets are "correlated" with each other axially and laterally. Therefore, one cannot eliminate atmospheric turbulence errors simply by taking lots of short-exposure measurements and averaging (as one does for vibration). According to the ergodic principle, the temporal variation along an optical path has the same statistical properties as the spatial turbulence. Thus, two measurements separated in time by less than the diffusion time are correlated, and therefore, averaging them will not yield a "zero" error. Rather, averaging correlated measurements yields a low-order error. The only way to eliminate atmospheric turbulence effects is to average measurements which are acquired at time intervals longer than the diffusion or correlation time. And the only way to obtain short diffusion times is a highly mixed, highly turbulent atmosphere.

5.2.3 Continuous Metrology Coverage

The old adage (and its corollary) is correct: "you cannot make what you cannot test" (or "if you can test it, then you can make it"). The key to implementing these rules is simple. Every step of the manufacturing process must have metrology feedback, and there must be overlap between the metrology tools for a verifiable transition. Failure to implement this rule typically results in one of two outcomes, either very slow convergence or negative convergence.

Overlapping metrology coverage requires tools which can precisely measure large dynamic ranges, for a range of surface textures during different fabrication processes, and over a range of different spatial frequencies. Regarding measurement precision and range, it is much easier to measure a 1 m radius of curvature to a precision of 10 μm than it is to measure a 10 m radius of curvature to a precision of 10 μm (or even to 100 μm).

The metrology tools designed to make such precision measurements can have range limitations. Also as distances become greater, all of the previously discussed problems such as mechanical stability and atmospheric turbulence affect precision. Another well-known but subtle effect is the Abbe sign error if the radius measurement is not being made directly on the optical axis of the component. Fortunately, the dimensional tolerances for large optics are frequently more relaxed than for small optics.

Large parts go through a variety of manufacturing processes, from machining to rough grinding to fine grinding to polishing and figuring. Each process has a different surface texture and different precision and dynamic range requirements. Typically, coarse metrology is done via a profilometer (mechanical stylus gage) for machining and grinding operations and fine metrology is done via an interferometer for polishing and figuring. The problem comes in when making the transition from grinding to polishing. Coordinate-measuring machines (CMMs) are great for machining and rough polishing. They have large dynamic ranges and work well with "mechanical" surfaces, that is, surfaces which are not smooth enough to reflect light. The primary issue for large optics is getting a CMM with a sufficiently large measurement volume. A secondary issues is that the larger the measurement volume, the more difficult it is to obtain high precision. High precision is what drives the overlap problem. A CMM with a 0.100 mm rms measurement uncertainty cannot provide a good metrology handoff to optical interferometry. To achieve good overlap with optical interferometry requires knowledge of the surface shape under test to an uncertainty of approximately 0.010 mm or 10 μm rms. Traditionally, this gap has been filled with infrared interferometry [12], but improvements in CMM precision will eventually allow for direct transition to optical interferometry. (CMMs capable of 8–10 m are expensive. So the choice may be aperture dependent.)

For JWST, Tinsley developed overlapping metrology tools to measure and control conic constant, radius of curvature, prescription alignment, and surface figure error throughout the fabrication process. During rough grinding, this was accomplished using a Leitz CMM (Figure 5.12). The CMM was the primary tool used to establish radius of curvature and conic constant. While these parameters can be adjusted in polishing, it is much easier to set them during grinding. During polishing, metrology was provided by a CoC interferometric test. Ordinarily, optical fabricators try to move directly from CMM to optical test during fine grinding. But, given the size of JWST PMSAs and the mid-spatial frequency specification, this was not possible. Bridge data were provided by a Wavefront Sciences Scanning Shack–Hartmann Sensor (SSHS) (Figure 5.13). Its infrared wavelength allowed it to test surfaces in a fine-grind state. And its large dynamic range (0–4.6 mrad surface slope) allowed it to measure surfaces which were outside the interferometer's capture range. The SSHS is an auto-collimation test. Its infrared source is placed at the focus for each PMSA prescription (A, B, or C) to produce a collimated beam. An infrared Shack–Hartmann sensor is then scanned across the collimated beam to produce a full-aperture map of the PMSA surface. The SSHS was only certified to provide mid-spatial frequency data from 222 to 2 mm. When this test was not used, convergence was degraded. Figure 5.14 shows an example of the excellent data agreement between the CMM and SSHS.

In additional to dynamic range and fabrication process stage, spatial sampling metrology overlap is also important. As the part becomes more and more perfect, it is necessary to control smaller and smaller features. High-resolution spatial sampling is needed to drive the polishing process. It is especially important if the optical component is an asphere just as a complex B-spline can be critical on machined parts. A common fabrication process for aspheric optics and for large optics is small tool computer-controlled polishing. But the size of the tool which can be used is limited by the spatial sampling of the metrology data [13].

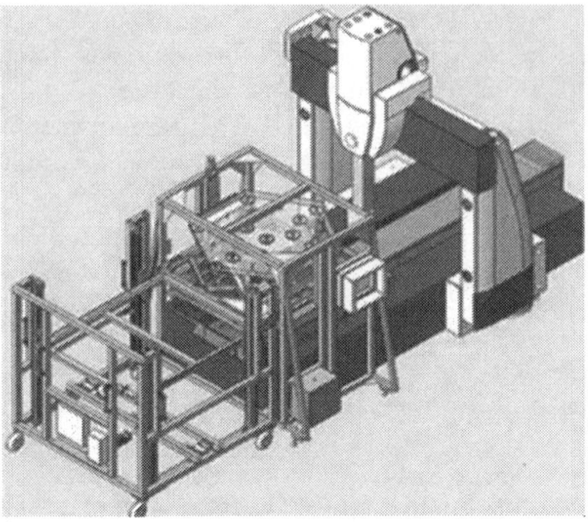

(a)

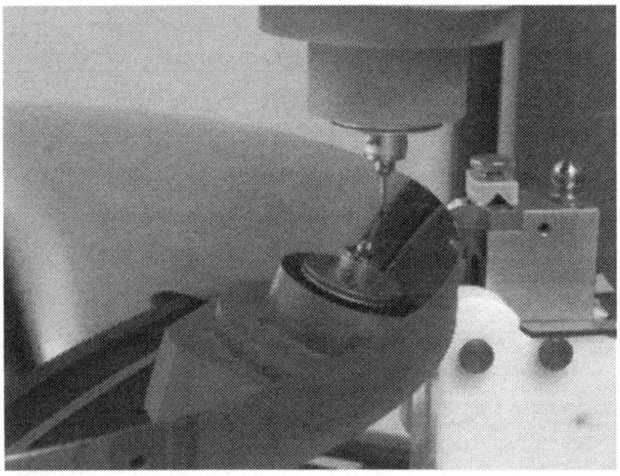

(b)

(c)

FIGURE 5.12
Leitz CMM (a) was used at Tinsley during generation and rough polishing to control radius of curvature, conic constant, and aspheric figure for PMSAs, secondary mirrors (b), and tertiary mirror (c). (From Stahl, H.P. et al., *SPIE Proc.*, 7790, 779002, 2010, DOI: 10.1117/12.862234.)

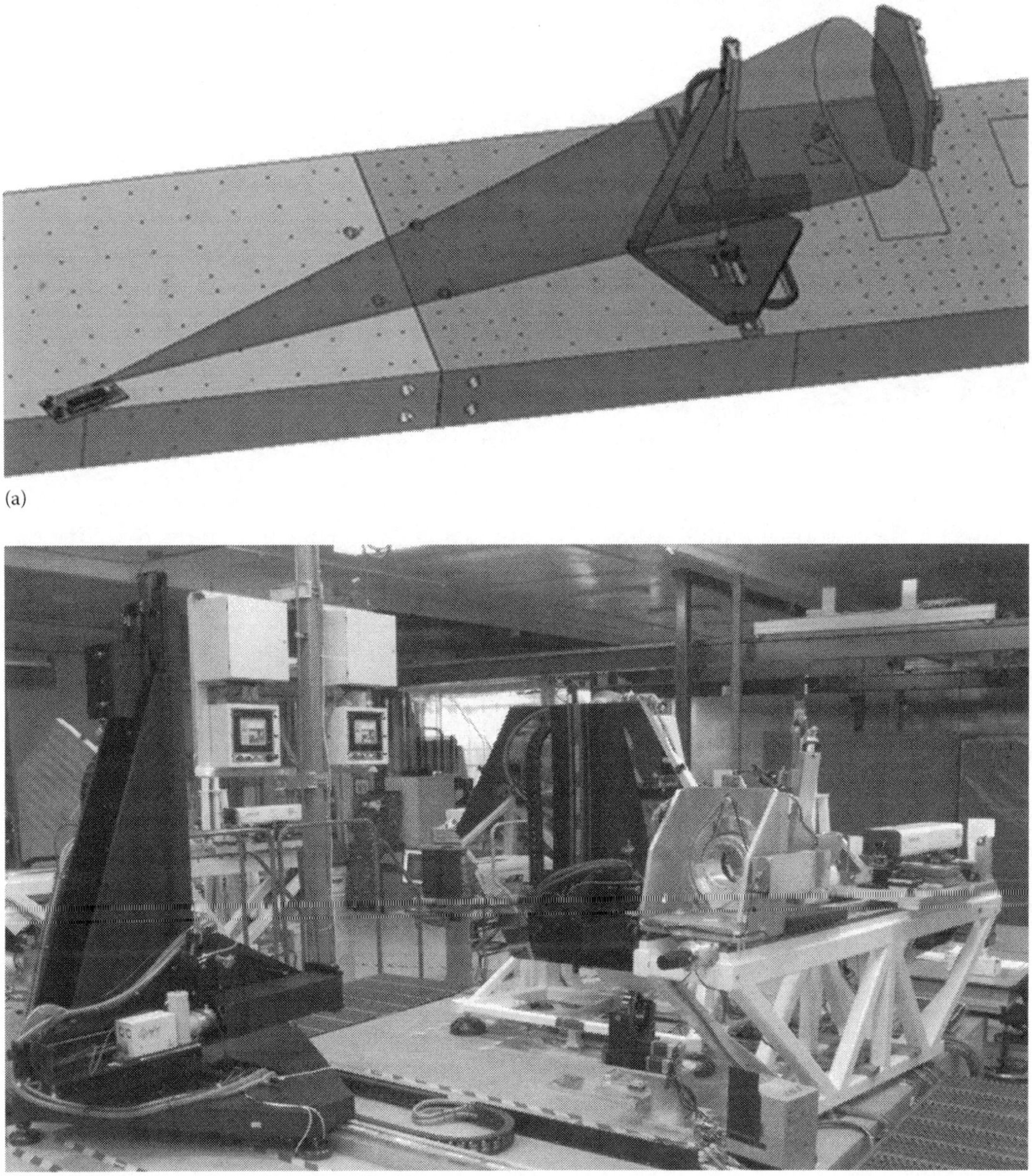

(a)

(b)

FIGURE 5.13
SSHS (manufactured by Wavefront Sciences) is an autocollimation test. (a) A 10 μm source is placed at focus and a Shack–Hartmann sensor is scanned across the collimated beam. There are three different source positions for the three PMSA off-axis distances. (b) Photo shows the sensor (white) mounted on the Paragon Gantry (black). (From Stahl, H.P. et al., *SPIE Proc.*, 7790, 779002, 2010, DOI: 10.1117/12.862234.)

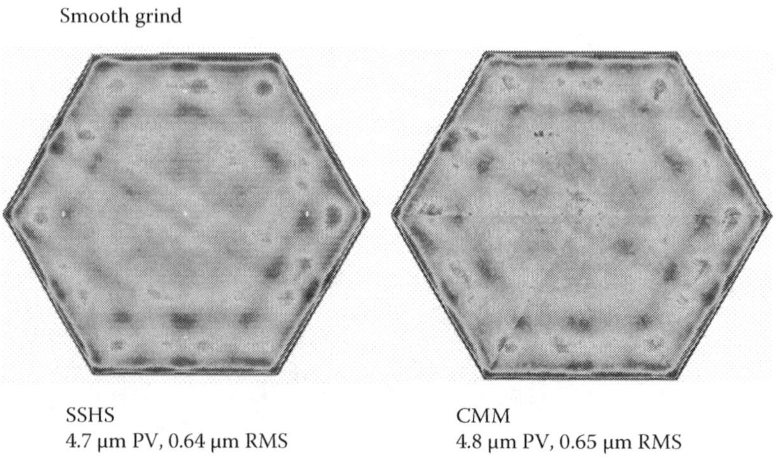

FIGURE 5.14
Comparison of CMM and SSHS data (for 222–2 mm spatial frequencies) after smooth out grind of the EDU. (From Stahl, H.P. et al., *SPIE Proc.*, 7790, 779002, 2010, DOI: 10.1117/12.862234.)

If one has an 800 pixel interferometer taking data on a 0.8 m component, then one has 1 mm spatial sampling. According to the Shannon sampling theorem, this should be sufficient to correct 2 mm spatial period errors, but in practice, it is only good enough for 3–5 mm spatial frequency errors. Extrapolating to larger apertures, an 800 pixel interferometer taking data on an 8 m mirror has 10 mm spatial sampling which can control 30–50 mm spatial frequencies. Depending upon the mirror's structure function specification, that is, its required surface figure versus spatial frequency, such a spatial sampling may or may not be sufficient. Additionally, segmented telescopes have edge requirements. On JWST, the polished optical surface needed to meet its specification to within 7 mm of the physical edge. While the JWST CoC interferometer had a projected pixel size of 1.5 mm and should have been able to resolve a 4.5–7.5 mm edge, it could not resolve to that level.

On JWST, grinding and polishing feedback was provided by a custom-built OTS (Figure 5.15). The OTS is a multipurpose test station combining the infrared SSHS, a CoC interferometric test with a computer-generated hologram (CGH) and an interferometric autocollimation test. This test simultaneously controls conic constant, radius of curvature, prescription alignment, and surface figure error. The CoC test pallet contains a 4D PhaseCAM, a Diffraction International CGH on a rotary mount, and a Leica ADM. The ADM places the test pallet at the PMSA radius of curvature with an uncertainty of 0.100 mm which meets the radius knowledge requirement. Please note that this uncertainty is an error budget built up of many contributing factors. Once in this position, if the PMSA were perfect, its surface would exactly match the wave front produced by the CGH. Any deviation from this null is a surface figure error to be corrected.

5.2.4 Know Where You Are

It might seem simple, but if you don't know where a feature is located on the part, you cannot correct it. To solve this problem, you must use fiducials. There are two

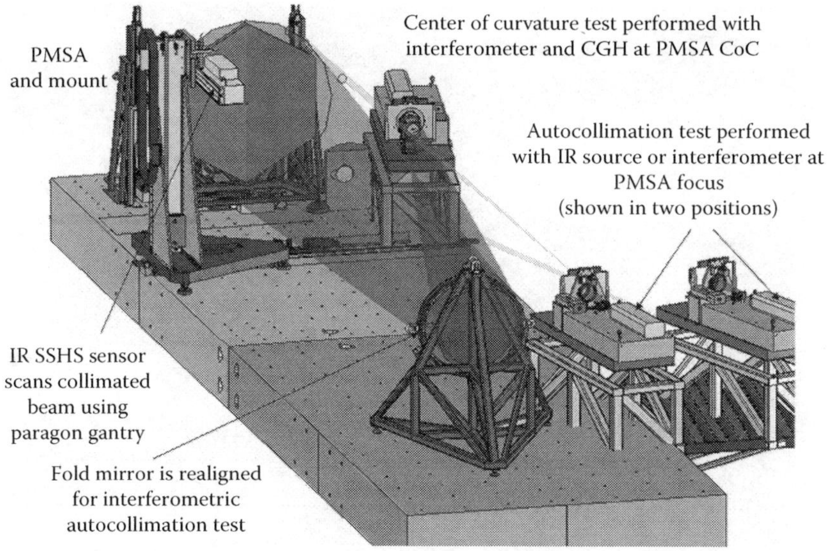

FIGURE 5.15
OTS is a multipurpose test setup with three different metrology tools: SSHS, CoC CGH interferometer, and autocollimation interferometer. (From Stahl, H.P. et al., *SPIE Proc.*, 7790, 779002, 2010, DOI: 10.1117/12.862234.)

types of fiducials: data fiducials and distortion fiducials. Data fiducials are used to define a coordinate system and locate the measured data in that coordinate system. Sometimes, this coordinate system is required to subtract calibration files; other times, it is required to produce hit maps. Distortion fiducials are used to map out pupil distortion in the test setup. Many test setups, particularly those with null optics, can have radial as well as lateral pupil distortion. Distortion can cause tool misregistration errors of 10–50 mm or more.

Fiducials can be as simple as a piece of tape or black ink marks on the surface under test or as sophisticated as mechanical "fingers" attached to the edge protruding into the clear aperture. Tape fiducials are acceptable for simple reproducibility or difference tests or to register a calibration alignment, but they are not recommended for computer-controlled process metrology. In these cases, fiducials define your coordinate system and need to be applied with a mechanical precision of greater accuracy than the required prescription alignment to the substrate. Additionally, because the interferometer imaging system might invert the image or because fold mirrors in the test setup might introduce lateral flips, an asymmetric pattern is highly recommended. A good pattern to use is one with fiducials at 0°, 30° (or 120), 90°, and 180°. The 0°/180° fiducials produce a central axis for the data set. The 90° fiducial defines left/right and the 30° fiducial defines top/bottom. Additionally, for test setups with null optics, pupil distortion can be a problem. In these cases, distortion fiducials are required. One option is to place multiple fiducial marks along a radius. For null tests with anamorphic distortion, a grid of fiducial marks is recommended. Finally, if one has a clear aperture requirement, make sure to place fiducial marks inside and outside of the required clear aperture distance; this way, it can be certified whether or not the requirement is achieved.

Another problem is software coordinate convention. Most interferometer analysis software assumes that the optical (Z-axis) positive direction points from the surface

FIGURE 5.16
PMSA mirrors with data and distortion fiducials are ready for loading into the MSFC XRCF.

under test toward the interferometer, such that a feature which is higher than desired is positive. But many optical design programs define the positive optical axis to be into the surface. The problem occurs because both programs will typically define the Y-axis as being up, so it is critical to understand which direction is X-axis. The problem is further complicated when interfacing with the optical shop. To avoid doubling the height or depth of a bump or hole because of a sign error, or adding a hole or bump to a surface because of a coordinate flip or inversion, a good metrologist must know the coordinate system of every computer-controlled grinding and polishing machine in the optical shop.

On JWST, the CoC null test simultaneously controls the PMSA conic, radius, figure, and prescription alignment. The key is knowing where the prescription is on the substrate and knowing where the prescription is in the test setup. Prescription alignment (off-axis distance and clocking) is controlled by aligning the PMSA into the test setup with an uncertainty which is smaller than the decenter and clocking tolerances. PMSAs are manufactured in Observatory Coordinate Space as defined by "master datums" on the back of each substrate. The optical surface figure is registered to the mirror substrate and to the observatory coordinate system via data fiducials placed on the front surface of each mirror. The CMM is used primary in establishing compliance with prescription alignment. Starting with the master datums, the CMM defines "transfer" fiducials on the side of the mirror. Then, the CMM establishes the data fiducials based on these secondary fiducials. Figure 5.16 shows fiducialized mirrors being loaded into the MSFC XRCF for cryogenic testing. Some of the mirrors have only the data fiducials. Others of the mirrors have both data fiducials and distortion fiducials (2D grid of dots). Distortion fiducials are necessary to compensate for anamorphic distortion introduced by the CGH.

5.2.5 Test Like You Fly

"Test like you fly" covers a wide range of situations, and of course, for ground applications, this rule could be "test like you use." Whenever possible, the part should be tested in its final mount, at its operational gravity orientation and at its operational temperature. While gravity is typically not a problem for small stiff optics, it can be a significant problem for large optics or, for that matter, any large precision part or structure (like a machine tool). Any optical component going into space needs to be tested in a "zero-g" orientation. This is typically accomplished by either averaging a cup-up/cup-down test to remove the concave/convex gravity sag contribution or by averaging a horizontal multiple rotation test to remove mount-induced bending [14].

Gravity sag can be very significant for very large ground-based telescopes. In this case, the best approach is to test the part in their final structure (or a suitable surrogate) at an operational gravity orientation. The one thing that a good metrologist should avoid is agreeing to test a very low stiffness mirror or part without a final support system. The reason is that it will be virtually impossible to achieve a stable, reproducible measurement. With such mirrors, simply picking it up and setting it back down on the metrology mount might result in unacceptable shape changes.

Finally, it is important to test a part under its intended atmospheric pressure and temperature conditions. If a lightweight mirror intended for use in vacuum does not have proper venting paths, it can result in a damaged mirror. A mirror intended for use at a cryogenic temperature can have very large CTE-induced figure changes. In such cases, it is necessary to characterize these changes and generate a cryogenic "hit" map to "correct" the surface figure for "at-temperature" operation.

Because JWST mirrors were fabricated at room temperature (300 K) but will operate in the cold of space (<50 K), it is necessary to measure their shape change from 300 to 30 K, generate a "hit-map," and cryo-null polish the mirrors such that they satisfy their required

FIGURE 5.17
MSFC XRCF, with its 7 m diameter and 23 m length can test up to 6 JWST PMSAs. Test equipment is located outside a window in ambient temperature and atmospheric conditions.

figure specification at 30 K. After coating, all mirrors underwent a final cryo-certification test of conic constant, radius of curvature, prescription alignment and surface figure error. These tests were performed at MSFC in the XRCF shown in Figure 5.17. Additionally, because JWST operates in the microgravity of space but is manufactured in the gravity of Earth, it is necessary to remove gravity sag from the measured shape. This is accomplished using a standard six rotation test.

5.2.6 Independent Cross-Checks

Probably the single most "famous" lesson learned from the Hubble Space Telescope is to never rely on a single test to certify a flight specification. Therefore, every component specification must have a primary certification test and a secondary confirming test. It is very important that these confirming secondary tests be performed early in the metrology process. The metrologist is always going to be under pressure to start in-process testing as soon as possible. While the argument will be made that precision is not required during the early fabrication phases, a good metrologist must insist on certifying and confirming the ability of their test setup to achieve the required error budget for each phase of the metrology process.

While technically not an independent cross-check test, it is recommended that a metrologist occasionally depart from their test routine and deliberately attempt to randomize the test. Metrologists tend to be highly structured, process-driven individuals—as is required by a profession which measures quantities to nanometers. But if by chance such an individual is unknowingly introducing an error into their measurement, then by being overly systematic, they will introduce that exact same error into the test every time they conduct the test. Examples of how to vary the metrology process include the following: deliberately misalign and realign the test setup, perform a settling vibration, and take data with different amounts of tilt or defocus.

As summarized in Figure 5.2, each JWST PMSA requirement has a verification test and at least one validation cross-check test. For example, the optical prescription has multiple cross-checks. The prescription is defined during fabrication at ambient temperature using the Tinsley CoC interferometer CGH test and confirmed with an independent autocollimation test. The PMSA prescription is further tested via an independent ambient test at BATC and the MSFC XRCF 30 K test. The prescription receives a final confirmation test at 30 K when the entire assembled primary mirror is tested at CoC with a refractive null corrector at Johnson Space Center.

5.2.7 Understand All Anomalies

Finally, of all the rules, this one maybe the most important and must be followed with rigor. No matter how small the anomaly, one must resist the temptation of sweeping a discrepancy under the metaphorical error budget rug. Any time that the actual data uncertainty for a given measured value is larger than its error budget, the reason for this discrepancy must be determined and understood. Do not eat into the contingency reserve because it will be needed at the end of the fabrication process or for the integration, alignment, and test (IA&T) process when, if something goes wrong, it is very difficult to fix an error. Similarly, if the actual data uncertainty for a measured value is less than its error budget, one can either adjust the total error budget to create margin for other more difficult parameters or increase the contingency reserve.

5.3 Conclusion

Testing large optics or large parts in difficult. This chapter has defined seven guiding principles that can be applied to any metrology application:

1. Fully understand the task
2. Develop an error budget
3. Continuous metrology coverage
4. Know where you are
5. "Test like you fly"
6. Independent cross-checks
7. Understand all anomalies

Although we have used specific examples from optical testing applications, the issues of error budgets, environmental issues, datum points, cross-checks, and understanding anomalies apply to measuring large part or structure – such as described in Chapters 3 and 4. Large sections on machine tools sag under gravity, girders holding up bridges will change with temperature, and many small errors in an engine will add up to a bad engine.

Many of these issues become most noticeable on large parts being made to high precision. A system like a turbine is made so precisely that a large engine able to move a jumbo jet can be easily turned by hand. However, as tolerances keep increasing for all manufacturing, more often than not, these considerations will hold true for smaller parts as well. The seven guiding principles therefore can be a valuable tool for any metrology application.

References

1. Stahl, H. P., Rules for optical metrology, *22nd Congress of the International Commission for Optics: Light for the Development of the World*, Puebla, Mexico, *SPIE Proceedings* 8011, 80111B, 2011.
2. Stahl, H. P. et al., Survey of interferometric techniques used to test JWST optical components, *SPIE Proceedings* 7790, 779002, 2010, DOI: 10.1117/12.862234.
3. Stahl, H. P., Testing large optics: High speed phase measuring interferometry, *Photonics Spectra* 12, 105, 1989.
4. Burge, J. H., D. S. Anderson, D. A. Ketelsen, and S. C. West, Null test optics for the MMT and Magellan 6.5 m F/1.25 primary mirrors, *SPIE Proceedings* 2199, 658, 1994.
5. Stahl, H. P., Phase-measuring interferometry performance parameters, *SPIE Proceedings* 680, 19, 1986.
6. Stahl, H. P. and J. A. Tome, Phase-measuring interferometry: Performance characterization and calibration, *SPIE Proceedings* 954, 71, 1988.
7. Hayes, J. B., Linear methods of computer controlled optical figuring, PhD dissertation, University of Arizona Optical Sciences Center, Tucson, AZ, 1984.
8. Montagnino, L. A., Test and evaluation of the Hubble Space Telescope 2.4 meter primary mirror, *SPIE Proceedings* 571, 182, 1985.

9. Stahl, H. P., Development of lightweight mirror technology for the next generation space telescope, *SPIE Proceedings* 4451, 1, 2001.
10. Smith, W. S. and H. Philip Stahl, Overview of mirror technology development for large lightweight space-based optical systems, *SPIE Proceedings* 4198, 1, 2001.
11. Stahl, H. P., J. M. Casstevens, and R. P. Dickert, Phase measuring interferometric testing of large diamond turned optics, *SPIE Proceedings* 680, 1986.
12. Stahl, H. P., Infrared phase-shifting interferometry using a pyroelectric vidicon, PhD dissertation, University of Arizona Optical Sciences Center, Tucson, AZ, 1985.
13. Mooney, J. T. and H. Philip Stahl, Sub-pixel spatial resolution interferometry with interlaced stitching, *SPIE Proceedings* 5869, 58690Z, 2005.
14. Evans, C. J. and R. N. Kestner, Test optics error removal, *Applied Optics* 35(7), 1015, 1996.

Part III

Optical Metrology of Medium Size Objects

6

Portable Metrology

Daniel Brown and Jean Francois Laurie

CONTENTS

6.1 Introduction on the Use of Optics in Metrology

Every day, optical technology plays a bigger and bigger role in metrology. No doubt optical metrology will soon be used to perform continuous, high-speed inspections of items of all shapes, sizes, and materials. Laser- and white light–based digitizing technologies are producing increasingly dense and specific data, while the x-ray tomography–based technologies already in use can inspect the internal geometry of the most complex equipment.

However, it will take some time to move from a world dominated by coordinate-measuring machines (CMMs) that use sensors to one where optical CMMs reign, especially in industries like the auto manufacturing or aeronautics, where every new device must prove its worth first. The stakes are sky-high in these industries. The financial consequences of any production shutdown, delay, or defect are enormous, meaning that change can only come about incrementally. This explains the success of solutions combining optics and sensors (laser trackers, optical trackers with sensors), which offer the benefit of optics (speed, portability, reduced sensitivity to the measurement environment, and higher measurement volume) while remaining compatible with known and proven procedures. In the same vein, while digital solutions are being rapidly developed for working with uneven surfaces that sensors cannot easily accommodate, sensor solutions still remain widely used for inspecting geometric components in processes such as boring or stamping.

Over the past 30 years, one of the biggest changes in metrology has been the development of portable measuring devices. This has brought inspection right into the production line, as close to the part as possible. The development of portable measuring arms in the early 1990s and the emergence of laser trackers shortly after made it possible to take measurements more quickly and more often, fostering huge improvements in response time and quality.

Over the past decade, the pace of change has been accelerated even more by the development of the arm-mounted scanners and self-positioning handheld 3D scanners that now dominate the market because of their portable digitization capabilities.

However, most portable measurement solutions still use mechanical technologies that present serious limitations in production environments. For example, these solutions require extreme stability throughout the measurement process, which means they require the use of costly, cumbersome, and generally inflexible control templates and heavy bases—including the granite tables long associated with CMMs.

Optical solutions, especially the ones using video cameras (technologies evolving out of photogrammetry/videogrammetry), sidestep these limitations by enabling automatic positioning and continuous device measurement (the self-positioning concept or dynamic referencing). While they are generally available with most optical technologies, there are some restrictions, as in the case of laser trackers.

Self-positioning handheld scanners are an example of this technology, since they make it possible to do away with external means of positioning.

6.2 History of 3D Measurement Systems (How We Evolve from CMMs to Optical Solutions)

The first CMMs appeared in the 1960s and consisted mainly of 3D tracing devices equipped with a digital readout (DRO) used to display X, Y, and Z coordinates. A number of companies claim to have invented the coordinate-measuring machine, commonly referred to as CMM. The first CMM was probably invented by the Italian company DEA (now part of the Hexagon Metrology Group), which introduced a Computer Numerical Control (CNC) portal frame CMM equipped with a probe and DRO in late 1950s.

Today, myriad different types of CMMs are available, including cantilever CMMs, bridge CMMs, gantry CMMs, horizontal arm CMMs, portal CMMs, moving table CMMs, fixed bridge CMMs, and articulated arm CMMs. Over 6000 CMMs are still produced every year. There are two main categories of CMMs: manually operated models where an operator guides the probe in contact with the part and models that are guided automatically by a program [1].

CMM measuring arms are equipped with either fixed probes (on manual CMM models, operators guide probe heads manually) or touch-trigger probes (which begins measuring automatically when the probe comes into contact with the part). Automated CMM technology was invented in the 1960s to be used for complex inspections of Concorde supersonic jet engines. This invention led to the creation of the company Renishaw in 1973, now the main supplier of CMM measuring heads [2].

Inspection software is needed in order to use a CMM, and CMM manufacturers originally provided the software along with the machine. Today, numerous types of inspection software are available from independent companies not affiliated with the machine manufacturers. They are compatible with most CMM brands and have enabled big companies with CMM fleets to standardize operation of their machines, reducing operator training costs considerably. Standardization also makes it possible to replace older machines with more recent models or newer technology without losing the many existing inspection programs or requiring operators to undergo lengthy training. The main software applications are all capable of inspecting prismatic and geometric entities in addition to lines or surfaces.

Operating a CMM requires a qualified technician and a controlled environment (vibrations, temperature, dust), and CMM accuracy inspections must typically be conducted every 12 months. Many standards have been established to regulate CMM accuracy, including B89 in the United States (American Society of Mechanical Engineers), VDI/VDE in Germany (The Society for Measurement and Automatic Control and Association for Electrical, Electronic & Information Technologies), and of course the ISO 10360 international standard.

Today, CMMs are most commonly used for 3D measuring of manufactured parts, reverse engineering, and digitizing prototype parts. However, conventional CMMs can be used only in very specific environments like a metrology lab or similar facility where temperature is very precisely regulated and vibrations are filtered using a vibration isolation pad.

The introduction of portable CMMs with measuring arms in the 1980s marked a new revolution in measurement: finally, it was possible to bring 3D measuring to the production line. This innovation eliminated the need to move manufactured parts to a dedicated, controlled environment, making this type of CMM very popular. However, because the measuring arm still used a classic technology based solely on precision mechanical parts, it was fragile. These types of CMMs remain very sensitive to the vibration environment, and considerable precautions are required when operating them. Optical solutions offer excellent portability, faster measurements, and superior reliability at an extremely competitive price. These variations of CMMs are shown in Figure 6.1.

For a long time, optical 3D measuring processes, which were limited to the use of theodolites, tacheometers, and conventional photogrammetry rooms (gelatin silver process), were the preserve of classic geodesy and cartography (see Chapters 1, 2, and 3 of this book).

The main reasons for this "limited" field of application had to do with the methods it involved: they did not deliver immediate measurement results because the data recorded by the sensors could not be directly transcribed in the desired XYZ coordinate points. Most

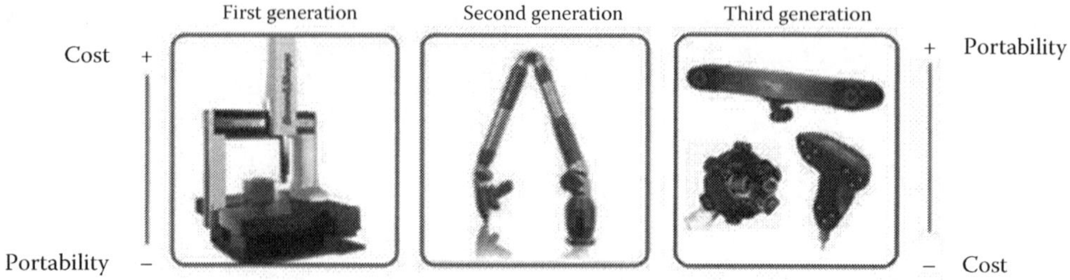

FIGURE 6.1
CMM evolution.

sensors provided angles that had to then be translated through geometric calculations into Cartesian coordinates, taking into account a certain number of data (triangulation, angle composition and distance, transfer of the sensor referential to the object referential, and other calculations). Consequently, and until such time computer tools became easily transportable, these methods remained "reserved" for fields where no other measuring methods were available.

In the mid-1970s, with the rapid development of computer technology, industry began turning to photogrammetry as it became possible to mathematically correct for the distortions in the optics used and perform complex calculations on desktop computers. But photogrammetry remained marginal because of the still significant delay between measurement taking and results and also because industry was skeptical of these unorthodox methods.

It was not until the mid-1980s that a series of new developments emerged to facilitate industry adoption of these new measuring methods, notably the arrival of the first electronic theodolites, the use of scanners to accelerate photogrammetry data processing, the advent of the PC as the primary computer tool, the subsequent arrival of the first electronic distance meters, and the miniaturization of video cameras. Industries where these methods were most invaluable (e.g., aerospace, automotive, shipbuilding, and nuclear plant maintenance) quickly integrated them into their production processes despite a few reserves and obstacles.

The main obstacle was the difficulty most users at the time had in mastering measurement uncertainties, which in a field of extremely stringent quality standards, could be a deal breaker. These methods have now reached maturity, and a certain number of studies and analyses have been conducted, but the obvious lack of normative documents still sometimes makes it difficult to implement these methods broadly in industry [3].

6.2.1 Diversity of Optical Solutions

6.2.1.1 Laser Trackers

Descendants of the theodolite laser trackers capitalize on clever, state-of-the-art technology: a stabilized laser beam aimed at a prism (which reflects the light back in the direction from which it came). The laser head thus records the distance from the prism (either by measuring the time it takes for the light to make its return trip or by combining this initial information with an interferometry measurement for greater precision) as well as the two angles (azimuth and elevation) corresponding to the target direction. Since the prism is placed on a spherically mounted retroreflector (SMR), it is possible to know the 3D position of the center of the SMR.

Laser trackers are currently the leading solution in the high-volume measurements market (particularly in aeronautics), thanks to their very high level of precision and good reliability. Their only serious competition for high-volume measuring is photogrammetry. However, laser trackers are quite expensive and very sensitive to environmental disruptions (vibrations, temperature, hygrometry, etc.). Moreover, the lack of a probe stylus makes it difficult to access hidden points, and the lack of information on the probing direction makes automatic compensation impossible. The company LEICA resolved this limitation by adding a probe around the prism. The orientation of the probe is determined by a camera located on the tracker head. Trackers are covered in more detail in Chapter 3 of this book.

6.2.1.2 Video CMMs

Video CMMs are offshoots of profilometers used to take 2D position measurements of the geometric features (contours) of a part placed on an X and Y displacement table. The Z measurement is most often determined via the optical focus distance. They are sometimes equipped with a probe for measurements that require greater accuracy or to probe points that cannot be seen in the image.

6.2.1.3 Optical Triangulation–Based Solutions

Introduced in the 1990s, these solutions are based on the principle of triangulation, which entails observing points in space from a number of viewpoints in order to determine the 3D position of these points. The typical number of viewpoints is two (matrix array cameras) or three (linear array cameras). A matrix array camera yields a bearing (azimuth and elevation) from the line passing through the camera's optic center and the point being measured, whereas a linear matrix camera only provides the azimuth of a plane passing through the point being measured. Two matrix array cameras are enough to obtain the 3D position of the point being measured, whereas three cameras are needed for the linear array method. Furthermore, a number of points in the space presenting different elevations may be projected onto the same spot in the image when using a linear array camera. In order to discriminate between them, they must be projected one after the other. This is why solutions involving three linear array cameras use light emitting diodes (LED) illuminated sequentially via a controller. Solutions using two matrix array cameras need only passive retroreflective targets, which eliminate all the wiring needed to power the LEDs and reduce the cost of accessories considerably. These two approaches are summarized in Figure 6.2.

These solutions are usually associated with a manual probe equipped with reference points (retroreflectors or LEDs) and a tip identical to that used in measuring arms or CMMs.

Optical triangulation–based solutions offer a key advantage over other solutions: they make it possible to measure the position of the part (through the placement of specific reference points) at any time in the measuring process. This makes them very robust solutions for measuring in unstable environments. This is particularly true for solutions using matrix array cameras, stereoscopic pairs with a fixed geometry, and passive retroreflective targets (e.g., Creaform's HandyPROBE™ system), because they measure the series of points simultaneously, safeguarding the system against the effects of even relatively high-frequency vibrations. These functionalities will be discussed in greater detail later.

It is possible to take continuous measurements of a number of 3D points with optical-based system at a relatively high speed, which paves the way for dynamic measurement

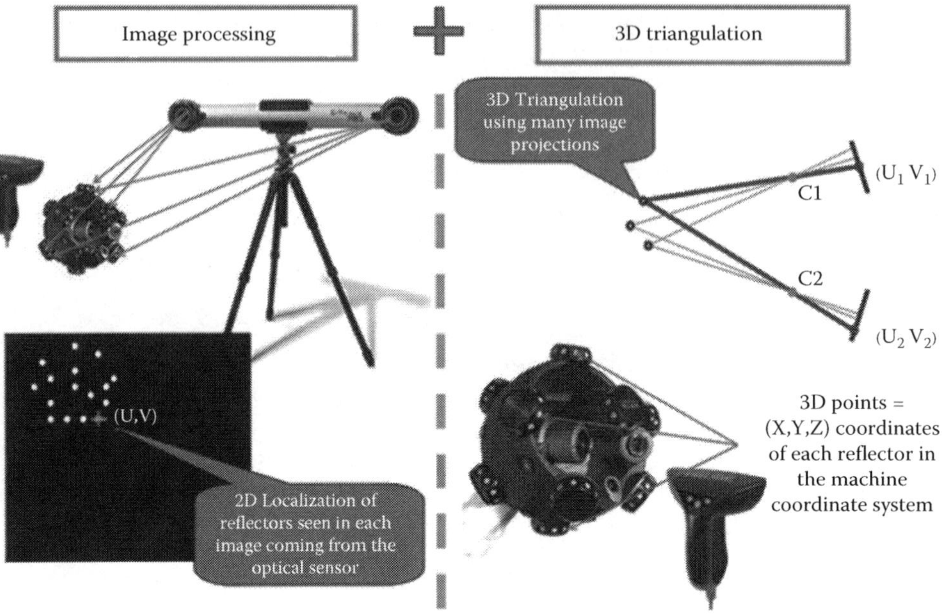

FIGURE 6.2
Basics of optical triangulation. Reflective targets are accurately detected in the images by image processing algorithms. Their 2D positions are then used in the 3D triangulation process to estimate the 3D position of each reflective target and to compute the position and orientation of the probe, the scanner or the part on which the reflective targets are affixed.

(displacement, distortion, trajectory, etc.). There is no question that these solutions—along with laser trackers, which are more accurate for high-volume measuring—are the future of portable 3D measuring. They are already gradually replacing measuring arms because of their greater accuracy and reliability in the operating conditions at most production sites.

6.2.2 Photogrammetry

Photogrammetry may be a state-of-the-art technology, but its technological foundations date back to the nineteenth century. Shortly after the emergence of photography, French mathematician François Arago presented the Academy of Science with a method using triangulation to determine the position of objects in space based on photographs taken from different viewing angles, without knowing the position of the shots beforehand. And *voilà*, photogrammetry was born.

It was not until nearly a century later in the late 1960s, however, that the first industrial applications appeared with the development of convergent photogrammetry.

American mathematician Duane Brown made a major contribution to the field by developing the first mathematical representation of the bundle adjustment process* and by establishing new camera calibration models. In 1977, he founded Geodetic Services, Inc., which successfully marketed the INCA and VSTAR systems (see Figure 6.3). Brown is considered one of the founding fathers of modern photogrammetry [4].

* Basic tool of photogrammetry used to find the optimal configuration of lines linking the points measured and the cameras' optical centers. Graphic representations of these lines can resemble bundles, hence the term "bundle adjustment."

FIGURE 6.3
Duane Brown with the CRC-1 camera (precursor of INCA photogrammetric cameras). (From Brown, J., *Photogramm. Eng. Remote Sens.*, 71(6), 677, 2005.)

With the emergence of computers, and later CCD image sensors, photogrammetry was finally able to take flight in industrial applications.

6.2.3 Scanning Solutions

The mechanical and optical solutions discussed thus far can only be used to measure 3D points, most often through contact with a probe, and sometimes by marking the point with a retroreflector or a LED. This process is not suited for surface inspections, because it is too slow when a large number of points are required, particularly for skew surfaces (the fastest CMMs can measure hundreds of points per second). So in the early 1980s, 3D scanning solutions appeared that were capable of creating files of points or "point clouds" (which are later processed into computer mesh data, such as Standard Template Library (STL) files) based on the measured surface. Certain scanners like the Handyscan 3D laser scanner from Creaform generate an STL file directly. This makes it possible to quickly generate a dense representation of shape imperfections (the fastest scanners can acquire hundreds of thousands of points per second), which has considerably advanced our understanding of certain fabrication processes (drawing, smelting, molding, etc.). Today, there are two main categories of scanners.

6.2.3.1 Structured Light Scanners

Structured light 3D scanners project a pattern of light onto a part and observe how the pattern is distorted such as shown in Figure 6.4. The light pattern is projected by either an LCD projector or a scanned or diffracted laser beam. One or more cameras are used to observe the projected pattern such as shown in Figure 6.5.

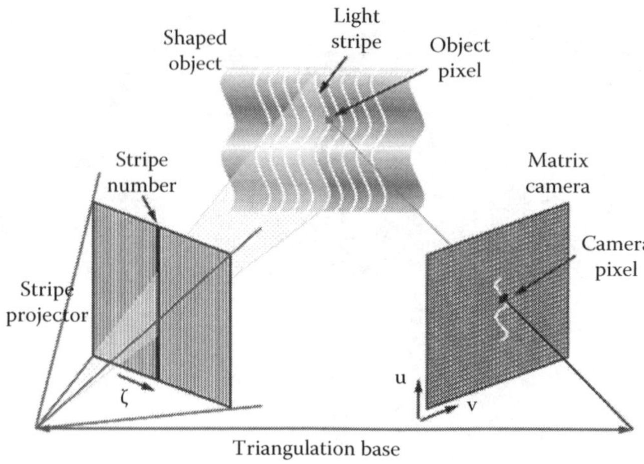

FIGURE 6.4
Basics of fringe scanner. (From Brown, J., *Photogramm. Eng. Remote Sens.*, 71(6), 677, 2005.)

FIGURE 6.5
Example of fringe projection on a part. (From Wikipedia [Web] on http://en.wikipedia.org/wiki/Structured-light_3D_scanner, visited on December 28, 2011.)

If only one camera is used, the projector's position in relation to the camera must be determined in advance; if two cameras are used, it is enough to calibrate the stereoscopic pair.

Speed is the biggest advantage of structured light, because it is possible to acquire thousands of points in a single acquisition rather than having to scan the points sequentially. Structured light also makes it possible to acquire unstable or moving objects, or even track distortions [6].

As with optical scanning methods, reflective and transparent materials are problematic. This obstacle is most often overcome by placing a powder on the material in question to yield a matte surface [7]. This type of 3D scanner is covered in more detail in Chapters 2 and 7 of this book.

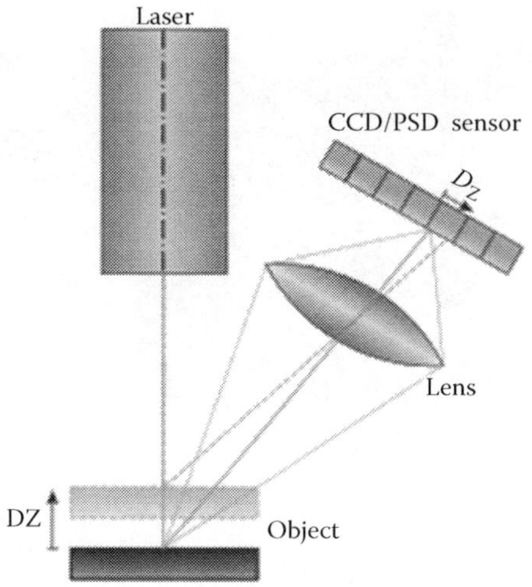

FIGURE 6.6
Distance estimation using the geometry of the pair camera laser. (From Wikipedia [Web] on http://en.wikipedia.org/wiki/File:Laserprofilometer_EN.svg, visited on December 28, 2011.)

6.2.3.2 Laser Scanners

Laser scanners project a point or a line on the object, but in this case, the pattern is a simple line. The point or line is observed by one or two cameras, making it possible to triangulate the 3D coordinates of the projected point shown in Figure 6.6. The surface is reconstructed progressively by moving the scanner in order to sweep the surface of the object with the projected line [8].

The National Research Council of Canada was one of the first institutes to develop this type of technology in 1978 [9].

6.3 Basics of Triangulation-Based Laser Scanners

If we look at the evolution of 3D measurement, we observe an improving trend in the speed and density of measurements. In a number of applications (measuring complex or free-form parts, for instance), the measurements needed to conduct the analysis or test must be information dense but should not unduly lengthen the required measuring time. This is particularly true when measuring parts as part of a first-article inspection or an in-depth tooling control process, because these types of metrology situations require the collection of dense data on all areas of the part.

This need is what sparked the invention of triangulation-based laser scanners, which make it possible to shift from point-by-point 3D measurements to fast measurements that can range in speed from thousands of points per second to millions. These scanners are particularly well suited to applications requiring comprehensive inspections.

As we mentioned earlier, these 3D scanners operate using the principle of triangulation (see Figure 6.7). A laser point or line is first projected on the object being measured. The resulting pattern is then recorded with optical components (usually one or two cameras) based on the principle of triangulation.

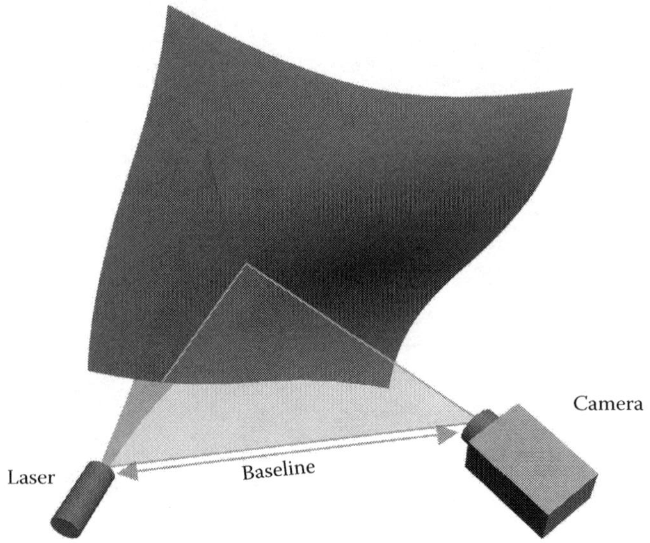

FIGURE 6.7
Basic soft triangulation–based laser scanning.

The main challenge with this type of laser scanner is to accurately follow scanner movements so as to place all reconstructed points in the same referential and align or record all of the points measured.

6.3.1 Laser Scanners Using an External Positioning Device

The natural solution to this challenge was to combine laser scanners with another system that had a referencing system, in this case, laser scanners using an external positioning device. With this type of system, each laser line is recorded by the laser scanner and placed in a 3D coordinate system defined by the external positioning system. The practice of combining scanning systems with CMMs or "portable CMMs" measuring arms illustrates this concept well (see Figure 6.8).

Other external referencing systems, such as a local GPS, may be used to obtain all the points measured in the same coordinate system. Even though these systems are all different, they nonetheless have one thing in common: to produce an accurate measurement, the referencing system they use must have a fixed position in relation to the object being measured. Consequently, special attention must be paid to the rigidity of the measuring setup when using the laser scanner with an external positioning reference (see Figure 6.9).

6.3.2 Self-Positioning Laser Scanners

In the mid-2000s, a major breakthrough took place in the laser scanner industry, with the appearance of self-positioning laser scanning systems. These 3D devices use no external positioning systems. Instead, they use retroreflective targets placed on the object as a positioning system.

These systems rely on two cameras that, in addition to triangulating the laser line, are also used for triangulating the object being measured—more specifically to determine the scanner's position in relation to specific points placed on the object (i.e., retroreflective targets) (see Figure 6.10).

FIGURE 6.8
Left: CMM. Right: 3D laser scanner mounted on a CMM.

FIGURE 6.9
Example of a rigid measuring setup when using a portable CMM.

6.4 Use of Self-Positioning Laser Scanners for Metrology

6.4.1 Self-Positioning 3D Laser Scanners

Laser scanning systems that use retroreflective targets as references have no need for external positioning systems, because the scanner's relative position in relation to the part or object being measured is determined by the principle of triangulation between the

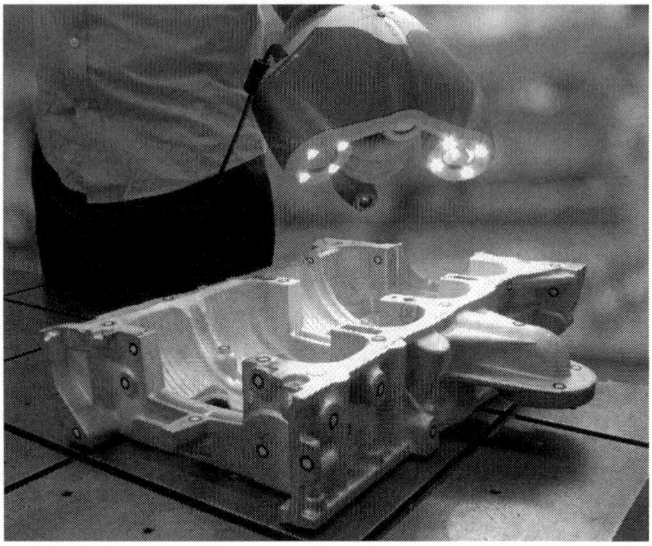

FIGURE 6.10
Self-positioning laser scanner measuring a casting part.

scanner's two cameras and the patterns of positioning targets, which is carried out in real time during data acquisition (see Figure 6.11b).

There are three main advantages to using these scanning systems in metrology. The first applies to all scanning systems: the ability to measure complex or free-form surfaces. Three-dimensional measurement by probing is the ideal way to measure geometric shapes but quickly proves ineffective and ill suited to measuring free-form surfaces that require a measuring speed of thousands of points per second. Scanning systems make it possible to quickly and effectively measure a very large number of points and fully characterize a complex or free-form surface.

The second advantage concerns self-positioning 3D scanners only: they are portable and their systems are simple to operate. Since these measuring systems do not require external references, they are much easier to use and move. As a result, they can be easily transported to the part being measured, no matter where it is.

The third advantage, and by no means the least important, is that these scanners do not require a fixed measuring setup during data acquisition—unlike other measuring systems where the system reference and part being measured must be set up on a rigid mount. For example, when measuring a part using a CMM or a measuring arm, you must make sure that

1. The part is securely fastened to a steel or marble table or any other support capable of providing a high level of rigidity and stability.
2. The CMM or measuring arm is also fixed in relation to the part (vibrations in the measuring environment must be taken into account).

With self-positioning scanners, the measurement reference is attached directly to the object, because the positioning targets are affixed to it (see Figure 6.11a). As a result, even if the object moves during data acquisition, the reference will remain fixed in relation to the part. These scanning systems use a reference that is measured in relation to the object on a continuous basis by the scanner's cameras during data acquisition.

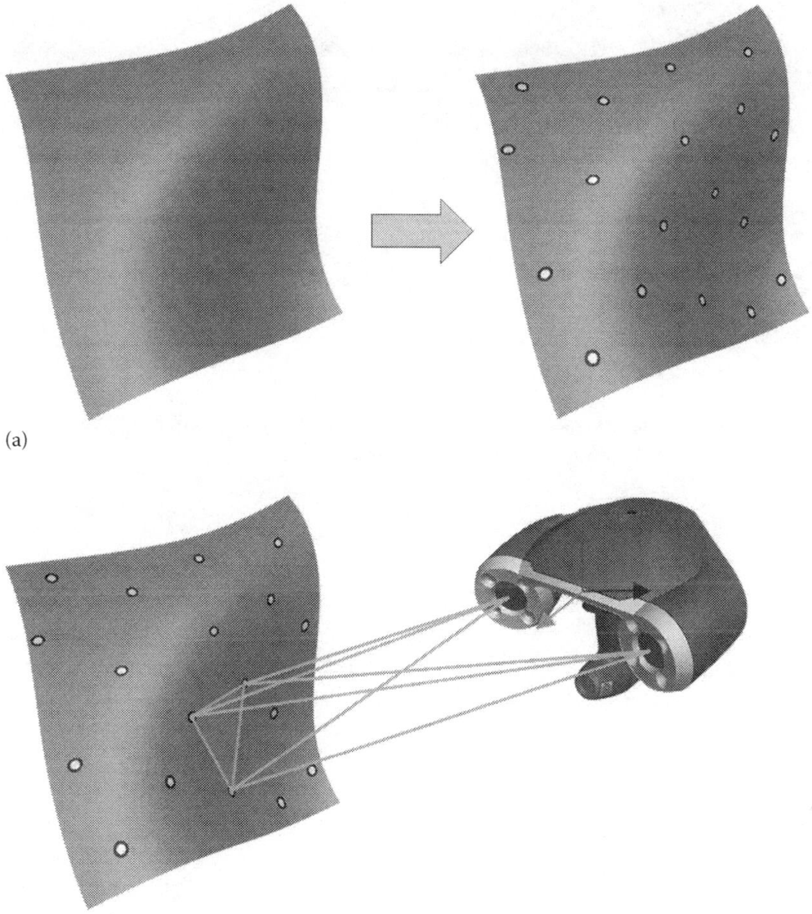

(a)

(b)

FIGURE 6.11
(a) When a free-form surface is measured using a self-positioning system, positioning targets are affixed to the part. (b) The scanner determines its relative position in relation to the object using the principle of triangulation between the scanner's two cameras and the random patterns of positioning targets.

6.4.2 Creating 3D Measurements

6.4.2.1 Laser Scanner Traditional Output

Once the scanner is positioned in space in relation to the part being measured, it must generate 3D measurements. At this stage, all laser scanners operate the same way by projecting one or more laser lines that conform to the shape of the object. The laser lines distorted by the object surface are then read and recorded by the scanning system camera(s) (see Figure 6.12), and the points extracted from this image are then placed in 3D space based on the reference or coordinate system created beforehand. Once data acquisition is complete, traditional laser scanners produce a 3D point cloud (see Figure 6.12), whereas the latest generation of portable scanners (which use automatic referencing via positioning targets) operate according to a similar process but produce more advanced and reliable measurements.

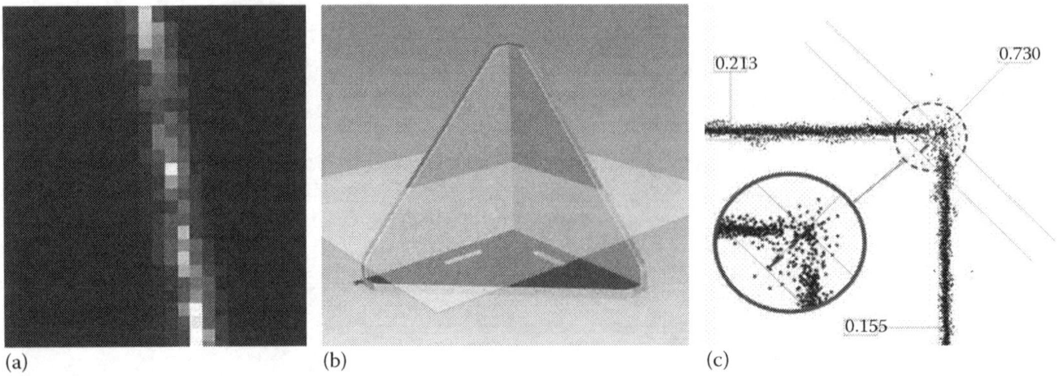

FIGURE 6.12
(a) Example of an image section of a laser trace. (b) Example of a 3D point cloud: output of a traditional laser scanner. (c) Distribution of the 3D points along the cross section.

6.4.2.2 Optimization of 3D Measurements

Self-positioning scanners also use a laser projection to characterize an object's surface. The laser is projected in the shape of a crosshair (see Figure 6.13a). Then, the surface is scanned by manually moving the scanner over the entire surface to be measured (see Figure 6.13b). However, with this type of scanner, the measurement produced is not a 3D point cloud but rather a more highly advanced and optimized 3D measurement shape: a mesh from an optimized surface that takes into account measuring conditions and measurement consistency (see Figure 6.13b).

6.4.3 Intelligent Measurement Process

This new generation of self-positioning laser scanners use not only an automatic reference system but also a new process for creating 3D measurements. Unlike conventional laser scanning systems, these systems are not limited to using simple data filtration methods. They instead use an intelligent measurement process that draws on the large quantity of measurements made available and analyze local distribution to evaluate consistency with the error model developed at the device calibration stage (see Figure 6.14). This serves as an additional data refinement step in comparison to conventional laser scanner 3D point generation methods. Moreover, the intelligent measurement process makes it possible to incorporate these steps in real time during data acquisition [10].

With this intelligent measurement approach, the measurements produced will be optimized and more accurate than those yielded from a simple 3D point cloud. This method therefore represents an advance in the way laser scanner data are used in metrology (see Figure 6.15).

6.5 Use of Optical CMM for Metrology

6.5.1 Main Causes of Measurement Uncertainty

Before discussing the advantages of optical CMMs, it is important to remember that a measurement can be affected by a number of factors, including

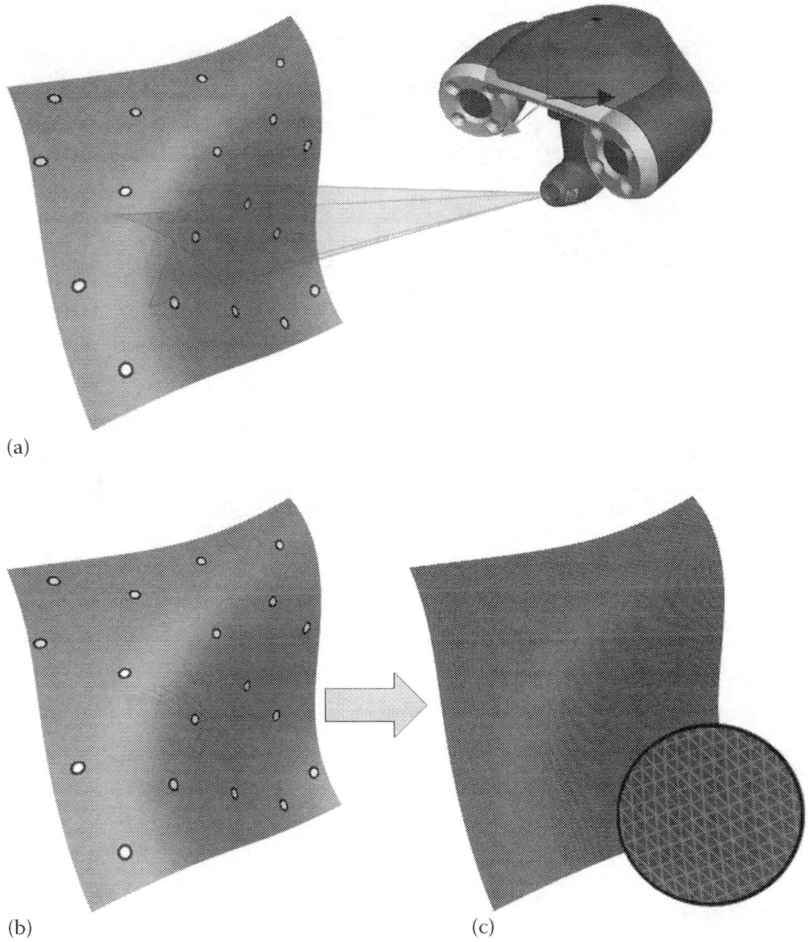

FIGURE 6.13
(a) The self-positioning scanner projects a laser crosshair onto the object's surface. (b) The object is scanned by manually moving the scanner over the entire surface to be measured. (c) An optimized mesh is generated.

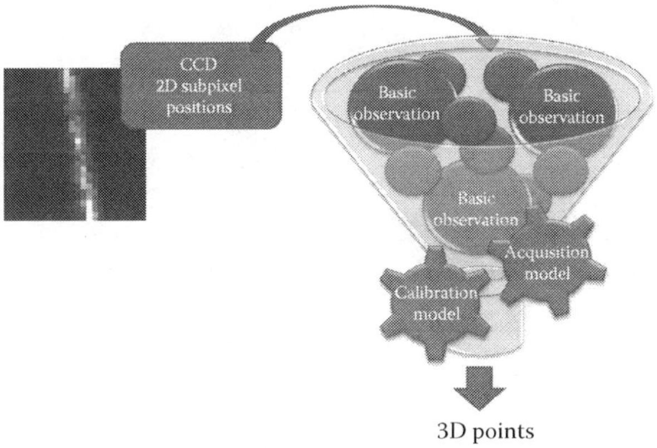

FIGURE 6.14
Creating 3D points in an intelligent measurement process.

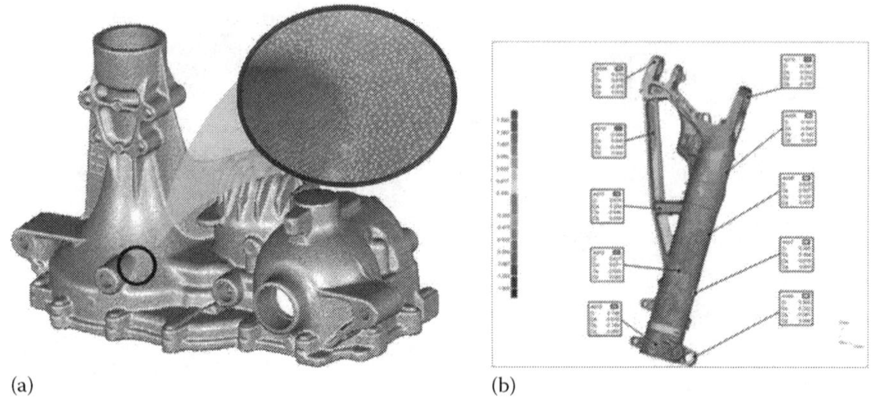

FIGURE 6.15
Examples of typical applications for intelligent measurement systems. (a) Example of the quality of the output generated. (b) Example of a typical application in the aerospace industry: inspection of a landing gear component. (Photo Courtesy of Messier-Bugatti-Dowty, Vélizy, France.)

1. Measuring machine
2. Measuring environment (temperature, vibrations, etc.)
3. Operator (skills, stress level, etc.)
4. Measuring method
   a. The part being measured

An easy way to remember this is to think of the 5 M illustrated in Figure 6.16: 1-Machine, 2-Man, 3-Measured part, 4-Method, 5-Measuring conditions. For example, conventional CMMs, measuring arms, and laser trackers are very sensitive to the environment (measuring conditions), whereas laser trackers require highly qualified operators, and measuring arms sometimes need leapfrog methods to measure larger parts, which can be a major source of errors.

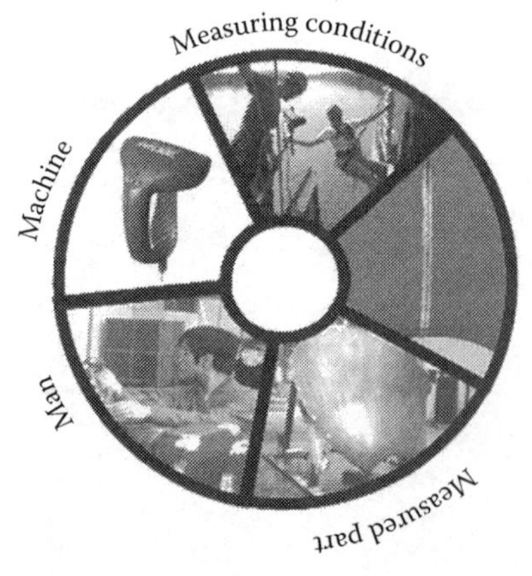

FIGURE 6.16
The 5M law.

So when deciding which measuring solution to use, it is important to take into account not only the accuracy of the machine but also its ease of use for a production operator and its overall performance under factory measuring conditions that are more restrictive than a lab.

Triangulation-based optical CMMs, for example, are easy to use for factory applications because, unlike measuring arms, they have no mechanical link. And unlike laser trackers, which require the implementation of a specific procedure to enable the tracker to lock back in on the prism when the laser beam is cut off by an obstacle, this type of CMM is tolerant of beam cutoff. With an optical tracker, even if the probe is accidentally obscured by an obstacle, it will be automatically detected once it is visible again.

But the biggest advantage of this solution is its robustness in typical shop floor environments—yielding a degree of measurement precision equivalent to that achieved in the lab using a conventional solution.

To achieve comparable levels of precision in a production line setting, some procedures that are beneficial include

- Quick and easy daily user calibration using a certified bar
- Continuous monitoring of the system's internal parameters (temperature, geometry)
- Dynamic referencing enabling the coordinate system to be locked onto the object being measured
- Automatic measurement range volume extension, making it possible to move the CMM in relation to the object without losing alignment
- Automatic alignment function using optical reflectors

Only optical CMMs offer all of these functionalities, a major breakthrough in the area of production line dimensional inspection. Let's take a closer look at these functionalities.

6.5.2 Dynamic Referencing

Dynamic referencing is a method that uses swivel reflectors rather than tooling balls. The main difference of dynamic referencing to traditional measurement lab methods is that conventional CMMs measure tooling balls only once at the beginning of the measuring phase, whereas optical CMMs measure optical reflectors continuously and simultaneously with each point (which is equivalent to realigning the part as each point is measured). Measurements are thus taken as if the part were attached to the optical CMM—in exactly the same way it would if the part were attached to the table of a conventional CMM (see Figure 6.17). The optical CMM recreates the stability of conventional CMMs with their granite tables and vibration isolation pads but in a virtual manner using a machine that is completely portable and can be used by a single operator anywhere in the production plant.

Instead of optical swivel reflectors, it is also possible to use simple adhesive retroreflective targets, which can be placed anywhere on the part. However, optical reflectors do present a number of advantages over these targets:

- Since the swivel reflector is centered on the axis of the swivel optical reflector, it is possible to move the CMM around the part without losing alignment. The reflectors are turned on their axes so they remain visible to the CMM.

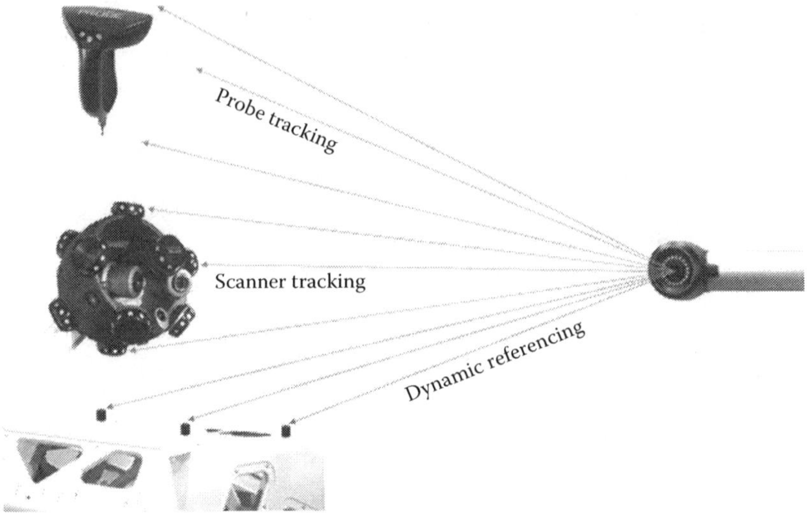

FIGURE 6.17
Using optical triangulation to dynamically track part and measuring tools.

- Optical reflectors can be used in place of mechanical reflectors for precise position locating (e.g., on a tool). This comes in very handy when it comes to automatically aligning the part or the tool, as we will see in the next section.

6.5.2.1 Automatic Alignment

We will once again use the optical reflectors described earlier, but this time, we will place them in precisely defined locations that have been measured beforehand. That means we know the 3D coordinates of these locations in the optical reflectors referential (measuring referential, CAD referential). Providing these coordinates to the CMM links the CMM's work referential to the part referential for the entire duration of the measurement (thanks to its dynamic alignment functions).

From that point on, all the points measured will be expressed directly in the part referential, eliminating the need to measure points by probing to align the part. This not only saves time but also, and more importantly, gets rid of all operator-based alignment errors on alignment points. These kinds of operator errors are very common and result in inspection defects that are often difficult to spot.

6.5.3 Calibration by the User

Another benefit of certain optical CMMs is the ability to quickly calibrate the CMM directly at the work site. With such capability, calibration takes minutes using a calibration bar that is certified annually by a laboratory that is a member of the International Committee for Weights and Measures. Operators position the calibration bar, which is equipped with several reflectors at known distances, at various locations within the CMM measurement volume. The CMM automatically identifies and uses the resulting data to immediately compensate for any measurement volume errors. This provides the operator with guaranteed stability and accuracy for the entire duration of use, whereas with other portable CMMs calibrated annually, highly restrictive shop floor environments can cause drift that can be difficult to identify.

6.5.4 Example of Application

Here's an overview of a typical inspection process using automatic alignment features based on an example from the aerospace industry.

The objective in this case is to control aeronautic parts before assembly by checking functional dimensions and edges for step and gap. These measurements have to be performed in unstable shop floor conditions by production line operators who are not metrologists. Temperature variations and vibrations from trucks and bridge cranes are always present.

Moving parts to the metrology lab for inspection on fixed CMMs would deliver accurate results but would take a lot of time, resulting in a bottleneck at the inspection process.

Using a portable CMM makes it possible to inspect parts directly on the shop floor. But conventional portable CMM solutions (arms, laser trackers) remain very sensitive to environment instability. They need fixed setups (jigs or tooling) which are not always practical and are expensive.

As probing remains the best way to measure geometrical features, an armless and wireless optical portable CMM can be easily moved to any location within the plant. It provides dynamic referencing feature that automatically locks the machine's coordinate system on to the part. This way, the alignment remains accurate during the entire measuring sequence.

6.6 How Does It Work?

The only preparation required is to position a few passive reflectors on the part or tool. As described earlier, the reflectors are measured continuously by the optical tracker.

Then, the measured points are directly computed in the local reference system attached to the part. The part or the tracker can move during measurement without having any impact on measurement accuracy. This makes it possible to move the part for better access to a particular area or simply to extend the measuring volume without having to do a painful leapfrog (NEXT).

Another feature is the extension of dynamic referencing. The 3D coordinates of certain specific elements on the part—the position of tooling balls on a jig, for example—are exactly known. Replacing the tooling balls with tooling reflectors and entering their coordinates in the optical CMM software make the machine reference system equivalent to the part reference system. Each point measured is now computed directly in the part reference system, eliminating the need for alignment. This saves time and drastically reduces the number of errors during a measuring process, as 75% of errors derive from the alignment procedure.

Let's look more closely at how this particular application works. The customer needs a solution to check the thrust reverser's pivot doors of an airplane engine. A total of 150 points have to be measured manually in less than 15 min, including the inverter door installation on the jig. The jig has to accommodate four types of doors, and the inspection report must be generated automatically.

The first step is to align the rotation axis of the door and the position of the lock. As points need to be taken on both sides of the door, the alignment has to be saved when rotating the door.

Two reflectors are fixed on the rotation axis of the door (see Figure 6.18). They are automatically measured by an optical tracker, generating two points used to compute a line.

Two other reflectors are placed on the plane support for the lock. The position of the reflectors is defined to ensure that the median point between the reflectors is at same height as the center point of the lock.

Three points are now available to create an automatic alignment. To make it work with different door types and in both door positions, reflectors are positioned in the different places reached by the lock.

Several reflectors are also added on the jig to continuously check jig conformity.

The control program primarily includes step and gap points all around the part and is repeated on both sides of the door. Using the metrology software probing assistant, the operator simply needs to probe points at the location indicated by a target in the CAD view. Only points taken inside a sphere centered on the target point are accepted.

When the door is rotated, the positions of the reflectors placed on the jig around the lock are automatically used for the alignment. They are measured at the beginning of the process so as to become invisible when rotating the door. The measurement process can thus continue.

Some specific tools have also been developed to obtain fast and accurate measurement of specific entities. For example, a tool with a target centered on it has been developed to measure the hole used for the hydraulic jack. The tool is positioned by the operator at the appropriate spot before starting the measurement process. By measuring the target, the center of the hole is automatically computed, eliminating the need to probe multiple points (on the supporting plane and inside the hole) and the potential errors associated with this process.

Thanks to automatic alignment, control time is reduced to only 15 min including the time required to put the door on the jig and to remove it. Previously, this would have taken around 2 h on a traditional CMM. Control time has been divided by a factor of 10.

Using automatic alignment and dynamic referencing, combined with probing assistants, 75% of errors are avoided. An automatic continuous verification of the jig is also available, thanks to the target permanently placed on the jig. The position of the targets can be verified on an annual basis using laser trackers and compatible tooling (available from major optical tooling manufacturers).

6.7 Summary Pros and Cons: When to Use a Portable CMM

Portable CMM technology offers a specific range of capability and convenience for metrology applications on medium and large parts. Typically, portable CMMs involved manual manipulation of a probe to measure the part but can be combined with a robotic system for automated operation. For controlled part inspection, the use of an optical probe such as a laser line triangulation probe, the inspection may be done using an optical probe on an articulated arm by a manual operator. However, for less-defined part and particularly for larger volumes of parts where detailed metrology is needed, a system with a remote head provides distinct advantages. Unlike many other metrology tools, a high degree of

(a)

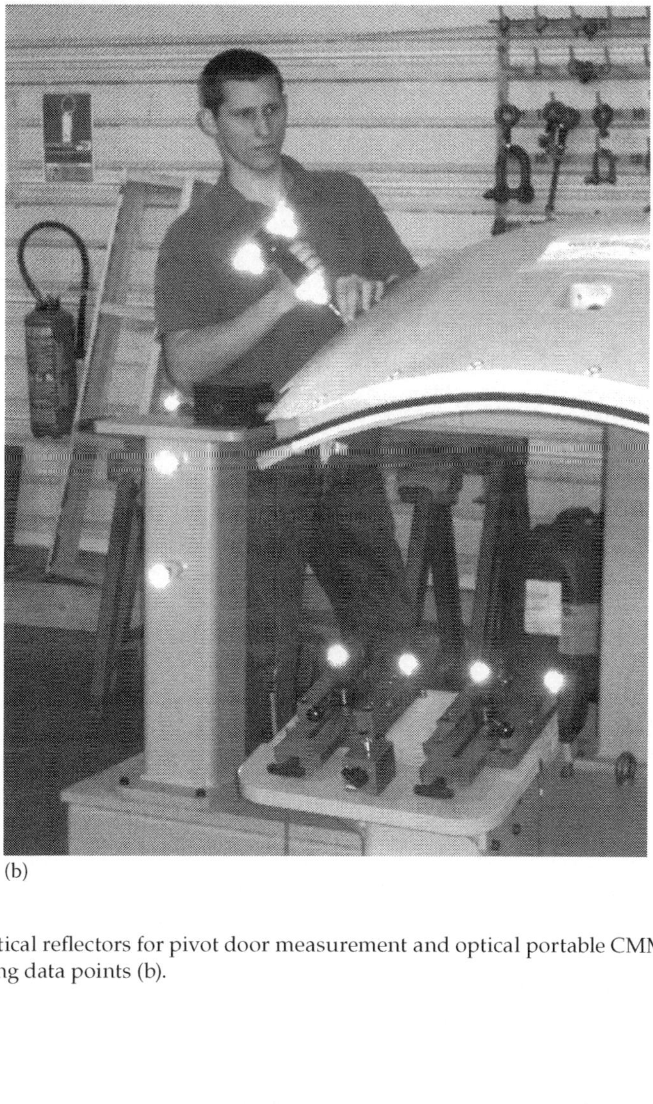

(b)

FIGURE 6.18
Jig equipped with optical reflectors for pivot door measurement and optical portable CMM. Overview of setup (a), and operator taking data points (b).

flexibility can make the measurement of volumes the size of a room with complexed features practical, that otherwise could be very time consuming. The advantages provided by a remote head optical triangulation probe include

- The ability to measure in an unstructured environment and undefined parts
- The ability to measure features from many directions, changing the direction of view as needed
- A high degree of flexibility in determining what and where to measure
- A working volume comparable to tracker systems but with more complete surface information collected at a faster rate

Application where remote head optical CMM application may not offer any advantage might include

- Volume production where a larger batch of the same part is measured repeatedly
- High-resolution measurement applications where micron-level measurements are needed
- Simple point-to-point measures such as larger structure alignment often done with trackers (covered in another chapter)

All in all, remote head, triangulation-based CMMs provide a unique range of capabilities not readily matched by other methods, thereby filling an important niche for a wide range of medium- to larger-size metrology applications.

References

1. CMMMETROLOGY [Web] on http://www.cmmmetrology.co.uk/history_of_the_cmm.htm (visited on December 28, 2011).
2. RENISHAW [Web] on http://www.renishaw.com/en/Our+company—6432 (visited on December 28, 2011).
3. AFNOR [Web] on http://www.bivi.metrologie.afnor.org/ofm/metrologie/ii/ii-80/ii-80-30 (visited on December 28, 2011, in French).
4. FERRIS [Web] on http://www.ferris.edu/faculty/burtchr/sure340/notes/history.htm (visited on December 29, 2011).
5. Brown, J., 2005, Duane C. Brown memorial address, *Photogramm. Eng. Remote Sens.*, 71(6):677–681.
6. M. C. Chiang, J. B. K. Tio, and E. L. Hall. Robot vision using a projection method, *SPIE Proc.* 449, 74 (1983).
7. *Fringe 2005, The 5th International Workshop on Automatic Processing of Fringe Patterns.* Berlin, Germany: Springer, 2006. ISBN: 3-540-26037-4, 978-3-540-26037-0.
8. J. Y. S. Luh and J. A. Klaasen. A real-time 3-D multi-camera vision system, *SPIE Proc.* 449, 400 (1983).
9. Mayer, R., 1999, *Scientific Canadian: Invention and Innovation from Canada's National Research Council.* Vancouver, Canada: Raincoast Books. ISBN: 1551922665, OCLC 41347212.
10. Mony, C., Brown D., Hebert, P., Intelligent measurement processes in 3D optical metrology: Producing more accurate point clouds. *J. CMSC*, October 2011.

7

Phase-Shifting Systems and Phase-Shifting Analysis

Qingying Hu

CONTENTS

7.1 Introduction

Phase-shifting surface profile measurement is a very important branch in optical metrology. When compared with other surface measurement techniques, it has many unique features such as varieties of configurations, high resolution, high accuracy, good repeatability, fast measurement speed, and superior surface finish tolerance. Especially in the past several decades, with the help of digital image devices and dedicated computer software, phase-shifting images were automatically processed at high speed over a full field of view (FOV), further enabling superfast 3D measurement without scanning. This chapter covers most of the aspects related to phase-shifting systems, including system configurations, phase-shifting algorithms, modeling and calibration of phase-shifting systems, and error analysis and compensation for accuracy improvement.

7.2 Phase-Shifting System and Its Benefits

Depending on how a fringe pattern is generated and how it is shifted, various phase-shifting configurations are available for use in optical metrology, aiming at different applications.

7.2.1 Fringe Patterns

A fringe pattern is a periodic grayscale pattern with alternative dark and bright areas. Based on pattern generation principles, the most common fringe pattern can be classified into three categories: interference pattern, moiré pattern, and projected pattern.

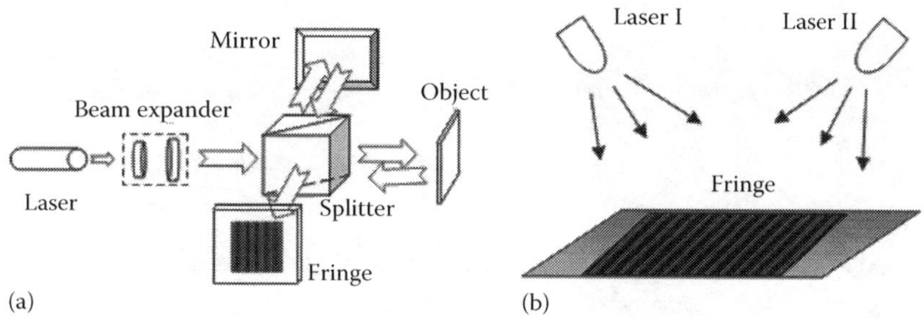

FIGURE 7.1
Interferometers: (a) Michelson interferometer and (b) AFI.

7.2.1.1 Interference Fringe Pattern

An interference pattern is well known in optical interferometers. When two coherent light beams with common polarization superpose in an area, at each point, the resulting light intensity, shown as grayscale in a camera, depends on the optical path difference (OPD) between these two light sources reaching this point. The OPD results in phase difference at a certain point, destructive or constructive, forming periodic pattern on the object surface. A good example is the famous Young's experiment in optics.[1]

In optical metrology, the two slots in Young's experiment are usually replaced by a splitter, either polarized or nonpolarized, to generate two wave fronts: one is a measurement wave front that is modulated by the geometric variation of the surface, and the other is a reference wave front under good control. When these two wave fronts superpose, the difference between them is revealed in an interference fringe pattern. A typical Michelson interferometer is shown in Figure 7.1a. Adjusting the tilting angle of the reference mirror will change the pitch of the fringe pattern. This interferometer has very high resolution, up to hundredth of wavelength, but the FOV is usually limited because the light beam diameter after expansion has to be slightly larger than the FOV so as to confirm to the related components. For large FOV measurements, the system will be too large and too costly.

To measure a large area with a small instrument footprint, a technique called Accordion Fringe Interferometry (AFI) was developed at MIT's Lincoln Laboratory in 1990s,[2] as shown in Figure 7.1b. AFI uses two-point lasers to illuminate the target divergently and a camera to record the interference fringe pattern that is modulated by the surface geometry of the sample under measurement. It also provides excellent accuracy performance with a large FOV but small footprint. Because the fringe pattern results from laser interference, the depth of focus for the fringe projection unit is infinite.

7.2.1.2 Moiré Pattern

A shadow moiré pattern looks like an interference pattern, but its geometric interference principle is very different.[3–5] Figure 7.2a shows a representative shadow moiré fringe image. It is generated by covering the measurement area with a physical grating while illuminating and viewing the area from an opposite direction, as shown in Figure 7.2b. When light passes through the grating at a tilted angle, a shadow of the grating will be generated on the sample surface. When this shadow is observed at a tilted angle from the opposite direction to the light source, a moiré pattern can be seen that represents the topology of the surface with the peak/valley rings representing the same height relative to the physical grating.

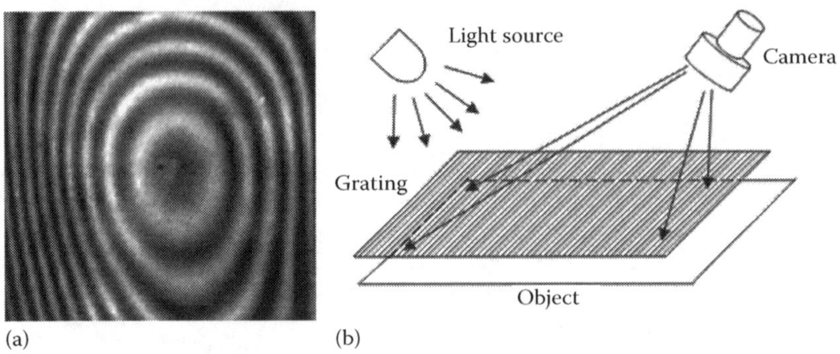

FIGURE 7.2
Shadow moiré: (a) moiré pattern and (b) setup.

7.2.1.3 Fringe Projection with a Physical Grating

When a transmission grating with a sinusoidal transmission profile such as holographic gratings is placed between a light source and a projection lens, the projected fringe pattern will also have a sinusoidal intensity profile, as shown in Figure 7.3a.[6] If a straight-line grating with a nonsinusoidal profile such as a ruled grating is used, the projection lens is usually defocused slightly so that a pseudo-sinusoidal pattern can be obtained. Figure 7.3b shows a projected fringe on an edge break.

Another technique called projection moiré requires a second physical grating to be placed before the camera lens.[7] The second grating can have a different pitch than that of the first grating used for fringe projection. In this configuration, a traditional moiré pattern will be captured in the imaging system. This technique is out of focus of this chapter but has been covered in Chapter 8.

7.2.1.4 Digital Fringe Projection

In digital fringe projection,[8–10] the fringe pattern can be generated with theoretically any intensity profile using computer software and projected to the object surface through an

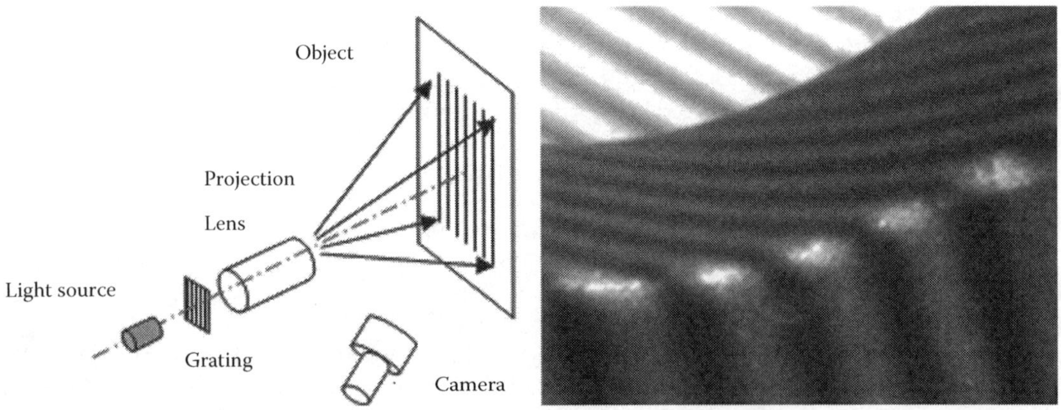

FIGURE 7.3
Projection moiré: (a) setup and (b) pattern.

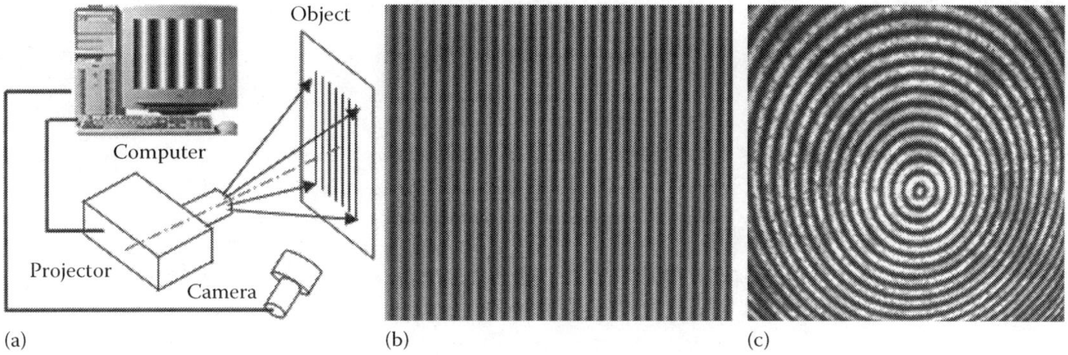

FIGURE 7.4
Digital fringe projection: (a) setup, (b) straight fringe pattern, and (c) circular fringe pattern.

off-the-shelf digital projector such as liquid crystal device (LCD), digital mirror device (DMD), and liquid crystal on silicon (LCOS) projectors. This provides a low-cost and flexible solution for fringe projection techniques. Figure 7.4 shows the typical set up and two projected fringe patterns.

7.2.1.5 Other Special Fringe Patterns

All fringe patterns discussed so far have sinusoidal or pseudo-sinusoidal intensity profiles. Sometimes, other special fringe patterns are also used for a specific purpose such as speed and simplicity considerations. These patterns include the trapezoidal pattern,[11–13] sawtooth,[14] and slope profile. Because this chapter focuses on traditional phase-shifting techniques related to sinusoidal patterns, these special patterns and related algorithms are not investigated further. Interested readers can find details in the corresponding references.

7.2.2 Fringe Pattern Analysis

7.2.2.1 Contour Analysis

To extract the geometric information in the fringe pattern, an appropriate analysis methodology has to be used. To better understand the challenges, it is necessary to take a look at the intensity profile of the captured fringe pattern image. Without losing generality, a representative cross section near the middle horizontal line of the fringe pattern depicted in Figure 7.2 is taken as an example shown in Figure 7.5.

Before the phase-shifting technique was invented, the only way to investigate the fringe pattern image was to count the peak and/or valley and follow the contour curve along the peaks and valleys,[3] as demonstrated in Figure 7.6. The calibration process was to find the factor that converted the peak/valley into the height dimension and was used to estimate the height variation over the FOV. The denser these fringes are, the steeper the slope magnitude is of the surface area. The resolution and accuracy of this analysis method was very low, and there is no way to identify the direction of the slopes from a single image without introducing a known tilt to the part, creating a bias fringe larger than any other expected slope. That is, by tilting the part, all slopes are made to be perturbations to that slope. The requirement for a bias greatly limits the use of such systems for measuring real parts.

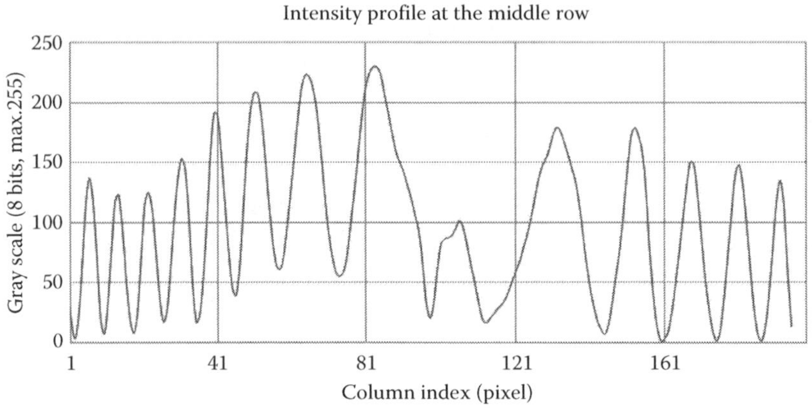

FIGURE 7.5
Intensity profile of the middle horizontal cross section in the moiré fringe.

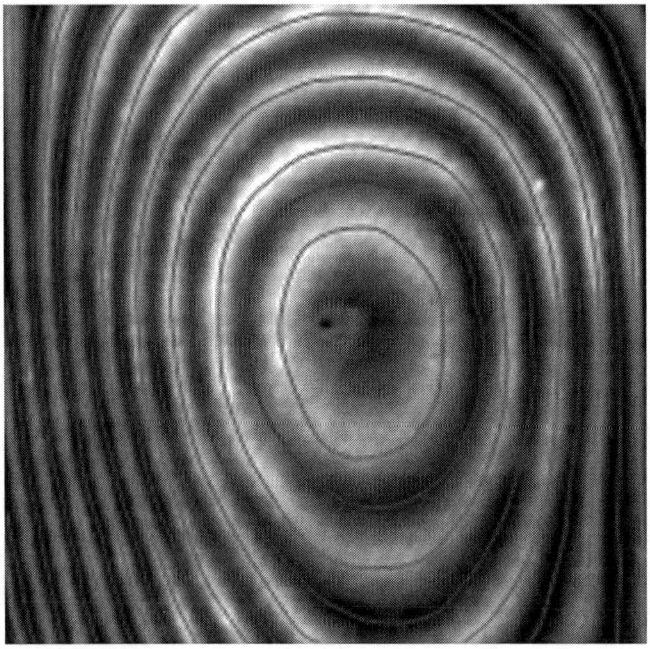

FIGURE 7.6
Contour showing peak (bright bands) and valley (dark bands).

7.2.2.2 Phase-Shifting Analysis

In the 1970s, thanks to the invention of digital cameras and computers, digital image analysis started to be used in optical metrology, and the phase-measuring methods became a reality that greatly improved the resolution, accuracy, speed, and repeatability of interferometers and moiré technology.[15] Over the years, various phase-measuring methods have been developed, with the phase-shifting method being the technique most widely used.[16,17]

An entire phase-shifting analysis process is demonstrated in Figure 7.7 using a three-step phase-shifting algorithm. In Figure 7.7, the three images in the top row are captured three fringe images of a master model with a 120° phase shift. The intensities of the three phase-shifted images at point (x, y) can be written as

$$I_1(x,y) = I'(x,y) + I''(x,y)\cos\left[\phi(x,y) - \frac{2\pi}{3}\right] = I'(x,y)\left\{1 + \gamma(x,y)\cos\left[\phi(x,y) - \frac{2\pi}{3}\right]\right\} \quad (7.1)$$

$$I_2(x,y) = I'(x,y) + I''(x,y)\cos\left[\phi(x,y)\right] = I'(x,y)\left\{1 + \gamma(x,y)\cos\left[\phi(x,y)\right]\right\} \quad (7.2)$$

$$I_3(x,y) = I'(x,y) + I''(x,y)\cos\left[\phi(x,y) + \frac{2\pi}{3}\right] = I'(x,y)\left\{1 + \gamma(x,y)\cos\left[\phi(x,y) + \frac{2\pi}{3}\right]\right\} \quad (7.3)$$

where
- $I'(x, y)$ is the average intensity
- $I''(x, y)$ is the intensity modulation
- $\phi(x, y)$ is the phase to be determined

By solving the earlier equations, phase $\phi(x, y)$ and image contrast $\gamma(x, y)$ can be obtained as

$$\phi(i,j) = \tan^{-1}\left(\sqrt{3}\,\frac{I_1 - I_3}{2I_2 - I_1 - I_3}\right) \quad (7.4)$$

$$\gamma(i,j) = \frac{I''(x,y)}{I'(x,y)} = \frac{\left[(I_3 - I_2)^2 + (2I_1 - I_2 - I_3)^2\right]^{1/2}}{I_2 + I_3} \quad (7.5)$$

This wrapped phase map includes the modulo 2π discontinuity, as shown in Figure 7.8. The continuous phase map $\Phi(i, j)$ can be obtained by use of a phase-unwrapping algorithm, as shown in Figure 7.9. How to implement a fast and robust phase-unwrapping process in computer software programming for a complex irregular 2D geometry with various slopes and discontinuity, such as holes, requires one to have both programming skills and an understanding of the phase-unwrapping principle. This is a complex issue that will require an entire book to discuss.[18]

Because the values of the unwrapped phase depend on the starting point of the unwrapping process, the obtained phase map is a relative phase map and cannot be used directly to represent the surface geometry although it contains the geometric information. For flat surfaces, the phase map can be either subtracted from a reference phase map or brought down to reveal the defects or qualitative geometric features, as shown in Figure 7.10. But for a complex geometry shape or quantitative dimension comparison with geometric tolerance, the difference among the actual surface geometry and phase map is obvious, as shown in the bottom row of Figure 7.7. However, when used with the appropriate model and phase-to-coordinate conversion algorithm,[19,20] an accurate 3D shape can be reconstructed from the wrapped phase map, as shown in the bottom-right picture of Figure 7.7.

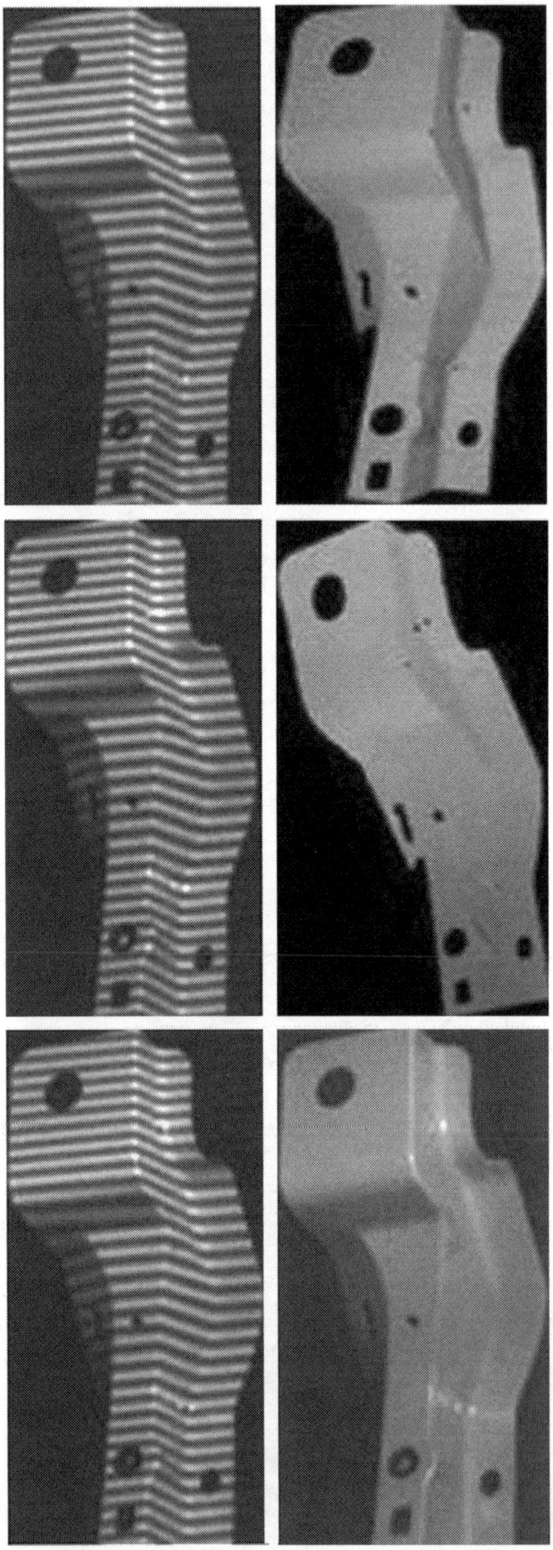

FIGURE 7.7
Phase-shifting fringe analysis process.

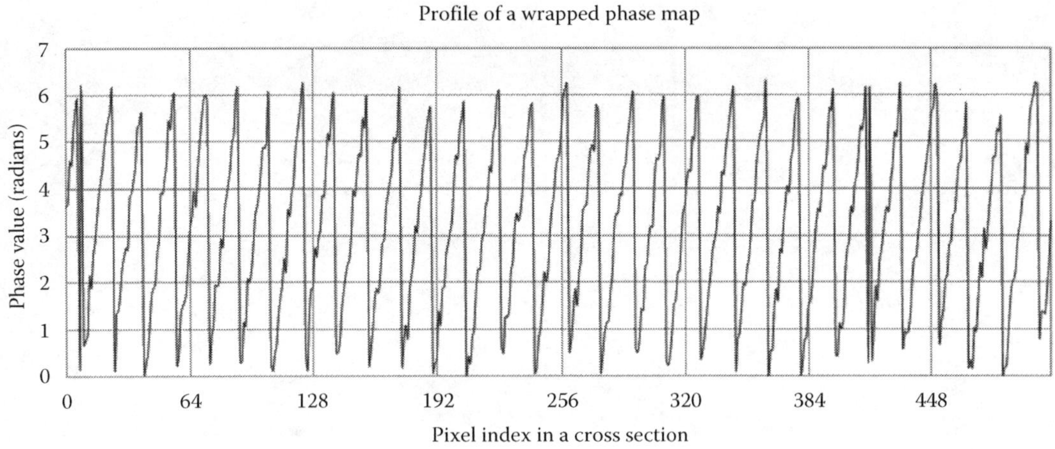

FIGURE 7.8
Profile of a wrapped phase map with 2π discontinuity.

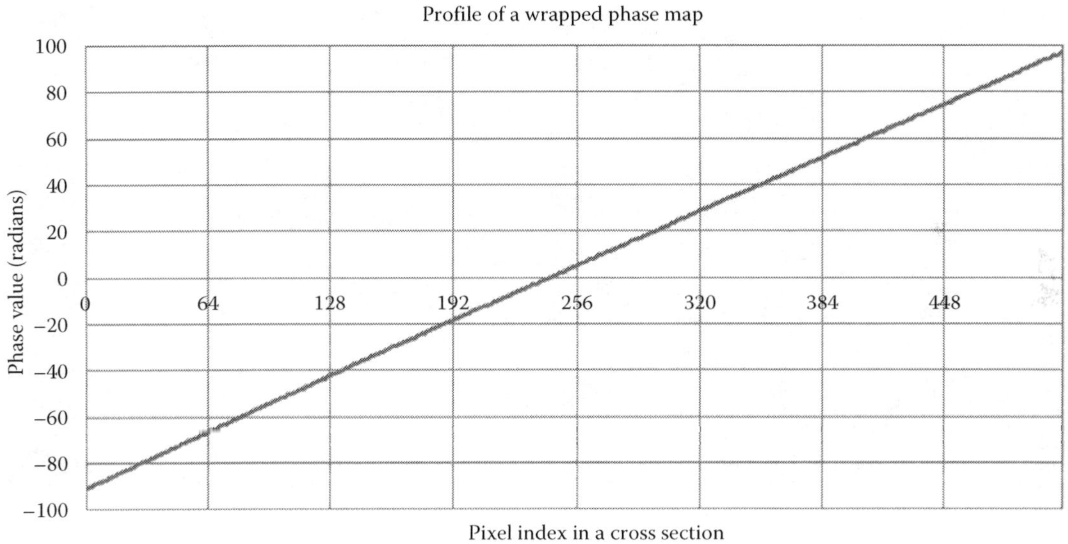

FIGURE 7.9
Profile of an unwrapped phase map without 2π discontinuity.

7.2.2.3 Benefits of Phase-Shifting Analysis

The phase-shifting analysis enables full-field analysis in areas because it provides geometric information for all sampled points between intensity peak/valleys. The obtained phase map provides directional information such as the positive or negative slopes and convex or concave local curvatures, along both lateral and vertical directions.

The phase-shifting analysis obtains phase information from image contrast, not intensity changes from peak to valley, thus enabling much higher accuracy and making the analysis tolerant to various surface finishes, including shiny surfaces on some parts. Figure 7.11 shows how the phase-shifting technique can be used to obtain dimensional

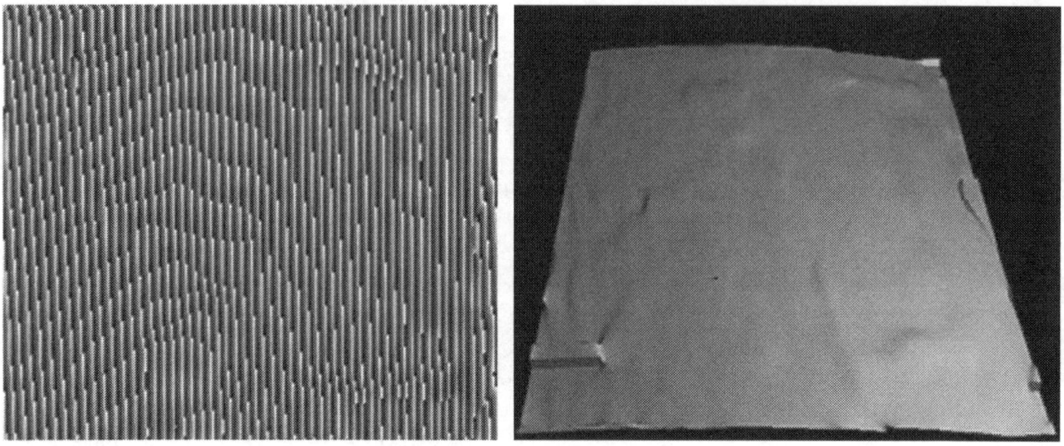

(a) (b)

FIGURE 7.10
Wrapped phase map (a) and unwrapped phase map (b) after bringing down.

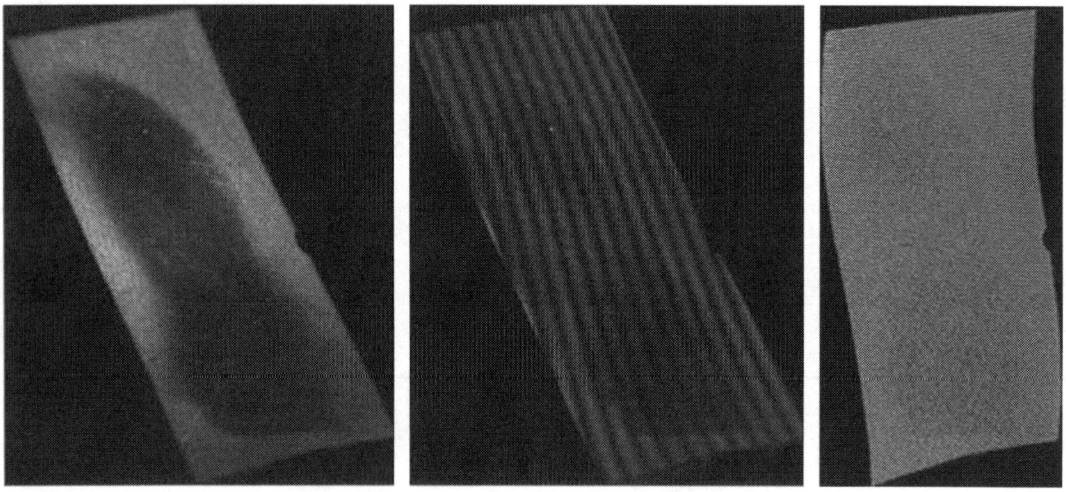

(a) (b) (c)

FIGURE 7.11
Shiny part measurement: (a) 2D picture, (b) fringe image, and (c) 3D point cloud.

information on a very shiny benched blade.[21] This surface tolerance feature makes the phase-shifting technique an excellent candidate for on-floor or in-line inspection in manufacturing because it eliminates the need for additional surface treatment of the parts to be measured.

7.2.3 Phase-Shifting Systems

There are various ways to do phase-shifting phase measurements. Phase-shifting systems can be classified into three categories: physical phase shifting requiring mechanical movement,[5,16,22–24] digital phase shifting through a digital projector without any movement,[8,9,12,21] and simultaneous phase-shifting techniques.

7.2.3.1 Physical Phase-Shifting System

In physical phase shifting, a translation stage such as a piezoelectric transducer (PZT) or other motorized stage is used to translate a component or a subsystem relative to others. In one phase-shifting Michelson interferometer,[25] the translated component is a mirror in the reference beam to introduce phase shifting. Many such interferometers have very high resolution with a small FOV. The relationship between the phase shift ∅ and the translation offset δ is calculated as (λ is the wavelength of the light source)

$$\varnothing = 4\pi \frac{\delta}{\lambda} \tag{7.6}$$

In a projection moiré, the grating is usually translated laterally in the grating plane in a direction perpendicular to the grating lines.[26] The relationship between the phase shift ∅ and the translation offset δ is calculated as (p is the pitch of the physical grating)

$$\varnothing = 2\pi \frac{\delta}{p} \tag{7.7}$$

In a shadow moiré system, either the grating[5] or the sample[27] can be translated. If the grating is translated, the translation is in the grating plane and Equation 7.7 is still valid. If the sample is translated, the translation direction is perpendicular to the grating and the amount of translation depends on the components and system configuration.

In a field shifting system,[28,29] the image-capturing unit and the fringe projection unit are translated relative to each other. The translation amount also depends on the system configuration and components for the required phase shift.

7.2.3.2 Digital Phase-Shifting System

In the digital phase-shifting systems as shown in Figure 7.4, a digital projector, for example LCD, DMD, and LCOS, is used to project software-generated fringe patterns with a certain intensity profile and to project a sinusoidal fringe onto the object surface. A high-resolution camera is used to capture the fringe patterns modulated by the object surface. Using phase-shifting algorithms, a relative phase map is obtained after phase wrapping and unwrapping. The x, y, z coordinates of the object surface with a corresponding pixel-level resolution are calculated from the phase map by use of a conversion algorithm.

For a straight-line sinusoidal fringe pattern, the equation used to generate the fringe image in the computer can be written as

$$I(u,v) = \frac{M}{2}\left[1+\cos\left(2\pi\frac{u}{p}+\theta\right)\right] \tag{7.8}$$

where

- $I(u, v)$ is the gray level at point (u, v) in the projector chip (LCD, DMD, or LCOS)
- p is the period of the fringe pattern in pixels
- M is the maximum grayscale the project supports
- θ is the phase shift

The fringe line is along the v direction.

For a circular fringe pattern centering at (u_c, v_c), the equation for fringe generation can be written as

$$I(u,v) = \frac{M}{2}\left[1 + \cos\left(2\pi\frac{r}{p} + \theta\right)\right] \tag{7.9}$$

where

$$r = \sqrt{(u - u_C)^2 + (v - v_C)^2} \tag{7.10}$$

in which r is the calculated radius from circular fringe center (u_c, v_c) and $I(u, v)$, p, M, and θ are the same as in Equation 7.8.

In any digital fringe pattern, the shifted phase θ depends on the phase-shifting algorithm to be used. These phase-shifting algorithms are discussed in detail in the next section.

7.2.3.3 Simultaneous Phase-Shifting Techniques

The physical phase-shifting and digital phase-shifting techniques need multiple fringe images in sequence with the assumption that the target will stay still during image capturing. However, sometimes, the target has to be measured in a vibrating environment or when the target is still moving. One such example is the measurement of facial expressions. Currently, there are two ways to address this issue: one method is to capture the images in a time period as short as possible such as using three red, green, and blue (RGB) channels. But this method has limitation. The other method is to use a simultaneous phase-shifting technique to capture multiple images or multiple sub-images at the same time.

A typical solution is to project an RGB color fringe with a phase shift of 120° between adjacent colors and use a color camera to capture the images.[8] From the color image, three monochromatic images can be extracted and used to calculate the phase map via the three-step phase-shifting algorithm. The color fringe method can also be applied to projection moiré.[30] In one color fringe projection research,[31] a projected grating that consisted of RGB-colored stripes was made, each with a separate set of lines for one color channel. The three line sets were identical in terms of the fringe pitch and the fringe width. They overlapped with an offset of one-third the line pitch, resulting in a 120° phase shift. A color camera captured the three fringes simultaneously. For these color fringe methods, channel balance is critical to obtain low-noise phase maps from the color fringe image because the camera and the projector might not have the same response for these three color channels. A more detailed discussion is provided in the following error analysis section.

A more advanced method is to make use of the projector hardware to make "fake" color fringes.[32,33] When a color fringe is sent to a DMD chip of a digital projector after removing its color wheel, the three channels of the projector will have three grayscale images with a 120° phase shift. Because these three images are projected in 10 ms, the black-and-white camera has to be synchronized with the projector to take all three phase-shifted images in this 10 ms time frame. The 3D shape of the object surface can then be reconstructed by phase-shifting algorithms. The 3D measurement speed can reach 100 Hz. This method does not need a color camera and thus has no color balance problem.

To avoid the color balance issue, some researchers also have used polarization splitting to generate multiple interferogram channels with, for example, a 90° phase shift and then used multiple cameras to capture the images.[34] Some other methods include wave front splitting with diffraction optics such as a holographic element[35] or a glass plate.[36] In the first case,[35] the test and reference beams pass through a holographic element that splits the beam into four separate beams, with each beam passing through a birefringent mask before entering the charge-coupled device (CCD) camera. The four mask segments introduce phase shifts between the test and reference beams. A polarizer is placed between the phase masks and the CCD sensor, resulting in the interference of the test and reference beams. In this setup, four phase-shifted interferograms are captured in a single shot on a single camera.

7.3 Phase-Shifting Algorithms for Phase Wrapping

Although there are different measurement principles and different ways to do phase shifting, phase-shifting systems all use multiple captured fringe images and share basic phase-shifting algorithms to extract the phase map from these fringe images.

7.3.1 General Phase-Shifting Algorithm

For both interferogram and projected fringes, the captured 2D fringe image can be written in the form of Equation 7.11 or 7.12:

$$I_k(i,j) = I_0(i,j)\left[1+\gamma(i,j)\cos\left(\phi(i,j)+\theta_k\right)\right], \quad k=1,2,3,\ldots,K \tag{7.11}$$

or

$$I_k(i,j) = I_0(i,j)+I'(i,j)\cos\left(\phi(i,j)+\theta_k\right), \quad k=1,2,3,\ldots,K \tag{7.12}$$

where

- k is the index number of the images used in the phase measurement method
- I_k is the intensity at pixel (i, j) in the captured image
- I_0 is the background illumination
- γ is the fringe modulation (representing image contrast)
- I' is the image contrast
- θ_k is the initial phase for the kth image
- K is the total number of the fringe images

In general, it is the phase term $\phi(i, j)$ in the fringe pattern Equation 7.11 or 7.12 that is to be calculated in the phase-shifting algorithms. In this section, attention is paid to the common discrete phase-shifting algorithms and their features. Readers interested in the development of various phase-shifting algorithms can refer to Refs. [17,37]. Also, the phase-shifting algorithms discussed in this section focus on the phase-wrapping process. Readers should keep in mind that the wrapped phase map includes the modulo 2π discontinuity, so phase unwrapping is needed to obtain a continuous phase map.

One very helpful feature in phase-shifting systems is the calculation of the image modulation γ. γ represents the image contrast and ranges between 0 and 1. The modulation γ can help generate a mask to avoid problems in phase unwrapping. In industrial applications, the shape of the parts and environmental lighting conditions vary a lot and may make some areas saturated or near saturated or too dark to analyze properly. At these areas, the signal-to-noise ratio is very low and the calculated phase information may not be correct. Therefore, these locations should be excluded in the following phase-unwrapping process. These locations can be detected by the modulation γ because it is much smaller in these areas. A common practice is to set a threshold for γ. If γ is smaller than the threshold in a pixel, the unwrapping process should bypass it.

7.3.2 Common Phase-Shifting Algorithms

7.3.2.1 Three-Step Phase-Shifting Algorithm

In the three-step phase-shifting algorithm,[17,38] the phase shift $\theta=-2\pi/3$, 0, and $2\pi/3$ is used for three fringe images, respectively. The intensities of the three phase-shifted images at pixel (i, j) are

$$I_1(i,j)=I(i,j)+I'(i,j)\cos\left[\phi(i,j)-\frac{2\pi}{3}\right] \tag{7.13}$$

$$I_2(i,j)=I(i,j)+I'(i,j)\cos\left[\phi(i,j)\right] \tag{7.14}$$

$$I_3(i,j)=I(i,j)+I'(i,j)\cos\left[\phi(i,j)+\frac{2\pi}{3}\right] \tag{7.15}$$

In these equations, there are three unknowns: I, I', and ϕ. By solving the earlier equations, phase $\phi(i, j)$ can be obtained as

$$\phi(i,j)=\tan^{-1}\left(\sqrt{3}\frac{I_1-I_3}{2I_2-I_1-I_3}\right) \tag{7.16}$$

The modulation can be calculated as

$$\gamma(i,j)=\frac{\left[(I_3-I_2)^2+(2I_1-I_2-I_3)^2\right]^{1/2}}{I_2+I_3} \tag{7.17}$$

The three-step phase-shifting algorithm only needs three fringe images and thus is among the fastest discrete phase-shifting algorithms. But this algorithm is vulnerable to errors in the system such as phase-shifting error, nonlinearity error, and noise.

7.3.2.2 Double Three-Step Phase-Shifting Algorithm

An improvement to the three-step phase-shifting algorithm is the double three-step phase-shifting algorithm, which can significantly reduce the error from system nonlinearity.

It has been proved that a second-order nonlinearity residual in the system can result in an error of $\Delta\varphi$ in the phase map:[39]

$$\tan(\Delta\phi) = \tan(\phi' - \phi) = \frac{\tan(\phi') - \tan(\phi)}{1 + \tan(\phi')\tan(\phi)} = -\frac{\sin(3\phi)}{\cos(3\phi) + m} \tag{7.18}$$

$$\Delta\phi = \arctan\left[-\frac{\sin(3\phi)}{\cos(3\phi) + m}\right] \tag{7.19}$$

where

- ϕ is the phase calculated with a traditional three-step algorithm when the system has perfect linearity
- ϕ' is the calculated phase with a traditional three-step algorithm when the system has a second-order nonlinearity
- m is a constant that depends on the system linearity

Equation 7.19 indicates that the frequency of the error pattern is three times that of the fringe pattern. If an initial phase offset is introduced in the phase-shifted fringe patterns, the phase of the error wave will vary correspondingly. When two phase maps are obtained with a relative initial phase difference of 60°, the phase difference between these two error patterns is approximately 180°. Therefore, when the two phase maps are averaged, the error will be significantly reduced. This means that we can do phase shifting twice with six fringe patterns with initial phases of 0°, 120°, 240° (group one) and 60°, 180°, 300° (group two), use the three-step algorithm twice to calculate the two phase maps from each fringe group, and then average the phase maps.

The effectiveness of the double three-step algorithm can be verified theoretically. In Equation 7.19, because the second-order nonlinearity residual ε is small, m will be large. If $m \gg 1$, Equation 7.19 can be simplified as

$$\Delta\phi = \arctan\left[-\frac{\sin(3\phi)}{m}\right] = -\arctan\left[\frac{\sin(3\phi)}{m}\right] \tag{7.20}$$

If we introduce another phase map with an initial phase offset of 60° for the fringe patterns, the phase error becomes

$$\Delta\phi' = \arctan\left[-\frac{\sin(3\phi + 180°)}{k}\right] = \arctan\left[\frac{\sin(3\phi)}{k}\right] \tag{7.21}$$

It is obvious that $\Delta\phi = -\Delta\phi'$. Therefore, if we average the two phase maps, the error will disappear.

7.3.2.3 Four-Step Phase-Shifting Algorithm

The four-step phase-shifting algorithm uses four fringe images with shifted phase θ as

$$\theta_i = 0, \frac{\pi}{2}, \pi, \frac{3\pi}{2}, \quad i = 1, 2, 3, 4 \tag{7.22}$$

The four images can be written as

$$I_1(i,j) = I(i,j) + I'(i,j)\cos\left[\phi(i,j)\right] \tag{7.23}$$

$$I_2(i,j) = I(i,j) + I'(i,j)\cos\left[\varphi(i,j) + \frac{\pi}{2}\right] \tag{7.24}$$

$$I_3(i,j) = I(i,j) + I'(i,j)\cos\left[\phi(i,j) + \pi\right] \tag{7.25}$$

$$I_4(i,j) = I(i,j) + I'(i,j)\cos\left[\varphi(i,j) + \frac{3\pi}{2}\right] \tag{7.26}$$

Using these trigonometric functions, the phase information can be calculated as

$$\phi(i,j) = \tan^{-1}\left(\frac{I_4 - I_2}{I_1 - I_3}\right) \tag{7.27}$$

The modulation can be calculated as

$$\gamma(i,j) = \frac{2[(I_4 - I_2)^2 + (I_1 - I_3)^2]^{1/2}}{I_1 + I_2 + I_3 + I_4} \tag{7.28}$$

The four-step phase-shifting algorithm has a 90° phase shift between adjunct frames and is easier to implement in some situations, making it the most useful algorithm in simultaneous phase-shifting systems.

7.3.2.4 Carré Phase-Shifting Algorithm

The Carré phase-shifting algorithm is a four-step phase shifting algorithm for use with an unknown phase shift. The four images can be written as

$$I_1(i,j) = I(i,j) + I'(i,j)\cos\left[\phi(i,j) - 3\theta\right] \tag{7.29}$$

$$I_2(i,j) = I(i,j) + I'(i,j)\cos\left[\phi(i,j) - \theta\right] \tag{7.30}$$

$$I_3(i,j) = I(i,j) + I'(i,j)\cos\left[\phi(i,j) + \theta\right] \tag{7.31}$$

$$I_4(i,j) = I(i,j) + I'(i,j)\cos\left[\phi(i,j) + 3\theta\right] \tag{7.32}$$

In this four-equation group, there are four unknowns. The phase ϕ can be calculated as

$$\phi(i,j) = \tan^{-1}\left(\frac{\sqrt{3(I_2 - I_3)^2 - (I_1 - I_4)^2 + 2(I_2 - I_3)(I_1 - I_4)}}{(I_2 + I_3) - (I_1 + I_4)}\right) \tag{7.33}$$

The obvious features of the Carré phase-shifting algorithm are that its constant phase step 2θ can be arbitrary and the measured phase is insensitive to all even harmonics.[17,40,41] Especially when the phase-shift system did not have linear response over the entire 2π range, the Carré phase-shifting algorithm still provides good results if phase shifting is performed within a small phase-shift range in a relatively linear segment.

The Carré phase-shifting algorithm has also proved to be adaptive to variation situations because of the flexibility in phase shift.[41] When there is a second-order phase-shift error, the average phase measurement error can be minimized when the phase step is 65.8°. When there is a systematic intensity error such as a nonlinearity in camera response, the best phase step is 103° for the minimum phase measurement error. For a high-noise image with a random intensity measurement error, a phase step of 110.6° will minimize the averaged phase measurement error.

Although the four-step phase-shifting method can be considered a special case of the Carré phase-shifting algorithm, it has features that the Carré phase-shifting algorithm does not have such as an easier and faster calculation of both the phase map and modulation.

7.3.2.5 Five-Step Phase-Shifting Algorithm (Hariharan Algorithm)

The five-step phase-shifting algorithm with an unknown but constant phase shift is also called the Hariharan algorithm.[17,42] The five fringe images are

$$I_1(i,j) = I(i,j) + I'(i,j)\cos\left[\phi(i,j) - 2\theta\right] \tag{7.34}$$

$$I_2(i,j) = I(i,j) + I'(i,j)\cos\left[\phi(i,j) - \theta\right] \tag{7.35}$$

$$I_3(i,j) = I(i,j) + I\,(i,j)\cos\left[\phi(i,j)\right] \tag{7.36}$$

$$I_4(i,j) = I(i,j) + I'(i,j)\cos\left[\phi(i,j) + \theta\right] \tag{7.37}$$

$$I_4(i,j) = I(i,j) + I'(i,j)\cos\left[\phi(i,j) + 2\theta\right] \tag{7.38}$$

When the phase shift $\theta = 90°$, the phase ϕ and modulation γ can be calculated as

$$\phi(i,j) = \tan^{-1}\left(\frac{2(I_2 - I_4)}{2I_3 - (I_1 + I_5)}\right) \tag{7.39}$$

$$\gamma(i,j) = \frac{3\sqrt{4(I_4 - I_2)^2 + (I_1 + I_5 - 2I_3)^2}}{2(I_1 + I_2 + 2I_3 + I_4 + I_5)} \tag{7.40}$$

This phase-shifting algorithm has good tolerance for phase-shift error because the first-order error terms cancel, even though the errors from second-order residuals still exist.

For an arbitrary phase shift, the phase ϕ and modulation γ can be calculated as[43]

$$\phi(i,j) = \tan^{-1}\left(\frac{\sqrt{4(I_2 - I_4)^2 - (I_1 - I_5)^2}}{2I_3 - (I_1 + I_5)}\right) \tag{7.41}$$

$$\gamma(i,j) = \frac{(I_2 - I_4)^2 \sqrt{4(I_2 - I_4)^2 - (I_1 - I_5)^2 + (I_1 + I_5 - 2I_3)^2}}{4(I_2 - I_4)^2 - (I_1 - I_5)^2} \tag{7.42}$$

As with the Carré phase-shifting algorithm, the five-step phase-shifting algorithm is insensitive to phase-shift errors.[44]

7.3.2.6 Phase-Shifting Algorithms with More Than Five Steps

Although phase-shifting algorithms using more than five image frames are seldom used, they do exist.[17,45,46] They require more computing resources and time to process the images, and it is sometimes impossible to calculate the modulation, but they are usually more resistant to some errors. For example, a seven-sample algorithm based on the Surrel six-sample algorithm using the averaging technique was demonstrated to be insensitive to linear and quadratic nonlinear phase-shift errors with linear compensation even when the fringe signal contains a second-harmonic distortion.[45] Higher-order phase-shifting algorithms (six-sample, eight-sample, and nine-sample algorithms) have also been investigated to show the effectiveness in compensating for a quadratic and spatially nonuniform phase-shift error.[46,47]

7.3.2.7 Spatial Carrier Phase-Shifting Algorithms

The spatial carrier method makes use of only one high-resolution fringe image from which multiple sub-images with lower resolution are extracted, so that a multiple-step phase-shifting algorithm can be used.[19,43,48–50] Because the spatial carrier method uses only one fringe image, it can be used in dynamic environments such as vibrating or moving object measurements while still having the benefits of the phase-shifting techniques. Because the sub-images are obtained by selecting every *N*th pixel in the original image (*N* is the number of steps used in the phase-shifting algorithms), these sub-images have a $1/N$ lateral resolution of the original image.

For surfaces with curvature such as edge breaks or sphere/cylindrical surfaces, the fringe pitch and thus the actual phase shifts between adjunct pixels can vary a lot over the entire surface. When the spatial carrier technique is used, it is critical to select phase-shifting algorithms with an unknown phase shift such as the Carré phase-shifting algorithm or the five-step phase-shifting algorithm with an arbitrary phase shift, as discussed in previous sections. For these algorithms, the local curvature has a decisive impact on the measurement accuracy.

The spatial carrier method is sensitive to random noise and is less tolerant to surface finish variation. Unlike traditional phase-shifting algorithms, where at corresponding pixel (i, j) the detected intensities I_k of these K fringe images come from the same physical location on the surface, in the spatial carrier technique, the intensity $I_k\ (i, j)$ corresponds to different physical locations (adjunct pixels in the original image). Therefore, any variation in the surface reflection or scattering angle, the surface finish, the illumination angle, and random noise may result in an error in the wrapped phase using traditional phase-shifting algorithms.

7.3.3 Selection of Phase-Shifting Algorithms

Many different phase-shifting algorithms have been developed in the past that can be used to reduce different types of errors. Because each algorithm has its own features and no single algorithm can meet all requirements, selection of the most appropriate phase-shifting algorithm for a specific phase-shifting measurement system needs careful analysis and trade-off considerations.

System dependence results from the fact that different systems have their own main error sources. For digital phase-shifting systems, phase shift is performed by software programming, and there is no phase-shifting error. The main error sources become the nonlinearity and noise from the camera and projector. Those phase-shifting algorithms that are insensitive to the system nonlinearity will provide the best measurement results. On the contrary, for physical phase shifting, the phase-shifting error is usually one of dominant error sources. When using a physical phase-shifting method, the algorithms that are insensitive to the phase-shift error will work best. In general, for incorrectly calibrated linear error and some nonlinear errors, algorithms that can work with an unknown phase shift such as the Carré phase-shifting algorithm and the five-step phase-shifting algorithm work the best. For simultaneous phase-shifting systems, the misalignment of multiple cameras or sub-images on the same camera may be the critical problem, and the selected algorithm needs to work in this situation.

Other considerations are more application oriented. Factors to consider include the part geometric (especially curvature variation and surface finish), measurement speed requirement, measurement environment, and accuracy requirements.

7.4 Phase-Shifting System Modeling and Calibration

Phase-shifting algorithms for phase wrapping result in a phase map with a 2π ambiguity. To remove the 2π ambiguity, a phase-unwrapping process is needed. A continuous phase map $\Phi(i, j)$ can be obtained after phase unwrapping. This phase map contains geometric information about the measured surface that sometimes looks like an unscaled, distorted 3D surface contour of the object. However, the phase value of the unwrapped phase map depends on the starting point of the unwrapping process. Thus, from a unique wrapped phase map $\phi(i, j)$, there may be many unwrapped phase maps $\phi(i, j)$ that can be obtained. Furthermore, a digitized geometry is usually represented by a point cloud with a set of 3D coordinates for each point, not the phase map. Converting from the continuous phase map to coordinates of the surface points is a critical process for accurate measurement, which requires a unique absolute phase map, system modeling, and system calibration.

7.4.1 Modeling of Phase-Shifting Measurement System

The phase map from the phase-shifting process contains information about an object profile and may be similar to its 3D shape, but the phase map and the profile are not the same. For industrial applications, it is the surface profile not the phase map that is desired. Once the phase map is obtained from the image(s), the coordinates at the sampled points on an object surface must be further calculated. This coordinate calculation at the sample points has to do with a conversion algorithm from phase map to coordinates through modeling.[19] Based on system configuration requirement and object complexity, current models for phase-coordinated conversion can be classified into three categories: linear model, partially linear model, and nonlinear model.

7.4.1.1 Unwrapped Phase Map and Absolute Phase Map

Before further discussing the model for phase-coordinated conversion, some clarification is needed for the term "absolute phase map." Absolute phase map was initially used by some researchers to stand for a continuous phase map without 2π discontinuity[51] such as the unwrapped phase map. Nowadays, it represents the phase map that is used to be converted to coordinates.

For a linear or partially linear model, the absolute phase map is usually obtained by subtracting a reference phase map from the measurement phase map. The measurement phase map is the unwrapped phase map after performing phase shifting on the object surface. Reference phase maps can be obtained by either performing phase shifting on a reference plane (usually flat) or by creating a flat phase plane determined by specific points (constituting a horizontal and a vertical line) on the measurement phase map. The coordinates calculated later will be referred to these planes. After subtraction, the measurement phase map is brought down. In the unwrapping process for both the measurement phase map and the reference phase map, the starting point for the unwrapping processing should be the same.

For a nonlinear model, an absolute phase map is linked to the coordinates system through either a projector or a camera. For physical phase shifting, special components with unique features can be used such as the seam of a folded mirror and overlapped points or seam or similar features. For digital phase shifting, an additional single-line segment or some kind of special pattern can be projected. These lines or patterns have known physical position in the projector and make it much easier to build the relationship between the phase value and physical location to further obtain the absolute phase map directly related to the system geometry.

One such example is the absolute phase map for the master gage shown in Figure 7.7. The digital phase-shifting system used to measure the master gage is similar to the configuration shown in Figure 7.4, which consisted of a black-and-white CCD camera, a digital light processing (DLP) projector with DMD technology, an image processor board (Matrox Genesis), a PC workstation, and windows-based software for system control and data processing. To obtain the absolute phase map, a vertical centerline through the center of the projector DLP chip was projected to the object surface. The captured centerline image is shown in Figure 7.12. The purpose is to correlate every pixel in the phase map to a point on the DMD chip of the projector.

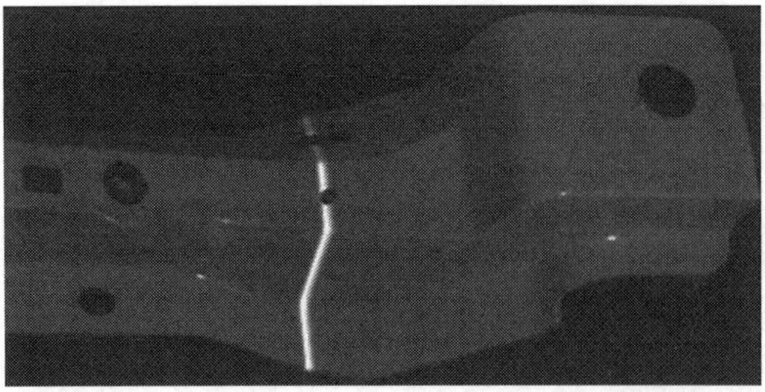

FIGURE 7.12
Image of the projected centerline.

The centerline image was used to identify the pixels in the phase map that correspond to the centerline in the projector chip. These pixels should have the same absolute phase as that of the centerline of the projection field where the project fringe patterns were programmed. With the absolute phase at these pixels known, the absolute phase map of the entire surface can be obtained by simply translating the relative unwrapped phase map $\Phi(i, j)$. Assume the absolute phase of the centerline to be Φ_0. The absolute phase map $\Phi'(i, j)$ can be obtained as follows:[38]

$$\Phi'(i,j) = \Phi(i,j) + \Phi_0 - \frac{1}{N}\sum_{k=1}^{N} \Phi_k \tag{7.43}$$

where

Φ_k's are the phases of the pixels that correspond to the centerline of the projection field

N is the total number of such pixels in a specified segment

The number N may be smaller than the total number of vertical pixels of the CCD sensor because the centerline may hit openings on the object surface and line centers near the openings should be excluded from calculation. Theoretically, the absolute phase at just one pixel is enough to obtain the entire absolute phase map of the object. However, by taking the average of the absolute phase values at multiple pixels, as is done in Equation 7.43, more accurate results can be obtained.

In digital phase shifting, some techniques other than the additional line projection have also been investigated such as embedded patterns or features in the fringe patterns.[52,53] Interested readers can find details in these papers.

7.4.1.2 Linear Model for Flat Surface Measurement

As a simple model that many researchers like to use, the linear model is very straightforward: the lateral dimensions are proportional to the pixel index while the vertical dimension is proportional to the absolute phase after reference phase subtraction. The calculation that converts pixel (i, j) with absolute phase Φ' to coordinates (x, y, z) can be represented by the following formulae:[53,54]

$$x = K_x(i - C_x) \tag{7.44}$$

$$y = K_y(j - C_y) \tag{7.45}$$

$$z = k_z\Phi' \tag{7.46}$$

where

K_x, K_y, K_z are scalars in the three coordinate directions

(C_x, C_y) are specified coordinate origin in the lateral directions

In practice, K_x and K_y are usually determined by calibration in the FOV and K_z is determined by step gage standards.

It is obvious that the linear model requires the camera viewing direction to be perpendicular to the object surface. This model is usually used in flat surface measurement only.

A good example is the shadow moiré technique used for flatness measurement in printed circuit boards (PCB), where the shadow moiré technique has been specified in several industrial standards as a warpage measurement tool.[55,56]

7.4.1.3 Partially Linear Model for Flat Surface Measurement

The partially linear model assumes that some dimensions (usually x and y coordinates) are proportional to the pixel index (i, j) on the image while the vertical coordinate is calculated from the absolute phase value using a nonlinear formula.

To deduce this kind of functions, some assumptions must be made such as assuming the camera is at the same height as the grating and/or assuming the optical axis of the camera/lens is perpendicular to the object surface. Using this method, the system configuration of a fringe projection system can be simplified as shown in Figure 7.13. Let (C_x, C_y) be the intersect of the camera sensing surface and the optical axis of the imaging lens; the coordinates (x, y, z) can be calculated as

$$x = K_x(i - C_x) \tag{7.47}$$

$$y = K_y(j - C_y) \tag{7.48}$$

$$z = h = \frac{L \times \Phi'}{\Phi' - 2\pi f D} \tag{7.49}$$

where K_x, K_y are scalars in the lateral directions determined by calibrating the FOV. In Equation 7.49, Φ' is the absolute phase, that is, phase difference at pixel (i, j) between the flat reference plane and the object plane, and f is the average frequency of fringe on reference plane. Φ' can be obtained by either subtracting the object phase map from the phase map of the reference plane or by removing the slope in the measurement phase map by bringing down the phase map. Equation 7.49 is not a universal function. Depending on system configurations, other similar functions may be derived.[57]

Under some situations, Equations 7.47 through 7.49 can provide reasonably good results, especially for measurements on flat surfaces. But there are so many assumptions in this particular analysis that they cannot provide the desirable accuracy for curved surface measurement. For example, in Figure 7.13, there is no way to ensure that the camera and fringe

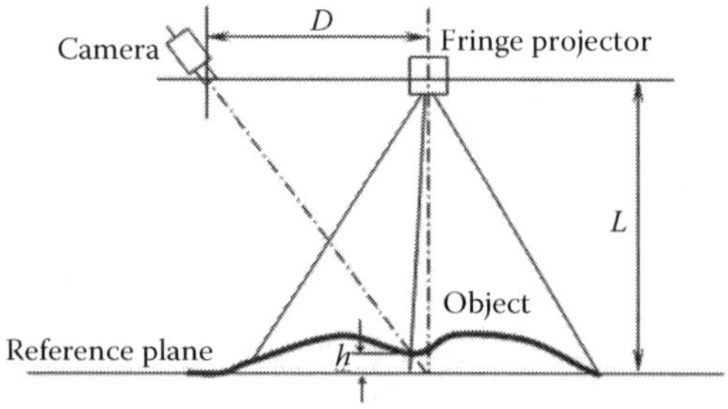

FIGURE 7.13
Simplified system configuration to calculate coordinate z.

projection unit are exactly at the same height. Furthermore, for complicated surface shapes with various curvatures, or in a configuration where there is no good means to ensure the object is perpendicular to the axis of the imaging system, due to magnification variation from point to point, the x and y coordinates are no longer proportional to the image index (i, j). The errors in the measured shape of complex surfaces can be very obvious when the measured result is compared with a model or other data measured from a good coordinate measurement machine (CMM).

When $L \gg h$, Equation 7.49 can be further simplified as

$$z = k_z \Phi' \tag{7.50}$$

where

$$k_z = -\frac{L}{2\pi f D} \tag{7.51}$$

In this case, the partially linear model is simplified to the linear model represented in Equations 7.44 through 7.46.

7.4.1.4 Nonlinear Model for Complex Shape

To obtain more accurate results, a more thorough nonlinear model than the linear and partially linear models is needed. To demonstrate the concept of the nonlinear model, the key dimensions of the digital fringe projection system of Figure 7.4 are shown in Figure 7.14. In this diagram, L is the imaging lens center, R is the projection lens center, and the global coordinate system XYZ has its origin at point O.

The 3D coordinates (x, y, z) of any corresponding object point P with an image pixel $Q(i, j)$ and absolute phase value $\Phi'(i, j)$ can be calculated uniquely in the 3D space. First, all points on each vertical light sheet RP of the fringe pattern have the same phase, so the positions of the fringe lines on the projector can be calculated from the absolute phase values Φ'. The equation of the light sheet RP with phase $\Phi'(i, j)$ can be stated as

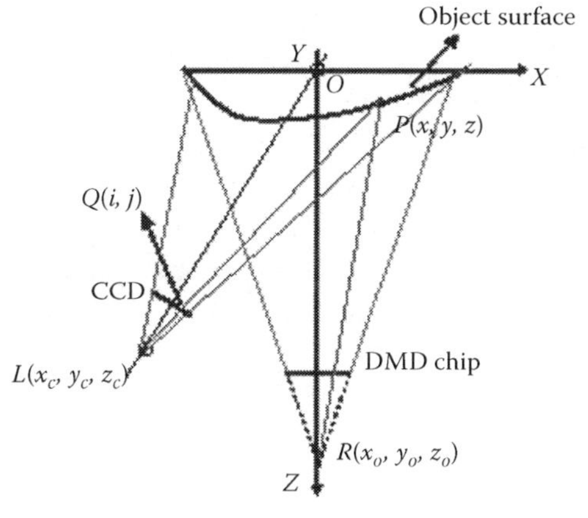

FIGURE 7.14
Diagram for a digital phase-shifting system with a DLP projector.

$$A(\Phi',i,j,c...)x + B(\Phi'',i,j,c...)y + C(\Phi',i,j,c...)z = D(\Phi',i,j,c...) \tag{7.52}$$

where A, B, C, and D define the function of the light sheet RP with phase value $\Phi'(i, j)$, image index (i, j), and system parameters c. On the other hand, in the imaging system, the object point P should lie at the line that connects image pixel $Q(i, j)$ and the camera lens center $L(x_c, y_c, z_c)$. The equation of this line LQ can be written as

$$\frac{x - x_c}{l} = \frac{y - y_c}{m} = \frac{z - z_c}{n} \tag{7.53}$$

where (l, m,n) is the direction vector of the line.

Calculating the intersection between the imaging ray LQ and projection light sheet plane RP by solving Equations 7.52 and 7.53 will provide coordinates (x, y, z) for any point P on the object surface.

This model works for all system configurations and complex surfaces. Unlike linear and partially linear models, it provides not only accurate shape but also space location of this surface in a global coordinate system determined in a calibration process, thus enabling data merging from multiple measurements[58] and accurate 360° shape reconstruction.[59] Figure 7.15 shows a reconstructed 360° shape of a flowerpot with complex surface textures from three measurements. Note that the steps between adjacent patches near the top edge result from magnification changes in the FOV between these measurements due to the tilting of the camera.

7.4.2 Phase-Shifting System Calibration

No matter which phase-to-coordinate conversion model is used, a calibration process has to be performed to determine the system parameters that are required by the conversion algorithms for the calculation of the object coordinates. This calibration process can be as simple as estimating the FOV and measuring a step gage with known step height as for a linear model and some partially linear models. However, in order to obtain accurate measurement results, a more complex calibration process is either desirable for lens error compensation or is desired for the nonlinear model.

FIGURE 7.15
Reconstructed 360° shape of a flowerpot with complex surface textures.

A calibration process is in fact an optimization process that finds a set of parameters to minimize the errors in the data collected for calibration. The ultimate goal of a calibration process is to find the systematic parameters that are used in the previous model for coordinate calculation of the object surface. These parameters are sometimes referred to as extrinsic parameters. Some calibration methods also have the capability to find intrinsic parameters to compensate for the imperfection in alignment, imaging lens, and cameras. Various calibration processes have been investigated,[38,60–66] among which Tsai calibration[60] using a well-aligned calibration target on a translation stage and Zhang calibration[61] using a check board placed in the 3D space with different orientations are most widely used and have many adaptive forms.

7.4.2.1 Camera Calibration

A camera model describes the mapping between points in a 3D space and a pixel in the 2D camera sensor chip. The parameters in a camera model can be classified into intrinsic parameters, which describe the geometry of the camera itself, and extrinsic parameters, which determine the camera's pose in the 3D space. Camera calibration is a process to find these intrinsic and extrinsic parameters. Thanks to high-quality imaging lens in the optical industry, a simplified pinhole camera model[67] can often meet the accurate calibration requirement, although more complex camera models have also been investigated.[68,69] The pinhole model is demonstrated in Figure 7.16. Assuming the focus length of the imaging lens is f, a point $P(x, y, z)$ in the 3D space can be projected into the 2D image plane at image point $Q(u, v)$ as

$$u = \frac{fx}{z} \tag{7.54}$$

$$v = \frac{fy}{z} \tag{7.55}$$

In this section, we briefly introduce Tsai and Zhang calibration methods, both of which are based on the pinhole camera model. Interested readers are referred to the referenced papers for more details about various camera models and related calibration techniques. Although there are many calibration toolkits available on the Internet, it is highly recommended that

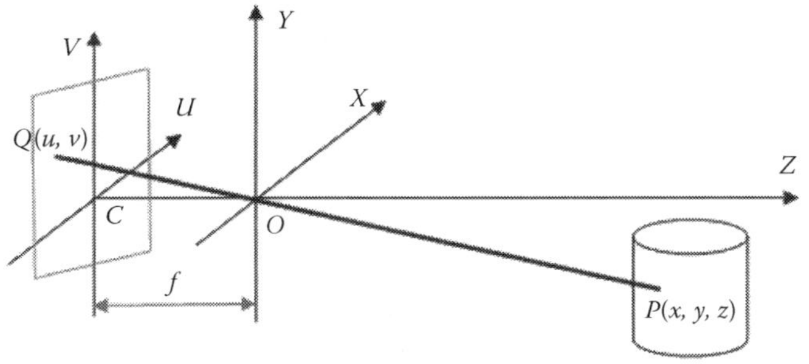

FIGURE 7.16
Pinhole camera model.

the related camera models for these toolkits be fully understood because different models may use the same terms for different meanings.

The Tsai camera calibration method can deal with coplanar and noncoplanar points. It is a two-step calibration method that can calibrate the intrinsic and extrinsic parameters separately. In Tsai calibration, there are 11 parameters that are to be optimized:

f: Effective focal length of the camera lens
k: Radial distortion coefficient of the camera lens
(C_x, C_y): Origin of the image coordinate (intersect of the lens axis and the image sensor)
S_x: Scale factor due to imperfections in hardware timing misstep
(R_x, R_y, R_z): Rotation angles between the global and camera coordinates systems
(T_x, T_y, T_z): Translation position between the global and camera coordinates systems

The rotation matrix R is deducted from the (R_x, R_y, R_z) as

$$R = \begin{bmatrix} r_1 & r_2 & r_3 \\ r_4 & r_5 & r_6 \\ r_7 & r_8 & r_9 \end{bmatrix} \tag{7.56}$$

where

$$r_1 = \cos(R_y)\cos(R_z) \tag{7.57}$$

$$r_2 = \sin(R_x)\sin(R_y)\cos(R_z) - \cos(R_x)\sin(R_z) \tag{7.58}$$

$$r_3 = \sin(R_x)\sin(R_z) + \cos(R_x)\sin(R_y)\cos(R_z) \tag{7.59}$$

$$r_4 = \cos(R_y)\sin(R_z) \tag{7.60}$$

$$r_5 = \sin(R_x)\sin(R_y)\sin(R_z) + \cos(R_x)\cos(R_z) \tag{7.61}$$

$$r_6 = \cos(R_x)\sin(R_y)\sin(R_z) - \sin(R_x)\cos(R_z) \tag{7.62}$$

$$r_7 = -\sin(R_y) \tag{7.63}$$

$$r_8 = \sin(R_x)\cos(R_y) \tag{7.64}$$

$$r_9 = \cos(R_x)\cos(R_y) \tag{7.65}$$

The translation matrix T is defined as

$$T = \begin{bmatrix} T_x \\ T_y \\ T_z \end{bmatrix} \tag{7.66}$$

A point (x, y, z) in the world coordinate system is transformed to the image coordinate system to be (x_i, y_i, z_i) through the translation matrix T and rotation matrix R as

$$\begin{bmatrix} x_i \\ y_i \\ z_i \end{bmatrix} = R \begin{bmatrix} x \\ y \\ z \end{bmatrix} + T \tag{7.67}$$

Further transformation to undistorted coordinates (x_u, y_u) to distorted (x_d, y_d) in image plane coordinates via the pinhole model is

$$x_u = \frac{fx_i}{z_i} \tag{7.68}$$

$$y_u = \frac{fy_i}{z_i} \tag{7.69}$$

$$x_d = \frac{x_u}{1+kr^2} \tag{7.70}$$

$$y_d = \frac{y_u}{1+kr^2} \tag{7.71}$$

where k is the lens distortion coefficient and $r = \sqrt{x_d^2 + y_d^2}$.

The transformation from distorted coordinates (x_d, y_d) to the final image index (x_f, y_f) is

$$x_f = \frac{s_x x_d}{d_x} + C_x \tag{7.72}$$

$$y_f = \frac{y_d}{d_y} + C_y \tag{7.73}$$

where (d_x, d_y) are camera pixel size in the X and Y direction.

In Tsai's calibration, a calibration setup is required for calibration data preparation. In this setup, a flat target with certain patterns is mounted to a translation stage with the target plane perpendicular to the translation direction. The target is translated in the calibration volume (usually as close to the measurement volume as possible). At each location, the image of the target is taken and saved with the translation reading. The X and Y axes origin is specified on the target (usually in the middle), while the translation direction defines the Z axis with one translation location set as $Z=0$. After processing the target images, for each feature on the calibration target, a unique correspondence between the real image index (x_f, y_f) in the image plane and its 3D coordinates (x, y, z) in the calibration volume is established. Applying this data set to the Tsai's calibration algorithm, together with known information about the camera and lens, all optimized intrinsic and extrinsic parameters can be obtained.

The Zhang camera calibration describes the relation between point (x, y, z) in the 3D space and image point (u, v) in the image plane in another form as

$$\underbrace{s\begin{bmatrix} u \\ v \\ 1 \end{bmatrix}}_{\tilde{m}} = \underbrace{\begin{bmatrix} \alpha & \gamma & u_0 \\ 0 & \beta & v_0 \\ 0 & 0 & 1 \end{bmatrix}}_{A} \underbrace{\begin{bmatrix} \mathbf{r}_1 & \mathbf{r}_2 & \mathbf{r}_3 & \mathbf{t} \end{bmatrix}}_{[\mathbf{R} \ \mathbf{t}]} \underbrace{\begin{bmatrix} x \\ y \\ z \\ 1 \end{bmatrix}}_{\tilde{M}} \quad (7.74)$$

where
 s is an arbitrary scale factor
 [R t] is the transformation matrix containing the extrinsic parameter
 scale matrix A contains the intrinsic parameters

$$A = \begin{bmatrix} \alpha & \gamma & u_0 \\ 0 & \beta & v_0 \\ 0 & 0 & 1 \end{bmatrix} \quad (7.75)$$

Physical meanings of these parameters are
 α: Effective focal length in the image u axis
 β: Effective focal length in the image v axis
 γ: Skewness factor of the u and v axes in the image plane
 R: Rotation matrix consisting of three column vectors $\mathbf{r}_1$, $\mathbf{r}_2$, and $\mathbf{r}_3$
 t: Translation matrix
 (u_0, v_0): Coordinates of the principal point
 (k_1, k_2): Two coefficients of the radial distortion

The distorted image coordinates (u', v') are related to the undistorted image coordinates (u, v) as

$$u' = u + (u - u_0) \times \left(k_1 r^2 + k_2 r^4\right) \quad (7.76)$$

$$v' = v + (v - v_0) \times \left(k_1 r^2 + k_2 r^4\right) \quad (7.77)$$

where

$$r = \sqrt{(u - u_0)^2 + (v - v_0)^2} \quad (7.78)$$

Zhang calibration first solves five intrinsic parameters and all extrinsic parameters using a closed-form solution and then estimates the coefficients of the radial distortion by use of a linear least-square algorithm (note: this estimate may be inexact as actual optical geometric distorting in a lens goes as a cube function).

In Zhang's calibration, there is no need for the translation stage. A flat target with certain patterns is placed at a minimum of three orientations: the first one at zero position for the *Z* coordinate, and then the others skewed relative to the first one. The motion of the target is flexible and need not be known. The *X* and *Y* origin is specified on the target. At each orientation, an image is taken and processed to obtain the calibration data set. By running the calibration codes with the data set and known parameters about the camera and lens, the parameters can be obtained.

In either Tsai calibration or Zhang calibration, an appropriate image processing is required. The calibration target needs to have features that are easy to identify and with known physical dimension. Common patterns include dot patterns, donut patterns, checkboard patterns, square grid patterns, and line grids. Either the center or edge corner on these patterns can be used as calibration features. Some researchers also use coded marks.[70]

In phase-shifting measurement systems, the calibrated parameters are used to calculate the 3D coordinates from each camera pixel and its phase information. Unlike calibration that starts from known 3D coordinates of target features in the 3D space, measurement has to start with the pixel index to calculate coordinates in the 3D space. This requires full understanding of the calibration model and properly using these parameters in a reverse way.

7.4.2.2 Projector Calibration

In general, projector calibration can be regarded as a reverse of the camera calibration using the same pinhole model. Although a projector cannot take images like a camera, it can project a pattern with known pixel information (u, v) on the projector chip such as a known grid or stripe. In Tsai calibration, the projector can be calibrated with the same setup as the camera calibration discussed in the previous section because the projected pattern on the calibration target can be captured by the camera. After camera calibration, the calibrated results can be used to calculate the coordinates (x, y, z) of the projected features such as grid points in the 3D space at some known *Z* locations. In this way, the correspondence between projector pixel (u, v) and related 3D coordinates (x, y, z) in the 3D space is established. The obtained data set can then be applied to the calibration algorithm to obtain the projector calibration parameters. It should be noted that, in this way, the projector and camera are calibrated in the same global coordinates system. Figure 7.17 shows one captured image of the projected grid pattern on a calibration target while Figure 7.18 shows the point clouds of the data set obtained at 5 Z positions for projector calibration.

Also, individual points can be projected as well to collect the data, instead of the grid. C. Sinlapeecheewa used the stereo vision method to obtain the 3D coordinates of the project point with a known pixel location in the projector chip.[71] Another researcher used CMM to place the target at specified 3D coordinates.[72]

A good use for the projector calibration with the gird data set is the masking on the projector.[21] A mask is a binary image that describes which pixel is valid. In the digital phase-shifting systems, both projected and captured images are programmable. To take advantage of this programmability to deal with double bounce light (light that reflects from one area on a part to another, causing confusing patterns), we can use masks to control where to project fringes and where to be measured. A fringe mask can be applied to the fringe projection unit so as to only illuminate a specific area. Because phase shifting–based fringe projection is a pixel-independent (each pixel is calculated independently) method, an image mask can be used to obtain measured data only on selected small patch of the part. By use of masks, we can easily divide the part into several measurement patches and

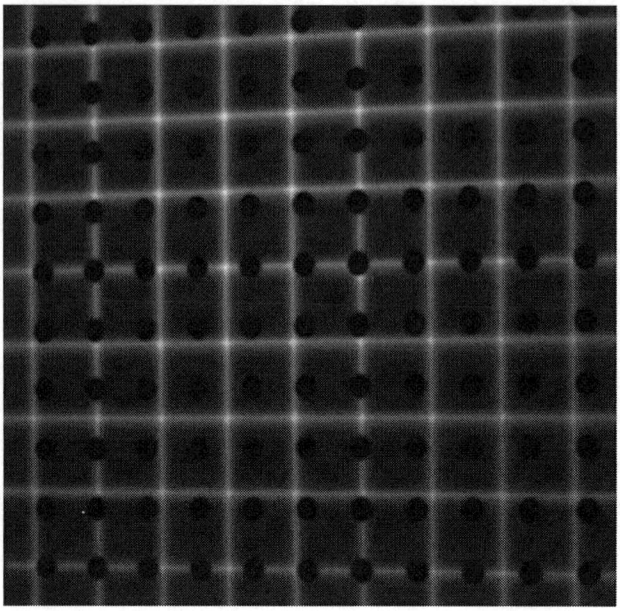

FIGURE 7.17
Projected grids on the calibration target with dot patterns.

FIGURE 7.18
Point cloud of the projected grid points on the target at 5 Z positions.

measure them separately. The fringe mask can be generated from a previous point cloud, manually selected mask border, or a 3D model at a known location.

Another way to calibrate the projector is to use a stripe data set instead of grid data set. When straight-line fringe patterns are used in the digital phase-shifting systems, in order to use the nonlinear model for phase-to-coordinate conversion, we need the light sheet orientation information that corresponds to an absolute phase value. It is known that a plane in 3D space can be interpolated from two known 3D planes called base planes. Assuming the two base planes have equations $A_1x+B_1y+C_1z+D_1=0$ and $A_2x+B_2y+C_2z+D_2=0$, for a phase value Φ', the pixel location δ in the projector chip is (p is the pitch of the digital fringe pattern)

$$\delta = \frac{p\Phi'}{2\pi} \tag{7.79}$$

whose corresponding light sheet plane in the global coordinate system can be determined as

$$A_1x + B_1y + C_1z + D_1 + \delta \times \left(A_2x + B_2y + C_2z + D_2\right) = 0 \tag{7.80}$$

To find the base plane equations, multiple line strips with known stripe location (determined by pixel number on the projector chip) can be projected onto the calibration target whose image is captured by the camera during calibration at each position. Similar to grid image processing, at each known Z position, the images of these stripes can be processed and the 3D coordinates of these strip lines on the target can be obtained after the camera calibration is performed. Figure 7.19 shows the captured stripe image, and Figure 7.20 the processed 3D coordinates in the 3D space. With this data set, the two base plane equation parameters (A_1, B_1, C_1, D_1) and (A_2, B_2, C_2, D_2) can be obtained through fitting.

When Zhang calibration is used in the camera calibration, no Z position is available for grid or strip image processing to obtain 3D point clouds. In this case, a link between the camera and the projector has to be established for the projector calibration.[63,73] The basic idea is to map the CCD image to the projector chip to form a so-called projector image. First, the vertical and horizontal fringes with additional centerline stripes are projected, and phase shifting is performed to obtain the absolute phase map in both the horizontal and the vertical directions as in regular phase shifting. The horizontal absolute phase map Φ'_x and vertical absolute phase map Φ'_y are saved to map the camera pixel to the projector pixel. At each orientation during calibration, the same checkerboard is used for the camera calibration and projector calibration, and grayscale images capturing and phase shifting

FIGURE 7.19
Projected stripes on the calibration target with dot patterns.

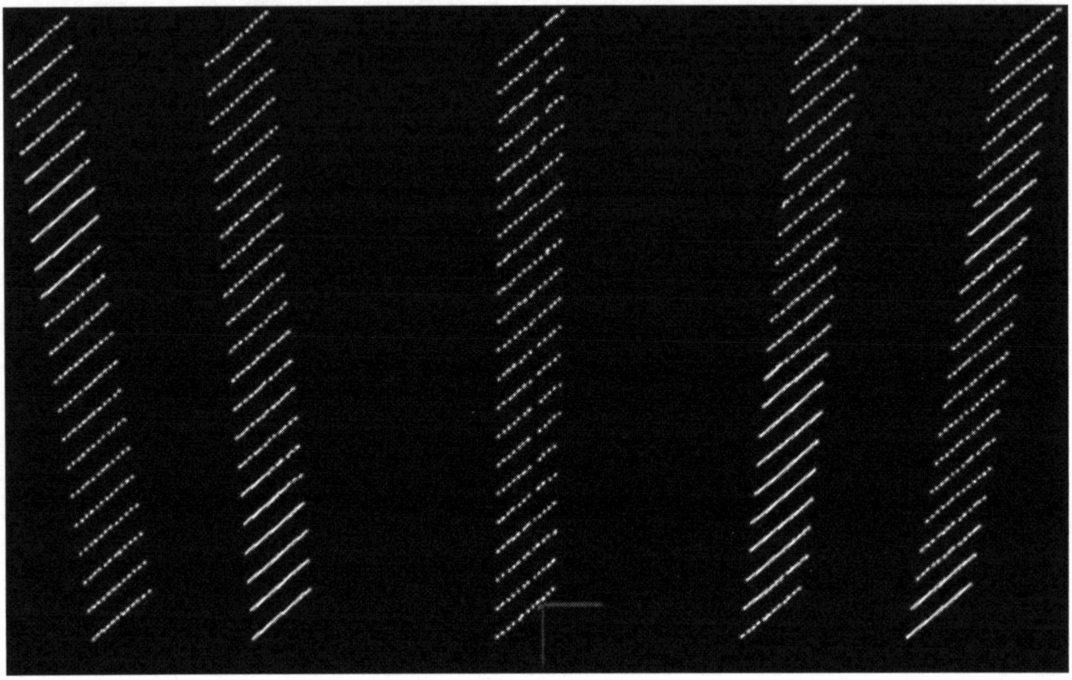

FIGURE 7.20
Point cloud of the projected five stripes on the target at 21 Z positions.

on the checkerboard are performed. First, the grayscale images of the checkerboard are processed for camera calibration using Zhang's calibration model. For each feature (corner of the squares) on the CCD image, its pixel index (i, j) can be mapped to the projector index (u, v) as (p_x and p_y are the pitches of the fringes)

$$u = \frac{p_x \Phi'_x(i,j)}{2\pi} \tag{7.81}$$

$$v = \frac{p_y \Phi'_y(i,j)}{2\pi} \tag{7.82}$$

The checkerboard's feature locations and their corresponding projector index (u, v) are then used for projector calibration just as in camera calibration. Because the same checkerboard is used for both phase shifting and calibration, a special colorful pattern is often used to ensure high-contrast images are available for both purposes.[63]

7.5 Error Analysis and Compensation for Phase-Shifting Systems

As in any other optical instrument, the phase-shifting measurement system should always use the most reliable and best-quality components if they are available and affordable. The reason is obvious: a carefully designed lens with very little distortion is more likely to provide better results than using a low-quality lens with software-based lens distortion correction; setting the projector gamma to linear is better than compensating for a nonlinear gamma curve in a projector. Moreover, coupling among different error sources may make the error compensation less efficient and more difficult. In addition, care has to be taken in system adjustment such as alignment and focusing/defocusing. With that said, error correction and compensation are still very useful as a last means of obtaining high-quality measurement results although it makes sense only after a "best" system is built.

7.5.1 Error Sources and Adjustment in the Phase-Shifting System

There are many error sources in an optical phase-shifting measurement system.[74,75] This section discusses the major and most common error sources and their behaviors.

7.5.1.1 Phase-Shifting Error

In digital phase-shifting systems, the phase shift is generated in a software program, and theoretically, there is no phase-shift error. In physical phase shifting, linear phase-shift error from stage miscalibration and nonlinear phase-shift error from poor stage response or control are one of the major concerns.[22]

A phase-shift error can sometimes be observed as ripples in averaged grayscale images. For example, in a three-step phase-shifting algorithm, the averaged image of the three fringe images should not have any ripple. Adding Equations 7.12 through 7.14, the averaged image $\bar{I}$ can be calculated as

$$\bar{I} = \frac{I_1 + I_2 + I_3}{3} = I(i, j) \tag{7.83}$$

which is a uniform background image whose brightness is about half of the maximum brightness.

If one image has a phase-shift error, the averaged image will have significant ripples. Figure 7.21 shows the averaged image of three simulated fringe images with a phase shift of 0°, 120°, and 245°.

One option to reduce the phase-shift error in physical phase-shifting systems is to use a very linear phase shifter and carefully calibrate the stage response to determine voltage or pulse signal. Another option is to select a phase-shift algorithm that is insensitive

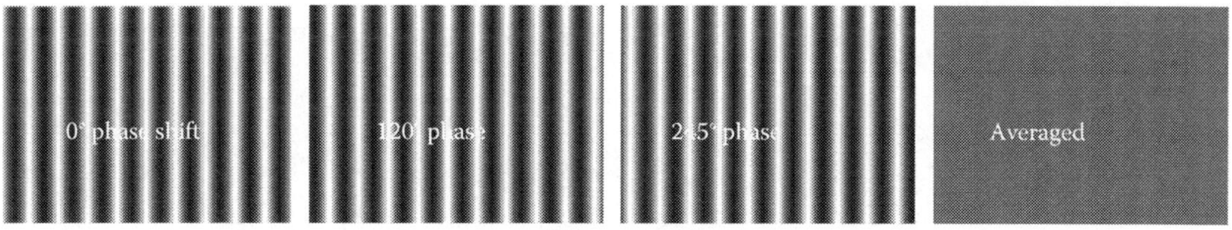

FIGURE 7.21
Ripples on the averaged image when the third fringe has a 5° phase-shift error.

to the phase-shift error such as the Carré phase-shifting and five-step phase-shifting algorithms.

7.5.1.2 Nonlinearity Error in the Detector/Projector

Nonlinearity errors may exist in both the camera and the projector. For industrial digital cameras, even though most have very good linearity unless the camera gain is set too low or too high, second-order nonlinearity may still exist. For presentation and home theater digital projectors, the default gamma setting is usually nonlinear because it is set for visual perception of nonlinear human eyes. Some projectors allow the users to reset gamma to be linear, but second-order nonlinearity may still exist. As one of the most severe error sources in digital phase-shifting systems, the nonlinearity will have to be compensated, which is discussed in more detail in the following sections.

A typical nonlinearity gamma curve is shown in Figure 7.22. This curve was obtained by inputting uniform grayscale images at 1 grayscale step up to 8 bits data limit (255 grayscale maximum) to a Canon SX50 LCOS projector. For each of the nine available gamma settings (from −4 to 4), a 12 bits digital QImaging camera was used to capture images of the projected uniform pattern on a white diffusive target for each projected grayscale image. The curves show nonlinearity from both the camera and the projector, mainly from the projector.

Some phase-shifting algorithms can deal with the nonlinearity error. For example, it has been proved that the double three-step phase-shifting algorithm is very efficient in eliminating the second-order nonlinearity error in the imaging/projecting system,[39] even if the second-order nonlinearity comes from the camera or the projector.

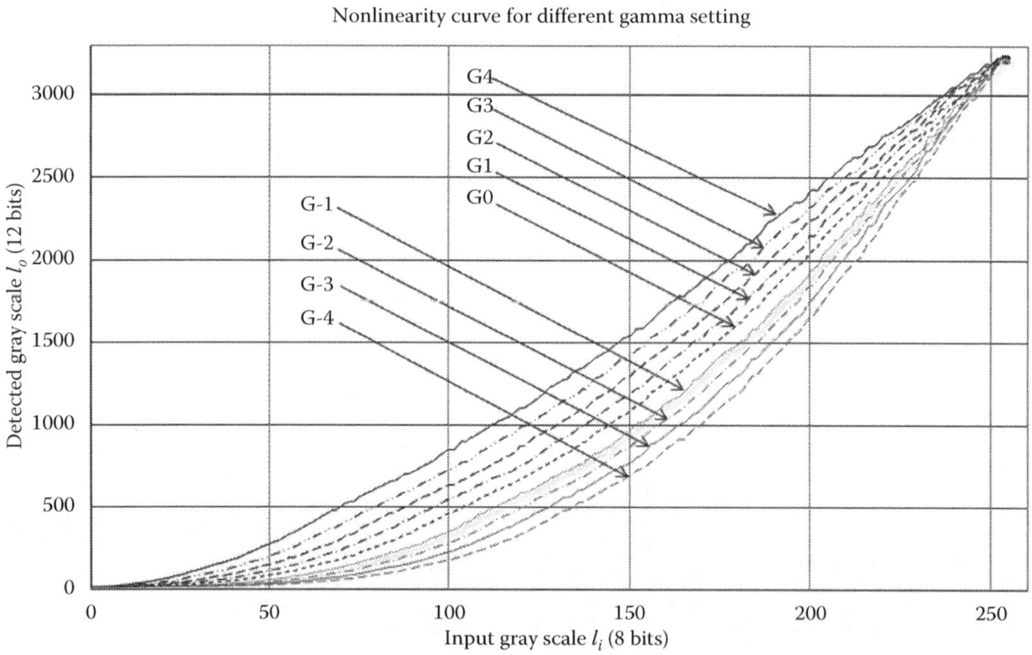

FIGURE 7.22
Nonlinearity curves of a Canon SX50 LCOS projector.

7.5.1.3 Modeling and Calibration Error

Selecting the correct modeling and calibration method is critical for accurate measurement. For a lens with large distortion (which may be a combination of magnification errors, field curvature, as well as geometric optical distortion), there is no way to obtain accurate results without correcting the lens errors using the calibration. Precision of the target, quality of the calibration setup, and calibration model including both selected intrinsic and extrinsic parameters all contribute to the calibration results.

The model used to convert an absolute phase map to a 3D point cloud is vital as well. For surfaces with curvatures, a simplified linear or partially linear model will result in very significant errors. In some cases that require multiview merging or extremely high accuracy, the light sheet from a grating or a digital projector cannot be taken as granted to be a flat plane. Instead, the light sheet has to be treated as a curved surface segment and needs to be fitted into a locally cylindrical surface.

7.5.1.4 Imbalance Error for Color Fringe Projections

Color fringe patterns have been used for phase shifting because they provide unique features—they enable three 120° phase shifts in one color fringe, allowing for fast measurement in a vibrating environment. However, for color fringe projection,[8,13,30,31,71] color balance is a big challenge. Because the human eyes have different sensitivity to different colors, most digital projectors have different gamma settings for RGB colors. The color camera may also have a nonuniform spectral response. In phase-shifting measurement, any captured brightness variation due to these imbalances among the three channels (corresponding to three fringe images) may contribute to error and noises. Also, colorful object surface may be a problem and needs special considerations.[76]

7.5.1.5 Quantization Error

Advancement in electronics has significantly reduced quantization error. Nowadays, 12 bits digital cameras are very common, which causes the digitization error to be in nanometer scale and thus negligible.[74] The digitization error in digital projectors can be reduced by use of a high-resolution projector with larger pitch and defocusing the projected fringes to act as a low-pass filter. Some projectors such as DLP can even accept 10 bits image input, which will reduce the digitization error significantly compared with the 8 bits data format.

7.5.1.6 Error and Noise from Environment

It has been demonstrated using simulation that the phase error due to vibration and air turbulence has a frequency of two times the fringe spatial frequency.[74] An obvious option is to add vibration isolation or to shield the instrument. Other means of removing vibration noise include selecting an appropriate phase-shifting algorithm that is less sensitive to the vibration, capturing data faster with less images and shorter shutter time, and using simultaneous phase shifting or color fringes.

The background and electronic noise of the camera can be reduced by averaging several images for each fringe pattern. For digital projectors, a reasonably long camera shutter can let the fringe be more stable while too short of a shutter time may cause some problems because the camera may capture the image at the moment of either the refreshing

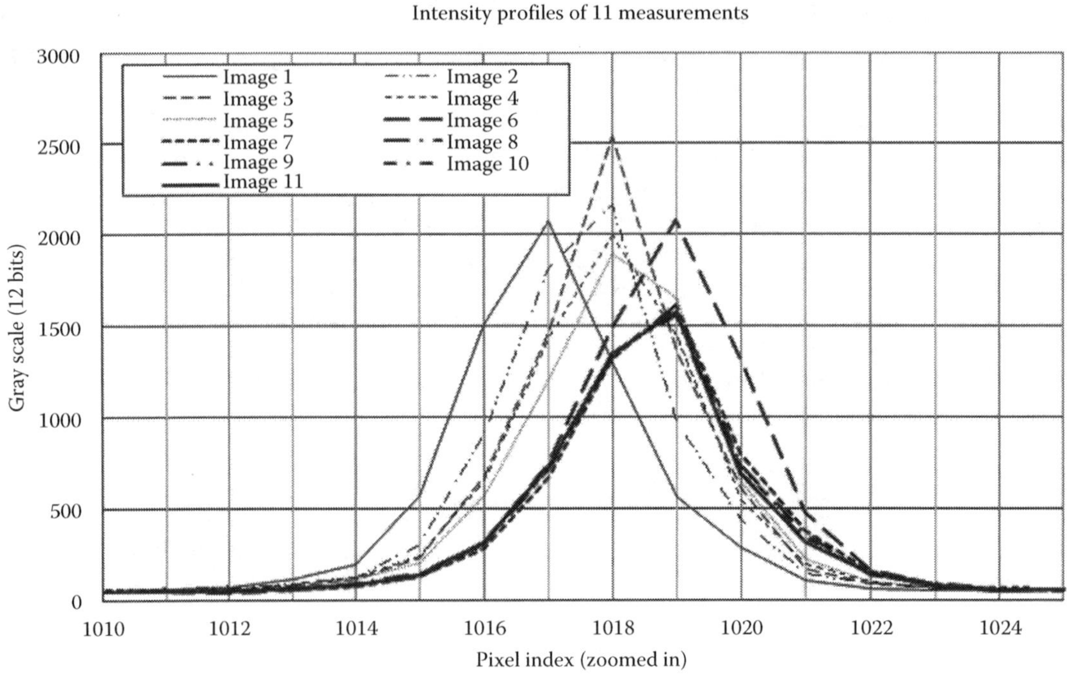

FIGURE 7.23
Drift in a Canon SX50 LCOS digital projector.

transition from one image frame to another or the dynamic binary on/off integration of the projector chip pixels during image formation.

For highly accurate, repeatable, and reliable measurement, thermal drift may be another serious problem. For large FOV measurement, a 0.1 pixel drift may cause 0.5 mm coordinate displacement in the 3D space. Figure 7.23 shows drifts in a Canon SX50 LCOS projector. These projected line images were taken every 10 min, and the intensity profiles at the same cross section were drawn. As can be seen in the figure, severe drift exists.

7.5.2 Nonlinearity Compensation with the Projector Gamma γ

The linearity of the phase-shifting system is so important in obtaining high-accuracy, low-noise point cloud that many papers have been published in this area. This section discusses the two most widely used gamma correction techniques in digital phase-shifting systems.

7.5.2.1 Gamma Correction with the Response Curve

As shown in Figure 7.22, nonlinearity in a digital projector can be very severe, and gamma correction is usually desirable. A first step is to pick up a gamma as close to linear as possible and then measure the system response to get a response curve like that in Figure 7.22. The response curve can be obtained by gradually changing the input gray level I_i and capturing the nonsaturated images with the fixed gamma setting and camera settings. A patch consisting of multiple pixels is used to reduce the noise by averaging their grayscales as the response I_o.

One gamma correction method using the gamma curve is through a compensation function. A polynomial function up to ninth order is usually used to fit the gamma curve such as

$$I_i = a_0 + a_1 I_o + a_2 I_o^2 + a_3 I_o^3 + a_4 I_o^4 + a_5 I_o^5 + a_6 I_o^6 + a_7 I_o^7 + a_8 I_o^8 + a_9 I_o^9 \tag{7.84}$$

Every intensity value calculated in Equation 7.8 or 7.9 needs to be substituted into Equation 7.84 as I_o to calculate the required input I_i so that the output fringe profile via the projector is sinusoidal.

For the gamma $\gamma = 4$ curve in Figure 7.22, a noncompensated gamma response will project a nonsinusoidal fringe pattern, although the input to the projector is sinusoidal, as shown in Figure 7.24. After processing the gamma curve compensation, the 10 coefficients in Equation 7.84 are listed in Table 7.1.

The compensation process is demonstrated in Figure 7.25 using the compensation coefficients given in Table 7.1.

An alternative way to the compensation function is to use a lookup table (LUT) and interpolation to modify the calculated intensity by ΔI from Equation 7.8 or 7.9 using

$$I' = I + \Delta I(I) \tag{7.85}$$

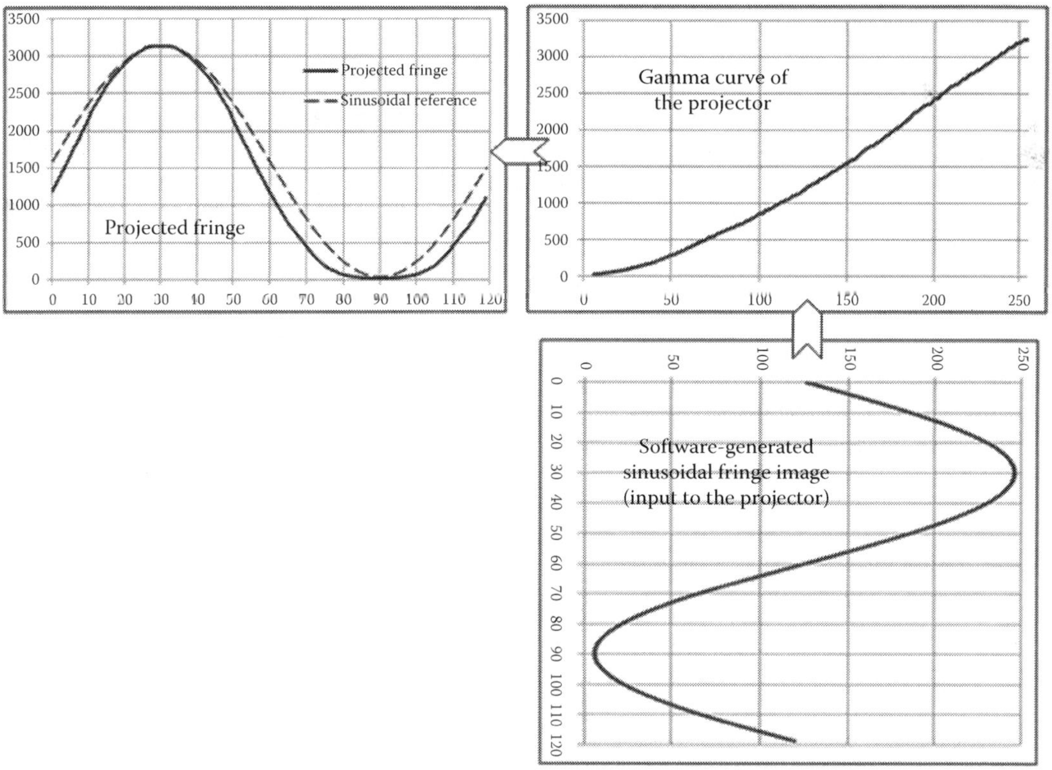

FIGURE 7.24
Projected fringe through a nonlinear projector (arrows indicate the data flow path).

TABLE 7.1

Ten Compensation Coefficients

Coefficient a_i	Value
a_0	8.005962569868930e+00
a_1	3.767418142271050e+00
a_2	−1.417556486538000e−01
a_3	4.004162099686760e−03
a_4	−6.551477558101120e−05
a_5	6.464338860749370e−07
a_6	−3.910659971627370e−09
a_7	1.417609255380700e−11
a_8	−2.823177656665190e−14
a_9	2.372943262233540e−17

where ΔI is a function of the calculated intensity I and is obtained by interpolating in the LUT.[77] The LUT is obtained by comparing the difference between the measured gamma curve and the ideal linear curve (upper curve), as shown in Figure 7.26. For each input grayscale value I, the system generates grayscale output g; for the system to be linear, the required output grayscale should be g', which needs input I'. In order to compensate for the nonlinearity, the calculated intensity I from Equation 7.8 has to be modified by ΔI:

$$\Delta I = I' - I \tag{7.86}$$

The LUT is obtained by recording all calculated ΔI for the input grayscale from 0 to 255 for 8 bits projector. Later when the fringe pattern is generated in the software using Equation 7.8, for each calculated I, the corresponding ΔI has to be calculated from the LUT through internal interpolation and the input image to the projector is then calculated using Equation 7.85. Because interpolation is involved, this method requires the projector's gamma curve to be monotonic.

7.5.2.2 Gamma Correction with a One-Parameter Gamma Model

The use of a one-parameter gamma function to estimate both the phase and gamma is also a hot topic in addressing the nonlinearity problem.[78–80] The gamma function that describes the relationship between input I_i and output I_o with a gamma γ can be modeled as

$$I_o = I_i^{\gamma} \tag{7.87}$$

Estimating the phase value for a linear system from the calculated phase map under a nonlinear system involves the phase shifting to obtain the phase and then an iterative process to estimate the gamma and phase alternatively. Some researchers also use the least-square fitting method with a few images to estimate the phase distribution[79] while others use many images for gamma estimation to apply a statistical method[78]or to reduce the error in gamma estimation.[80] Due to an error in the measurement and processing, the estimated gamma may vary from pixel to pixel in the image and the average of the estimated gamma over the entire image is taken as the "global" gamma. Using the gamma function (model)

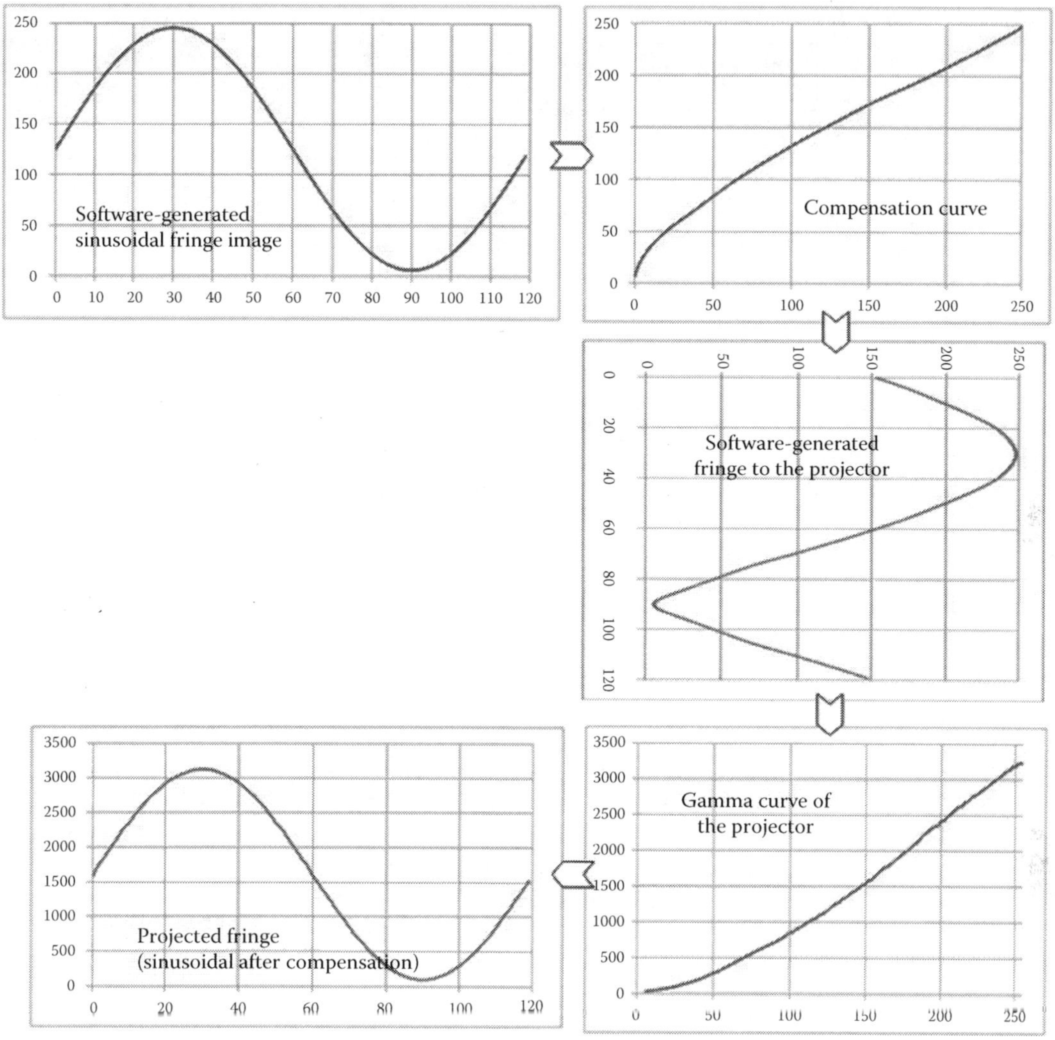

FIGURE 7.25
Nonlinearity compensation process (arrows indicate the data flow path).

for phase estimation does not need the gamma curve and one-time compensation but it usually needs additional fitting or iteration, and thus is time consuming.

7.5.3 Phase Error Compensation

As an intermediate between the captured images and desired coordinates (point cloud), a phase map can also be compensated, a method that has some advantages. Compared with direct coordinate correction, phase compensation can be faster and easier to implement as long as a good correction mechanism can be built that is reliable and not dependent on the measurement settings during image capturing.

Over the past several years, such phase compensation techniques have started to emerge.[77,81–83] It has been demonstrated that the phase error due to the system nonlinearity

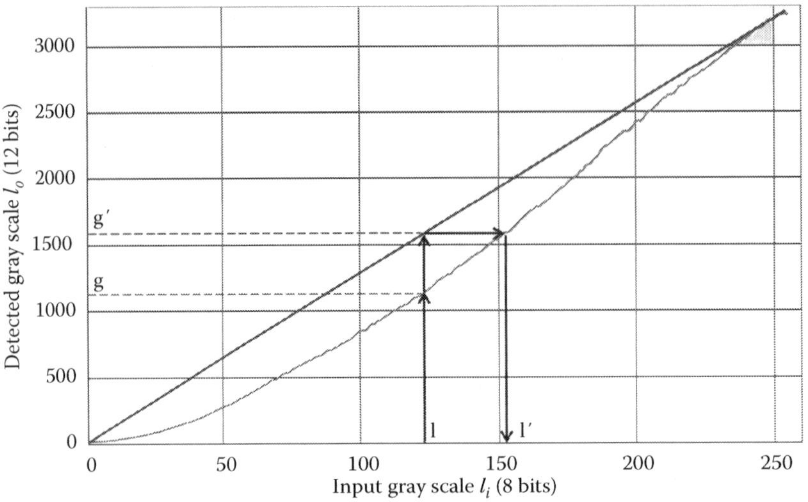

FIGURE 7.26
Method for the LUT generation.

is independent of the pitch used in the fringe generation. The phase error LUT only needs to be built on one 2π phase cycle. The required LUT is generated by performing a phase-shifting measurement with a large pitch on a white flat surface. The phase with and without nonlinearity issues is compared to construct the error map. The established phase error LUT can be applied to the wrapped phase map when a part is measured.

Some researchers[84] have also investigated the use of a phase compensation function to correct the distorted phase map directly through an inverse function. The inverse function is a polynomial function obtained through an iterative fitting process. After original phase shifting is performed, the calculated phase map is then modified by this correction function and the new phase map will have much less error from system nonlinearity.

Compared with the gamma correction discussed in the previous section, the phase error compensation technique takes more computing effort. The gamma correction can be done once before measurement while the phase correction has to be performed pixel by pixel after measurement.

7.5.4 Coordinate Compensation

Although the effects of many error sources can be reduced by various means, the achieved accuracy may still be limited because there are some errors that cannot be reduced by these measures completely. This limitation makes direct coordinate error compensation a very important technique for measurement systems to reach higher accuracy without significantly increasing manufacturing cost.

7.5.4.1 Coordinate Error Map

The first step for coordinate error compensation is to obtain an error map in the 3D space. This is performed by measuring a feature (point or surface) and comparing the measured data with a reference data point. Two methods were investigated to collect data for the error map construction. One method is to use a flat surface as a reference. The measured data of the surface are fitted into a plane, and the deviation from the fitted plane at each point can be used as an error map at that location. Moving the target to various positions in the 3D space and obtaining the error map at each location provide an error map in the entire measurement volume.

The other method to obtain the error map is to use CMM to provide the coordinate reference.[85] In this setup, a small target with a dot in the center is mounted on a CMM probe. CMM moves the target to predetermined points in the measurement volume. At each point, the target is measured and the coordinates of its center dot are extracted. An error map in the 3D space can be obtained by comparing the measured coordinates with the CMM coordinates.

Once the error map is obtained, the errors in the measurement system can be compensated at measured points through either error functions or interpolation.

7.5.4.2 Coordinate Error Compensation

In some cases, it is possible to find an error compensation function rather than use a LUT, especially when the system measurement volume and error map are symmetric, in which case the error functions are the easiest to reconstruct. A traditional method to reconstruct the error function Δ is to fit the error map into a function of coordinates (x, y, z) and errors (e_x, e_y, e_z) such as

$$\Delta_x = f\left(x, y, z, e_x\right) \tag{7.88}$$

$$\Delta_y = f\left(x, y, z, e_y\right) \tag{7.89}$$

$$\Delta_z = f\left(x, y, z, e_z\right) \tag{7.90}$$

The coordinates (x', y', z') after compensation can be calculated as

$$\left(x, y, z\right) = \left(x + \Delta_x, y + \Delta_y, z + \Delta_z\right) \tag{7.91}$$

In general cases, the error map will not be symmetric, and it might be impractical to reconstruct an error function with high accuracy. More often, a LUT can be built to compensate the coordinate errors through interpolation. A 3D interpolation technique called Shepard's method[86] is used for error compensation in some research. The interpolated value s is given by a function

$$s(v) = \sum_{i=1}^{N}\left[w_i(v)\Delta(v_i)\right] \tag{7.92}$$

where

v is a vector representing a point
N is the number of points used in the interpolation
$\Delta(v_i)$ is the error at point v_i

The weighting function $w_i(v)$ has the form

$$w_i(v) = \frac{\left\| v - v_i \right\|^{-2}}{\sum_{j=1}^{N} \left\| v - v_j \right\|^{-2}} \tag{7.93}$$

Obviously, if $v = v_i$, $s(v) = \Delta(v_i)$. If a point v_i is closer to the point v to be interpolated, it is given a larger weight. The Euclidean normal is defined as

$$\left\| v \right\| = \sqrt{\sum_{k=1}^{K} \xi_k^2} \tag{7.94}$$

where ξ_k is the element of vector v. In most cases, $K = 3$ and s can be the interpolated error in any coordinate direction depending on $\Delta(v_i)$. Since the error map is in the form of a 3D grid, the number of points used in the interpolation N can be set to be eight. For every measured point, the data set of the error map is searched to find the eight points that are the closest to the measured point and calculate their weight functions $w_i(v)$ ($i = 1$–8) according to Equation 7.93. The interpolated error for the measured point is then calculated by Equation 7.92. This error is then subtracted from the measured coordinates of the point to improve measurement accuracy.

7.6 Summary

In this chapter, we tried to detail the science and methods behind the field of phase shifting–based methods. These methods have been used in a wide range of commercial systems made for such applications as reverse engineering of part geometry, process control of formed parts like airfoil and sheet metal structures, as well as small area mapping of features like edge breaks on machined part out to dental impressions of people's teeth. The use of phase-shifting methods has been made more widely practical by high-speed computers and larger memory chips that allow a typical phase-shift measurement to be made in a few seconds in most cases and at the frame rates of cameras using some dedicated hardware.

In Chapter 1, we discussed some of the challenges in applying 3D technology for industrial metrology purposes. In subsequent chapters, we will present a wide range of examples of applications of this technology.

References

1. E. Hetcht, *Optics*, 3rd edn., Addison-Wesley, Reading, MA (1998).
2. M. S. Mermelstein, D. L. Feldkhun, and L. G. Shirley, Video-rate surface profiling with acousto-optic accordion fringe interferometry, *Optical Engineering*, 39, 106 (2000); doi: 10.1117/1.602342.
3. H. Takasaki, Moiré topography, *Applied Optics*, 9(6), 1467–1472 (1970).

4. F. P. Chiang, Moiré methods for contouring, displacement, deflection, slop and curvature, *Proceedings of SPIE*, 153, 113–119 (1978).
5. Y. Wang. and P. Hassell, Measurement of thermally induced warpage of BGA packages/substrates using phase-stepping shadow moiré, *Proceedings of the First Electronic Packaging Technology Conference*, Singapore, pp. 283–289 (1997).
6. K. G. Harding and S. L. Cartwright, Phase grating use in moire interferometry, *Applied Optics*, 23(10), 1517 (1984).
7. K. Creath and J. C. Wyant, Moiré and fringe projection techniques, in *Optical Shop Testing*, D. Malakara (ed.), 3rd edn., Chapter 16, pp. 559–652 John Wiley & Sons, New York (2007).
8. P. S. Huang, Q. Hu, F. Jin, and F.-P. Chiang, Color-encoded digital fringe projection technique for high-speed three-dimensional surface contouring, *Optical Engineering*, 38, 1065 (1999); doi: 10.1117/1.602151.
9. P. S. Huang, F. Jin, and F.-P. Chiang, Quantitative evaluation of corrosion by a digital fringe projection technique, *Optics and Lasers in Engineering*, 31(5), 371–380 (1999).
10. Y. Y. Hung, L. Lin, H. M. Shang, and B. G. Park, Practical three-dimensional computer vision techniques for full-field surface measurement, *Optical Engineering*, 39, 143–149 (2000).
11. P. S. Huang, S. Zhang, and F.-P. Chiang, Trapezoidal phase-shifting method for three-dimensional shape measurement, *Optical Engineering*, 44, 123601 (2006).
12. L. C. Chen, X. L. Nguyen, and Y. S. Shu, High speed 3-D surface profilometry employing trapezoidal HSI phase shifting method with multi-band calibration for colour surface reconstruction, *Measurement Science and Technology*, 21(10), 105309 (2010).
13. S. Lina, Y. Shuang, and W. Haibin, 3D measurement technology based on color trapezoidal phase-shifting coding light, *IEEE 9th International Conference on the Properties and Applications of Dielectric Materials* (ICPADM 2009), Harbin, China, pp. 1094–1097 (2009).
14. L. Chen, C. Quan, C. J. Tay, and Y. Fu, Shape measurement using one frame projected sawtooth fringe pattern, *Optics Communications*, 246(4–6), 275–284 (2005).
15. K. Creath, Phase-measurement interferometry techniques, in *Progress in Optics*, Vol. 26, Chapter 5, Elsevier Science Publishers B.V., Amsterdam, the Netherlands (1988).
16. M. Kujawinska, Use of phase-stepping automatic fringe analysis in moiré interferometry, *Applied Optics*, 26(22), 4712–4714 (1987).
17. H. Schreiber and J. H. Bruning, Phase shifting interferometry, in *Optical Shop Testing*, 3rd edn., John Wiley & Sons, New York, pp. 547–655 (2007).
18. D. C. Ghiglia and M. D. Pritt, *Two-Dimensional Phase Unwrapping: Theory, Algorithms, and Software*, 1st edn., John Wiley & Sons, New York (1998).
19. Q. Hu and K. G. Harding, Conversion from phase map to coordinate: Comparison among spatial carrier, Fourier transform, and phase shifting methods, *Optics and Lasers in Engineering*, 45, 342–348 (2007).
20. Q. J. Hu, Modeling, error analysis, and compensation in phase-shifting surface profilers, *Proceedings of SPIE*, 8133, 81330L (2011).
21. Q. Hu, K. G. Harding, X. Du, and D. Hamilton, Shiny parts measurement using color separation, *Proceedings of SPIE*, 6000, 60000D (2005).
22. C. Ai and J. C. Wyant, Effect of piezoelectric transducer nonlinearity on phase shift interferometry, *Applied Optics* 26(6), 1112–1116 (1987).
23. M. P. Kothiyal and C. Delisle, Polarization component phase shifters in phase shifting interferometry: Error analysis, *Optica Acta: International Journal of Optics*, 33(6), 787–793 (1986).
24. Y.-B. Choi and S.-W. Kim, Phase-shifting grating projection moiré topography, *Optical Engineering*, 37(3), 1005–1010 (1998).
25. J. Y. Cheng and Q. Chen, An ultrafast phase modulator for 3D imaging, sensors, cameras, and systems for scientific/industrial applications VII, in *IS&T Electronic Imaging, Proceedings of SPIE*, Vol. 6068, M. M. Blouke (ed.), 60680L (2006).
26. B. F. Oreb, I. C. C. Larkin, P. Fairman, and M. Chaffari, Moire based optical surface profiler for the minting industry, in *Interferometry: Surface Characterization and Testing, Proceedings of SPIE*, Vol. 1776 (1992).

27. J. Pan, R. Curry, N. Hubble, and D. A. Zwemer, Comparing techniques for temperature-dependent warpage measurement, *Global SMT & Packaging*, 14–18 (February 2008).
28. A. J. Boehnlein and K. G. Harding, Field shift moire, a new technique for absolute range measurement, *Proceedings of SPIE*, 1163, 2–13 (1989).
29. L. H. Bieman, K. G. Harding, and A. Boehnlein, Absolute measurement using field shifted Moiré, in *Optics, Illumination, and Image Sensing for Machine Vision VI, Proceedings of SPIE*, Vol. 1614, pp. 259 264 (1991).
30. K. G. Harding, M. P. Coletta, and C. H. VanDommelen, Color encoded Moiré contouring, in *Optics, Illumination, and Image Sensing for Machine Vision III, D. Svetkoff (ed.), Proceedings of SPIE*, Vol. 1005, pp. 169–178 (1988).
31. M.-S. Jeong and S.-W. Kim, Color grating projection moire´ with time-integral fringe capturing for high-speed 3-D imaging, *Optical Engineering*, 41(8), 1912–1917 (2002).
32. P. S. Huang, C. Zhang, and F.-P. Chiang, High-speed 3-D shape measurement based on digital fringe projection, *Optical Engineering*, 42, 163 (2003).
33. S. Zhang and P. S. Huang, High-resolution, real-time three-dimensional shape measurement, *Optical Engineering*, 45(12), 123601 (2006).
34. C. L. Koliopoulos, Simultaneous phase shift interferometer, in *Advanced Optical Manufacturing and Testing II, Proceedings of SPIE*, Vol. 1531, V. J. Doherty (ed.) pp. 119–127 (1992).
35. J. C. Wyant, Advances in interferometric metrology, in *Optical Design and Testing, Proceedings of SPIE*, Vol. 4927, pp. 154–162 (2002).
36. L.-C. Chen, S.-L. Yeh, A. M. Tapilouw, and J.-C. Chang, 3-D surface profilometry using simultaneous phase-shifting interferometry, *Optics Communications*, 283(18), 3376–3382 (2010).
37. D. W. Phillion, General methods for generating phase-shifting interferometry algorithm, *Applied Optics*, 36(31), 8098–8115 (1997).
38. Q. Hu, P. S. Huang, Q. Fu, and F. P. Chiang, Calibration of a three-dimensional shape measurement system, *Optical Engineering*, 42, 487 (2003).
39. P. S. Huang, Q. Hu, and F. P. Chiang, Double three-step phase-shifting algorithm, *Applied Optics*, 41(22), 4503–4509 (2002).
40. P. S. Huang and H. Guo, Phase shifting shadow moiré using the Carré algorithm, in *Two- and Three-Dimensional Methods for Inspection and Metrology VI, Proceedings of SPIE*, 7066, 70660B (2008).
41. Q. Kemao, S. Fangjun, and W. Xiaoping, Determination of the best phase step of the Carré algorithm in phase shifting interferometry, *Measurement Science and Technology*, 11, 1220–1223 (2000).
42. P. Hariharan, B. F. Oreb, and T. Eiju, Digital phase-shifting interferometer: A simple error-compensating phase calculation algorithm, *Applied Optics*, 26, 2504–2506 (1987).
43. K. G. Larkin, Efficient nonlinear algorithm for envelope detection in white light interferometry, *Journal of the Optical Society of America A*, 13(4), 832–843 (1996).
44. J. Novak, Five-step phase-shifting algorithms with unknown values of phase shift, *Optik-International Journal for Light and Electron Optics*, 114(2), 63 68 (2003).
45. H. Zhang, M. J. Lalor, and D. R. Burton, Robust, accurate seven-sample phase-shifting algorithm insensitive to nonlinear phase-shift error and second-harmonic distortion: A comparative study, *Optical Engineering*, 38, 1524 (1999).
46. K. Hibino, B. F. Oreb, D. I. Farrant, and K. G. Larkin, Phase-shifting algorithms for nonlinear and spatially nonuniform phase shifts, *Journal of the Optical Society of America A*, 14(4), 918–930 (1997).
47. K. G. Larkin and B. F. Oreb, A new seven-sample symmetrical phase-shifting algorithm, in *SPIE Conference on Interferometry Techniques and Analysis, SPIE Proceedings*, Vol. 1755, San Diego, CA pp. 2–11 (1992).
48. K. H. Womack, Interferometric phase measurement using spatial synchronous detection, *Optical Engineering*, 23, 391–395 (1984).
49. M. Kujawinska, Spatial phase measurement methods, in *Interferogram Analysis: Digital Fringe Pattern Measurement Techniques*, D. W. Robinson and G. T. Reid (eds.), Institute of Physics, Bristol, U.K. pp. 141–193 (1993).

50. J. Xu, Q. Xu, and H. Peng, Spatial carrier phase-shifting algorithm based on least-squares iteration, *Applied Optics*, 47, 5446–5453 (2008).
51. W. Osten, P. Andrae, W. Nadeborn, and W. Jiiptner, Modern approaches for absolute phase measurement, *Proceedings of SPIE*, Vol. 2647, *International Conference on Holography and Correlation Optics*, O. V. Angelsky (ed.), Chernovtsy, Ukraine, pp. 529–540 (1995).
52. H. Cui, W. Liao, N. Dai, and X. Cheng, A flexible phase-shifting method with absolute phase marker retrieval, *Measurement*, 45(1), 101–108 (2012).
53. K. Creath, *Phase-Measuring Interferometry Techniques* [*Progress in Optics XXVI*], Elsevier Science Publishers B.V., Amsterdam, the Netherlands, pp. 349–393 (1988).
54. L. Tao, K. Harding, M. Jia, and G. Song, Calibration and image enhancement algorithm of portable structured light 3D gauge system for improving accuracy, in *Optical Metrology and Inspection for Industrial Applications*, *Proceedings of SPIE*, Vol. 7855, 78550Y-1, K. Harding, P. S. Huang, and T. Yoshizawa (eds.), 78550Y (2010).
55. Y. Wang and P. Hassell, Measurement of the thermal deformation of BGA using phase-shifting shadow Moiré, *Electronic/Numerical Mechanics in Electronic Packaging*, 2, 32–39 (1998).
56. J. Pan, R. Curry, N. Hubble, and D. Zwemer, *Comparing Techniques for Temperature Dependent Warpage Measurement*. Plus 10/2007, pp. 1–6.
57. G. Mauvoisin, F. Brémand, and A. Lagarde, Three-dimensional shape reconstruction by phase-shifting shadow Moiré, *Applied Optics*, 33(11), 2163–2169 (1994).
58. Q. Hu, K. G. Harding, D. Hamilton, and J. Flint, Multiple views merging from different cameras in fringe-projection based phase-shifting method, *Proceedings of SPIE* 6762, 676207 (2007).
59. Q. Hu, P. S. Huang, and F. P. Chiang, 360-degree shape measurement for reverse engineering, *Proceedings of the International Conference on Flexible Automation and Intelligent Manufacturing (FAIM 2000)*, University of Maryland, College Park, MD (June 2000).
60. R. Y. Tsai, A versatile camera calibration technique for high-accuracy 3-D machine vision metrology using off-the-shelf TV cameras and lenses, *IEEE Journal of Robotics and Automation*, RA-3(4), 323–344 (1987).
61. Z. Zhang, A flexible new technique for camera calibration, Microsoft Research Technical Report MSR-TR-98-71.
62. A. M. McIvor, Nonlinear calibration of a laser stripe profiler, *Optical Engineering*, 41(01), 205–212 (2002).
63. S. Zhang, and P. Huang, Novel method for structured light system calibration, *Optical Engineering*, 45(08), 083601 (2006).
64. J. Heikkilä and O. Silvén, A four-step camera calibration procedure with implicit image correction, *IEEE Proceedings of the 1997 Conference on Computer Vision and Pattern Recognition (CVPR '97)*, San Juan, PR, pp. 1106–1112 (1997).
65. R. Legarda-Sáenz, T. Bothe, and W. P. Jüptner, Accurate procedure for the calibration of a structured light system, *Optical Engineering*, 43(2), 464–471 (2004).
66. S. Q. Jin, L. Q. Fan, Q. Y. Liu and R. S. Lu, Novel calibration and lens distortion correction of 3D reconstruction systems, *International Symposium on Instrumentation Science and Technology*, *Journal of Physics: Conference Series*, 48, 359–363 (2006).
67. K. M. Dawson-Howe and D. Vernon, Simple pinhole camera calibration, *International Journal of Imaging Systems and Technology*, 5(1), 1–6 (1994).
68. J. Kannala and S. S. Brandt, A generic camera model and calibration method for conventional, wide-angle and fish-eye lenses, *IEEE Transactions on Pattern Analysis and Machine Intelligence*, 28(8), 1335–1340 (2006).
69. T. Rahman and N. Krouglicof, An efficient camera calibration technique offering robustness and accuracy over a wide range of lens distortion, *IEEE Transactions on Image Processing*, 21(2), 626–637 (2012).
70. Y. Yin, X. Peng, Y. Guan, X. Liu, A. Li, Calibration target reconstruction for 3-D vision inspection system of large-scale engineering objects, in *Optical Metrology and Inspection for Industrial Applications*, *Proceedings of SPIE*, Vol. 7855, K. Harding, P. S. Huang, and T. Yoshizawa (eds.), Beijing, China, 78550V (2010).

71. C. Sinlapeecheewa and K. Takamasu, 3D profile measurement by color pattern projection and system calibration, *IEEE International Conference on Industrial Technology (IEEE ICIT'02)*, Bangkok, Thailand, Vol. 1, pp. 405–410 (2002).
72. T. S. Shen and C. H. Menq, Digital projector calibration for 3-D active vision systems, *Journal of Manufacturing Science and Engineering*, 124(2), 126–134 (2002).
73. M. Kimura, M. Mochimaru, and T. Kanade. Projector calibration using arbitrary planes and calibrated camera. *2007 IEEE Computer Society Conference on Computer Vision and Pattern Recognition (CVPR 2007)*, IEEE Computer Society, Minneapolis, MN, June 18–23 (2007).
74. K. Creath, Error sources in phase measuring interferometry, *Proceedings of SPIE*, Vol. 1720, *International Symposium on Optical Fabrication, Testing, and Surface Evaluation*, J. Tsujiuchi (ed.), Tokyo, Japan, pp. 428–435 (1992)
75. K. Creath and J. Schmit, Errors in spatial phase-stepping techniques, *Proceedings of SPIE*, Vol. 2340, *Interferometry '94: New Techniques and Analysis in Optical Measurements*, M. Kujawinska and K. Patorski (eds.), Warsa, Poland, pp. 170–176 (1994).
76. L. C. Chen et al., High-speed 3D surface profilometry employing trapezoidal phase-shifting method with multi-band calibration for colour surface reconstruction, *Measurement Science and Technology*, 21(10), 105309 (2010).
77. S. Zhang and P. S. Huang, Phase error compensation for a 3-D shape measurement system based on the phase-shifting method, *Optical Engineering*, 46, 063601 (2007); doi: 10.1117/1.2746814.
78. H. Guo, H. He, and M. Chen, Gamma correction for digital fringe projection profilometry, *Applied Optics*, 43, 2906–2914 (2004).
79. T. M. Hoang, Simple gamma correction for fringe projection profilometry system, *SIGGRAPH 2010*, Los Angeles, CA, July 25–29 (2010).
80. K. Liu, Y. Wang, D. L. Lau, Q. Hao, and L. G. Hassebrook, Gamma model and its analysis for phase measuring profilometry, *Journal of the Optical Society of America A*, 27(3), 553–562 (2010).
81. S. Zhang and S.-T. Yau, Generic nonsinusoidal phase error correction for three-dimensional shape measurement using a digital video projector, *Applied Optics*, 46(1), 36–43 (2007).
82. H. Cui, X. Cheng, N. Dai, T. Yuan, and W. Liao, A new phase error compensate method Of 3-D shape measurement system using DMD projector, *Fourth International Symposium on Precision Mechanical Measurements, Proceedings of SPIE*, Xinjiang, China, 7130, 713041 (2008).
83. X. Chen, J. Xi, and Y. Jin, Phase error compensation method using smoothing spline approximation for a three-dimensional shape measurement system based on gray-code and phase-shift light projection, *Optical Engineering*, 47, 113601 (2008).
84. Y. Liu, J. Xi, Y. Yu, and J. Chicharo, Phase error correction based on inverse function shift estimation in phase shifting profilometry using a digital video projector, in *Optical Metrology and Inspection for Industrial Applications, Proceedings of SPIE*, 7855, K. Harding, P. S. Huang, and T. Yoshizawa (eds.), 78550W (2010).
85. P. S. Huang, Q. Hu, and F. P. Chiang, Error compensation for a three-dimensional shape measurement system, *Optical Engineering*, 42, 482 (2003).
86. P. Alfeld, Scattered data interpolation in three or more variables, in *Mathematical Methods in Computer Aided Geometric Design*, T. Lyche and L. Schumaker (eds.), Academic Press, Boston, MA, pp. 1–33 (1989).

8

Moiré Measurement

Toru Yoshizawa and Lianhua Jin

CONTENTS

8.1 Introduction

The French term "moiré" originates from a type of textile, traditionally of silk, with a grained or "watered" appearance. Nowadays, moiré is generally used to describe a fringe that is created by superposition of two (or more) patterns such as line gratings and dot arrays. Moiré phenomenon has been well known because it can be observed in our daily life, like the pattern seen in folded netting. Lord Rayleigh made the first scientific description of moiré phenomena in 1874. Since then, a lot of research papers have been reported in various scientific fields. The first generation of moiré research was mainly focused on the physical interpretation of moiré.[1] The famous research papers by Oster and Nishijima suggested future availability of moiré techniques in scientific and industrial fields. At this point in time, we can review these works through a book compiled from selected papers on moiré.[2] The second generation was focused on in-plane moiré, that is, 1D or 2D applications of moiré. Researchers in the field of experimental mechanics intensively investigated moiré applications to strain measurement. This technique is highlighted by the fact that both noncontact and full-field measurements are attainable.

Many excellent papers have been reported, and a few popular books have been widely read.[3,4] Chiang's review[5] is a nice introduction to strain analysis by moiré technique. The third generation of moiré research applies moiré technique to out-of-plane moiré used to capture 3D deformation and profilometry.[6,7] These two research papers pioneered a new technique for 3D shape measurement, or profilometry,[8] and inspired people in the medical, dental, and anthropological fields. Moiré topography has been established as a key method for measuring the shape of the human body and the diagnosis of regional change, including for apparel and cosmetics purposes.

Currently, moiré applications have reached a stage where digital techniques and devices can be incorporated to find new principles based on moiré and to attain higher sensitivity measurement and real-time analysis. Nowadays, moiré methods are found in a variety of different scientific and engineering systems, and their technological applications cover industrial fields such as optical, mechanical, electrical, and chemical engineering mainly emphasizing metrological purposes. Typical applications have been described in a detailed book[9] that contains detection of temperature effect to electronic package, thermal stress, and strains and loading effect to various complex-shaped samples in addition to newly developed techniques using digital devices.

Some moiré patterns need filtering methods to eliminate undesired artifacts,[10] for example, the patterns produced during scanning a halftone picture. Some moiré patterns, on the other hand, act as very useful phenomena for measurements in different application fields. For instance, in the textile industry, designers intentionally generate beautiful moiré patterns with silk fabrics; in the healthcare field, moiré is applied to a diagnostic test of scoliosis,[11] which is more common in teenage females (young female's muscle is not as strong in comparison with young male's, and girl's spines are apt to be deformed).

In this chapter, applications of the moiré phenomenon to the optical metrology are described. In moiré metrology, the moiré fringe results from the superposition of two periodic grating structures with 1D or 2D lines: one is called reference grating and the other object grating, which is to be distorted by a structure whose deformation or shape is represented by the resulting moiré fringes. This object grating may be called the signal grating that is modulated by the deformation or profile of the sample to be measured. In the case of moiré topography, the object pattern is sometimes called the deformed grating pattern. The moiré fringe created by these two superposed gratings in the same plane is termed in-plane moiré, which is used to detect 1D or 2D deformation, displacement, and rotation, and the moiré caused by two gratings in different planes is named out-of-plane moiré, which is used for capturing 3D deformation or shape of the object.

In the following sections, we introduce the principles of pattern formation of in-plane and out-of-plane moirés and describe basic applications to strain analysis and profilometry.

8.2 In-Plane Moiré Method and Measurement of Strain

8.2.1 Pattern Formation of In-Plane Moiré

An in-plane moiré pattern is obtained by superimposing the reference and the signal gratings by direct contact or by optically overlapping one onto the other. In general, the reference grating consists of constant and equally spaced period lines with a fixed

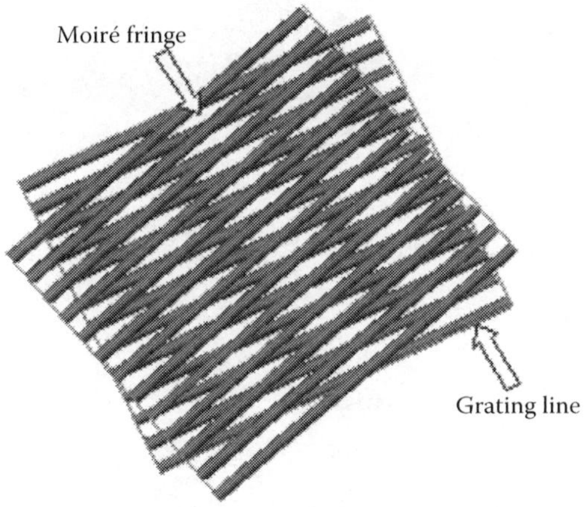

FIGURE 8.1
Superposition of two gratings and moiré fringe.

spatial orientation, and the same pattern is either printed or attached to the object or sample surface to be measured. Before deformation of the sample object, the period and orientation of this grating are identical to that of the reference grating. After deformation or displacement, this grating pattern is deformed or modulated to be the signal grating that contains information caused by deformation or displacement. In other words, this can be called as object grating in a manner analogous to holography. Here, let us consider the signal/object grating generated due to rotation by an angle θ with respect to the reference grating, as shown in Figure 8.1. When seen from a distance, we can no longer resolve the original grating lines with high frequency, and we only see dark and pale bands with low frequency, that is, the moiré pattern. The pale bands correspond to the lines of nodes, namely, broad lines passing through the intersections of many lines in the two gratings.

How are the orientation and interval of these moiré patterns in Figure 8.1 determined? Let p and p' be the periods of the reference and signal/object gratings, φ and d the orientation and interval of moiré pattern, respectively. From the geometry, in Figure 8.2,

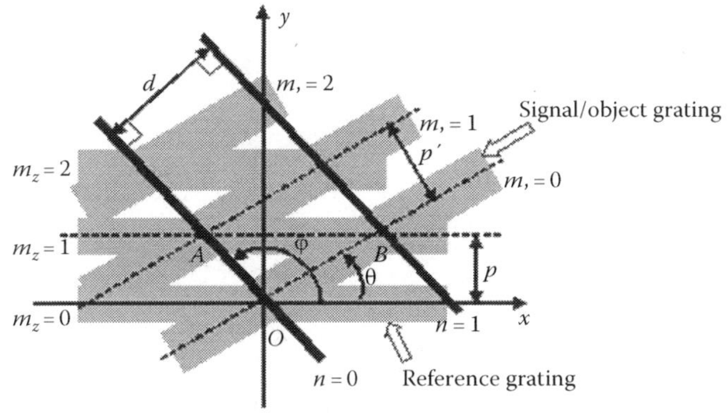

FIGURE 8.2
Order n, orientation φ, and interval d of in-plane moiré pattern. $n = mr - mo$ (mr, mo are the number of reference and signal/object grating lines, respectively.)

$$OA = \frac{p}{\cos(\phi - \pi/2)} = \frac{p}{\sin\phi} \tag{8.1}$$

$$OA = \frac{p'}{\cos(\phi - \pi/2 - \theta)} = \frac{p'}{\sin(\phi - \theta)} \tag{8.2}$$

Therefore,

$$p'\sin\phi = p\sin(\phi - \theta) \tag{8.3}$$

Rearranging this by using geometric functions, we obtain

$$\phi = \tan^{-1}\frac{p\sin\theta}{p\cos\theta - p'} \tag{8.4}$$

From Figure 8.2, we have

$$OB = \frac{p}{\sin\theta} \tag{8.5}$$

and

$$d = OB\cos\left(\phi - \frac{\pi}{2} - \theta\right) = OB\sin(\phi - \theta) \tag{8.6}$$

Substituting Equations 8.3 and 8.5 into Equation 8.6 leads to

$$d = \frac{p'\sin\phi}{\sin\theta} \tag{8.7}$$

Rearranging Equation 8.4 by using geometric relationships and then substituting into Equation 8.7 yield

$$d = \frac{pp'}{\sqrt{p^2\sin^2\theta + (p\cos\theta - p')^2}} \tag{8.8}$$

From Equations 8.4 and 8.8, it is obvious that once p, p', and θ are given, the orientation φ and interval d of the moiré fringe are uniquely decided. Conversely, from the measured values of φ and d, and a given period p of the reference grating, we can obtain the period p' and orientation θ of the signal/object grating whose structure may be deformed by strains.

8.2.2 Application to Strain Measurement

Strain is the geometric expression of deformation caused by the action of stress on a sample object. Strain is calculated by assuming change in length or in angle between the initial and final states. If strain is equal over all parts of the object, it is referred to as homogeneous strain; otherwise, it is known as inhomogeneous strain. Here, we apply in-plane moiré to the measurement of homogeneous linear strain and shear strain.

8.2.2.1 Linear Strain Measurement

The linear strain ε, according to Eulerian description, is given by

$$\varepsilon = \frac{\delta \ell}{\ell_f} = \frac{\ell_f - \ell_o}{\ell_f} \tag{8.9}$$

where ℓ_o and ℓ_f are the original and final lengths of the object, respectively. According to the Lagrangian description, the denominator is the original length ℓ_o instead of ℓ_f; for the application of moiré methods, here we use Eulerian description. When the deformation is very small, the difference between the two descriptions is negligible. The extension δl is positive if the object has gained length by tension and is negative if it has reduced length by compression. Since ℓ_f is always positive, the sign of linear strain is always the same as the sign of the extension.

In measuring the uniaxial linear strain, the object/signal grating will be printed onto the object surface with its lines perpendicular to the direction of the strain, and the reference grating superimposed on it with the same orientation, as shown in Figure 8.3. Figure 8.3c shows the resulted moiré patterns across the object deformed by the tension. In this case, the period of the object grating will be changed from p to p', and its orientation will remain unchanged, namely, the rotation angle θ in Figure 8.2 is zero.

Substituting $\theta = 0$ into Equation 8.4, the interval d of the moiré fringe becomes

$$d = \left| \frac{pp'}{p - p'} \right| \tag{8.10}$$

Thus, from Equations 8.9 and 8.10, we obtain the linear strain

$$|\varepsilon| = \left| \frac{\ell_f - \ell_o}{\ell_f} \right| = \left| \frac{p - p'}{p'} \right| = \frac{p}{d} \tag{8.11}$$

Since both p and d are positive magnitude, the sign of absolute value is introduced to the strain, which means that the appearance of moiré fringe will not explain whether it is a result of tensile or compression strain. To determine the sign of the moiré fringe as well as the resulting strain, various techniques are available such as the mismatch method, which introduces an initial bias (equivalent to either compression or tension by having a different grating period in the reference grating), or the fringe-shifting method, by which translating the reference grating causes the moire pattern to move, with one direction indicating compression and the other tension.

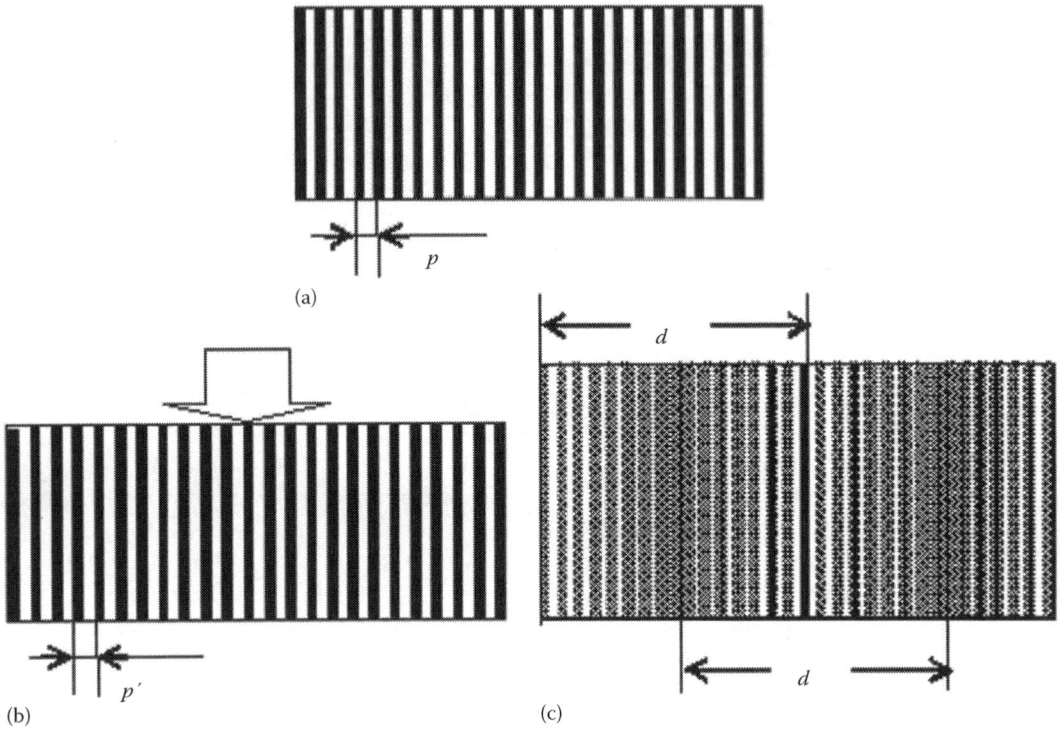

FIGURE 8.3
Reference grating (a) before and (b) after deformation of the object by linear tension and (c) resulted moiré pattern.

8.2.2.2 Shear Strain Measurement

The shear strain γ is defined as the angular change between any two lines in an object before and after deformation, assuming that the lengths of the lines are approaching zero. In applying the in-plane moiré method to measure the shear strain, the object grating should be so oriented that its principal direction is parallel to the direction of the shear, as shown in Figure 8.4a. Figure 8.4b shows typical moiré patterns across the object deformed by the shear. In this deformation, the period p' of the object grating is assumed to be the same as p before deformation, and the orientation changes from zero to θ (the quantity of θ is very small). Thus, from Equation 8.8, the interval of the moiré pattern is

$$d = \frac{p}{\theta} \tag{8.12}$$

Comparing Figure 8.4c with Figure 8.2, one can see easily that the shear strain γ is the resulting rotation angle θ of the object grating relative to the reference grating; hence, it can be expressed by

$$\gamma = \frac{p}{d} \tag{8.13}$$

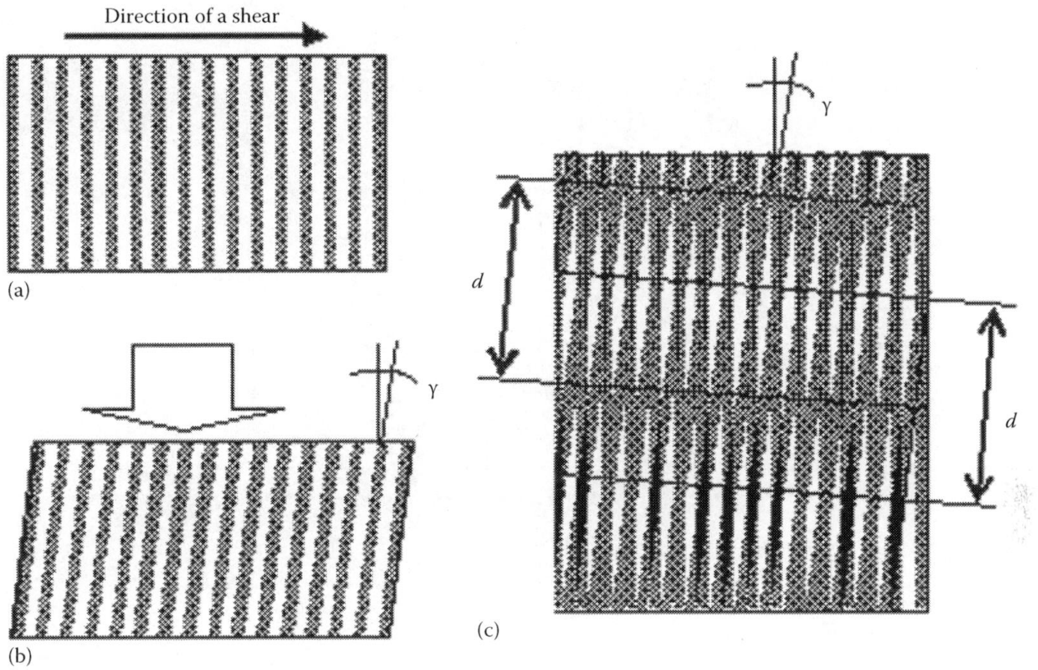

FIGURE 8.4
Reference grating (a) before and (b) after deformation of the object by shear and (c) resulted moiré pattern.

Although linear and shear strain are discussed independently, in a general 2D deformation, the object/signal grating undergoes rotation as well as change of period that is not uniformly distributed, as depicted in Figure 8.5. From this 2D moiré pattern, we obtain the linear and shear strain values given by

$$\varepsilon_x = p\frac{\partial n_x}{\partial x} \tag{8.14a}$$

$$\varepsilon_y = p\frac{\partial n_y}{\partial y} \tag{8.14b}$$

$$\gamma = p\left(\frac{\partial n_x}{\partial y} + \frac{\partial n_y}{\partial x}\right) \tag{8.14c}$$

where n_x, n_y are the fringe order at the point P along x, y direction, respectively.

The gratings used for strain analysis with an in-plane moiré method have usually such period as 20–40 lines/mm and are formed by optical ruling, holographic interference techniques, e-beam writing, x-ray lithography, or similar pattern formation methods. The original reference grating is transferred to the sample object (usually metal) to be measured by lithography using, for example, a photosensitive coating, photoresist exposure, or dichromate gelatin.

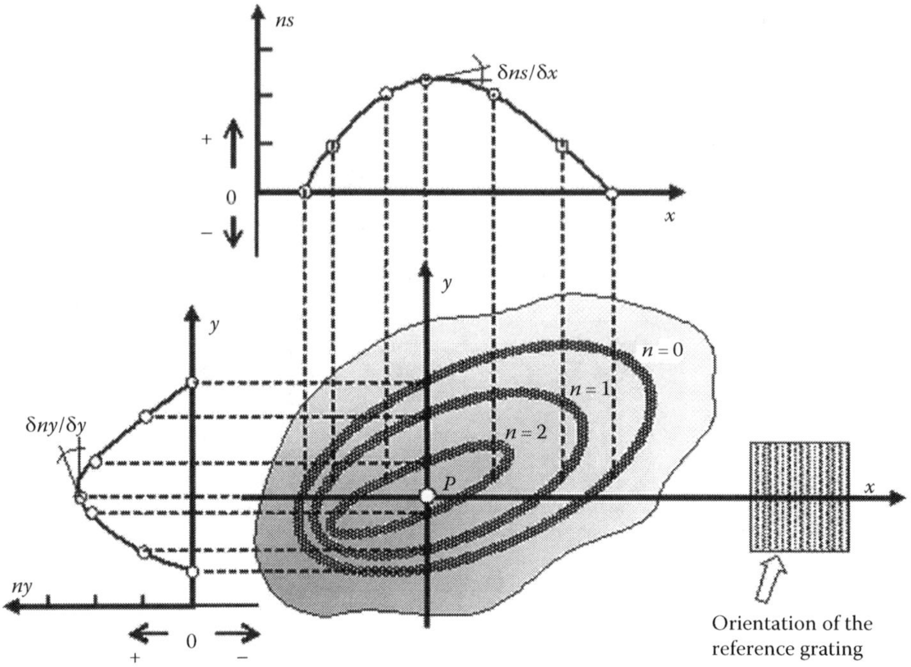

FIGURE 8.5
Strain analysis from 2D moiré fringe.

8.2.3 Practical Applications of In-Plane Moiré Methods

In-plane moiré has been applied in a variety of industrial fields, and its applications have been spreading more and more in combination with other optical methods such as diffraction and the use of digital devices. As suggested in Figure 8.4, if two gratings with the same period p overlap with the inclination angle θ, moiré fringe appears with the larger period d. In cases where this angle of intersection θ is sufficiently small, d is given by p/θ, that is, $d = p/\theta$. This means the period of the moiré fringe becomes broad enough to be captured by a detector even if the grating pattern cannot be seen. If one grating is displaced by such a small distance as p, the resultant moiré moves as much as d. That is to say, a small displacement p is optically augmented or leveraged to $d = p/\theta$. In addition to this fundamental principle, moiré brings averaging effects that reduce error due to non-uniformity of the period of the grating pattern. When one grating with a slightly rotated angle θ overlaps on the other grating, moiré fringes that are quite the same phenomenon as shown in Figure 8.4 appear.

Higher sensitivity is attained by using diffraction effects due to finely ruled grating scale and by precise arrangement of photodetectors to capture moiré fringes. If signals with mutually different phase values (usually 1/4 period of the moiré fringe) are available, it is easy to identify the direction of movement.

The basic description of these techniques and some applications to length or displacement measuring machines, coordinate measuring machines (CMMs), and positioning systems can be found in a number of books.[12–14] One example is shown in Figure 8.6,[15] where the displacement of the moving grating (main grating) relative to the fixed reference grating (index grating) is determined by detecting moiré fringes. This is a classical arrangement

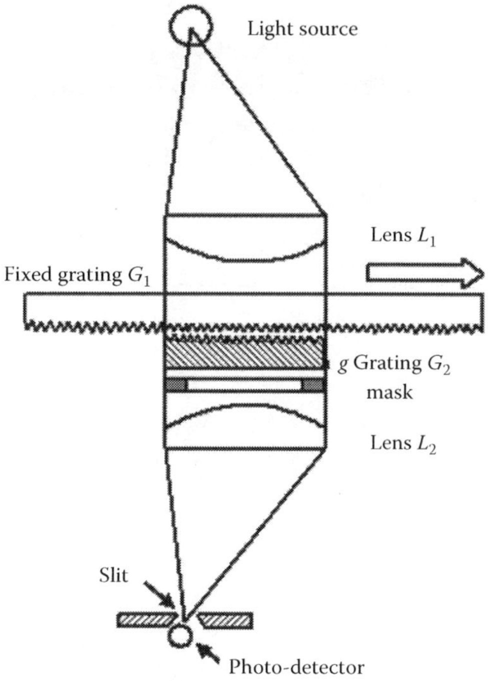

FIGURE 8.6
Displacement detection using moiré fringe detection. (From Shimizu, K., *J. Jpn Soc. Precis. Eng.*, 32(12), 857, 1966, in Japanese; Shimizu, K., *Sci. Mach.*, 16, 4, 1964.)

(based on Shimizu's original figure)[15,16] for checking the feed displacement of a cutter in a machine tool or for detecting length in length measurement machines. Here, the moiré scale optically or photographically ruled on the glass or metal substrate was used.

At the earlier stage (1960s) of this kind of moiré applications, moiré gratings and diffraction gratings with such periods as 100–250 lines/mm were manufactured. The length of the grating varied from 200 to 2000 mm, and ruling accuracy was expected to be ±0.5–±1.0 μm depending on the scale length. The highest resolution attained was 0.2–1.0 μm in the case of a 400–1000 mm scale length. Since then, remarkable progress has been made in length and displacement measurement, alignment, and positioning using various scales (diffraction gratings)[17] and hologram/laser scales including magnetic scales.

Recently, hologram/laser scales have been popularly used in measurement and control that achieve high resolution less than 1 nm. Sinusoidal wave patterns of approximately 138 nm signal period are realized by a hologram scale with high diffraction efficiency and a high-resolution detecting head based on the grating interferometric method (http://www.mgscale.com/mgs/language/English/product/, Magnescale Co., Ltd.). This potentially leads to 17 pc resolution after passing through an electric interpolator. Grating interferometric principle linear encoders generate signals of 0.14 mm period that are 1/140 of conventional linear encoder with a 20 mm signal period.

One example of an industrial application of these moiré methods is the accurate positioning techniques for proximity printing in x-ray lithography that have been introduced.[18,19] This automatic and precision alignment technique is used for proximity printing in x-ray lithography. In Figure 8.7, a laser beam, divided into two beams, is diffracted by two pairs of gratings (A, C and B, D). When relative displacement occurs between A, C and B, D,

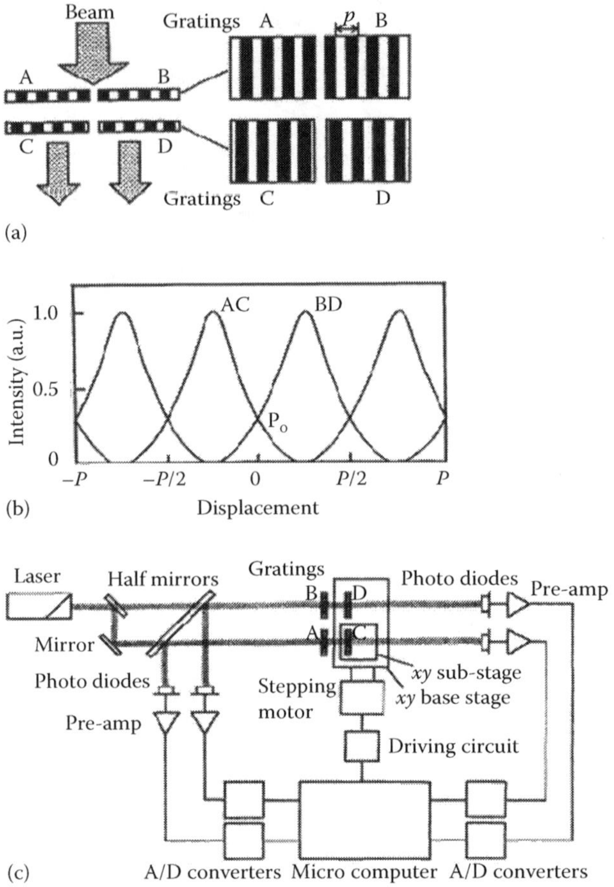

FIGURE 8.7
Principle and schematic diagram for moiré alignment system. (Courtesy of Uchida.): (a) phase-shifted gratings, (b) moiré signals of zeroth-order beam, and (c) experimental arrangement.

moiré fringes with phase difference of 180° are produced. The zeroth order of the moiré signals in transmission and reflection is detected by photodiodes. By analyzing these signals, high automatic alignment sensitivity and stability as high as 0.5 μm over times of 20 min became attainable in 1990.

The most widely used application of in-plane moiré is found in photomechanics from the viewpoint of methodology. These methods are mainly based on optical techniques using moiré, holography, speckle, and other optical principles. A tremendous amount of research results has been published in the experimental mechanics field, and elaborate books have been published and widely read.[3,4,20] Hence, further description is not necessary on this application. However, let us introduce one unusual technique that makes use of the electron moiré method.[21,22]

One interesting trial is referred to here that measures the deformation of the strain around holes in a polyimide resin substrate, that is, thermal strain of electronic packaging component, and tensile creep around grain boundary in a pure copper specimen.[23] Here, a grating with a high frequency of up to 5000 lines/mm was produced on a single deposited metal layer and on double deposited metal layers that had sufficient heat resistance.

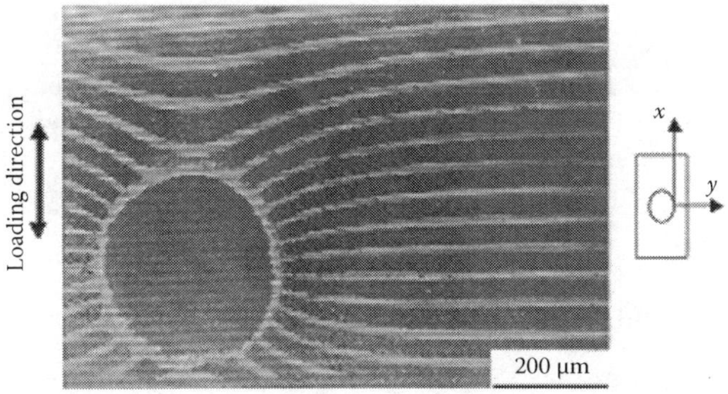

FIGURE 8.8
Electron moiré (u field) around the hole in the polyimide substrate. (From *Opt. Lasers Eng.*, 34(1), Kishimoto, S., Xie, H., and Shinya, N., Electron moiré method and its application to micro-deformation measurement, pp. 1–14, 2000; From *Opt. Eng.*, 32(3), Kishimoto, S., Egashira, M., and Shinya, N., Microcreep deformation measurement by a moiré method using electron beam lithography and electron beam scan, pp. 522–526, 1993, Copyright 2010, with permission from Elsevier.)

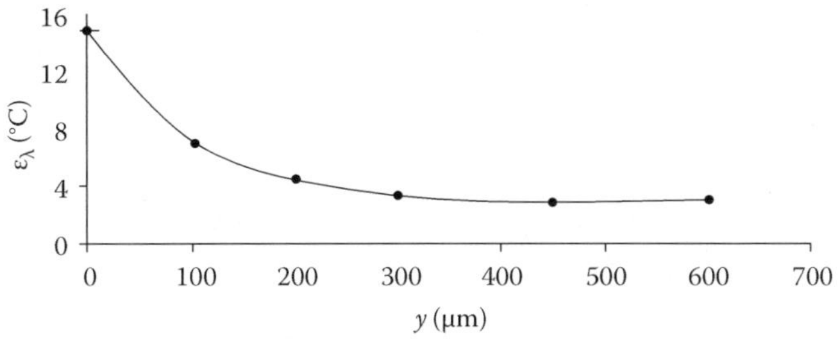

FIGURE 8.9
Distribution of the normal strain εx in y axis. (From *Opt. Lasers Eng.*, 34(1), Kishimoto, S., Xie, H., and Shinya, N., Electron moiré method and its application to micro-deformation measurement, pp. 1–14, 2000; From *Opt. Eng.*, 32(3), Kishimoto, S., Egashira, M., and Shinya, N., Microcreep deformation measurement by a moiré method using electron beam lithography and electron beam scan, pp. 522–526, 1993, Copyright 2010, with permission from Elsevier.)

A rectangular polyimide sheet with an elliptical hole (with long/short diameter 1.19, short diameter 305 μm) was tested under the conditions of a grating period: 6.6 mm and tensile stress of 46 MPa. Figures 8.8 and 8.9 show the results where the normal strain is inhomogeneous near the hole and the strain concentration appears due to the hole.

8.3 Out-of-Plane Moiré Method and Profilometry

As mentioned in the previous section, in-plane moiré patterns are generated by superposing the reference and object gratings in the same plane. In applying out-of-plane moiré method to the contour mapping, the object grating formed across the object is distorted in accordance with the object profile. This out-of-plane moiré method is also termed moiré topography. Moiré topography can be mainly categorized into two methods: shadow moiré and projection moiré. The shadow moiré method was the first method to apply the moiré phenomenon to 3D measurement. Coincidentally, in 1970, Takasaki[6] and Meadows et al.[7] published their papers on moiré applications to photogrammetry, and this principle (shadow moiré) attracted a lot of attention as it was successfully used to observe the contours of an object's surface.

In addition to describing the principle of a shadow moiré pattern, the following sections also introduce its application to field measurements. Later, the projection moiré method is also described.

8.3.1 Shadow Moiré and Contour

A shadow moiré pattern forms between the reference grating and its shadow (the object grating) on the object. The arrangement for shadow moiré is shown in Figure 8.10. The point light source and the detector (the aperture of detector lens is assumed to be a point) are at a distance l from the reference grating surface, and their interseparation is s. The period of the reference grating is p ($p \ll l$ and $p \ll s$). Without loss of generality, we may assume that a point O on the object surface is in contact with the grating. The grating lying over the object surface is illuminated by the point source, and its shadow is projected onto the object. The moiré pattern observed from the detector viewpoint is the result of the superposition between the grating elements contained in OB of the reference grating and the elements contained in OP of the objected grating, which is the shadow of elements contained in OA of the reference grating. Assuming that OA and OB have i and j grating elements, respectively, from geometry

$$AB = OB - OA = jp - ip = np \quad n = 0, 1, 2, 3, \ldots \tag{8.15}$$

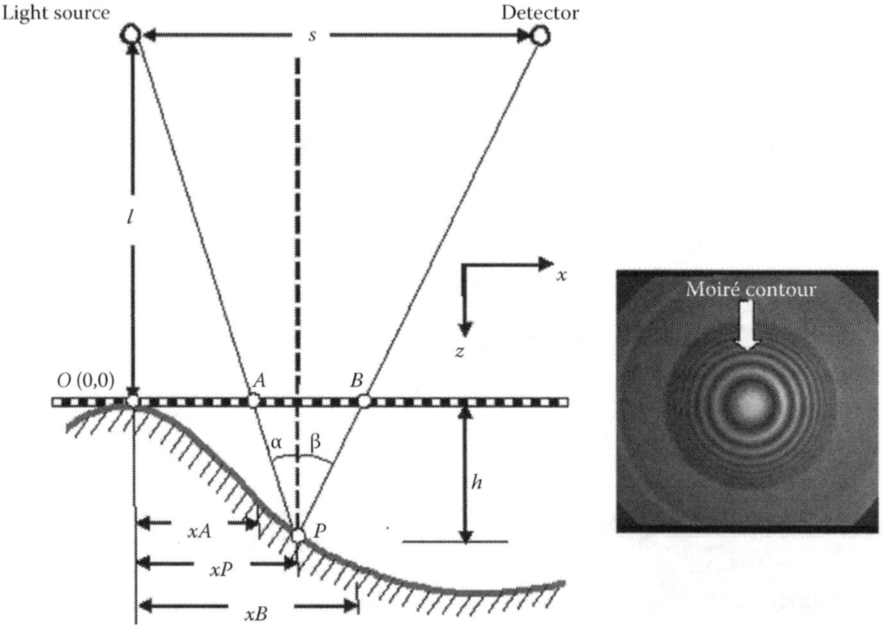

FIGURE 8.10
Optical arrangement for shadow moiré.

$$AB = h_n(\tan\alpha + \tan\beta) \tag{8.16}$$

where

n is the order of the moiré pattern

h_n is the depth of nth-order moiré pattern as measured from the reference grating in Figures 8.10 and 8.13

Hence,

$$h_n = \frac{np}{\tan\alpha + \tan\beta} \tag{8.17}$$

From Figure 8.10, we also have

$$\tan\alpha = \frac{P_x}{l + h_n} \quad \text{and} \quad \tan\beta = \frac{s - x_P}{l + h_n} \tag{8.18}$$

where x_P is the x component of OP.

Substituting earlier equations in Equation 8.17 and rearranging lead to

$$h_n = \frac{npl}{s - np} \tag{8.19}$$

From Equation 8.19, it is seen that the nth order of moiré fringe lies in the contour plane of equal depth measured from the reference grating like a contour line on a topographic map. The interval between two adjacent contours $\Delta h = h_n - h_{n-1}$ is given by $\Delta h = pl/s$ (=constant) if $s \gg np$ or if telecentric optics is used in Figure 8.10. We should note that Δh is not always constant, but it depends on the fringe order n.

Figure 8.10b shows moiré contour lines on a convex object. Besides these contours, the intensity distribution of these moiré fringes also provides interesting information. In practice, knowing the order n of a contour line, we can approximately guess the location of points on the object, and knowing further about the intensity of that fringe, we can exactly plot the measurement points in X, Y, and Z.

8.3.2 Intensity of Shadow Moiré Pattern

Before mathematical development of intensity of moiré fringes, let us review the square wave grating shadow moiré description. In Fourier mathematics, it is known that all types of periodic functions (including square wave function) can be described as a sum of simple sinusoidal functions. In shadow moiré, the amplitude transmittance of the square wave grating is considered as that of a sinusoidal grating:

$$T(x,y) = \frac{1}{2} + \frac{1}{2}\cos\left(\frac{2\pi}{p}x\right) \tag{8.20}$$

Then, the resulting intensity at the point P is proportional to the product $T_A(x_A, y) \cdot T_B(x_B, y)$:

$$I(x,y) = \left[\frac{1}{2} + \frac{1}{2}\cos\left(\frac{2\pi}{p}x_A\right)\right]\left[\frac{1}{2} + \frac{1}{2}\cos\left(\frac{2\pi}{p}x_B\right)\right] \tag{8.21}$$

From Figure 8.10, it is seen that $x_A = lx_P/(l+h_P)$, $x_B = (sh_P + lx_P)/(l+h_P)$. Substituting these equations in Equation 8.21 and rearranging, we have the following normalized intensity:

$$I(x,y) = 1 + \cos\frac{2\pi}{p}\left(\frac{lx_P}{l+h_n}\right) + \cos\frac{2\pi}{p}\left(\frac{sh_n + lx_P}{l+h_n}\right) + \frac{1}{2}\cos\frac{2\pi}{p}\left(\frac{2lx_P + sh_n}{l+h_n}\right) + \frac{1}{2}\cos\frac{2\pi}{p}\left(\frac{sh_n}{l+h_n}\right) \tag{8.22}$$

The last term in Equation 8.22 is solely dependent on height and is termed the contour term. The other three cosine terms representing the reference grating, although height dependent, are also dependent on x (location) and, hence, do not represent contours. The patterns corresponding to these three terms can obscure the contours, as shown in Figure 8.11a. The intensity distribution along a cross-sectional line clearly shows the disturbance of the reference grating itself. To remove these unwanted patterns, Takasaki proposed, during exposure, to translate the grating in azimuth. The resultant intensity is proportional to

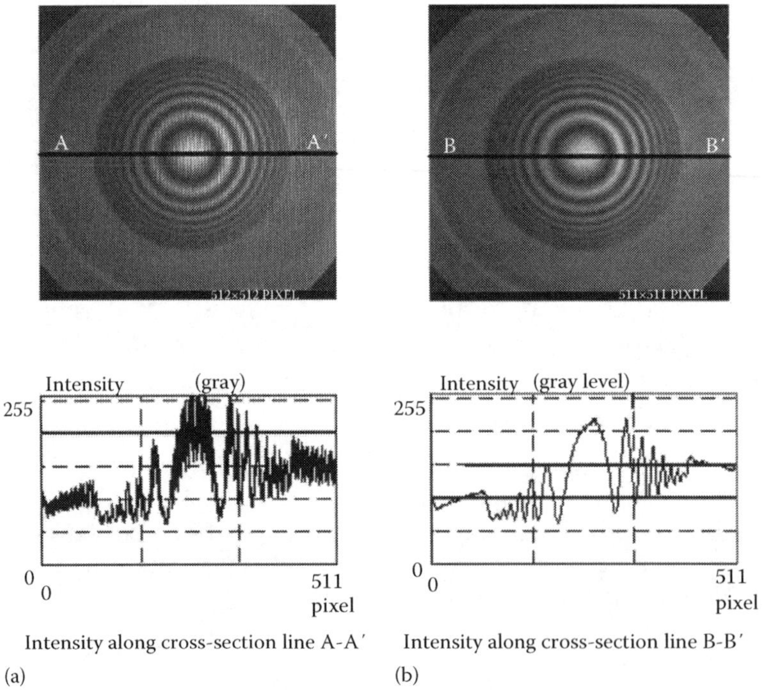

FIGURE 8.11
Shadow moiré pattern (a) before and (b) after translating the grating.

$$I(x,y) = K(x,y)\left[1+\frac{1}{2}\cos\frac{2\pi}{p}\left(\frac{sh_n}{l+h_n}\right)\right] = a(x,y)+b(x,y)\cos\left[\frac{2\pi}{p}\left(\frac{sh_n}{l+h_n}\right)\right]$$
$$= a(x,y)+b(x,y)\cos\phi(x,y) \tag{8.23}$$

where
- $a(=K)$ is the intensity bias
- $b\ (=K/2)$ is the amplitude
- ϕ is the phase related to the temporal phase shift of this cosine variation

The resultant moiré fringes of this equation are shown in Figure 8.11b. Compared with Figure 8.11a, it is obviously seen in Figure 8.11b that the unwanted noise patterns are "smoothed out" due to the averaging effect of the reference grating movement. As an added benefit, the periodical error of the grating is also averaged out as the resulting pattern is averaged over multiple grating lines. The importance of the removal of unwanted patterns and noise from moiré topography is clearly described by Allen and Meadows.[23]

When an object moves while we take pictures because of exposure, we will get unclear pictures. Here, this effect helps us, on the contrary, to have clear contour image. In addition, in in-plane moiré method applications, translating the reference grating is one of the important techniques to shift the moiré fringe and then determine the sign and order of the fringe as described previously.

Another example is shown in Figure 8.12, where the effect of this translation technique is clearly demonstrated. The original grating lines on the surface of a medal (right) are removed by translating the grating during the exposure (left). This technique, first invented by Takasaki,[6,24] plays an important role in developing precise measurement by moiré topography and especially contributes by giving us beautiful moiré photos.[25,26]

Figure 8.13 shows the fundamental arrangement for shadow moiré. This optical arrangement is so simple that it is easy to set up the precise alignment of devices such as the light source (halogen lamp), the camera and the grating (which must be as large as the objective to be measured), and the reference plane is which are all defined by the grating. This figure shows the relation between the phase φ and the moiré contour.

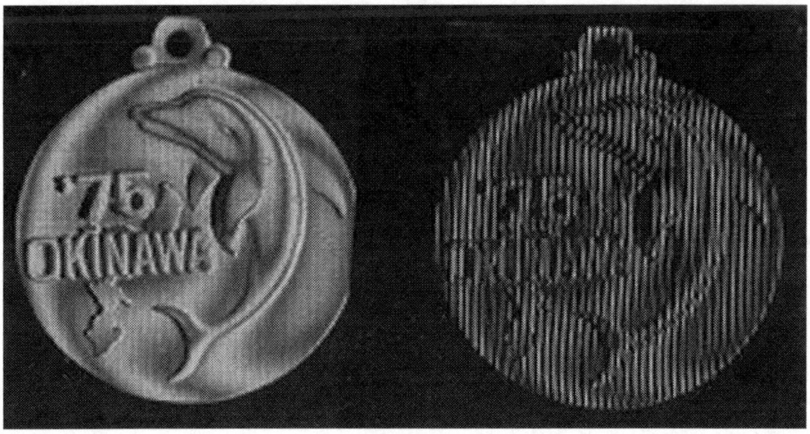

FIGURE 8.12
Effect of grating translation: with translation (left) and without translation (right).

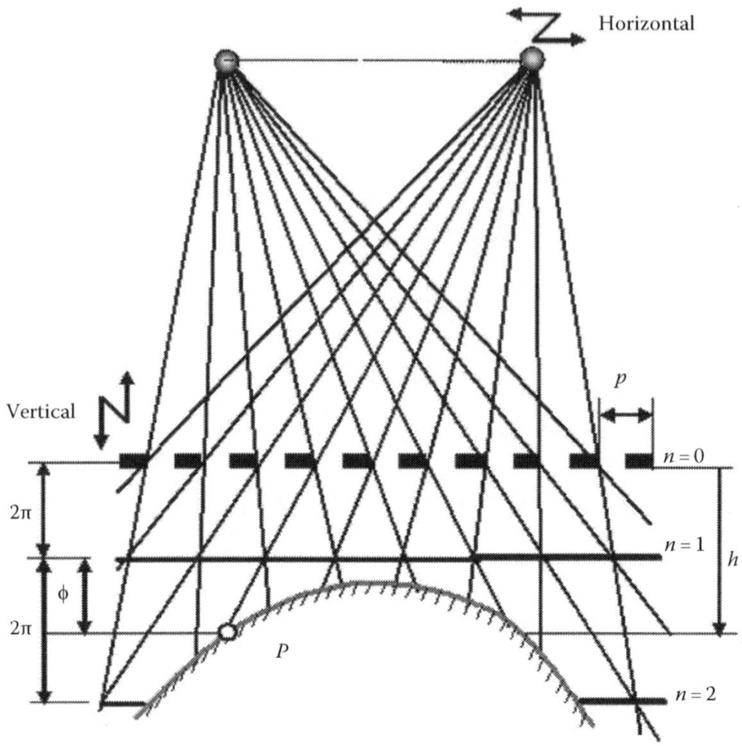

FIGURE 8.13
Relation between phase and shadow moiré pattern.

The depth $h(x, y)$ of any point $P(x, y)$ on the objective surface, then, can be obtained from the following equation:

$$h(x,y) = h_n + (h_{n+1} - h_n)\left[\frac{\phi(x,y)}{2\pi}\right] \quad (8.24)$$

We should note that the contour interval $\Delta h = h_n - h_{n-1}$ is not equal but depends on the contour order n. This fact means the phase-shifting method (described in more detail later) is not easily applied to this type of moiré measurement system.

8.3.3 Application to Three-Dimensional Profile Measurement

To map 3D profiles of objects with shadow moiré topography, the phase distribution encoded in the intensity distribution can be retrieved from the more pattern by using a phase-shifting method.

8.3.3.1 Phase-Shifting Method

In electromagnetic waves interference, the phase-shifting method is well used to get the phase information from a modulated intensity. The four-step algorithm is particularly well known for phase calculation in image processing. Here, we apply this algorithm to mechanical (geometrical) interference moiré. The four-step algorithm to obtain the phase φ in intensity Equation 8.23 is given by

$$\phi = \tan^{-1}\frac{I_3 - I_1}{I_0 - I_2} \tag{8.25}$$

where I_k is the intensity of the moiré fringes across the object and is described as follows:

$$I_k = a + b\cos\left(\phi + \frac{\pi}{2}k\right), \quad k = 0,1,2,3 \tag{8.26}$$

In shadow moiré, a vertical movement of the grating results in a change of moiré pattern as well as a shift in the phase (see Figure 8.13). The distance Δh between adjacent moiré fringes (i.e., $\Delta n = 1$) can be deduced from Equation 8.19

$$\Delta h_{n,n-1} = h_n - h_{n-1} = \frac{dpl}{(s-np)[s-(n-1)p]} \tag{8.27}$$

Hence, when the quantity of a vertical movement of the grating is Δl, the shifted phase can be expressed as follows:

$$2\pi\frac{\Delta l}{\Delta h_{n,n-1}} = \frac{2\pi\Delta l(d-np)[d-(n-1)p]}{dpl} \tag{8.28}$$

It is seen that this quantity is not constant but decreases with the order n of moiré contours. Therefore, it is impossible to attain a constant phase change merely by vertical movement of the grating. To solve this problem, a few methods have been reported. Yoshizawa and Tomisawa proposed to move the grating vertically and at the same time to translate the light source vertically.[27] In the arrangement in Figure 8.14, the interval Δh (0.65 mm) is too large to measure the surface profile of the coin (0.25 mm in maximum height). Therefore, the phase-shifting technique[28] (that was not popular until 1974 due to computing needs) is required to be applied to realize higher sensitivity. They moved the grating vertically by Δl and the light source horizontally by Δd, then four images with a mutual phase difference of $\pi/2$ were acquired as shown in Figure 8.15. Let us note that one or less than one contour is observed in this case. However, computer processing of these four images (a–d) using the four-step phase-shifting algorithm can produce a representation of the result shown in Figure 8.15e. This method was applied to check damage or wear of a score cap (Figure 8.16a), which is used for die stamping on the top of cans. For the calibration of the measurement (Figure 8.16b), cross-sectional profiles obtained by this method (Figure 8.16c) are shown with the result obtained by a mechanical contact profilometer. The two results agree well despite the fact that these measurements have to be performed differently.

To keep the phase shift in every order as constant as possible, another technique by Jin et al. is also available that rotates the reference grating in addition to moving it vertically.[29] Figure 8.17 shows the reference grating rotated, and this rotation results in variation of the measurement period by

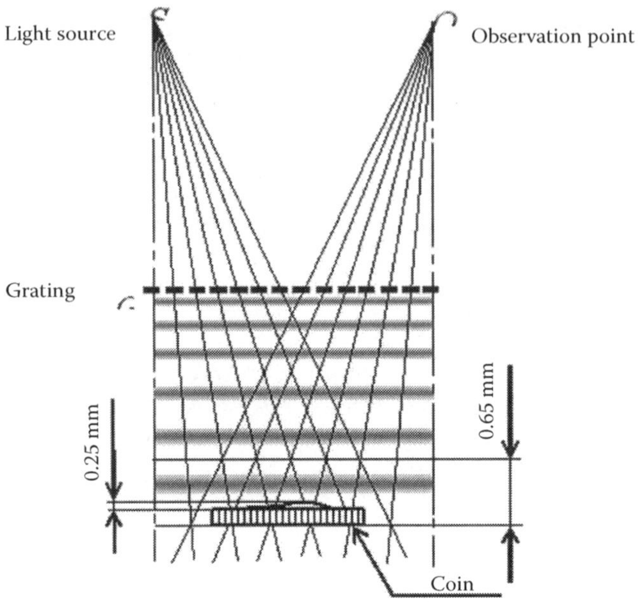

FIGURE 8.14
Arrangement for surface measurement of a coin.

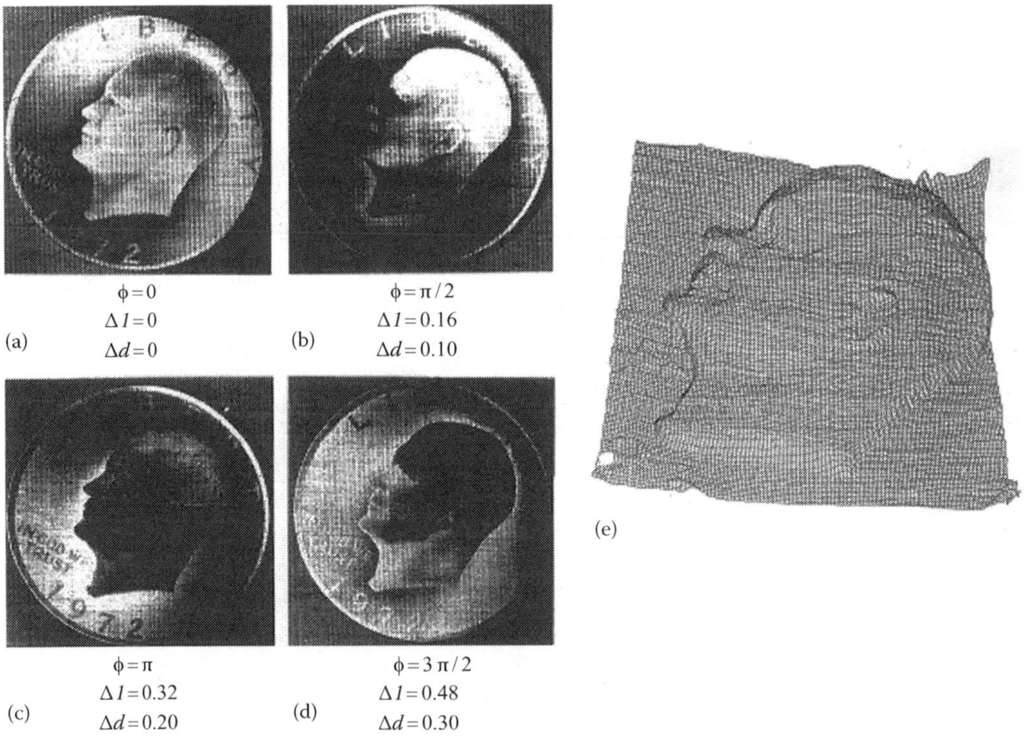

FIGURE 8.15
(a–d) Images with different phases produce the result and (e) 3D expression produced by the four images.

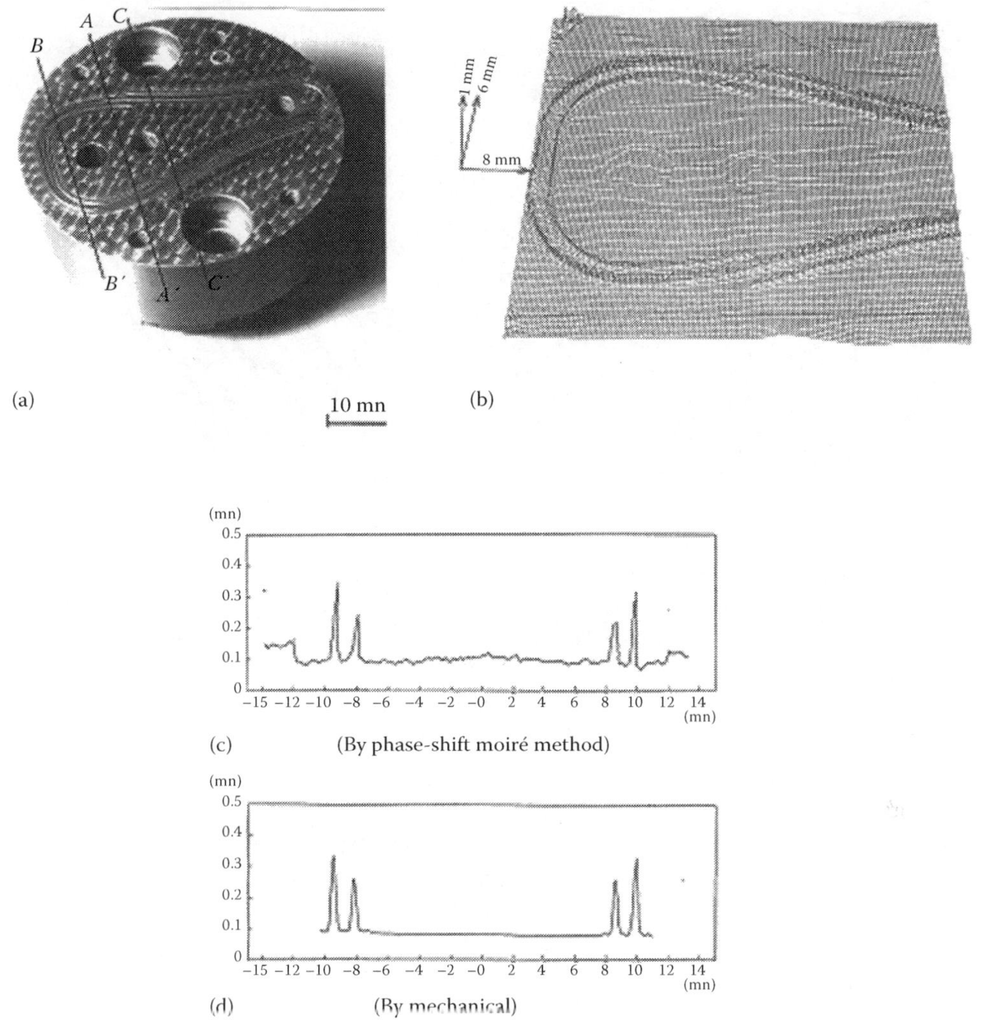

FIGURE 8.16
Inspection of score cap: (a) score cap, (b) measurement result, (c) cross-section AA′ by moiré, and (d) cross-section AA′ by contact method.

$$p' = \frac{p}{\cos\theta} \tag{8.29}$$

where θ is the rotation angle of the reference grating.

The combination of the vertical movement and rotation of the reference grating result in two equations: Firstly, the distance h'_n between the moiré contour of nth order and the reference grating plane, which is transformed from h_n, is as follows:

$$h'_n = \frac{n(l + \Delta l)[p/\cos\theta]}{s - n[p/\cos\theta]} \tag{8.30}$$

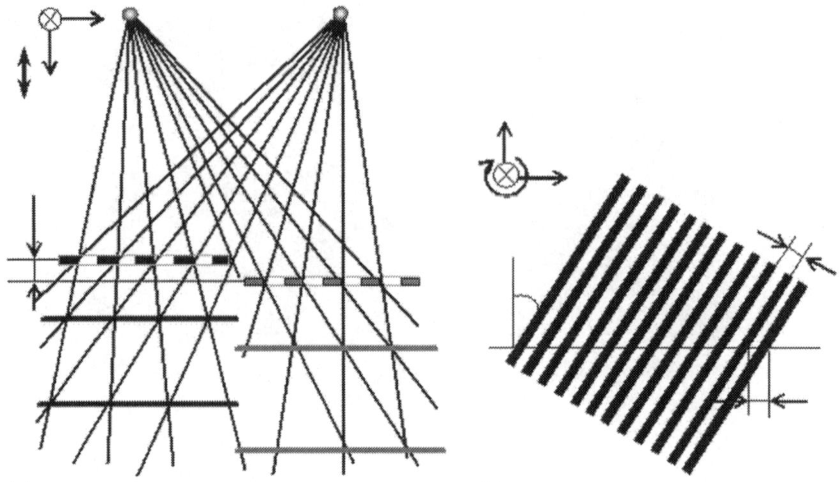

FIGURE 8.17
Up and down moving of the grating and the resultant moiré pattern shift (a) and rotation of the grating (b).

Secondly, the depth h at the point $P(x, y)$ can be expressed with h'_n and Δl under the condition that h'_n exists between h_n and h_{n+1}:

$$
\begin{aligned}
h &= h_n + (h_{n+1} - h_n)\phi/2\pi \\
&= h'_n + \Delta l
\end{aligned}
\tag{8.31}
$$

From Equation 8.31, to shift the phase ϕ by $\pi/2$, π, $3\pi/2$, the needed quantities of vertical movement Δl and rotation angle θ of the reference grating are

$$\Delta l = \frac{\phi p l}{2\pi(s - p)} \tag{8.32}$$

$$\theta = \cos^{-1}\left(\frac{l}{l + \Delta l}\right) \tag{8.33}$$

Figure 8.18 shows an example of the four images with phase-shifted moiré patterns, and the analyzed results expressed with a wire-frame representation.

The limitation of this phase evaluation method is that it cannot be applied to those objects with discontinuous height steps and/or spatially isolated surfaces, because the discontinuities and/or the surface isolations hinder the unique assignment of fringe orders and the unique phase unwrapping.

8.3.3.2 Frequency-Sweeping Method

In the laser interferometric area, the wavelength-shift method is used to measure 3D shapes of objects. A similar concept can be applied to moiré pattern analysis. With this concept, named frequency sweeping, the distance of the object from the reference grating plane can be measured by evaluating the temporal carrier frequency instead of the phase.[30]

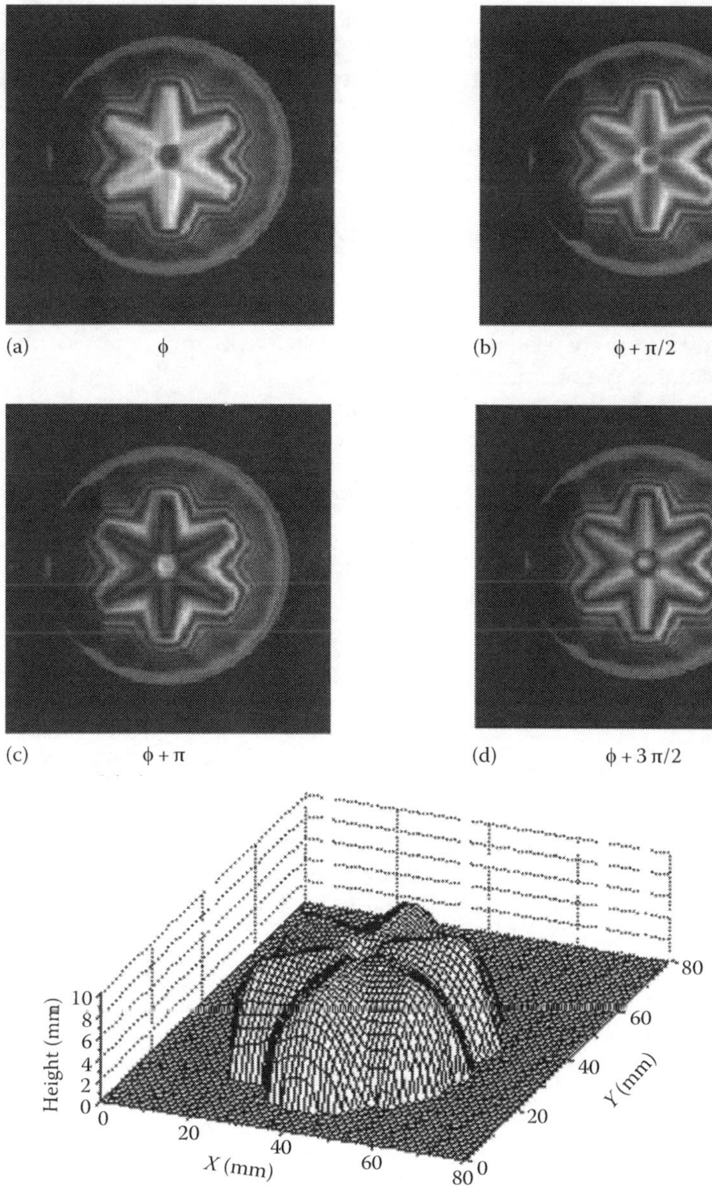

FIGURE 8.18
Images with modulated intensity distribution (a–d) and analyzed result.

Different from the wavelength-shift method in optical interferometry, this technique changes the grating period p by rotating the grating in different intervals and produces the spatiotemporal moiré patterns.

Let us consider Equation 8.23 again:

$$I(x,y) = a(x,y) + b(x,y)\cos\left[\frac{2\pi}{p}\left(\frac{sh(x,y)}{l + h(x,y)}\right)\right] \tag{8.23a}$$

where $2\pi/p$ is defined as the virtual wave number that is analogous to the wave number $k = 2\pi/\lambda$ (λ: wavelength).

As mentioned earlier, when the reference grating is rotated, the measurement grating period is changed, namely, the virtual wave number $g = 2\pi/p$ is changed with time t. By controlling the amount of the rotation angle θ, we can get the following quasi-linear relationship between the different virtual wave number and time:

$$g(t) = g_0 + C \cdot t \tag{8.34}$$

where

C is a constant showing the variation of the virtual wave number
g_0 is the initial virtual wave number $2\pi/p$

Then, Equation 8.23a can be rewritten in the following time-varying form:

$$I(x,y;t) = a(x,y;t) + b(x,y;t)\cos\left[g(t)H(x,y)\right] \tag{8.35}$$

$$H(x,y) = \left[\frac{dh(x,y)}{l + h(x,y)}\right] \tag{8.36}$$

Substituting Equation 8.35 into Equation 8.23a results in the following equation:

$$I(x,y;t) = a(x,y;t) + b(x,y;t)\cos\left[CH(x,y)t + g_0H(x,y)\right]$$

Here, let us define the temporal carrier frequency $f(x, y)$ as

$$f(x,y) = \frac{C \cdot H(x,y)}{2\pi} \tag{8.37}$$

and the second-term initial phase as

$$\varphi_0 = g_0 \cdot H(x,y) \tag{8.38}$$

Then,

$$I(x,y;t) = a(x,y;t) + b(x,y;t)\cos[2\pi f(x,y)t + \phi_0(x,y)] \tag{8.39}$$

As the virtual wave number varies with time t, the intensities at different points vary as shown in Figure 8.19. In this sinusoidal variation, it is obvious that the amounts of the modulated phase φ_0 and the temporal carrier frequency $f(x, y)$ depend on the distance $h(x, y)$ to the object. This means, for the latter, the further the distance, the higher the frequency. Therefore, the height distribution of an object can be obtained from the temporal carrier frequency:

$$h(x,y) = \frac{l}{(dC/2\pi f(x,y)) - 1} \tag{8.40}$$

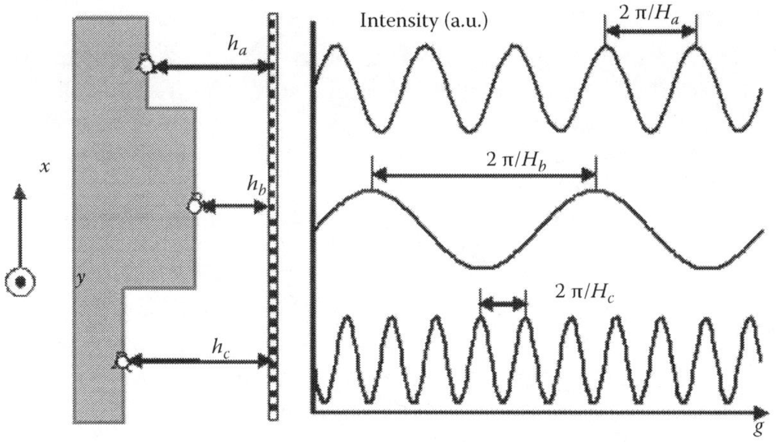

FIGURE 8.19
Relation between the distance h and spatial frequency g.

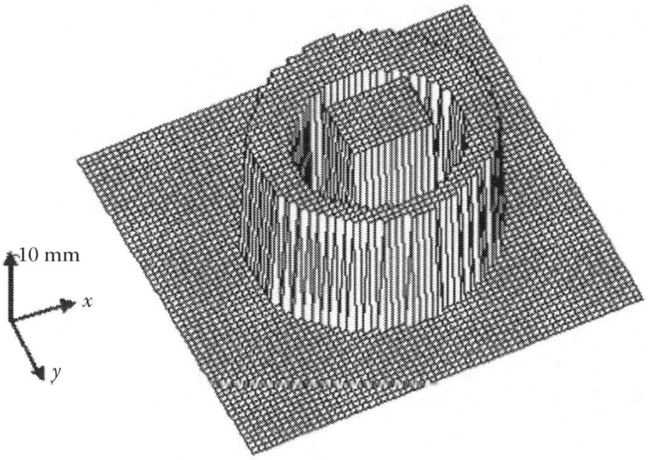

FIGURE 8.20
Measurement result by means of the frequency-sweeping method.

The frequency f in Equation 8.40 is available by applying the "fast Fourier transform" method. Since this frequency-sweeping method does not involve the phase φ, it is not necessary to carry out the phase-unwrapping process.

Figure 8.20 shows the measurement result by using the frequency-sweeping method on two objects (ring and rectangle in shape) separated from each other. Note that many moiré contours are used in this example.

The problem of applying the shadow moiré method is that a big grating is necessary for the measurement and the size is dependent on the objects. The projection moiré method can solve this problem very flexibly by using two gratings.

8.3.3.3 Practical Application of Shadow Moiré

A lot of moiré applications to profilometry exist. At the first stage of moiré topography, the demonstration of the superiority of moiré to measure irregularities of the human body shape without contact and at one instant in time stimulated investigators in the medical and clinical engineering fields. The possibility of defining the relationships of the body surface to underlying anatomical states as seen in scoliosis diagnosis became a big topic. Many examples of measurement techniques and applications to a variety of body surface measurements have since been reported.

8.3.4 Projection Moiré

Projection moiré uses two identical gratings: one for projection and the other for the reference, as shown in Figure 8.21. The projection grating is projected across the object, and the detector captures an image through the reference grating. From Figure 8.21, it is obvious that the principle of moiré pattern formation is similar to the principle of shadow moiré.

The concept of phase-shift and frequency-sweeping methods is also valuable for projection moiré. In order to apply these methods to the projection method, one of the two gratings has to be moved. The gratings can be easily made photographically, using liquid crystal panel plates or similar pattern-generating devices, for example, for movie projection. Recently, without using any physical gratings, grating patterns were projected, which were designed with the aid of application software, through a liquid crystal projector or a digital mirror device (DMD) projector based on Digital Light Processing (DLP) technology. The resulting images of the patterns deformed on the surface of a part are often superimposed on the same grating patterns, inside of the analysis system computer. All process such as shifting the grating and modulating the period can be easily carried out with computer programming. During the developing stages of moiré topographic technology, projection was tried especially in the industrial field and various types of

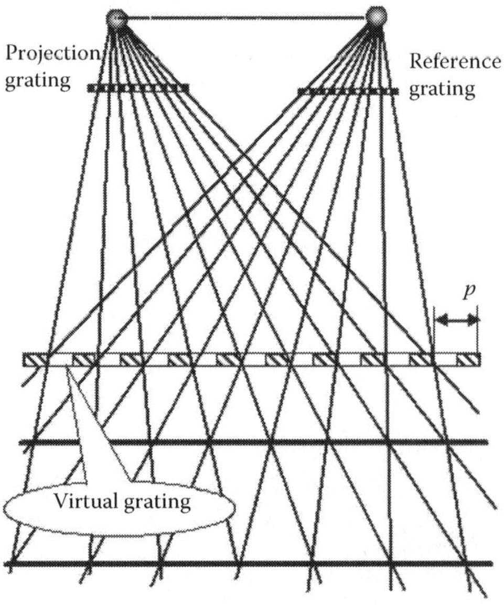

FIGURE 8.21
Principle of projection moiré.

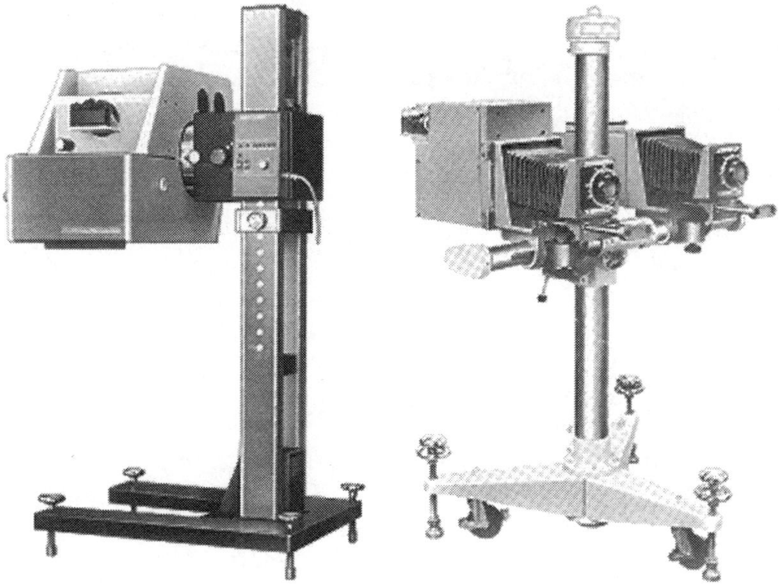

(a) (b)

FIGURE 8.22
Commercialized moiré measurement systems: (a) shadow moiré type and (b) projection moiré type. (Courtesy of Fuji Photo Optical Co., Ltd. Tokyo, Japan.)

measuring systems were manufactured. In Figure 8.22, two "moiré cameras" (Fuji Photo Optical Co., Ltd.) are shown: FM-3011 (a) incorporates the shadow-type moiré and (b) FM-80 is based on the projection-type moiré.

A good example of application to the car industry is shown in Figure 8.23. This trial was made during the early stages of practical applications of moiré to solve the problems inherent to the press forming of a car body. One of the important things to take care during car body fabrication is checking of growth and removal of buckles that can cause trouble in

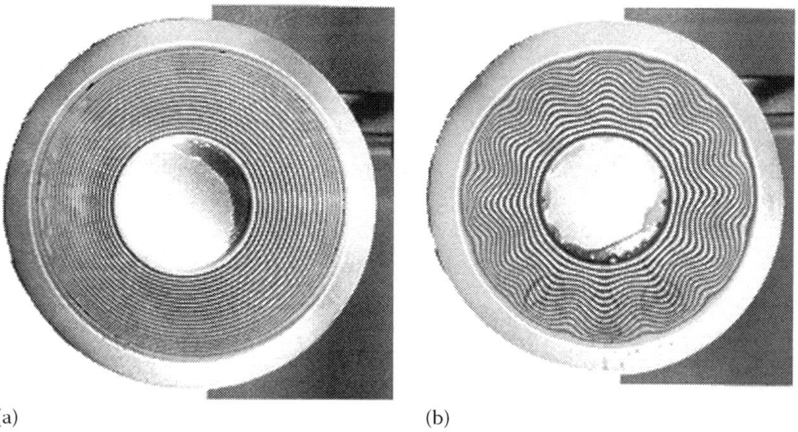

(a) (b)

FIGURE 8.23
Measurement of buckles in press forming of a mild steel sheet using a circular truncated die. The contour interval: 2 mm (Hayashi), $\Delta \doteq 2$ mm, height of dies: 30 mm (a), 40.5 mm (b).

making finely formed large-sized and geometrically complicated car parts. Buckles of a flat plate due to the nonuniform stretching and excessive draw-in of a shrink-flange into a conical die were investigated using the projection-type moiré topography method. Figure 8.23 shows the press formed result using a circular truncated die.[31]

Another example is the measurement of thermally induced warpage of electric devices such as ball grid array (BGA) packages, which is one of the most appropriate applications of moiré. Wang and Hassell indicate[32] that "The thermomechanical behavior of the BGA substrate, the silicon die, and the encapsulated package is particularly interesting to engineers as adverse behavior, or warpage, can significantly increase rework costs and effect product reliability." Usually, deformation of BGAs is of the order of dozens of micrometers, but this amount is too large to be captured by traditional interferometric methods and too small for conventional moiré methods. For a 27 mm BGA, if the maximum deformation is 25 μm, too many fringes, approximately 80 fringes, will appear in the case of interferometry. Conventional moiré techniques, whether shadow type or projection type, have difficulty in solving this problem. The authors were successful in measuring thermally induced warpage in real time (when the sample is driven through a simulated reflow process) by using the phase-shifting shadow moiré technique.

One measurement example of warpage of the BGA during reflow[33] is shown in Figure 8.24.

With respect to the in-plane deformation of electric packages, Quad Flat Package (QFP) and multi-chip module (MCM), the moiré interferometry method was also successfully applied by Arakawa et al.[34] Here, the displacement fields (in u and v directions) due to thermal loading, that is, thermal deformations, were experimentally measured, and at the same time, the result was simulated by finite element method (FEM) to check the validity of the results (Figures 8.25 and 8.26).

In the case of projection moiré, the phase-shifting technique can be easily applied by moving one of the two separated gratings. To get better quality of the moiré images, one pair of gratings should be translated at the same time during the exposure. A primitive proposal was made by Yoshizawa and Yonemura in 1977,[35] but a refined system was reported recently by Dirckx et al.[36,37] by incorporating such digital devices as a liquid crystal grating and charge-coupled device (CCD) camera. In their arrangement shown in Figure 8.27, the first grating using the green channel of the projector (reference grating) is projected on to the sample through the projector lens, and the deformed grating image is formed on the blue channel

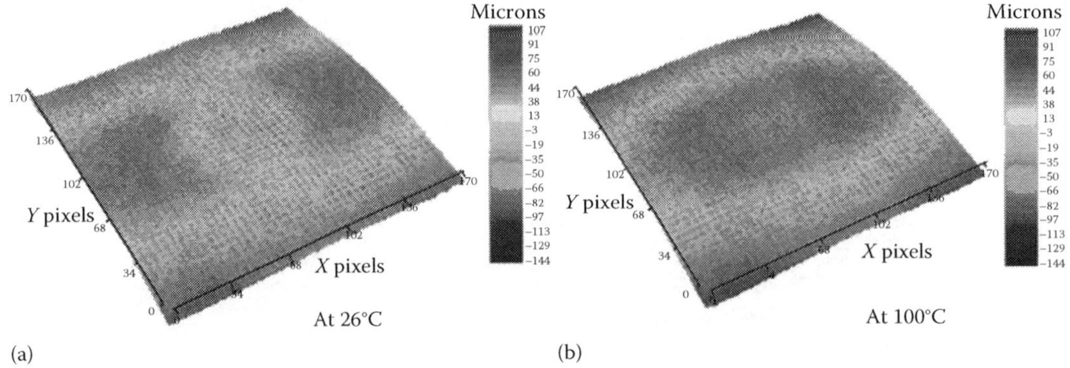

FIGURE 8.24
Warpage variation during reflow: (a) 26°C and (b) 100°C. (From Pan, J., Presented at the *ICCES (International Conference on Computational & Experimental Engineering and Science)'11*, Nanjing, China, April 19, 2011.)

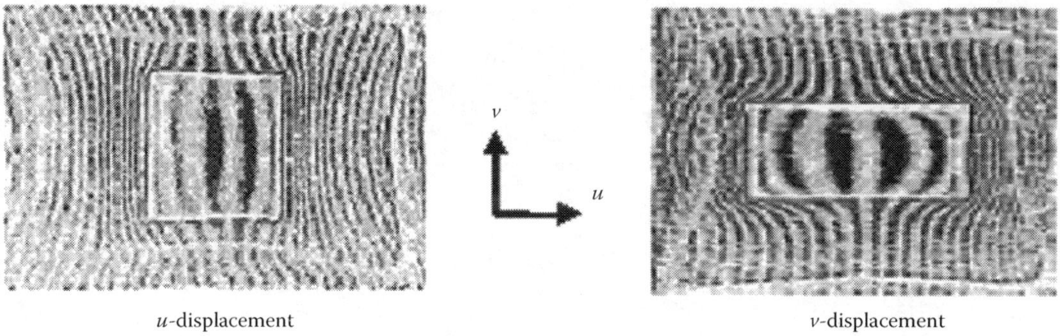

FIGURE 8.25
Moiré patterns formed on QFP (one silicon tip). (From Arakawa, K. et al., Measurement of displacement fields by moiré interferometry, Application to thermal deformation analysis of IC package, *Proceedings of the Japanese Society for Experimental Mechanics*, pp. 113–116, June 26, 2001, in Japanese.)

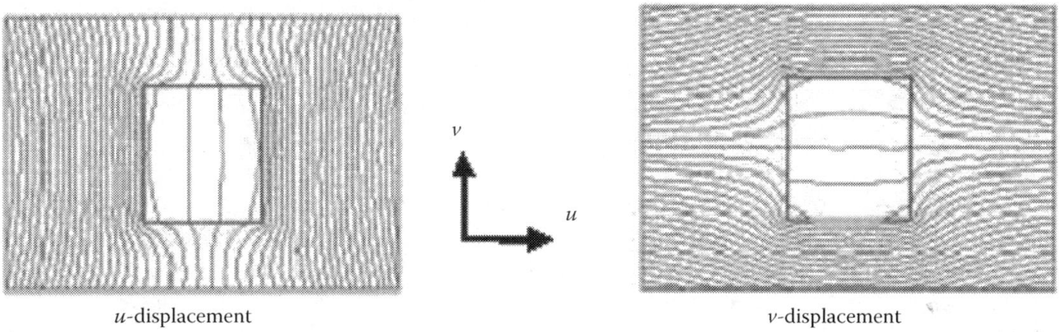

FIGURE 8.26
FEM analysis for QFP (one silicon tip). (From Arakawa, K. et al., Measurement of displacement fields by moiré interferometry, Application to thermal deformation analysis of IC package, *Proceedings of the Japanese Society for Experimental Mechanics*, pp. 113–116, June 26, 2001, in Japanese.)

liquid crystal light modulating matrix placed outside of the projector with two orthogonal polarizers (second grating). The produced moiré pattern is captured by the CCD camera. In addition to phase shifting easily realized by moving one grating, these paired gratings are translated simultaneously to remove grating noise and averaging of the "ruling error" (small spacing errors induced by the picture elements of the projector) of the gratings. This sophisticated system solved various problems inherent to the classical moiré measurement method.

8.4 Reflection Moiré

In the preceding sections, the moiré technique is performed for strain or profile measurement of diffuse objects. The reflection moiré method can be applied on objects with mirror-like surfaces. According to how the reflection moiré fringe is obtained, there are several types of reflection moiré methods. Figure 8.28 shows examples of reflection moiré.[3,38,39] The mirror-line surface of the object makes the virtual mirror image of the grating. Then, the mirror image of the grating is observed through the lens. The moiré fringe can be

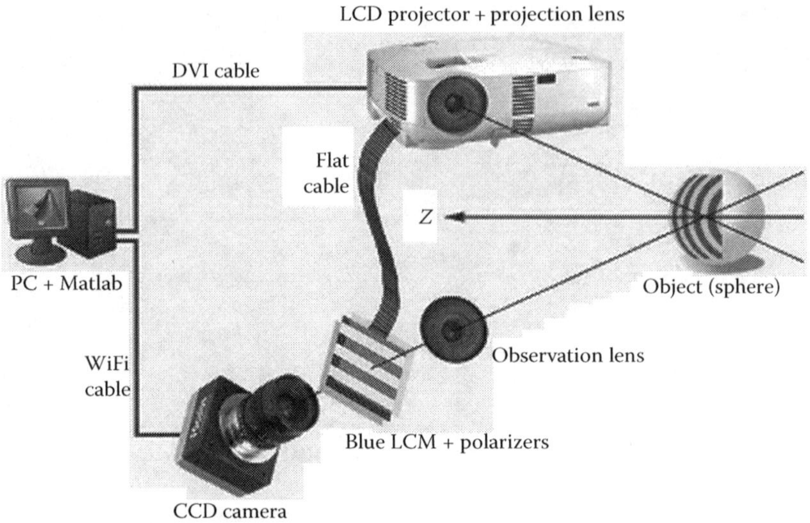

FIGURE 8.27
Setup for digital moiré topography using an LCD projector. (From *Opt. Lasers Eng.*, 48, Dirckx, J.J.J., Buytaert, J.A.N., and Van der Jeught, S.A.M., Implementation of phase-shifting moiré profilometry on a low-cost commercial data projector, pp. 244–250, 2010, Copyright 2010, with permission from Elsevier.)

obtained by exposing two times, before and after deformation of the object, or by placing the reference grating in the image plane of lens. The resulting reflection moiré pattern can be applied to analysis of slope deformation. The general method known as Ronchi testing used in testing optical elements during fabrication is based upon this approach.

When a rectangular grating with a fine periodical spacing is illuminated by a monochromatic parallel light beam under an angle of incidence θ_0, between the grating and the object, light beams are generated of zeroth and diffracted higher orders, as shown in Figure 8.29. That is, the grating acts as a diffraction grating. Each beam then interferes with all other beams. In practice, due to the principles of diffraction, only the zeroth and ± first orders interfere to form the fundamental frequency sine wave pattern. The interference pattern formed by the zeroth and first diffraction order has the same period as that of the diffraction grating p. It extends in a direction θ:

$$\theta = \frac{\theta_0 + \theta_1}{2} \tag{8.41}$$

where $\theta_1$is the propagation direction of the first diffraction order beam.

According to diffraction theory, the angle θ_1 is given by

$$\theta_1 = \sin^{-1}\left(\frac{\lambda}{p} + \sin\theta_0\right) \tag{8.42}$$

where λ is the wavelength of the incident light.

As depicted in Figure 8.30, the interference pattern is reflected at the surface of the object. The reflected interference pattern will be deformed by the flatness of the object. Through a detector, one can observe moiré fringes formed between the reflected interference pattern

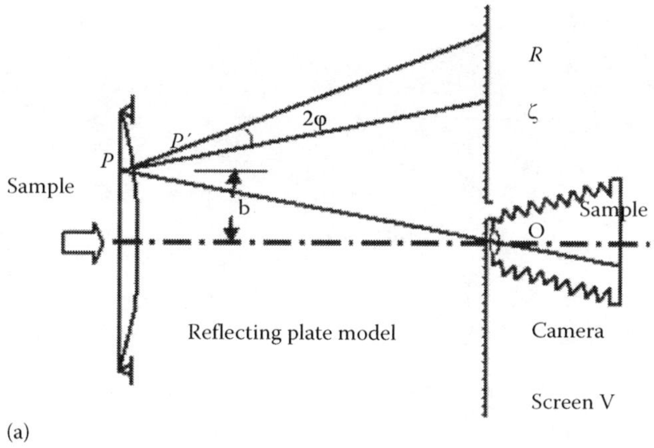

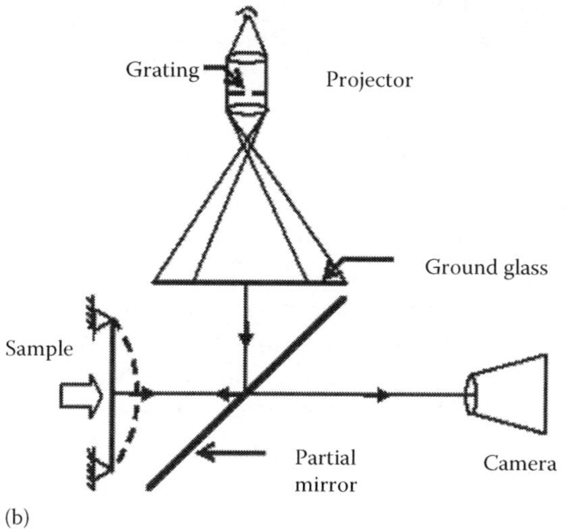

FIGURE 8.28
Principle of reflection moiré: (a) Ligtenberg's arrangement and (b) Chiang's arrangement.

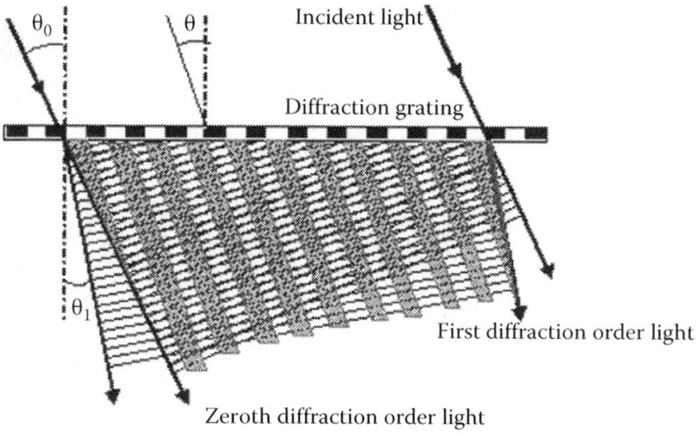

FIGURE 8.29
Interference pattern by diffracted light waves.

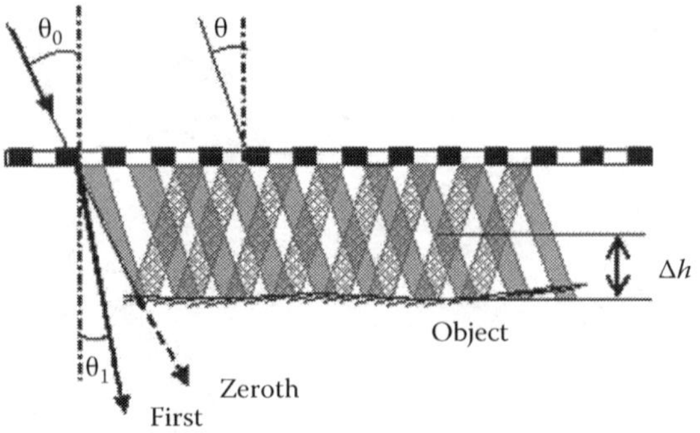

FIGURE 8.30
Generation of moiré pattern.

and the diffraction grating (reference grating). From Figure 8.30, for an optical arrangement with parallel illumination and parallel detecting (so-called collimated beams), the distance Δh between two contours can be expressed as

$$\Delta h = \frac{p}{2 \tan \theta} \tag{8.43}$$

The sensitivity of this method (refer to Equation 8.43) will be changed by the incidence angle. This method is also applicable for coherent and incoherent monochromatic light sources, and due to its high sensitivity, it is very attractive for the flatness measurement field for such objects as computer disks, wafers, and glass substrates that are highly processed. Figure 8.31 shows a system applying the UV moiré method to measure the flatness of a soda glass substrate.[40] The moiré pattern can be shifted by moving the grating perpendicularly to the reference plane, that is, the grating plane.

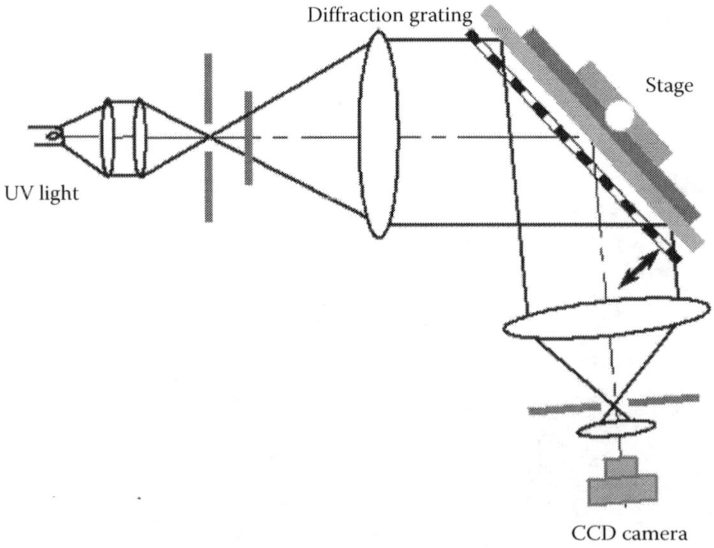

FIGURE 8.31
Flatness measurement system using the UV moiré method.

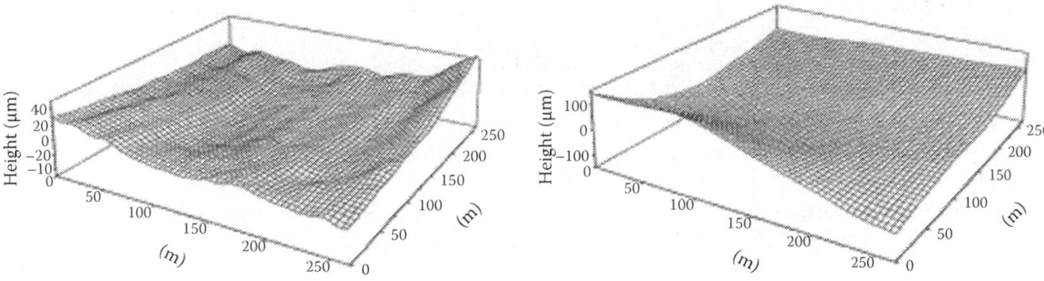

FIGURE 8.32
Measurement result of an LCD glass substrate (left: not polished; right: polished). (From Fujiwara, H. et al., *Proc. SPIE*, 2862, 172, 1996.)

The phase-shift method introduced in the previous section can be applied for pattern analysis. In this system, to cancel the influence of the light reflected from the back surface of the substrate, an ultraviolet (UV) light ($\lambda = 313$ nm) is used as the light source. Part of the UV light will be reflected on the front surface of the soda glass, and the residual part of the beam transmitted into the soda glass will be absorbed while passing through the glass. Figure 8.32 shows the measurement result using this system. The grating employed was 10 lines/mm, and the incident angle was 60°.

8.5 Conclusions

Moiré metrology and its fundamental principles and applications have been briefly described in this chapter. Moiré phenomenon has been known for a long time, and a variety of applications have been reported. However, this technique has been an interesting topic in academic and applicative fields and has shown the potential for use in many more applications. For instance, in 3D profilometry, such techniques as pattern/fringe projection method are very popular. But this technique has the same principle as the moiré topographic method.[41] Moiré is still an attractive topic in the field of optics and will be utilized more in practical industrial applications in the future.

Finally, the authors recommend readers, especially engineers and scientists in industrial fields, to read a series of Kevin Harding's reports[42–44] that are written from the viewpoint of the practical use of moiré technology in industry. What is needed or should be revised, not only in moiré but also in other optical methods, is also described in his recent paper[45] from the viewpoint of industrial applications.

This chapter was rewritten with revision on the basis of "Chapter 9: Moire Metrology" taken from the *Handbook of Optical Metrology: Principles and Applications* (CRC Press, 2009).

References

1. I. Amidror, *The Theory of the Moiré Phenomenon*, Kluwer Academic Publishers, Dordrecht, the Netherlands, 2000.
2. G. Indebetouw and R. Czarnek, *Selected Papers on Optical Moiré and Applications*, SPIE Milestone Series, SPIE Vol. MS 64, 1992.

3. P. S. Theocaris, *Moiré Fringes in Strain Analysis*, Pergamon Press, New York, 1969.
4. A. J. Durelli and V. J. Parks, *Moiré Analysis of Strain*, Prentice-Hall, Englewood Cliffs, NJ, 1970.
5. F. P. Chiang, Moiré method of strain analysis, in *Manual for Engineering Stress Analysis*, 2nd edn., A. Kobayashi, ed., Society for Experimental Mechanics, Bethel, CT, 1978.
6. H. Takasaki, Moiré topography, *Applied Optics*, 9(6), 1467–1472, 1970.
7. D. M. Meadows, W. O. Johnson, and J. B. Allen, Generation of surface contours by moiré patterns, *Applied Optics*, 9(4), 942–947, 1970.
8. T. Yoshizawa, Surface profilometry, in *Handbook of Optical Metrology: Principles and Applications*, Chapter 19, T. Yoshizawa, ed., CRC Press, Boca Raton, FL, 2009.
9. C. A. Walker, *Handbook of Moiré Measurement*, Institute of Physics Publishing, Bristol, PA, 2004.
10. D. N. Sidorov and A. C. Kokaram, Suppression of moiré patterns via spectral analysis, *Proceedings of SPIE*, 4671, 895–906, 2002.
11. I. A. F. Stokes, J. R. Pekelsky, and M. S. Moreland, eds., *Surface Topography and Spinal Deformity*, Gustav Fischer Verlag, Stuttgart, New York 1987; M. D'Amico, A. Merolli, and G. C. Santambrogio, eds., *Three Dimensional Analysis of Spinal Deformities*, IOS Press, Washington, DC, 1994.
12. H. Walcher, *Position Sensing Angle and Distance Measurement for Engineers*, Butterworth-Heinemann, New York, pp. 66–69, 1994.
13. F. T. Farago and M. A. Curtis, *Handbook of Dimensional Measurement*, 3rd edn., Industrial Press, New York, pp. 321–324, 1994.
14. J. R. Rene Mayer, Optical encoder displacement sensors, in *The Measurement, Instrumentation, and Sensors*, J. G. Webster, ed., CRC Press, Boca Raton, FL, pp. 6-98–6-119, 1999.
15. K. Shimizu, Optics of moiré fringes and applications, *Journal of the Japan Society of Precision Engineering*, 32(12), 857–864, 1966 (in Japanese).
16. K. Shimizu, Moiré fringes and applications, *Science of Machine* 16, 4 and 5, 1964 [the best short review on moiré fringes and applications, unfortunately in Japanese].
17. J. Guild, *Diffraction Grating as Measuring Scales*, Oxford University Press, London, U.K., 1960.
18. Y. Takada, Y. Uchida, Y. Akao, J. Yamada, and S. Hattori, Super-accurate positioning technique using diffracted moiré signals, *Proceedings of SPIE*, 1332, 571–576, 1990.
19. K. Hane, S. Watanabe, and T. Goto, Moire displacement detection by photoacousic technique, *Proceedings of SPIE*, 1332, 577–583, 1990.
20. D. Post, B. Han, and P. Ifju, *High Sensitivity Moiré: Experimental Analysis for Mechanics and Materials*, Springer-Verlag, New York, 1994.
21. S. Kishimoto, M. Egashira, and N. Shinya, Microcreep deformation measurement by a moiré method using electron beam lithography and electron beam scan, *Optical Engineering*, 32(3), 522–526, 1993.
22. D. Read and J. Dally, Theory of electron beam moiré, *Journal of Research of the National Institute of Standards and Technology*, 101(1), 47–61, 1996.
23. S. Kishimoto, H. Xie, and N. Shinya, Electron moiré method and its application to micro-deformation measurement, *Optics and Lasers in Engineering*, 34(1), 1–14, 2000.
24. J. B. Allen and D. M. Meadows, Removal of unwanted patterns from moiré contour maps by grid translation techniques, *Applied Optics*, 10(1), 210–212, 1971.
25. His best photo (one pair of stereo photos) is seen in T. Yoshizawa ed. *Handbook of Optical Metrology: Principles and Applications*, CRC Press, Boca Raton, FL, p. 440, 2009.
26. H. Neugebauer and G. Windischbauer, *3D-Fotos Alter Meistergeigen* (*3D-Photos of Antique Master Violins*), Verlag Erwin Bochinsky, 1998.
27. T. Yoshizawa and T. Tomisawa, Shadow moiré topography by means of the phase-shift method, *Optical Engineering*, 32(7), 1668–1674, 1993.
28. J.-i. Kato, Fringe analysis, in *Handbook of Optical Metrology: Principles and Applications*, Chapter 21, T. Yoshizawa, ed., CRC Press, Boca Raton, FL, 2009.
29. L. Jin, Y. Kodera, Y. Otani, and T. Yoshizawa, Shadow moiré profilometry using the phase-shifting method, *Optical Engineering*, 39(8), 2119–2213, 2000.
30. L. Jin, Y. Otani, and T. Yoshizawa, Shadow moiré profilometry using the frequency sweeping, *Optical Engineering*, 40(7), 1383–1386, 2001.

31. H. Hayashi, *Handbook of Sheet Metal Forming Severity, Sheet Metal Forming Research Group*, 3rd edn., Nikkan Kogyo Shinbun, Ltd., Tokyo, Japan p. 259, 2007 (in Japanese); See also K. Yoshida, H. Hayashi, K. Miyauchi, Y. Yamamoto, K. Abe, M. Usuda, R. Ishida, and Y. Oike, The effect of mechanical properties of sheet metals on the growth and removal of buckles due to non-uniform stretching, *Scientific Papers of Physical and Chemical Research*, Saitama, Japan 68, 85–93, 1974.
32. Y. Wang and P. Hassel, Measurement of thermally induced warpage of BGA packages/substrates using phase-stepping shadow moiré, *Proceedings of the Electronic Packaging Technology Conference*, pp. 283–289, 1997.
33. J. Pan, Presented at the *ICCES (International Conference on Computational & Experimental Engineering and Science)'11*, April 19, 2011, in Nanjing, China; See also J. Pan, R. Curry, N. Hubble, and D. Zwemer, Comparing techniques for temperature-dependent warpage measurement, Production von Leiterplaten und systemen, 1980–1985, 2007.
34. K. Arakawa, M. Todo, Y. Morita, and S. Yamada, Measurement of displacement fields by moiré interferometry (Application to thermal deformation analysis of IC package), *Proceedings of the Japanese Society for Experimental Mechanics*, pp. 113–116, June 26, 2001 Tokyo, Japan (in Japanese); See also Y. Morita, K. Arakawa, and M. Todo, Experimental analysis of thermal displacement and strain distributions in a small outline J-leaded electronic package by using wedged-glass phase-shifting moiré interferometry, *Optics and Lasers in Engineering*, 46(1), 18–26, 2008.
35. T. Yoshizawa and M. Yonemura, Moire topography with lateral movement of a grating, *Journal of the Japan Society for Precision Engineering*, 43(5), 556–561, 1976 (in Japanese).
36. J.A.N. Buytaert and J.J.J. Dirckx, Phase-shifting moiré topography using optical demodulation on liquid crystal matrices, *Optics and Lasers in Engineering*, 48, 172–181, 2010.
37. J.J.J. Dirckx, J.A.N. Buytaert, and S.A.M. Van der Jeught, Implementation of phase-shifting moiré profilometry on a low-cost commercial data projector, *Optics and Lasers in Engineering*, 48, 244–250, 2010.
38. F.K. Ligtenberg, The moiré method: A new experimental method of the determination of moments in small slab models, *Proceedings of SESA*, 12, 83–98, 1955.
39. F.P. Chiang and J. Treiber, A note on Ligtenberg's reflective moiré method, *Experimental Mechanics*, 10(12), 537–538, 1970.
40. H. Fujiwara, Y. Otani, and T. Yoshizawa, Flatness measurement by reflection moiré technique, *Proceedings of SPIE*, 2862, 172–176, 1996.
41. M. Suganuma and T. Yoshizawa, Three-dimensional shape analysis by use of a projected grating image, *Optical Engineering*, 30(10), 1529–1533, 1991.
42. K. Harding, Optical moiré leveraging analysis, *Proceedings of SPIE*, 2348, 181–188, 1994.
43. A. Boehnlein and K. Harding, Large depth-of-field moiré system with remote image reconstruction, *Proceedings of SPIE*, 2065, 151–159, 1994.
44. K. Harding and Q. Hu, Multi-resolution 3D measurement using a hybrid fringe projection and mire approach, *Proceedings of SPIE*, 6382, 63820K-1–63820K-8, 2006.
45. K. Harding, Optical metrology: Next requests from the industrial view point, *Proceedings of SPIE*, 7855, 785513-1–785513-13, 2010.

Part IV

Optical Micro-Metrology of Small Objects

9

Automation in Interferometry

Erik Novak and Bryan Guenther

CONTENTS

9.1 Introduction

Interferometry provides the highest quantitative vertical resolution and repeatability of any surface measurement technique. As such, it has found application in a variety of research groups as well as industry for quantification of feature heights, surface texture, relative angles, radii of curvature, and other metrics which require precision distance or surface measurements. Interferometric methods are also typically quite rapid, with distance-measuring interferometers sensing thousands of times per second, phase-shifting measurements capturing entire fields of view in a fraction of a second, and coherence-sensing techniques (such as white-light interferometry) typically taking under 10 s. This combination of speed, accuracy, and repeatability has led to wide adoption of interferometry in a variety of situations where automated results are important.

Automation in interferometry can be divided between automation within a measurement, automating multiple measurements on a sample for greater areal coverage, and automating measurements across many samples. Automation within a measurement is designed to achieve the best possible individual measurement through automatic changes to the instrument hardware, software, or part alignment to optimize results for that measurement. Automating multiple measurements on a single test piece is needed to perform one or more of the following operations: combine multiple measurements together to get larger coverage area, sample multiple locations to achieve statistical significance, or sample multiple locations in order to determine systematic variations across a sample (such as center-to-edge height changes on a semiconductor wafer). Automating measurements across samples is typically employed in production where consistency and throughput are important or where parts should not be manually handled due to contamination, breakage, or due to size constraints.

This chapter systematically explores the key automation functions which are available on various interferometers and discusses key aspects of each function and its effect on the quality of results. In addition, fixturing of parts is discussed as this is a key element to successfully automating any interferometric measurements.

9.2 Automation within a Measurement

When measuring a part on an interferometer, a variety of variables can affect the quality of the results. While interferometric methods generally provide high-quality data with minimal adjustments, there are many situations where getting the highest quality results is necessary. In such cases, ensuring all variables are optimized on the system is critical to achieving the required performance. One example of this is in the data storage industry, where single-system repeatability and reproducibility on roughness and height measurements must be significantly below 0.1 nm and even system-to-system matching must be below 1 nm for key metrics. Tight manufacturing tolerances which require optimal measurements are also found in the semiconductor, automotive, optics, tooling, and medical device industries as well as in many other precision machining applications. To ensure consistency for such measurement cases, and to obtain best quality for normal operation, many interferometers offer a variety of automation features.

9.2.1 Focus Automation

One of the most critical items which can be automated is focus of the instrument on the part. For both laser interferometers that measure larger parts and microscope-based interferometers, focus can dramatically impact the results. An interferometer must generally achieve two functions simultaneously: proper focus on the part to maximize the optical resolution and localization of the interference fringes at that focus position. For laser interferometers, the localization of the fringes is typically not a consideration due to their long coherence length sources. However, for microscope-based interferometers, proper fringe localization is important and is typically adjusted by the manufacturer at time of installation. However, when new parts are measured, they are often at a different distance from the instrument due to fixturing, thickness variations, or other factors, and thus the instrument must be focused on those parts.

For laser interferometers, focus is generally a weak variable with respect to measurement quality. Defocus will cause a quadratic phase shift, usually noticeable only near the edge of the field of view.[1] Blurred edges from defocus can reduce the usable field of view by a few camera pixels with large defocus, which typically translates into a few millimeters of potentially incorrect data near the edges. Edge effects from blurring can also lead to phase unwrapping errors in the measurement, usually noticeable as sudden jumps in the data, which can affect the data quality. Improper focus may also increase distortion or cause other systematic errors in the measurement, which can be from a few nm to a few tens of nm.

Automated focus adjustment for laser interferometers typically involves the user choosing a region of interest with a sharp feature, such as an edge, scratch, dust, or other noticeable marks. The instrument will then adjust focus until the sharpness of that feature is maximized. The user usually may select the distance over which to look for best focus and the sensitivity of the focus method; the higher the sensitivity, typically, the slower the focus procedure. Laser interferometers generally have a long depth of focus, so for a given setup, focus is often adjusted once and only adjusted again when major changes are made in the measurement path.

For white-light interferometers, or other interferometers with fringes highly localized near focus, the focus setup operation is typically very similar, with the user again selecting sensitivity, the region on which to focus, and the distances over which to search. In this case however, sharpness of a feature is not maximized, but rather the contrast of the interference fringes is maximized. Thus, focus can be achieved when no sharp features are present in the field of view.

Low-magnification microscope objectives typically have a large depth of focus and many high-contrast fringes, and thus focus is not very critical. However, for large magnification objectives (20× and higher), focus is often quite critical to ensure proper signal levels for phase-shifting methods. For coherence-sensing methods, automated focus often allows one to minimize the scan range of the system and thus improve throughput. Figure 9.1 shows a focus setup screen from a white-light interferometer. In this case, five regions of interest can be selected so that if focus fails (for instance, due to a defect in a certain location as one moves around on a part), another location can be tried.

Variables that are available for focus adjustment often include the size of the region of interest, the speed at which the focus should be adjusted, how many locations on which to attempt to focus in the case of a failure, and what to do if automatic focus cannot be achieved; the options for the latter may be to move on to the next measurement location, flag a system error and wait for operator intervention, or to fail the measurement.

9.2.2 Intensity Automation

To obtain the best possible results on any instrumentation, the signal-to-noise ratio must be maximized. For an interferometer, this typically involves ensuring the maximum light level appropriate for a given measurement, such that the instrument is using the highest dynamic range of the sensor. For phase-based interferometry, this usually involves setting the intensity such that no pixels in the desired measurement field saturate the camera; saturation is caused when too much light is returned to the camera and will cause errors in the phase calculations. For coherence-sensing interferometry, some interferometer manufacturers can achieve good data even in the presence of saturation. This can be used, for

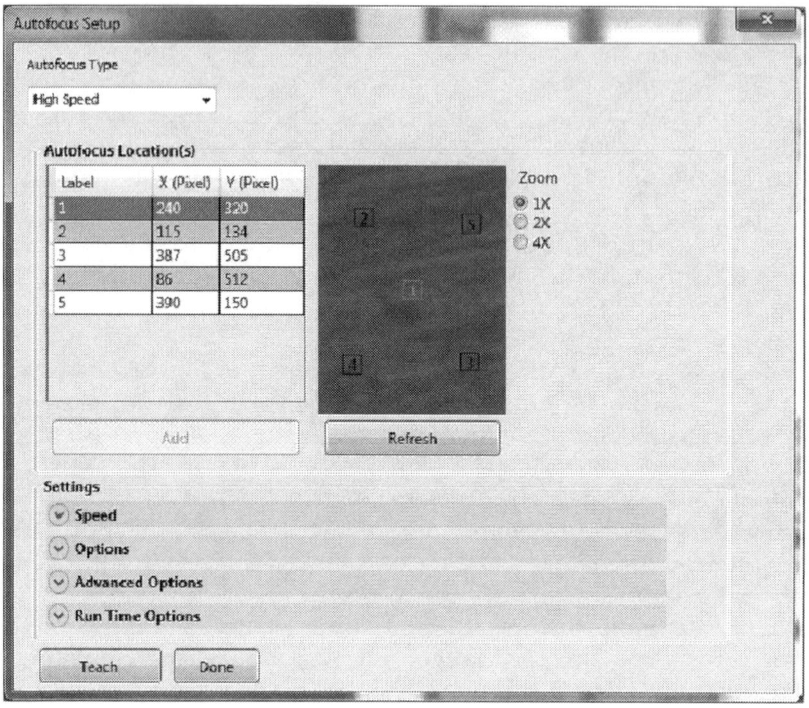

FIGURE 9.1
Example focus automation screen showing multiple regions on which autofocus will be attempted and some of the common categories of settings used to optimize the focus procedure.

instance, if one has an area with poor surface quality (and thus poor signal to noise) in the same measurement field as an area with high surface quality or reflectivity; one may choose to saturate the brighter area in order to achieve better data from the area with worse signal to noise.

Intensity automation typically involves performing a measurement, examining the signal level for each pixel, and then adjusting the light level appropriately. Typically, the only user-settable parameter is the amount of camera saturation to allow in those cases in which it is desired. The only time intensity automation is not straightforward is if one is performing many measurements on a sample or across multiple samples. In that case, one can adjust intensity each time, or one must choose the best balance of intensity against the expected sample variation. If a part is fairly uniform in reflectivity and shape, however, intensity optimization is generally not required for good measurement quality even when automating multiple measurements.

9.2.3 Scan Length Automation

Interferometers that perform only phase-shifting operations, such as laser Fizeau or digital holography systems, perform small scans, typically only a few micrometers in length. For such systems, the scan length is automatically determined by the phase algorithm employed. White-light interferometers, however, must scan vertically such that every location on the part passes through best focus of the instrument. For highly variable surfaces, or surfaces which are tilted with respect to the interferometer, one may need to change the vertical scan range from location to location as one moves about the sample, or one must use the worst-case scan range, which can slow down measurements.

Systems often therefore include an option to automatically stop a scan once the image is known to have passed through focus at a given location, or after a certain final distance after enough data is acquired. When this is done, the worst-case scan length can be specified in the automation, but throughput is maximized since each individual measurement will stop once the desired data has been measured. For instance, a user may tell the instrument to look for when 80% of the data is acquired and then scan only ten more μm before halting; on a flat surface, this will usually ensure all of the data that needs to be acquired is gathered, and it keeps the system from scanning long distances when no further information can be gathered.

Automating scan length can have an enormous impact on measuring complex surfaces. It works well in conjunction with automatic focus adjustment to reduce the amount of time the system spends acquiring data. Typically, an optimized automation sequence using focus and scan length automation can be completed in less than 25% the time compared to a sequence which does not automate those factors.

9.2.4 Measurement Averaging

As mentioned previously, interferometers have extremely good vertical resolution and have a unique advantage against most other surface measurement systems since the vertical resolution is independent of the field view of the measurement. Therefore, a single measurement is usually sufficient at any location to achieve good data quality. However, for very smooth surfaces (Sa < 0.5 nm) or where gage capability of the instrument is challenged by low part tolerances, it may be desirable to average multiple measurement results together. The vertical noise in interferometers is typically limited by the camera used to acquire the data, and this noise is mostly random. Thus, averaging results together should reduce the noise by the square root of the number of averages; averaging four measurements should halve the noise, nine measurements should reduce it to 1/3, etc. Therefore, most interferometric instruments have an option of enabling averaging. The user enters in the number of measurements to average at each location, and the instrument will automatically take multiple scans and combine them into a single result.

Figure 9.2 shows the result of a single measurement on a SiC mirror and the same location measured using 64 averages. Surface roughness reduces by about 30%, and fine details such as polishing marks under 0.2 nm in height can be seen in the data taken using averaging. The drawback of averaging is the increase in measurement time. Eventually, the longer measurement time will allow environmental drift to affect the measurements and begin to degrade the results again. Typically, any measurement that takes over 5 min will suffer from such effects except in very controlled environments, so this presents a practical limit to the benefit that can be achieved.

9.2.5 Tilt Automation

The last key item for single-measurement automation discussed is tilt of the sample with respect to the measurement system. This is sometimes accomplished by tilting the part and at other times is accomplished by tilting the instrument. Minimizing the relative tilt is important for three primary reasons: less tilt increases the signal to noise of measurements because optics collect more light from the sample, reduced tilt minimizes off-axis errors associated with the system optics, and reduced tilt allows for shorter scan lengths

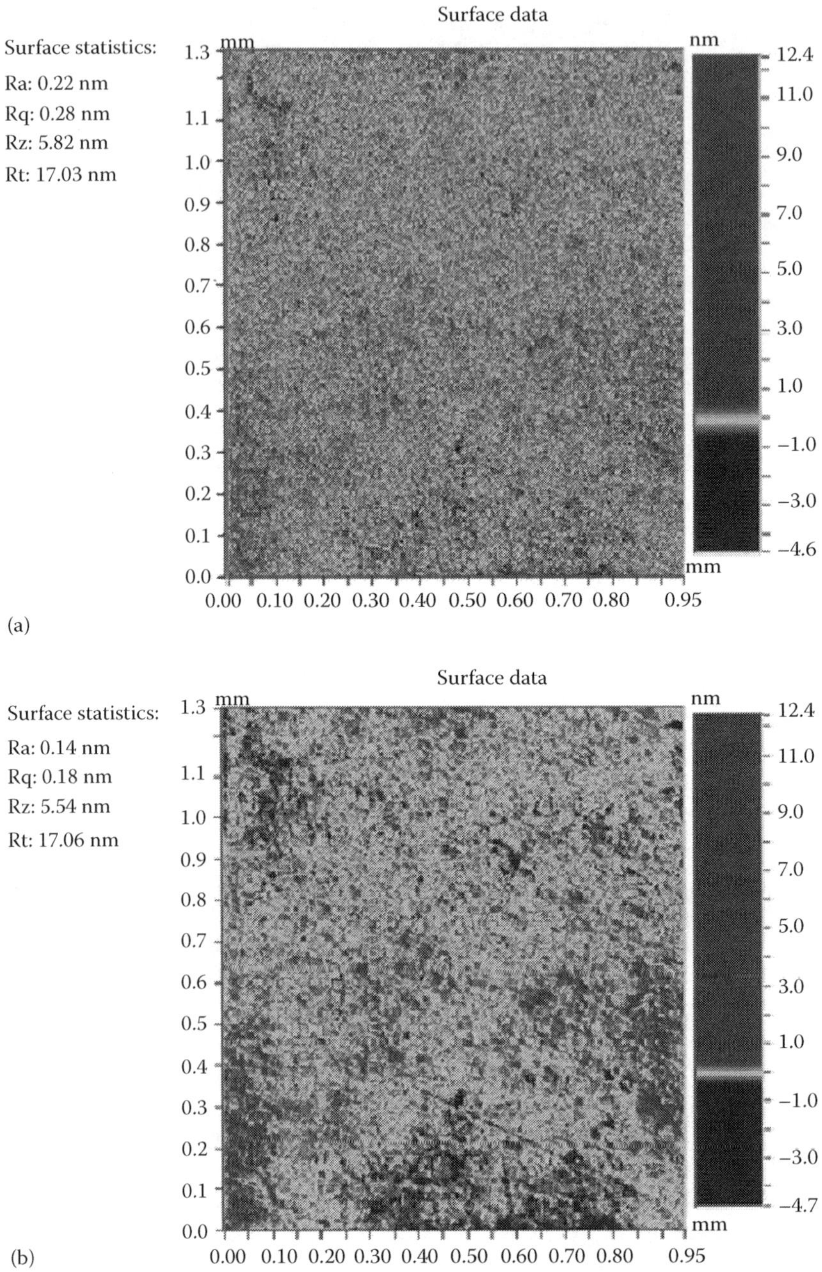

FIGURE 9.2
(a) Non-averaged surface data with random noise. (b) Surface measurement using multiple averages showing a sub-nm deep polishing marks on a SiC mirror.

for coherence-based measurements where every point on the sample must be scanned through focus.

Tilt adjustment does not typically introduce noticeable errors for rough surfaces (Sa > 1 μm) measured with white-light interferometers. However, errors can be on the order of a few tens of nm, and thus for phase-shifting measurements of smooth surfaces, the errors can be significant. Thus, for smooth surfaces or where best gage capability is desired, adjusting tilt may be required for proper results. One must keep in mind as well that since adjusting tilt changes the amount of light collected by an interferometer, intensity adjustment may need to be performed after adjusting tilt in order to avoid saturating the detector. Also, as noted earlier, for coherence-scanning interferometers, having the sample perpendicular to the optical axis means the instrument can minimize the amount of focus adjustment needed to focus through the entire part.

9.2.6 Summary of Automation within a Measurement

Typically, the automated processes earlier are used in conjunction to optimize measurement results on a sample. This results in the fastest, most accurate, and most repeatable results. The order of operations is important to ensure that the best results are achieved; for instance, intensity should be adjusted after tilt is optimized, since tilt affects the light level as seen by the instrument. For a fully automated measurement of a single location, the operations performed by an interferometer would be as follows:

1. Perform quick through-focus scan to determine focus position.
2. Perform any necessary fine adjustment of focus.
3. Check light level to ensure it is sufficient for measurement.
4. Perform measurement to determine tilt.
5. Optimize tilt of the sample relative to the instrument.
6. Optimize intensity.
7. Perform final scan where part is measured.
8. Repeat scan for the desired number of averages.
9. Analyze results.

Of course, depending on the sample under test and the required performance, one or more of the steps earlier may be eliminated.

9.2.6.1 Repeated Measurements

With any measurement system, it is important to examine the stability of the results over time. Both short-term and long-term measurement stability are important. Short-term stability is often used as a basic measure of the noise floor. Some systems offer a measurement sequence called a "difference measurement" where the instrument performs two successive measurements and displays only the subtracted result. Ideally, such a result, which represents the residual roughness as seen by the instrument, should look like random noise. This would indicate there are no systematic errors. In the presence of vibration, the result will often contain structure at twice the period of the interference fringes, indicating there may be an instrument, environment, or fixturing issue. Figure 9.3 shows a difference measurement showing mostly random noise on the left and in the presence of vibration on the right where the interference fringes have printed through into the measurement result.

For most interferometric systems, the average residual roughness in an ideal situation should be less than 1 nm as in Figure 9.3. Such measurements are very useful in assessing the environment and also in optimizing parameters such as the ideal number of averages to reduce noise. Figure 9.4 shows how the subtraction of two successive measurements

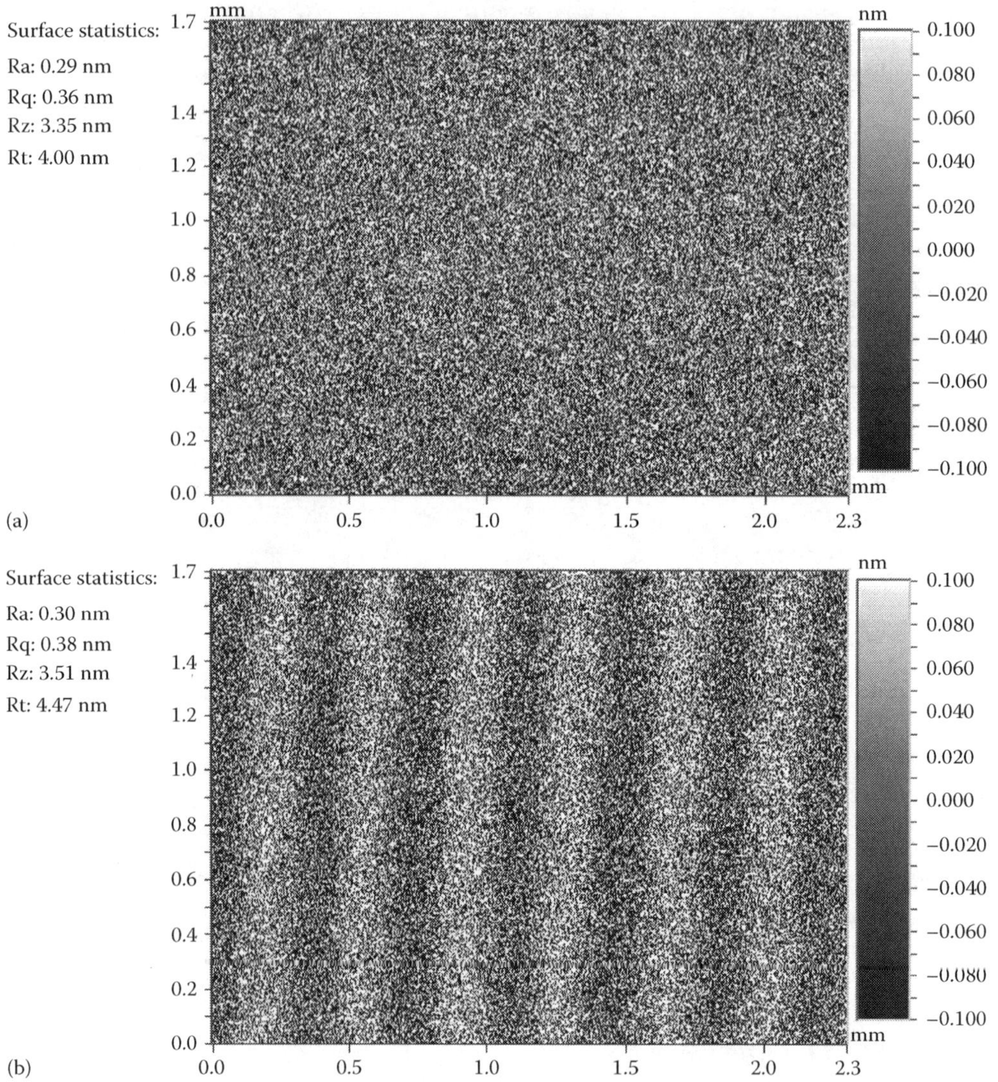

FIGURE 9.3
(a) A difference measurement taken without vibration showing mostly random noise. (b) A difference measurement in the presence of large vibration, showing errors from the interferometric fringes.

improves as the number of averages increase. In this case, using 25 averages improves the measurement repeatability by approximately a factor of 5, as expected from theory.

9.2.7 Automating Multiple Measurements

As mentioned earlier, multiple measurements on a sample have a variety of purposes. One is to combine overlapping measurement regions into one larger dataset that combines both large area and high lateral resolution; this is commonly referred to as data stitching. Another purpose is to sample multiple locations on a part to determine consistency of a parameter across a sample. This could be measuring different areas on

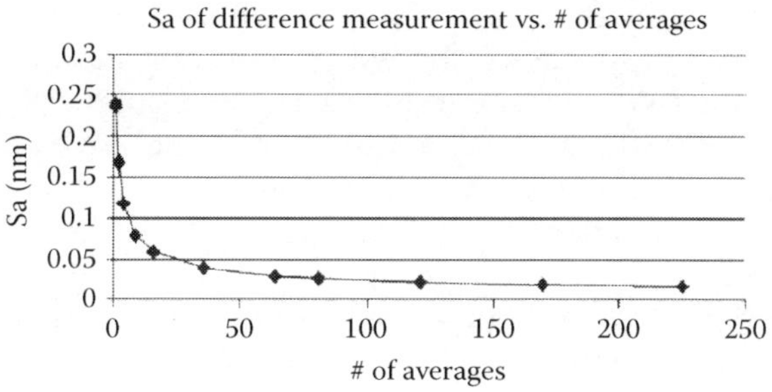

FIGURE 9.4
Sa of difference measurements using different numbers of averages, showing the reduction of residual noise as averaging increases.

a solar cell to ensure texture uniformity (which correlates to efficiency),[2] measuring multiple dies on a wafer to ensure each has the same properties, or measuring several distinct portions of a large sample to determine its properties, such as multiple points on an engine block. Any time, the sample or the interferometric system is moved to take multiple measurements; however, sample fixturing becomes critical for achieving high-quality data. Therefore, this is discussed first, and the different automation schemes are explored later.

9.3 Part Fixturing

Interferometry can produce highly detailed and accurate measurements due to the sensitivity of the technique. This sensitivity, however, makes results susceptible to relative movement between the sample and the instrument. Any such movement will cause a displacement of the interferometric fringes which in turn will lead to errors or degradation of the measurement. In a typical installation, there are many potential causes of relative movement between the instrument and the sample. Seismic (floor) vibration, acoustic (noise) disturbances, and both thermal and humidity changes can affect the instrument stability.

Seismic vibration can transmit from the floor up into the instrument. Sources of seismic vibration include building movement (especially upper floors or raised floors), heavy machinery operating nearby, traffic at nearby highways, or even earthquakes. Isolation tables—either mechanical, pneumatic, or active—are usually required to isolate the instrument from these vibration sources. In addition, placing instruments on solid floors, near support pillars, and away from the worst sources of floor vibration will greatly improve measurement results.

Acoustic sources of vibration can often be more subtle and difficult to combat. The typical interferometer is a complex piece of equipment with numerous parts and subassemblies. Each of these parts will tend to vibrate or "ring" at its resonant frequency if excited by an appropriate source. Often, this source takes the form of acoustics. Air conditioners, fans, loud music, nearby machinery, or even people talking can create acoustic disturbances of the correct frequency to excite components in the instrument to vibrate.

These vibrations are usually too small to see but can manifest themselves as a vibration or unwanted movement of the fringes. Solving this problem can sometimes be difficult and can include removing the source of vibration (often impossible, especially in a clean room where air movement is necessary), moving the instrument to reduce the noise influence, stiffening the instrument to increase the rigidity of the problem component, or isolating the instrument using a shroud or enclosure designed to attenuate acoustic disturbances.

Changes in the air temperature or humidity can similarly affect the performance of the instrument. Aluminum is a common material used in the construction of interferometric systems but has a relatively high coefficient of thermal expansion. Changes in the ambient temperature will cause the structure to shrink or grow resulting in movement of the fringes. If this occurs during a measurement, the accuracy can be affected.

Humidity changes can cause similar effects. Plastics are sometimes used in the construction of the instruments or in special fixtures. Many plastics absorb moisture from the air and can change shape significantly. Again, the most dramatic effects occur when there are rapid changes in humidity. To avoid these effects, it is important to maintain stable temperature and humidity around the instrument. If this is impossible, it may be necessary to time the taking of measurements with the cycling of the air conditioning systems to avoid rapid changes.

9.3.1 Sample Holding General Requirements

The operating principle of an interferometer makes the orientation of the sample under test very important. Generally speaking, the surface under test must be held perpendicular to the optical path of the instrument for best results. Most instruments incorporate a leveling stage to make fine tilt adjustments, either to the optical head or the sample. However, if the sample under test has angled faces that exceed the travel of the leveling stage, a special fixture may be required to bring the test face into alignment.

In addition to achieving proper part orientation, fixtures must hold the part securely. Often, large or heavy parts are simply placed on the staging associated with the interferometer, and it is assumed that gravity will hold the part securely with respect to the stage when the stage is moved. This is often a bad assumption, and errors in combined datasets after stage motion can be severe even with part motion of only a few μm as the stage is moved. Vacuum fixturing, clamping the part, or merely holding using clay along the edges of the part where it contacts the stage may be needed to accomplish stable and repeatable motion.

It is also important to consider how the sample will be loaded into the instrument, particularly for microscope-based systems. Many optical objectives have a very small working distance, so the clearance between the part and optics is small, typically between 1 cm and several millimeters. It is important to avoid contact between the sample and the objective as either may be damaged. Interferometric objectives can be very expensive and may require time-consuming adjustments to focus and align if replacement is necessary. Most interferometric instruments incorporate a motion axis which can be used to retract the optics during sample loading, but it is sometimes preferable to design a fixture that will position the sample without requiring retracting the objective. This is often faster and may help to reduce setup time.

In some cases, especially in a production setting where parts will be frequently installed and removed, it is useful to position the part in a repeatable way. The technique of kinematic mounting locates the part with the minimum number of contact points necessary to uniquely locate it. Vukobratovich and others describe these techniques in detail.[3]

9.3.2 Nonadjustable Fixtures

Sample fixtures designed to hold parts statically are generally the simplest kind to design and fabricate. However, they can range in complexity from a simple flat plate to a complex kinematic mount made from low coefficient of expansion material. As mentioned previously, it is often useful to design fixtures such that they hold the sample in a repeatable fashion using kinematic design. For a flat sample, this can consist of three balls or small pads upon which the sample rests. Side-to-side alignment can similarly be provided by two balls or pads on one edge and another on a second edge. Many other schemes are available for mounting other shapes and can be found in the references.

When designing sample holding fixtures, it is important to consider how the sample itself will deform when mounted. This is especially important with thin or flexible samples. An interferometric measurement is sensitive enough to "see" deformation of a sample as a result of incorrect mounting. Semiconductor wafers or thin plastic samples mounted on a three-ball mount can easily deform to an unacceptable level due to gravitational sag. In these cases, it may be necessary to fully support the sample with a precision ground flat surface. Figure 9.5 shows a mirror measured where it is properly mounted and unstressed and a second measurement where the mounting holds the mirror too tightly and has an astigmatic shape due to stress-related deformation.

When it is necessary to hold a thin sample in place, it is common to employ a vacuum fixture. Grooves or holes cut into the top surface can be connected to a vacuum source. Flat samples will usually form a sufficient seal to be held in place against the top surface provided the surfaces are clean and the fixture is flat and smooth. The sample will conform to the shape of the vacuum fixture, however, and thus, the flatness should be properly specified against the measurement needs. Very thin samples can also be deformed locally in the areas over the vacuum grooves or holes. In these cases, porous metal, ceramic, or plastic vacuum fixtures can be employed. These materials distribute the holding force over a larger area and can avoid the local distortion common with vacuum grooves. This is particularly useful for very pliant samples such as plastic films or for samples with numerous holes that would not seal well with a conventional vacuum fixture. Figure 9.6 shows an example of these chucks.

9.3.3 Adjustable Fixtures

It is sometimes useful to design fixtures that incorporate some type of motion. Some samples may require a leveling adjustment in excess of what the standard instrument can accommodate because of their shape. Or, samples may require additional linear movement if their length exceeds the travel of the standard stage. Figure 9.7 shows one type of fixture, designed to hold a part at multiple orientations with respect to the instrument so that the front face and back face of the object under test may be imaged. For two-position fixtures such as this, adjustable hard stops are easily used, leading to highly repeatable positioning.

In these cases, it is important to pay careful attention to the design and construction of the fixture to avoid unwanted movement that could degrade the measurement. Excessive backlash or play in the fixture can cause drift of the fringes or unwanted movement of the sample. Depending on the type of actuator used, there are many techniques for removing backlash. Preload springs can be employed to keep the moving portion of the fixture pressed against the actuator. If lead screws are employed, there are many types of anti-backlash drive nuts available. Gravity can also be used if the motion axis can be inclined

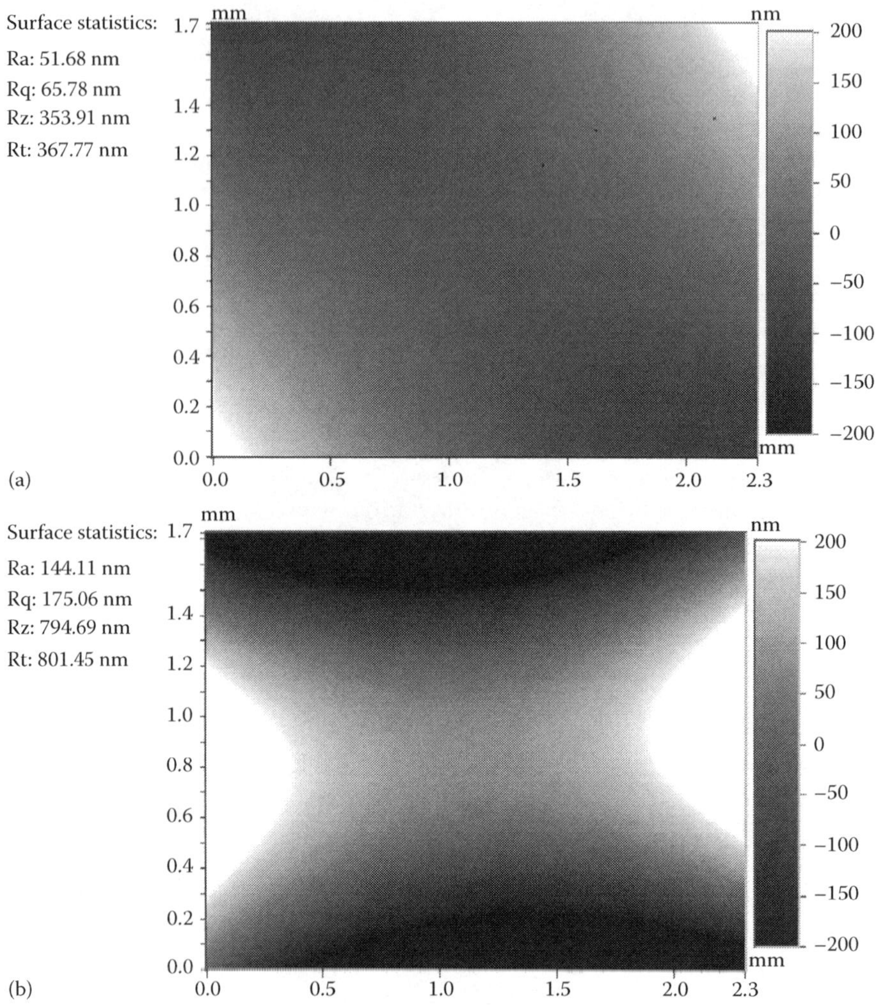

FIGURE 9.5
(a) Unstressed part showing slight amount of bow (66 nm). (b) Stressed part with more than three times the bow and strong astigmatic (potato chip) shape.

such that the weight of the moveable portion will be pressed against the actuator. In all cases, it is critical that the sample is firmly fixed in position during a measurement. Any unwanted motion or drift will negatively impact the measurement.

Most part fixtures will require some adjustment, either on initial setup to align the fixture to the instrument or ongoing adjustments related to variations in the parts themselves or how they are mounted within the fixture; such ongoing adjustments typically allow parts to be properly oriented and features of interest placed within the measurement area of the instrument. There are numerous actuator types available for use in fixtures. It is important to consider the resolution of the motion and use an actuator that meets the requirements. For many laser interferometers, fine lateral adjustments are only necessary for specialized testing such as for aspheric parts. However, for microscope-based interferometers, objectives may have submillimeter fields of view and therefore require at least that resolution from the actuator.

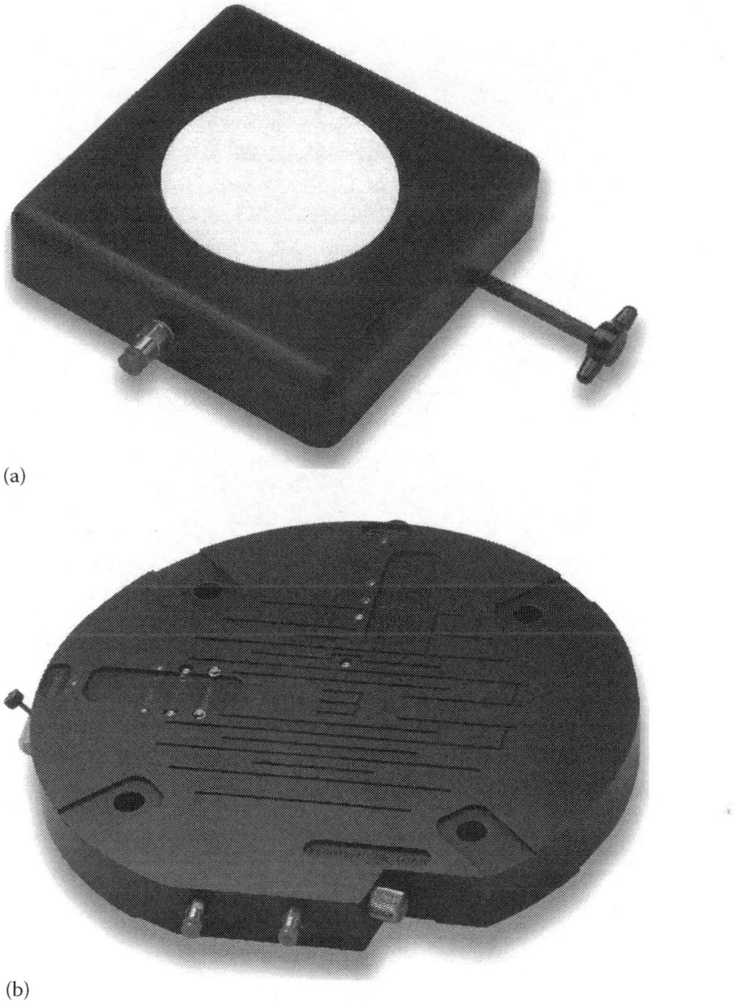

(a)

(b)

FIGURE 9.6
(a) Porous ceramic vacuum chuck. (b) More traditional grooved vacuum chuck for wafers, with retractable mounting pins to accommodate different size wafers.

There are so many different types of actuators that space does not permit listing them all. Manual actuators can be as simple as a threaded screw, lever, or knob. Manual threaded adjusters can provide extremely fine motion depending on the thread pitch selected. If the actuator is a manual threaded adjuster, select the thread pitch such that no less than 1/8 turn, at a minimum, is required to make the smallest adjustment. If the actuator is motorized, consult the manufacturer's information to make sure they guarantee that it can make the necessary minimum adjustment. Figure 9.8 shows an example of a simple stage employing threaded screws. If finer adjustment is required than can be achieved by a simple threaded adjuster, consider using a differential micrometer. They achieve extremely fine motion by using two threads, very close to the same pitch. The result is an extremely fine motion. The effective pitch is given by $1/P_{eff} = 1/P_1 - 1/P_2$, where P_1 and P_2 are the thread pitches of the two screws.

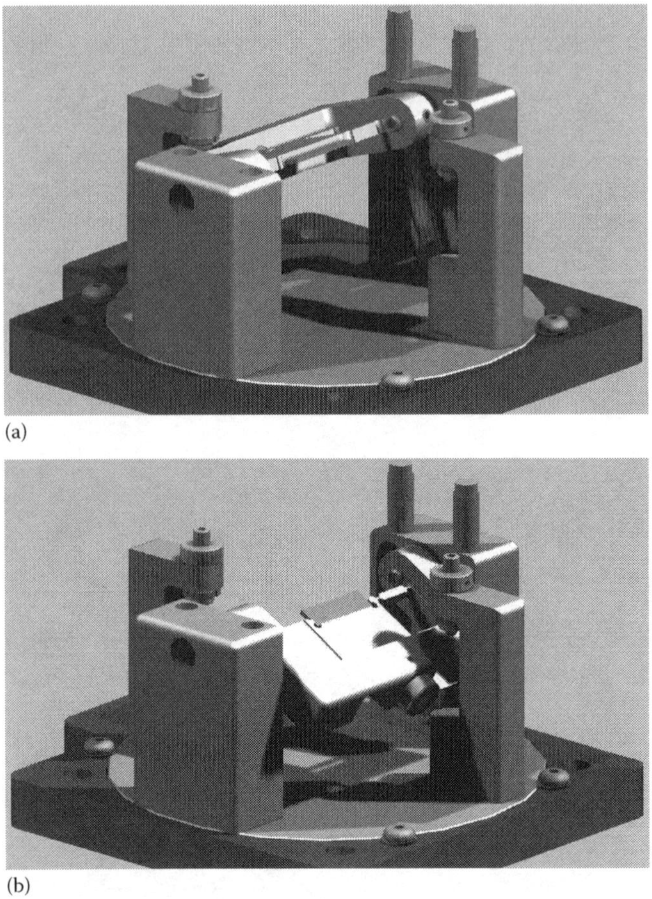

FIGURE 9.7
(a and b) Example of moveable fixture at two different positions. Fixture can swivel through different angles to position sample.

Automated actuators range from motors and pneumatic cylinders to piezoelectric transducers, shape memory alloys, and magnetostrictive materials. Selecting the simplest and easiest method to achieve the necessary motion will almost always provide the best solution. There are numerous bearings available for use in adjustable fixtures. Simple sliding dovetail slides will provide low-cost linear motion. This may be satisfactory if there is a means for locking the slide in place after adjustment as the play in the slide can cause unwanted motion of the sample. For more precise linear adjustment, there are ball and crossed roller bearings that can provide very accurate motion without side-to-side play. These bearings are commonly available from a number of suppliers and can be purchased in travels from millimeters to many meters in length. At the very high end of performance and cost are air bearing linear slides that offer the lowest friction and the highest straightness of travel. These slides require a compressed air supply (and some a vacuum supply) in order to function. Make sure to test the particular air bearing slide on an instrument prior to incorporating in a design. Some can cause a slight movement due to the fluctuation of air pressure.

Often, it is easier and less expensive to provide rotary motion instead of linear motion. Rotary bearings are very common, relatively inexpensive, and available in a nearly limitless

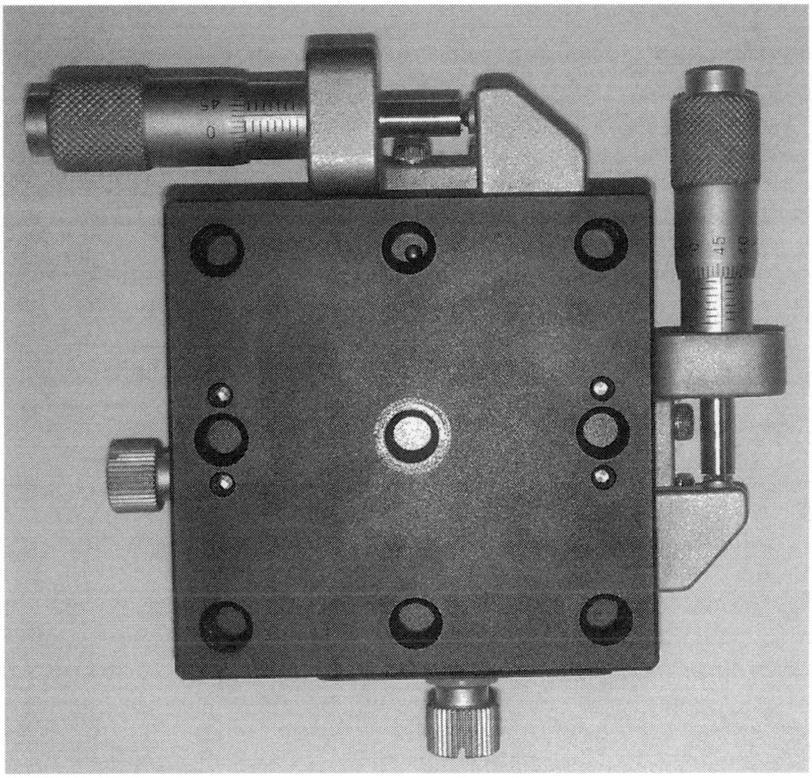

FIGURE 9.8
Manual translation stage employing micrometers.

number of sizes and types. The least expensive are simple sleeve bearings. Often used in pairs, these bearings can provide simple rotation about an axis. As with the dovetail slide mentioned earlier, sleeve bearings usually are designed to have some clearance and thus may not control the position of the fixture adequately. Simple ball bearings are a step up from sleeve bearings and can provide very accurate motion provided they are loaded per the manufacturer's recommendation. These bearings will not usually support large axial loads (loads applied along the axis of the rotating shaft).

When very high axial loads need to be supported, angular contact bearings can be employed. These bearings are designed to support higher axial loads due to their construction. The bearings are usually used in pairs and are preloaded axially to remove all play. Without preloading, angular contact bearings will generally exhibit too much play to accurately control the position of the sample.

For very accurate motion, flexures can be a good choice. This is a category of devices that rely on an element to bend in order to provide motion. Their advantages are low cost (although there are high-end flexures that can be more expensive), simple construction, no play or backlash, and low force over small distances. Flexures do not excel, however, at providing large amounts of travel. Flexures can be fabricated using spring steel cut to appropriate shapes or machined into a variety of materials as part of a larger device. There are also commercially available flexure pivots that can provide nearly friction-free rotation over short distances. These devices use a pair of flexures at right angles to one another to maintain a constant rotation center.

An important consideration for many sample types is to provide pins or features against which samples can be aligned. Often, more than one sample size can be accommodated on a single fixture by providing removable pins that can be repositioned as required. Care must be taken to prevent the alignment pins or features from protruding above the surface of the sample by more than the minimum working distance of the objective being used.

Another important consideration is to provide features, usually grooves in the fixture surface, that will allow users access to the sample so that they may manipulate them either by hand or with tweezers. Many samples are delicate and cannot be touched on the top surface, so providing any easy means of handling them is an important consideration.

9.3.4 Special Considerations

With any sample fixture, it is important to take into consideration the ease with which samples can be loaded and unloaded. It is imperative to avoid contact between the sample and interferometer. Not only can samples be damaged, but the optics in an interferometer are delicate and expensive items, and accidental contact can damage optical surfaces or knock the focus out of position.

Automated interferometers typically incorporate a focusing stage which can be moved to a safe position for sample loading. Often, however, it is desirable to leave the focus stage in place to speed up the process, or if the sample is too bulky. In these cases, it may be beneficial to incorporate a mechanism that will allow the sample to be moved horizontally to a loading/unloading position. Many of the techniques discussed earlier regarding adjustable fixtures can be employed in such a mechanism. Of paramount importance are the stability and repeatability of the fixture in the measurement position. The fixture should return to the same position every time it is loaded, and it should not have any play or slop such that it could be affected by vibration.

Safety is an important consideration with any fixture, especially those that have motorized motion. Care should be taken to avoid sharp edges or areas where fingers or hands can become trapped or pinched. Not only can such hazards exist on fixtures but attention must also be paid to how close moving parts come to portions of the instrument. For example, moving a fixture to its load position may create a pinch point with the instrument's sample stage. Often, simple guards can be included in a fixture design to cover such hazards. If hazards cannot be eliminated, they should be marked. There are a number of commercially available labels for this purpose.

There are cases where pinch points cannot be entirely eliminated and present such a substantial hazard that some sort of interlock must be included. One method is to enclose the entire instrument and provide switches on the access doors that will shut off the motion if opened. Other methods include light curtains, through-beam sensors, or dual switches that both must be engaged (one with each hand) in order to operate the system.

Finally, any good fixture design will take into account ergonomics. The ease with which an operator can load, unload, and manipulate a fixture is very important. Further, most companies have specifications regarding ergonomics. It is important to consider how much load must be lifted, how much force must be applied, and how far the operator must reach in order to operate the fixture. There are limits on all of these in most ergonomic specifications, and these considerations may dramatically alter the ultimate design of the fixture.

9.3.5 Automated Positioning Stages

Proper fixturing of the part is important, but equally important is how the part under test is moved relative to the instrument or how the instrument is moved relative to the part.

The most common positioning stages are linear stages which convert the motion of an actuator into straight-line motion. It is common to stack two linear stages at right angles to create a system that will position a sample in two orthogonal directions. These are commonly referred to as "XY" stages. Rotary stages are appropriate for many rotationally symmetric parts such as wafers or optics and convert the motion of an actuator into angular motion. Rotary stages can be stacked with linear stages to provide total coverage of round samples. These are sometimes referred to as "R-Theta" stages.

Either type of stage can be equipped with position feedback devices commonly known as encoders. These are available in various accuracies down to a micron and below. Encoders can be connected to the motor drive electronics to allow them to position the stage to very high accuracies. Encoders can be very expensive, however, and their cost must be weighed against their benefit.

Stage actuators are available in numerous types and resolutions. Stepper motors are capable of rotating in discrete steps and can be an inexpensive way to achieve high accuracy without expensive position feedback devices. Servo motors can achieve very smooth motion but require an encoder in order to move accurately to a given position. Piezo actuators can perform extremely small moves and are available in a variety of styles. Because they can operate at very high frequencies, piezoelectric transducers can also achieve relatively high speeds.

The stability of any sample positioning is very important. Virtually any unwanted movement of the sample will degrade the interferometric measurement. Properly preloaded, rolling element (crossed roller or ball bearing) bearings are usually preferred. Sliding bearings can be problematic unless all play is locked out during measurements. Some stepper motors can actually generate enough vibration when stopped to disturb the measurement; they must be properly adjusted to ensure stable operation. Some brands of stepper motors and drive electronics are better than others when it comes to their vibration. It may be necessary to evaluate offerings from several suppliers in order to find the quietest version.

9.4 Automation Multiple Measurements on a Sample

9.4.1 Data Stitching

The simplest and most common form of automation across a sample in interferometry is to combine multiple measurements of overlapping fields of view into a single, larger field of view measurement. This process is referred to as data stitching. There are a variety of situations where such stitched measurements are desirable or necessary. Usually, stitched measurements are employed because one needs to maintain a high lateral resolution but also to measure a large area; this might be because one needs to measure fine features, quantify defects, or determine roughness on a high spatial scale, but the part of interest is larger than the field of view offered by the interferometer at the required resolution. An example of such a case is shown in Figure 9.9, where a coin measuring more than 20 mm in diameter would be inspected for defects on the order of a few micrometers wide; even with today's high-resolution cameras, such a combination of field of view and resolution could not be achieved in a single measurement.

A second case where stitching would be required is for measuring parts with steep slopes over large distances. Example applications might include examination of precision

FIGURE 9.9
Stitched dollar coin and close-up showing high lateral resolution even over a data area of 25 mm in diameter.

screw threads, as shown in Figure 9.10, where the angles of some surfaces with respect to the instrument are up to 70°. In this case, to obtain good data, a fairly high-magnification measurement (typically 20× or higher) is needed to properly resolve the slopes. To obtain data over multiple threads, multiple measurements can be combined such that the pitch and angles can be calculated along the screw, rather than merely seeing a single or partial thread measurement.

The basic idea behind stitching interferometric measurements together is simple, but there are many details which can affect the quality of the final result. The primary variable to be controlled governs the percent of overlap between the individual scans. Typically, stitched measurements rely either on known stage accuracy and merely average any overlapped region together to create a continuous surface, or they use some kind of correlation technique in the overlap region to precisely align one measurement to another independent of the stage accuracies or other positional errors. The greater the overlap region, generally the better the alignment between adjacent locations and the lower the overall surface errors because there are more points on which to align. An overlap of 10%–20% is common. The best surface stitches can achieve surface accuracies of about 1 part in 10^6 in terms of height errors with respect to lateral stitched size. Thus, a stitch of a 10 mm in length involving high magnification objectives may have a shape error on the order of 10 nm or so across the entire stitch. The pattern of stitching may also differ based on the interferometer and part geometry. Two stitching patterns are highlighted in Figure 9.11. Figure 9.12 shows the individual measurements and combined stitch resulting from a stitched pattern as in Figure 9.11b.

For the highest quality stitched result on smooth surfaces, often the systematic errors of the interferometric system must be removed. Most systems allow the option of generating a reference surface, where a high-quality part is measured at multiple locations and the resulting averaged surface measurement is then said to represent the fixed errors associated with the measurement system. Once generated, this data can be subtracted from future measurements to achieve greater levels of accuracy.

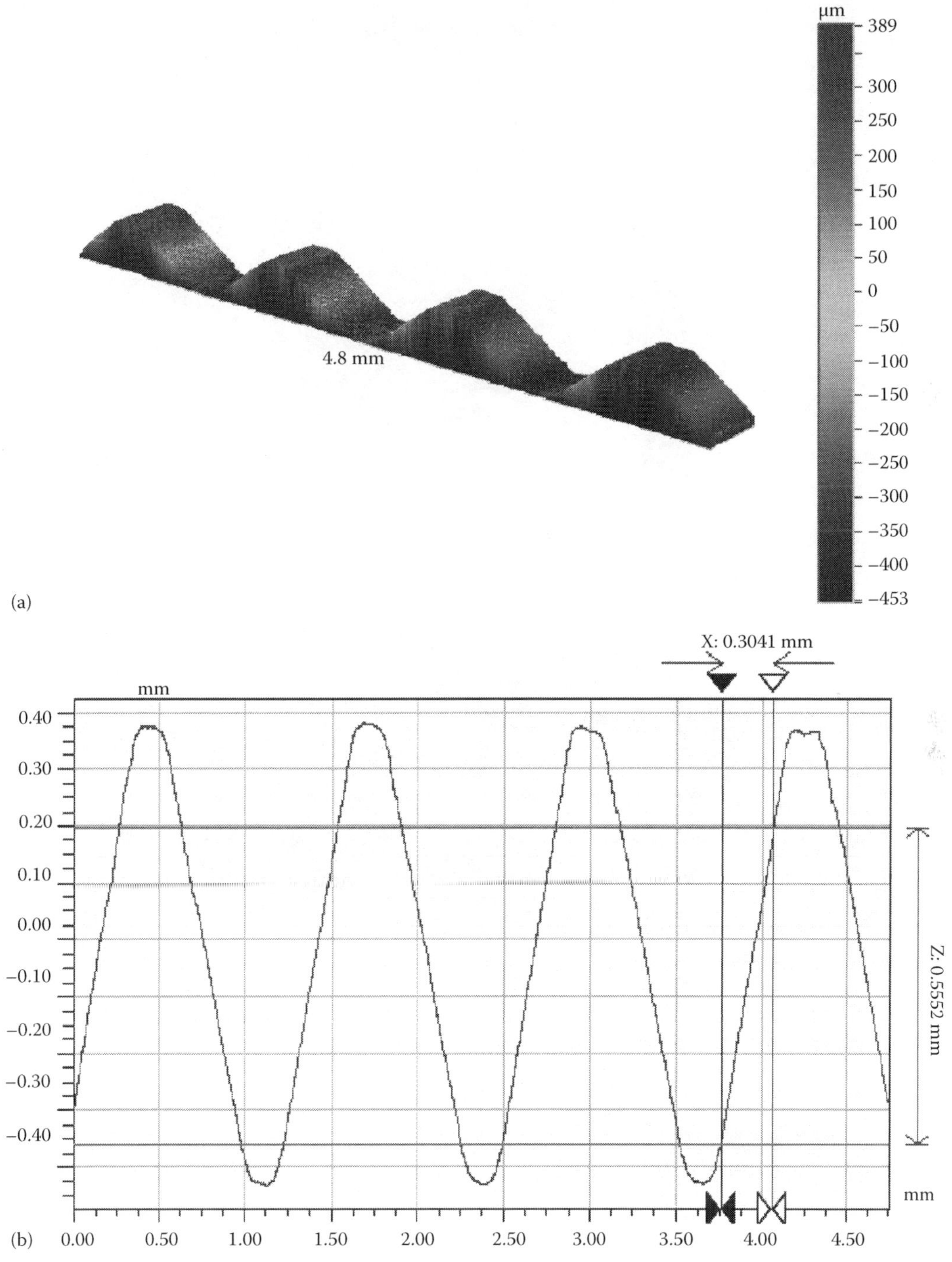

FIGURE 9.10
(a) 3D view and (b) 2D cross-section of steep screw threads measured with an interferometric profiler.

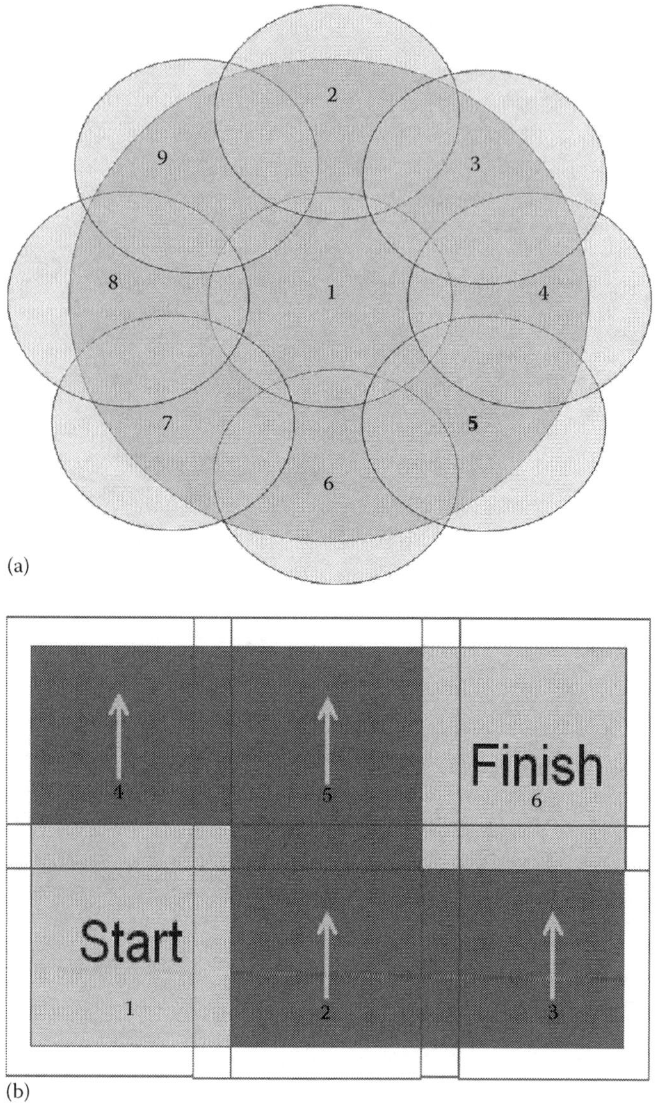

FIGURE 9.11
(a) Stitched pattern for a laser Fizeau system measuring a large part via stitching and (b) a typical stitch pattern for an interferometric microscope.

9.5 Automating Measurements across Samples

9.5.1 Multipoint Automation

It is often desirable to measure multiple locations on one part which are not contiguous or multiple parts on an instrument. This can include multiple single fields of view or multiple stitched measurement sets for each part. For instance, one might wish to sample a wafer for roughness at various points on the wafer, or one might wish to stitch together

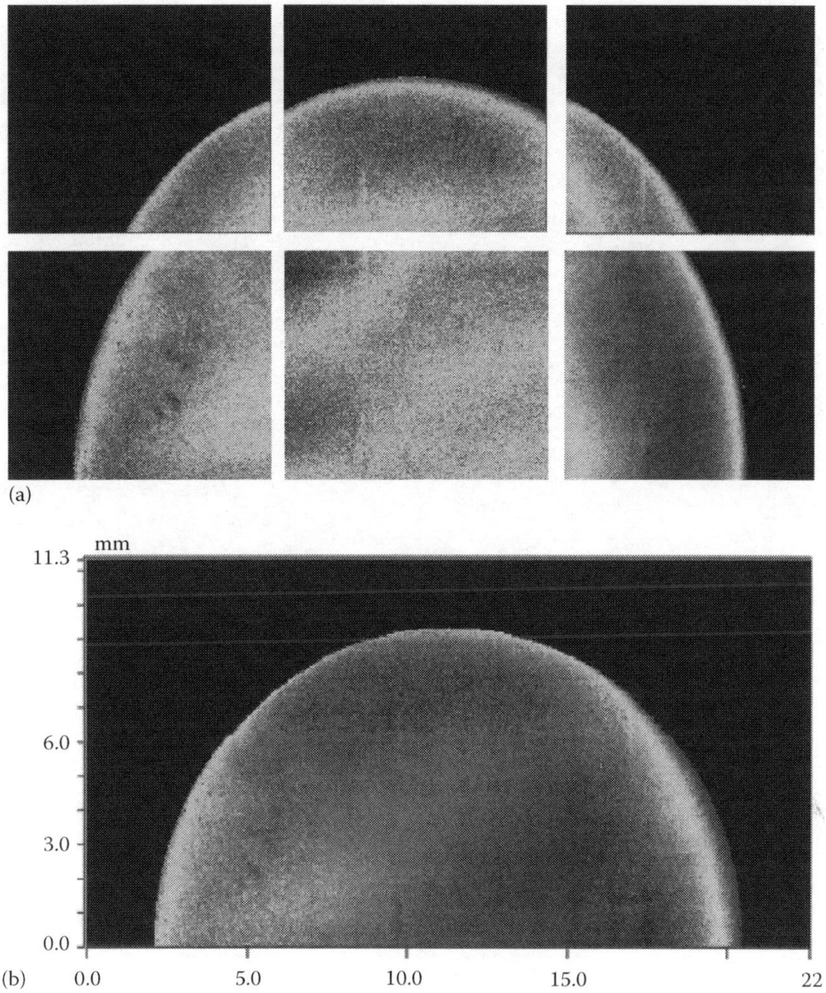

FIGURE 9.12
(a) Individual measurements and (b) final stitched result of a mirror.

entire dies at multiple locations in order to attain enough data to understand the variations across the wafer. Additionally, one might then wish to measure those multiple locations on a variety of samples placed underneath the interferometer.

Most interferometers offer some form of stage automation. They may support measurement grids for wafers or other parts and may also allow one to measure nonperiodic locations on a part through manual or automated selection. Figure 9.13 shows a possible map used for wafer automation, where the system will cycle through any or all of the grid locations depending on which automation cells are activated. Choosing the order in which the cells are traversed, shown in the right side of the image, can be important to improve throughput or to ensure that the instrument measurement order corresponds to a desired order based on other considerations in the process. For instance, if the output of the instrument is to be used to control another point in the process, it may be important that data are always taken and presented with each row being traversed left to right, even though moving in a serpentine fashion reduces movements and therefore improves throughput.

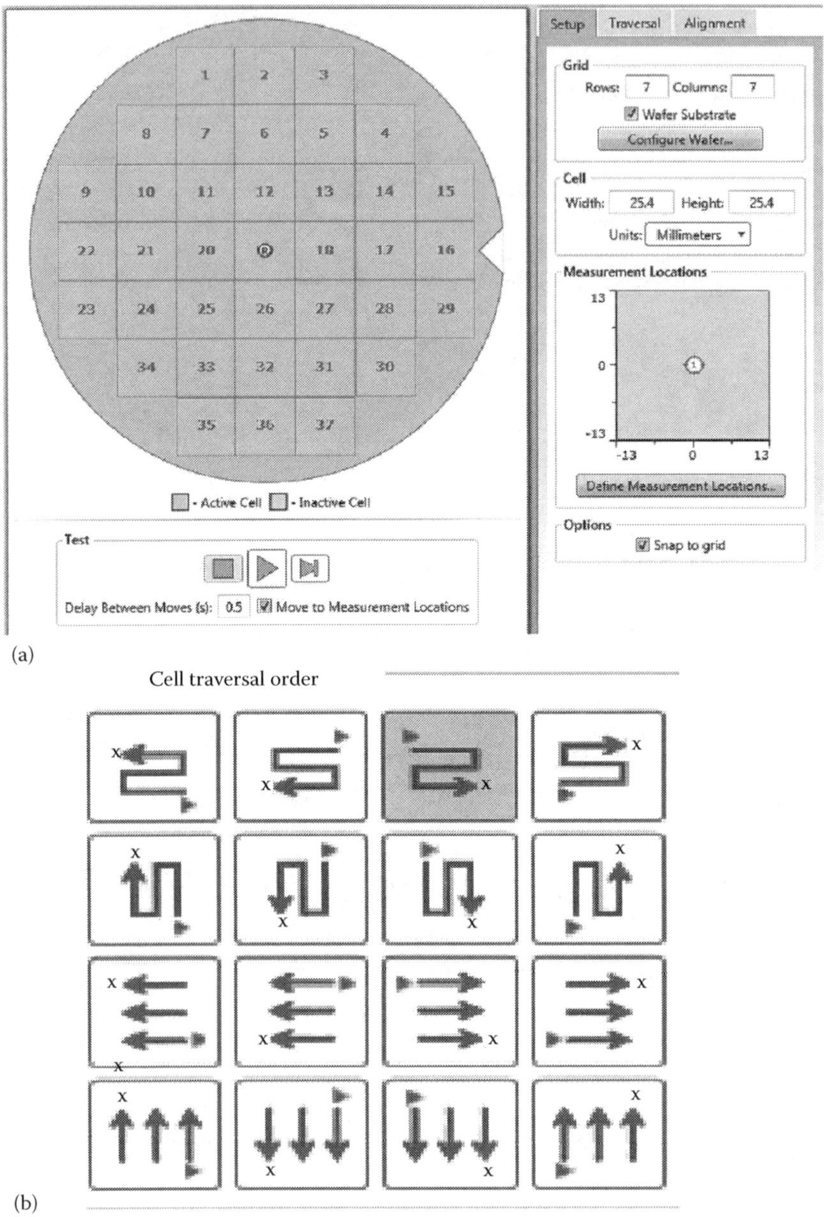

FIGURE 9.13
(a) Wafer grid automation and (b) the stage traversal options used to optimize measurement speed and results.

When automating multiple locations, it is especially important to employ the automation features associated with single measurements. Even a nominally flat sample such as a wafer may bow by hundreds of microns if it is stressed. Such bow will require a different focus on the interferometer at each location, may require light level adjustment due to tilt of the sample, and may even require adjustment of tilt to ensure the proper quality data.

A key consideration for all multipoint automation is determining the orientation of the sample beneath the instrument. Parts may vary in terms of their dimensions or feature placement within the sample. Additionally, even the most careful loading of a sample will have some error in positioning. If one merely wishes to sample approximate locations, this may not be an issue. However, to measure specific features, the part location relative to the interferometer must be known. Determination of alignment may be semi-manual, where the user drives to multiple known locations, centers them, and then the system will automatically adjust the stage file, removing any tilt in the part.

9.5.2 Part Loading

In some applications, it is desirable to load parts automatically to minimize contamination, damage, and to ensure consistent placement with respect to the interferometer. This is particularly true for applications involving wafers, such as in semiconductor and LED manufacturing. However, placement of copper panels, glass displays, lenses, data storage read heads, optical filters, and other precision components has been automated on various interferometers. Such automation reduces operator error, improves throughput, and generally results in more consistent measurement results due to repeatable placement of samples underneath the measuring instrument.

It is important to understand what exactly is required from an automated system to be successful. Some systems take parts from a tray, place them onto your fixture, measure them, and then return them to their original locations. At other times, it may be desirable to sort parts based on the results of the interferometer. The interferometer may also need to communicate with factory automation software when employed in production environments. This may be as simple as sending out results via a file or simple communication protocol such as TCP/IP or may require direct interface via GEMS/SECS protocols or with third-party software which controls other instrumentation. For the more sophisticated automation interfaces, a commercial integrator is typically employed who specializes in incorporating instrumentation into a production environment.

Figure 9.14 shows an example of an automated wafer handling system designed to feed parts to two separate interferometric 3D microscopes. Part handling automation is a highly

FIGURE 9.14
Two optical profilers sharing a single wafer automation system to help minimize cost.

expensive addition to a system, and often the handling is much faster than the measurements. Thus, employing one part handler to feed multiple measurement systems can be highly cost-effective as well as minimizing the space requirements for a given throughput.

While wafer handling systems are fairly standard, most other automated handling mechanisms are customized for the specific part being manipulated. As such, it is hard to provide specific guidance about key considerations. However, to optimize throughput, part placement should be consistent to within at least half the field of view of the interferometer such that fiducial finding software or other means of determining the exact part location can minimize any search area. In addition, handlers should not heat or cool the devices to be tested while picking up and placing them to avoid distorting the part shape. In addition, they should not overly stress the components being handled as the relaxation time of some materials, particularly polymers, may be several minutes and thus can affect the data quality.

The intelligence of the handler should be carefully specified when purchasing any system which incorporates automated part loading. It is important to understand what the handler should do with missing and/or damaged components when trying to load them. Also, if there is a power outage or a system hardware or software malfunction, the handler must be able to recover sufficiently to load or unload any parts in transit to the interferometer without damage. A universal power supply is often incorporated into fully automated systems such that power drops will not affect the system performance.

Automated part handling can dramatically improve consistency of results, reduce part damage and contamination, and increase the throughput of the interferometer being used to measure components. However, the complexity of such systems and their relatively high cost must be weighed against these advantages. Typically, only a small fraction of production applications fully automate part handling as opposed to manually loading/unloading samples and merely automating the rest of the measurements.

9.5.3 In-Line Instrument Considerations

In some cases, it is necessary to mount the interferometer head directly in-line with another manufacturing process so that samples can be measured without removing them from the process line. Some suppliers offer their interferometer heads separate from the mounting structure for this purpose. In order to successfully mount the head assembly in-line, it is necessary to consider several important factors.

The primary consideration is the rigidity of the mounting structure. It must be sufficiently stiff to prevent unwanted relative motion between the optical head and the sample. As discussed earlier, even minute motions can disturb the measurement. In general, it is preferable to attach the optical head to the same structure that supports the sample. Attaching to a wall or ceiling mount probably will not afford the necessary structural rigidity. If it is not possible to mount to the same base as the sample, a solid floor mount can be used as long as the floor is very solid such as a concrete slab on grade.

Mounting structures for the optical head must be well designed for stiffness. Cantilever structures should be avoided and instead "bridge" structures (supported on both ends) should be used, if possible. It is difficult to describe exactly how stiff the structures must be, but the structure, with the interferometer attached, should have a resonant frequency above 180 Hz as a guideline. Consult the instrument supplier for assistance with your specific application.

In some cases, it may be possible to place the instrument directly on the sample by designing special feet that contact the sample and hold the optical objective at the proper

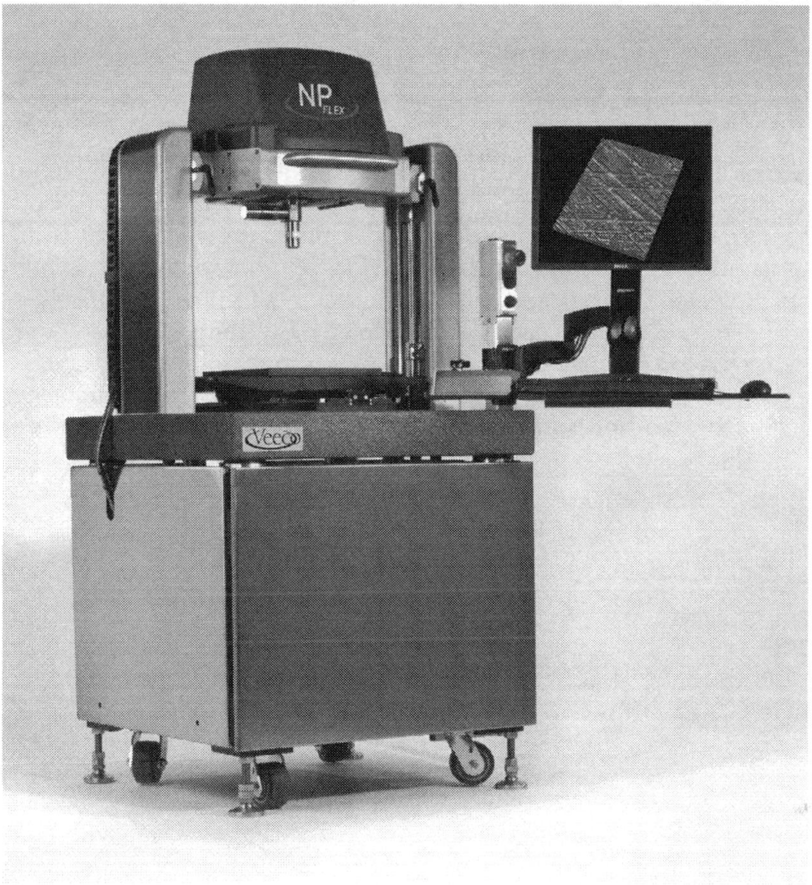

FIGURE 9.15
Interferometric profiler using a bridge design for increased stability and measurement capability on large samples.

distance from the sample. If the sample is a sufficiently damage-resistant material, this approach can provide a very rigid connection between the sample and instrument. Feet must be of a hard material but not so hard as to damage the sample. Often, aluminum or brass feet can be used if the sample is steel, for example.

Before embarking on the development of a custom instrument mount, consult your instrument supplier to see if they offer an instrument that will meet your needs. The interferometer can often be packaged into different platforms, and many such platforms have been developed as the technology has matured in various production and research applications. The Bruker-Nano NPFlex interferometer shown in Figure 9.15 is an example of a surface metrology instrument tailored for large or irregularly shaped samples. It has 12 in. × 12 in. × 12 in. sample capability and a tilting head design for measuring angled surfaces.

Finally, consult with your instrument supplier on techniques for determining whether your instrument is performing correctly. Some instruments contain software that can use the interferometric fringes to measure vibration between the sample and head. This can be used to evaluate different locations or environments to insure that the instrument performs as well as possible.

9.6 Conclusions

Automating interferometric measurements can be a powerful tool to improve system functionality. Automated results are typically more consistent, avoid sample damage, are faster, and are often more accurate than those performed manually. Automation routines can be used to find focus, adjust light levels, and adjust relative orientation of the sample to the instrument. Multiple measurements can be combined into large, high-resolution datasets, and automated staging can also be used to sample multiple locations on the test piece. Part handlers may also be married with interferometers to provide entirely operator-free operation, where the instrument can be controlled by and communicate with factory automation software. Last, systems may be incorporated directly into production lines to provide in-line process control. Each flavor of automation has its own unique concerns, and generally, many of the earlier automation elements will be combined to provide the fastest, highest quality data.

References

1. Murphy, P., Brown, T., Moore, D., Measurement and calibration of interferometric imaging aberrations, *Applied Optics* 39(34), 6421–6429 (2000).
2. Blewett, N., Novak, E., Photovoltaic cell texture quantitatively relates to efficiency, optics for solar energy, *OSA Technical Digest*, paper SWC3 (2010).
3. Yoder, P., Vukobratovich, D., Paquin, R., *Opto-Mechanical Systems Design*, 3rd edn., CRC Press, Boca Raton, FL, 2005.

10

White-Light Interference 3D Microscopes

Joanna Schmit

CONTENTS

3D microscopes employing white-light interferometry (WLI) are easy to use and provide unbeatable surface topography measurements of engineered surfaces. These systems deliver vertical resolution down to a fraction of a nanometer while maintaining the sub-micron lateral resolution measurements found in any typical microscope.

These instruments, used both in research labs and on the production floor, are based on a digital microscope equipped with an interferometric objective (from 1× to 115× magnification) and a computer-controlled scan through focus. The precision of the height measurement, levels only possible by using the interference of light, is independent of the numerical aperture of the microscope objective and can reach vertical resolutions down to a fraction of a nanometer. A WLI 3D microscope (see Figure 10.1) functions as regular microscope but delivers height measurements at hundreds of thousands to over a million points simultaneously creating a 3D surface topography measurement.

In a single vertical scan, areas from 50 μm × 50 μm up to 20 mm × 20 mm can be measured depending on the magnification and the size of the camera chip used. The measurement

(a) (b) (c)

FIGURE 10.1
White-light interference 3D microscope (a), bifocal contact lens to be measured (b), surface measurement result of the lens (c).

is highly repeatable and reproducible, and because only light contacts the surface to be examined, no damage occurs during evaluation. Light allows for very fast and precise measurements; heights up to 10 mm are possible using commercial systems. Only the working distance of the objective limits the heights that can be measured.

This 3D surface metrology has become an established method in many industries, including the precision machining, MEMS, solar, medical, and biological fields. This method is the basis of a future ISO (International Organization for Standardization) standard for surface metrology under the name of coherence scanning interferometry (ISO/CD 25178-604). Automation of measurement including sample handling, focusing, pattern recognition data processing, and specialized analysis makes these instruments very suitable for continuous, 24-h surface topography testing with excellent gage repeatability on different surface features. Different chapters in this book discuss measurement automation in 3D microscopy.

10.1 White-Light Interference

Because the interference of light is the way that surface topography is determined, we begin with an explanation of how interference fringes are produced. White-light interferometric 3D microscopes often use Köhler illumination with superluminescent diodes (SLDs) or light-emitting diodes (LEDs) to create interference fringes around the zero optical path difference (OPD), which is set to coincide with the best focus position of the microscope. A SLD light source has a broadband visible spectrum with a bandwidth of around 100 nm. SLD combines the high power and brightness of laser diodes with the low coherence of conventional LEDs. In addition, SLDs and LEDs last a long time, consume little power, and stay relatively cool as compared to the halogen lamps previously employed in systems.

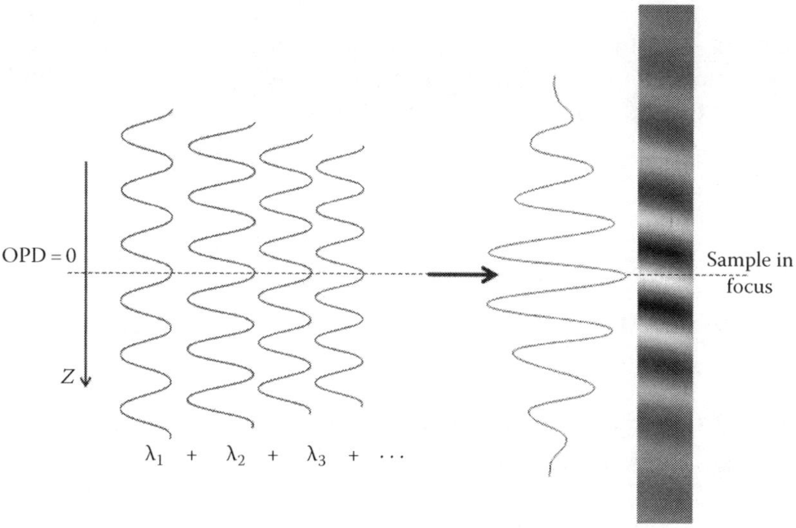

FIGURE 10.2
White-light fringe formation along axial scan z.

The different wavelengths λ from the source spectrum are mutually incoherent, and the superposition of interferometric fringes for individual wavelengths creates white-light fringes as shown in Figure 10.2. A solid-state CCD (charge-coupled device) or CMOS (complementary metal–oxide–semiconductor) detector observes the sum of all the fringe irradiances at each moment of the objective's scan through focus. Because each wavelength of the source creates fringes of different spacing, the maxima of fringes will align around only one point where the OPD in the interferometer is zero for all wavelengths. Away from this zero OPD position, along the axial scan z of the objective, the observed sum of the white-light fringe irradiances quickly decreases. In Figure 10.2, white-light fringes along the axial scan are presented in two ways: as a function observed by a single camera pixel, as well as grayscale image observed by a few pixels. The maximum of the fringe envelope marks the best focus position. In practice, the optical paths in an interferometer are not perfectly balanced creating slight shifts of fringes maxima.

White-light interference observed by a single pixel during an axial scan z can be mathematically described as the integral of all the fringes for all wavelengths for the full bandwidth of the spectrum and for different incident angles depending on the numerical aperture of the objective (de Groot and de Lega 2004). This summation results in the following mathematical form describing the detected fringes around the position h of the point on the object:

$$I(z) = I'\left[1 + \gamma(z)\cos(k_0(h - z) + \varphi)\right] \tag{10.1}$$

where

I' is the background irradiance
$\gamma(z)$ is the fringe visibility function or coherence envelope along axial scan z
$k_0 = 2\pi/\lambda_0$ is the central wave number for fringes under the envelope
φ is the phase offset of fringes maximum from the envelope's maximum due to dispersion in system

The broader the bandwidth of the source spectrum, the narrower is the width of the envelope. For a white-light source, this width is on the order of 1–2 μm. Determining the position of each of the envelopes on each point on the object is equivalent to finding the best focus position *h-z* at each point with great precision due to the narrowness of the envelope of the fringes. Best focus position at each point in turn determines the object's topography.

10.2 Measurement Procedure

The measurement procedure is simple, and no special sample preparation is needed. The sample is first placed underneath the objective. Then the sample needs to be brought into focus by moving either the objective or the sample plus its stage vertically as with regular optical microscopes, the main difference being that on portions of the sample which are in focus a few fringes will appear. Then the top or the bottom of the surface may need to be found manually or automatically by bringing it to focus. Once found, measurement can begin. The automatic scan will bring the sample beyond the focus and then scan the sample through its desired height while a camera will be collecting images at constant rate. The images will include fringes required for the topography measurement with this method. A scan has to start from the position at least a few microns above the highest point on the surface to be measured to a few microns below the lowest point to be measured; thus, a scan starts and ends with no fringes in the field of view.

Figure 10.3 presents four camera frames with interferograms from a scan through focus for a semispherical surface. Each interferogram shows the part of the object that is in focus for a given scan position.

The shape of these fringes already provides information about the spherical topography of the object. Additionally, a grayscale image of the fringe signal as seen by a few pixels during scan through focus is shown under the objective in Figure 10.3. For mirrorlike surfaces without steps or scratches, only these fringes are signaling when the sample is in the best focus position since no other object features "come to focus." Thus, 3D white-light interferometric microscopes are ideal for measuring smooth, mirrorlike finished surfaces.

Measurement of the best fringe contrast position results in a nanometer level precise measurement of surface topography. Methods for finding the position of best contrast fringes were described by many authors (Ai and Novak 1997, Caber 1993, Danielson and Boisrobert 1991, de Groot and Deck 1995, Kino and Chim 1990, Larkin 1996, Park and Kim 2000).

Unlike the monochromatic interferometric techniques, white-light interferometric techniques can be used for measuring not only smooth but also rough surfaces. Applications for this technique are enormously varied; for example, surface roughness that affects the cosmetic appearance of glossy paper or car finishes can be measured including the food roughness which apparently affects our eating experience. But 3D profilers are also commonly used to measure roughness of engine components, other machined parts (see Figure 10.4), bearings, or medical implants and contact lenses for which roughness and its character determines functionality of the part.

White-light fringes for rough surfaces can vary from locally well-defined fringes to fringes in the form of speckles. Figure 10.5 shows four through-focus images for a rough sample. Two images taken from above and below the sample's height range show no fringes, and two images taken at two different focus positions on the sample show clear fringes. The fringes indicate rather smooth islands along with rougher valleys among the islands.

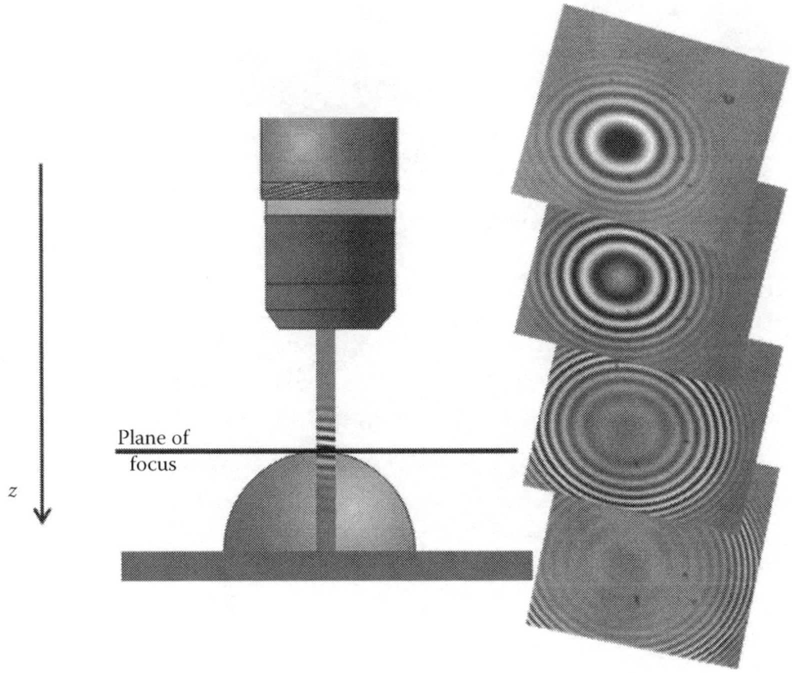

FIGURE 10.3
Four interferograms from a scan through focus for a semispherical surface.

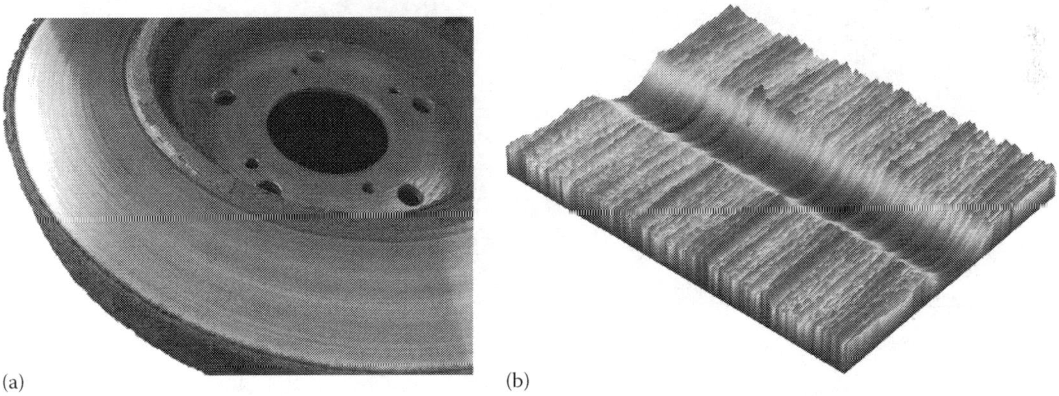

FIGURE 10.4
A metal machined part (a) and the measurement result of wear mark on this part (b).

10.3 Interferometric Objectives

3D WLI microscopes use interferometric objectives that are based on infinity-corrected bright-field objectives with a beam splitter and a reference mirror. The interferometer portion enhances determination of the focus position by creating fringes around the best focus allowing for object shape measurement with sub-nanometer resolution for high as well

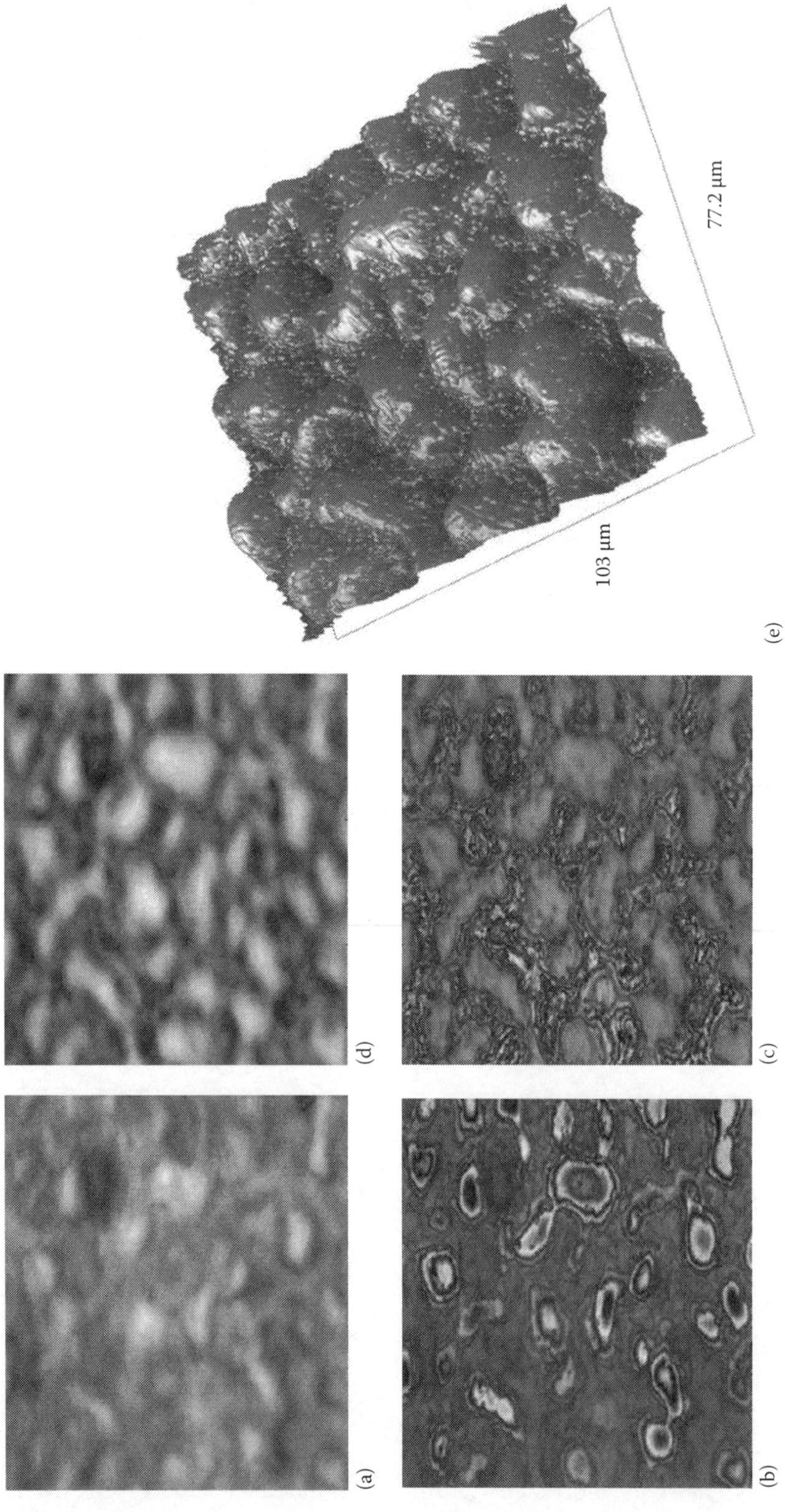

FIGURE 10.5
3D WLI microscope images for rough surface taken (a) above and (d) below the sample height range that show no fringes. (b) and (c) show two different focus positions on the sample and show fringes. (e) is the measurement result with average roughness Ra = 419 nm.

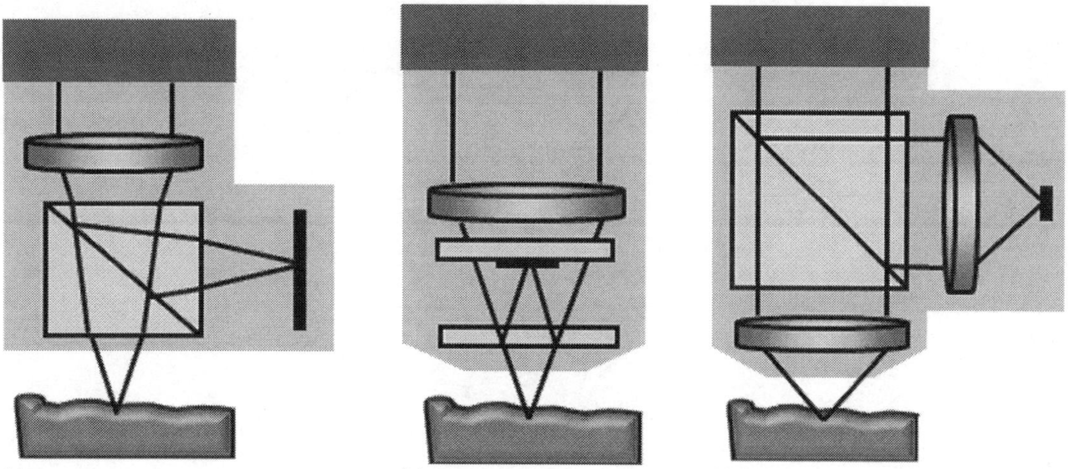

(a) (b) (c)

FIGURE 10.6
Interferometric objectives: (a) Michelson (b) Mirau and (c) Linnik.

as low magnification. Three basic interferometric objectives are used in these systems: Michelson, Mirau, and Linnik.

A Michelson interferometric objective uses a beam splitter cube placed underneath the objective and a reference mirror placed to the side (see Figure 10.6a); this type of interferometer is used only with low-magnification objectives having a long working distance and thus low numerical apertures. The Mirau interferometric objective (Figure 10.6b) uses two thin glass plates underneath the objective in objectives with shorter working distances and thus medium to high magnifications. The lower plate acts as a beam splitter, and the plate above contains a small reflective spot that acts as the reference surface. Mirau setups are not very useful at magnifications of less than about 10× because at these lower magnifications, the reference spot needs to be bigger as the field of view gets bigger, and then spot obscures too much of the aperture.

The third type of interferometric objective, a Linnik objective, is built for a few different reasons. The Linnik system (Figure 10.6c) allows an interferometer to be set up for any magnification objective from two identical bright-field objectives and beam splitter before objectives. Thus, such objectives can be built in the lab for any magnification. On one hand, the Linnik systems may be the only solution for very high-magnification, high-numerical aperture objectives that have short working distances that could not facilitate a beam splitter plate. On the other hand, the Linnik objective may be the only solution if a long working distance needs to be maintained for the measurement of large samples.

However, these objectives are quite bulky, and for higher magnifications, they may be very difficult to adjust since not only the position of the reference mirror needs to be adjusted but also the position of the objective and the reference mirror separately. In addition, the two objectives need to be matched with a beam splitter to provide a wavefront with minimum aberration and maximum fringe contrast.

10.3.1 Alignment of Interferometric Objectives

In order to obtain fringes at best focus, the position of the reference mirror needs to be set also at the best focus of the objective. This is done in three steps (see Figure 10.7): first, the

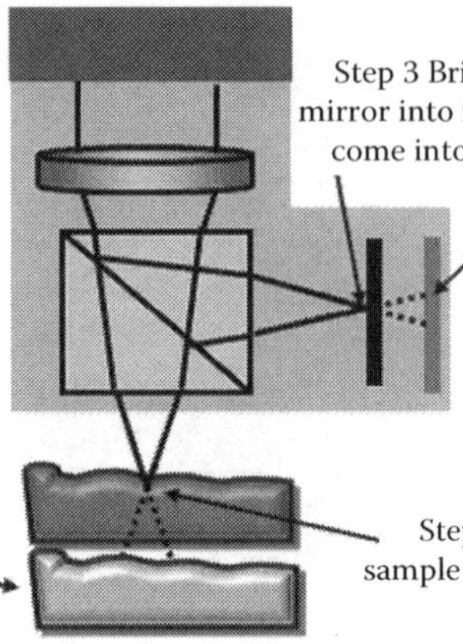

FIGURE 10.7
Procedure of interferometric objective alignment.

reference mirror is moved a few or tens of microns away from focus; second, the objective is focused on the object with some features like the edge of a sharp but not too tall step (fringes are not visible at this moment); and third, the reference mirror is brought to focus and stopped when best contrast fringes are obtained. When best contrast fringes and best sample focus are achieved, the zero OPD between arms of the interferometer coincides with best focus. This procedure is done in the factory, and the position of the reference mirror is then locked before shipping the system.

10.4 Calibration of WLI 3D Microscope

In order to provide precise and accurate measurements of surface topography, each microscope needs to be calibrated not only in the lateral but also vertical dimensions. Calibration is commonly done using special "artifacts" with certified heights and lateral features. Calibration artifacts need to be handled with care and stored and cleaned in accordance with manufacturer's instructions.

10.4.1 Lateral Calibration

Lateral calibration in x and y direction serves as a magnification calibration and is done for each objective and field of view lens. A calibrated pattern on the artifact is measured, and its lateral scale is set to match the lateral size of features via a magnification value setting. The pattern may be a grating composed of parallel lines or concentric circles.

Nowadays, systems are designed for the infinity-corrected objectives. It is worth noting that the magnification of the objective is not the value assigned to the objective but the combination of objective and the microscope's tube length. The tube length may vary between 160 and 210 millimeters, and thus, if the nominal magnification of the objective assigned by the manufacturer is based on a 160 mm tube length, then the magnification of this objective on a system with 210 mm tube length will be about 30% greater, as magnification equals tube length divided by the focal length of the objective.

10.4.2 Vertical Calibration

The vertical calibration relies on the proper settings of scanner speed, the cameras collected number of frames per second, and the known distance the scanner has traveled. This distance may be determined by the height of a calibrated artifact or by the interference fringes of a stable source wavelength.

Typically, such calibration is done via the measurement of a known step-height standard (SHS). Calibration is done by means of changing the scanner speed to achieve the value of the NIST (National Institute of Standards and Technology) or PTB (Physikalisch-Technische Bundesanstalt) traceable artifact to be within its allowable determined uncertainties. These standards are called secondary standards because their value is measured against a primary step-height standard established by NIST or PTB. Step-height standards come in different heights from nanometer through micron ranges. This type of calibration determines the average scanner speed and assumes that the scanner's speed remains constant both over the whole range of the scanner and across many measurements. For faster measurement of tall samples, scanner speed can be set even to approximately 100 μm/s (Deck and de Groot 1994, Schmit 2003). Often, the overall variations in scanner performance are mapped at the factory and corrected, but this correction may not remain stable over time.

Because of the number of possible different sources of uncertainty such as scanner inconsistency, nonlinearities, ambient temperature changes, and variances in secondary standard values, a primary standard can be embedded into WLI 3D microscopes that evaluates each moment of every measurement and continuously self-calibrates the system for maximum precision (Olszak and Schmit 2003). The primary standard is based on the wavelength of an additional embedded laser interferometer, which continuously monitors scanner motion, measures the wavelength of the detected fringes, or gives a trigger signal to collect images at constant distances. The repeatability of step-height measurement can be achieved on the order of only a few nanometers over a long period of time (days) and regardless of the changes in ambient temperature.

The need to rapidly inspect components by white-light interferometry has become increasingly important especially for high-volume manufacturing environments. For example, data storage firms use multiple machines continuously to screen hundreds or thousands of magnetic heads per hour in a reliable, automated way with a minimum of user control. In situations where multiple machines measure the same parts with sub-angstrom precision, repeatability and system to system correlation play a crucial role in production. In these cases, not only must the same measuring procedure and analysis be applied across all the systems but also the alignment and calibration of the systems must be well established and controlled.

10.5 WLI 3D Microscope Settings and Surface Properties to Be Measured

In order to correctly measure the sample's surface over the area, one has to have a strong knowledge of the system's abilities and limitations as well as the properties of the surfaces to be measured in order to assess if the system will be acceptable for the required measurement.

There is a wide range of options available when it comes to 3D measurement systems, but systems based on microscopes will most likely be best for measuring objects with small features, and an interferometric method of measurement is highly desirable for its unparalleled vertical precision and measurement speed. In other words, a WLI 3D microscope will be the best choice when roughness measurements, lateral resolution, vertical precision, and speed are important. When only form needs to be measured and lateral resolution larger than 10 μm is sufficient, then other methods not based on the WLI microscope should be considered.

Once the WLI 3D microscope has been chosen as the measurement instrument, one will have to make some decisions within the system. For example, the choice of objective will depend on the size of the total area to be measured and the lateral features, roughness, as well as the angle of local smooth slopes. While the magnification of the objective (plus intermediate optics and tube length) and size of the CCD camera determine the measurable area, the numerical aperture (NA) of the objective determines what lateral features on the object can be measured and the maximum measurable smooth slopes. In this section, we discuss a wide range of sample features that can be successfully measured with a white-light interferometric microscope. In addition, we discuss if the presence of film and dissimilar materials on the sample, or even the cover glass above the sample, will require some special consideration and solutions.

10.5.1 Vertical Size

We often think of microscopes as examining very small samples, a smear of blood on a glass slide slipped under the aperture like in high school biology. However, nowadays, small features needing measurement are often part of much larger systems. In these cases, the setup of the microscope needs to be much more flexible. One solution is to mount the WLI 3D microscope on a large frame platform, so the sample can fit under the objective; alternately, the WLI 3D microscope must be designed to be portable so that it can be brought to large samples like printer rollers or engines. These systems also require the ability to tilt at large angles so that the objective is roughly perpendicular to the surface as shown in Figure 10.8.

10.5.2 Vertical Resolution

Samples that require high-precision measurement should be tested using an interferometric technique because of this method's ability to deliver results with nanometer or a fraction of a nanometer vertical resolution. Interferometric objectives, in contrast to methods based on bright-field objectives, deliver excellent vertical resolution regardless of their magnification (numerical aperture).

The quality of the vertical measurement with bright field objectives will vary depending on the numerical aperture/magnification of the objective. The precise measurements can only be achieved using the highest, 50× or 100×, magnifications since only for these objectives is a very small portion of object imaged with sharp focus. Clearly, different parts of the object will come into focus with the vertical motion of the objective providing the information to make a measurement. These high-magnification bright-field objectives have a small depth of field on the order of a few microns. For low-magnification objectives, 5× magnification, for example, the whole sample may be in sharp focus at once. In this case, determining the best focus plane is impossible because of the large depth of field.

The quality of the vertical measurement with interferometric objectives is determined by the width of the fringe envelope and not the depth of field of the objective. The width of the fringe envelope, which is based on the bandwidth of the light source, typically on the

FIGURE 10.8
WLI 3D microscope system on large platform adapted to the measurement of significantly titled surfaces on large objects.

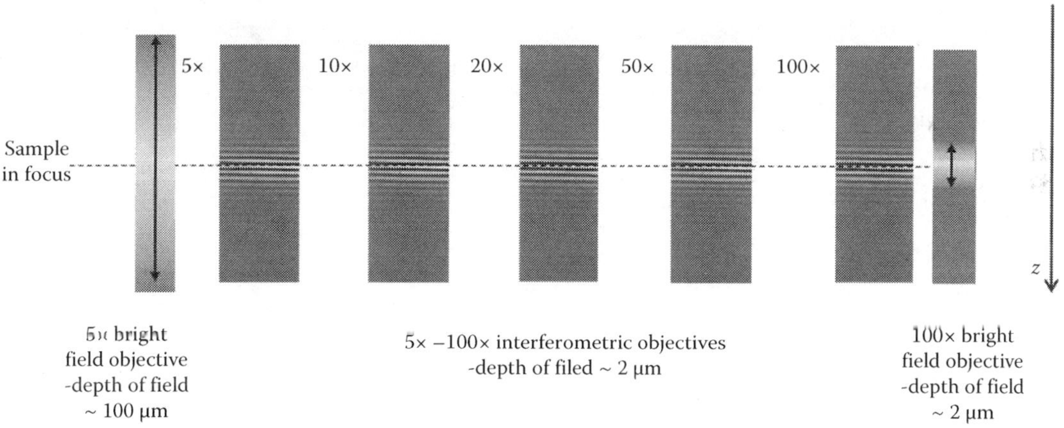

FIGURE 10.9
White-light fringe envelope spans a few microns around the best focus for any magnification objective, while the depth of field of non-interferometric objectives can be as large as 100 μm for 5× magnification and only can go down to a few microns for the highest magnifications.

order of 1–2 μm, is the same for any objective as shown in Figure 10.9; this allows for the same great height precision measurements regardless of the objective.

Nanometer and sub-nanometer level measurement precision requires stable environmental conditions; an anti-vibration air table is required to achieve optimal precision. The low noise of the measurement allows for the excellent vertical resolution but creates some susceptibility to vibrations in the form of ripples following the fringes. Non-interferometric measurements typically have a noise level that is higher than the level of ripples coming from vibrations, and thus, these ripples are not noticeable in the measurement noise of non-interferometric systems but at a cost of vertical resolution.

With white-light interferometric methods, sub-nanometer vertical resolution can be achieved but only for smooth, close to mirrorlike quality surfaces. For these smooth surfaces, the phase of the fringe maximum carries information about object height; measuring this phase position enables the sub-nanometer precision. For stepped, smooth surfaces in addition to measuring the envelope peak, the phase of fringes can also be measured (de Groot et al. 2002, Harasaki et al. 2000); this allows for the measurement of tens of microns tall steps and at the same time sub-nanometer level measurement on step surfaces. Figure 10.10 shows an example of locally smooth surface measurement, first with fringe envelope peak detection and then with combined fringe envelope and phase detection. Measured roughness from a few nanometers reduces to a fraction of nanometers. Figure 10.11 shows the result of 4 nm tall binary grating measured with a 20× objective and a fringe envelope peak and phase detection method. The noise level of this method easily allows for the measurement of a single nanometer tall grating.

Rougher surfaces (above roughness Ra = 50 nm) can only be measured using peak envelope position because the phase information loses its meaning on these rough surfaces (Dresdel et al. 1992). Measurement of optically smooth surfaces (roughness Ra below 20 nm) may be

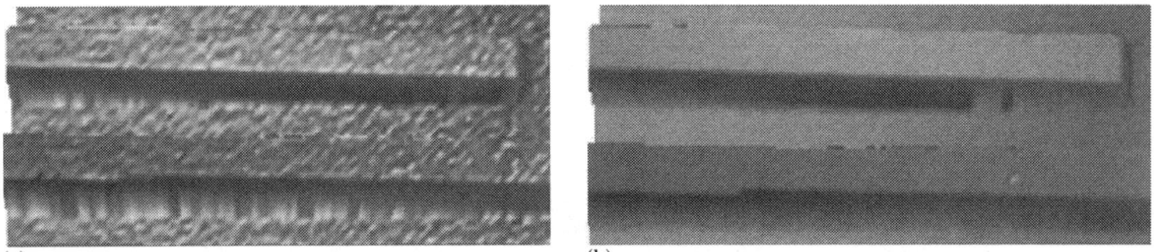

(a) (b)

FIGURE 10.10
Surface roughness of smooth surface calculated with algorithm based on (a) fringe envelope peak detection and (b) combined fringe envelope peak and phase detection.

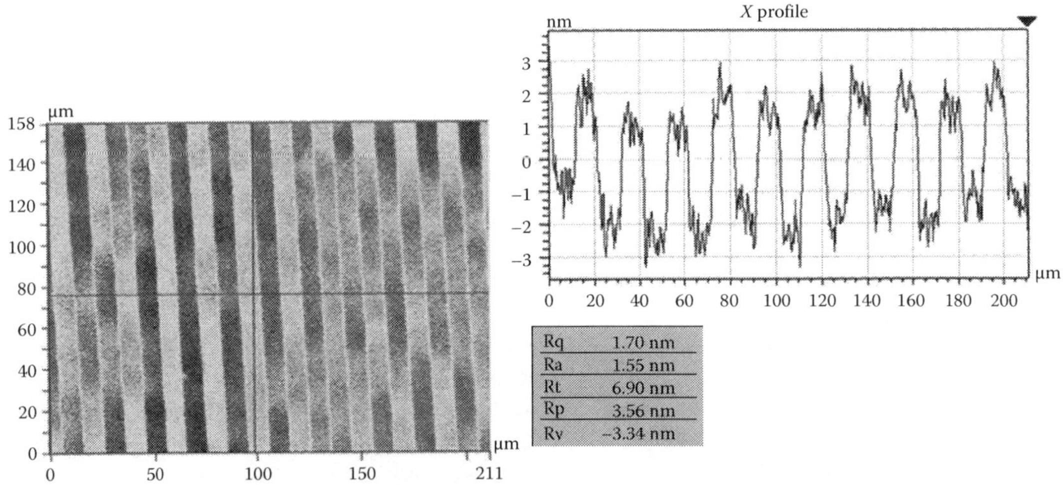

FIGURE 10.11
Four nm tall binary grating measured using a fringe envelope peak and phase detection method that yields results sub-nanometer vertical resolution (meaning very low noise).

based only on the phase of the fringes; this method is called phase shifting interferometry (PSI) and does not require fringes with decreasing amplitude of the envelope. However, the limitation of PSI is that only objects that have discontinuities less than about quarter of the wavelength (typically around 150 nm) can be measured. A significant advantage of PSI is that measurement time is only on the order of a fraction of a second and vertical resolution easily reaches sub-angstroms. Many article and book chapters have been written on phase shifting interferometry (i.e., Schreiber and Bruning 2007)

10.5.3 Lateral Size

The size of the measured area depends on the magnification of the system and the camera. In general, objectives of lower magnification can measure samples up to 20 mm × 20 mm, while the highest magnifications (100× objective) have fields of view of about 50 μm × 50 μm. A stitching procedure based on multiple measurements with a slight lateral overlap allows for the measurement of larger areas. Often, the stitching procedure is used if a high-magnification objective is required to resolve high slopes and small lateral features on the sample and larger areas also need to be measured. Figure 10.12 shows the measurement of roughly a 25 mm diameter Arizona quarter coin measured with a 5 mm field of view objective. Profiles show height changes from which lateral dimensions can be measured.

10.5.4 Lateral Resolution

Optical resolution determines the resolvable size of the lateral features on the object. This resolution depends only on the wavelength and the numerical aperture of the microscope objective and ranges from a few microns to a fraction of a micron for higher magnification objectives. Sparrow and Rayleigh (Born and Wolf 1975) cite slightly different criteria for an incoherent system imaging two radiating points as an object:

$$\text{Sparrow optical resolution criteria} = \frac{0.5 * \lambda}{\text{NA}} \tag{10.2}$$

$$\text{Rayleigh optical resolution criteria} = \frac{0.6 * \lambda}{\text{NA}} \tag{10.3}$$

Although these criteria are a good approximation for a WLI microscope, we have to remember that two lines can be better distinguished than two points, and for the WLI 3D microscope, height changes also influence the lateral resolution (de Groot and de Lega 2006). In addition, registering an image with a CCD camera requires at least three pixels in order to resolve an image of two points. Systems with low-magnification objectives may be limited by detector sampling. Thus, lateral resolution will depend on the larger of the optical resolution and the lateral sampling of the detector. Typically, microscope systems are optically limited for magnifications 20× and higher and detector limited for magnifications 5× and lower. Lateral resolution beyond the diffraction limits is difficult but possible to achieve on systems with sub-nanometer vertical resolution, an excellent scanner, and good knowledge of the system applied in the algorithm. Images in Figure 10.13 show the improvement in lateral resolution beyond the diffraction limits of 200 nm period grating, measured with an enhanced surface interferometric system and algorithm, over the

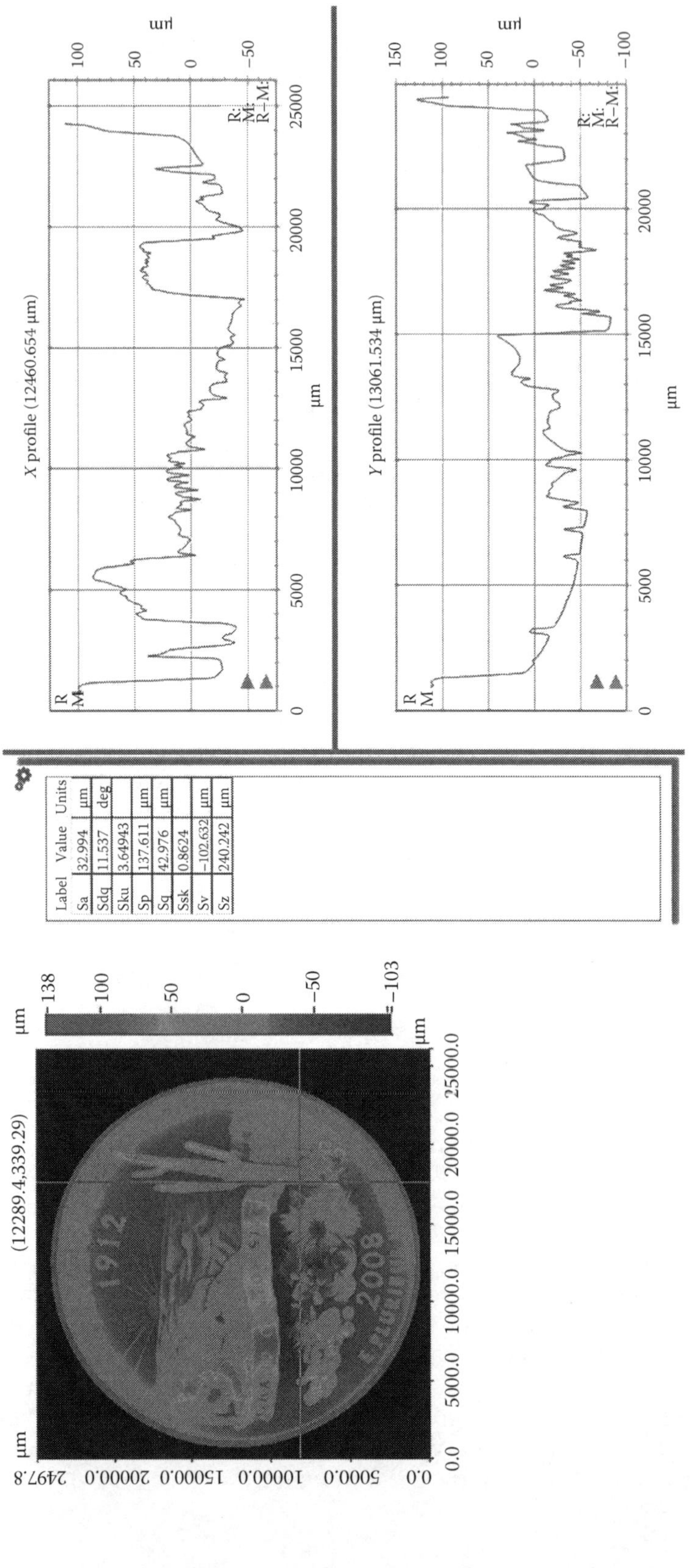

FIGURE 10.12
Topography of Arizona quarter coin measurement including two profile traces.

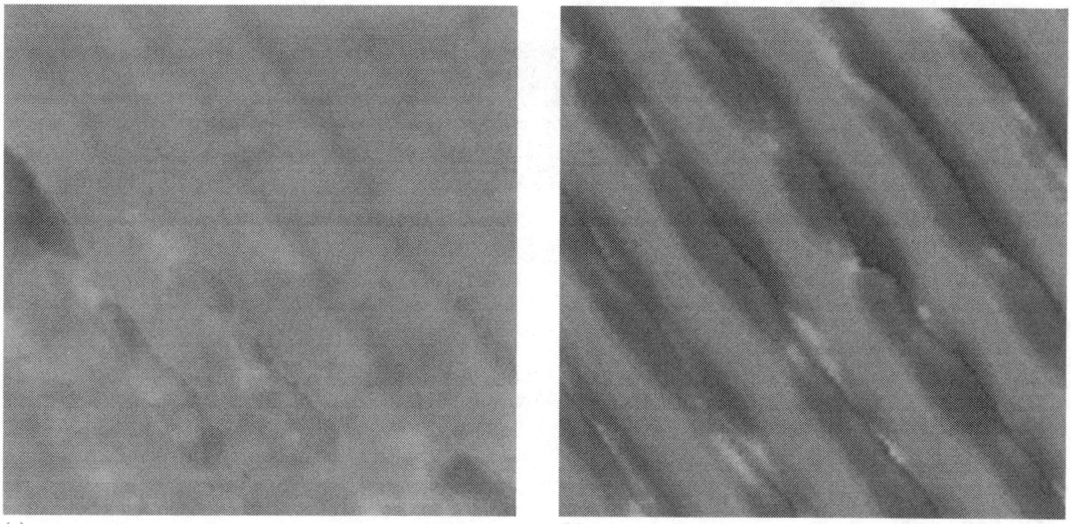

(a) (b)

FIGURE 10.13
Two hundred nanometer period binary grating as measured using (a) white-light interferometric measurement resulting in hardly resolvable features and (b) enhanced interferometric measurement with improved lateral resolution.

regular measurement (Novak and Munteanu 2011). Excellent vertical and lateral resolution allows for measurement of fine structures as well as small defects and scratches on solar cells, plastic films, and any other high-quality surfaces.

10.5.5 Lateral Dimensions

With information about object height, it is easy to distinguish, for example, traces, and automatically determine their widths, lengths, and distances as shown in Figure 10.14. In order to improve the precision of the measurement, the height map of each sample can be aligned to the same orientation, which makes measuring the x and y dimensions easy and repeatable. Lateral and vertical dimensions along with other parameters like curvature or roughness can be calculated for each of the identified features on the object.

10.5.6 Surface Slopes

Surface topography refers to the different heights, slopes, and roughness present on a sample. Regardless of the sample topography, light must come back to the objective in order for the surface features to be measured. The amount of light gathered by the objective depends on its NA. The limits on the maximum measurable slope differ depending on whether the surface being measured is smooth or rough. A smooth, mirrorlike sloped surface reflects light in mainly one direction according to the specular reflection angle, while a rough surface reflects light in a range of directions dependent on the specific properties of the surface. As shown in Figure 10.15, if light that is reflected from the sloped smooth surface totally misses the objective, the slope cannot be measured. However, some light that is reflected from a similarly sloped but rough surface will make its way to the objective and measurement of the overall shape and slope is easy.

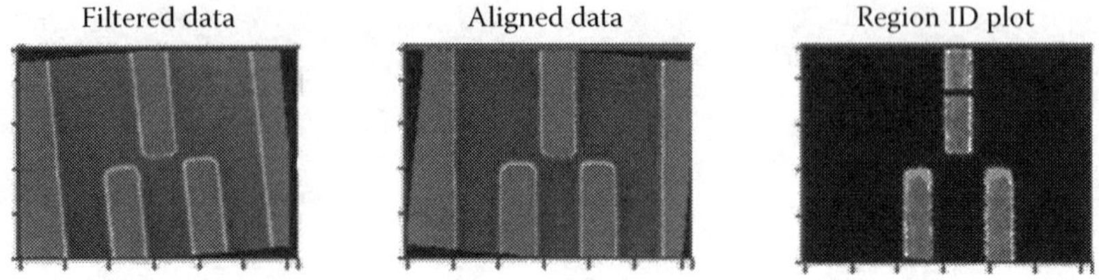

Height		Lateral	
Very top avg height	381.62 nm	Very top Y width	4.35 μm
Top avg height	385.11 nm	Very top X width	2.98 μm
Left avg height	386.08 nm	Top Y width	6.32 μm
Right avg height	382.96 nm	Top X width	3.03 μm
		Left X width	3.09 μm
		Left Y width	9.89 μm
		Right Y width	10.21 μm
		Right X width	3.06 μm

FIGURE 10.14
Measured traces before and after rotational alignment and with identified traces regions and their vertical and lateral parameters calculated.

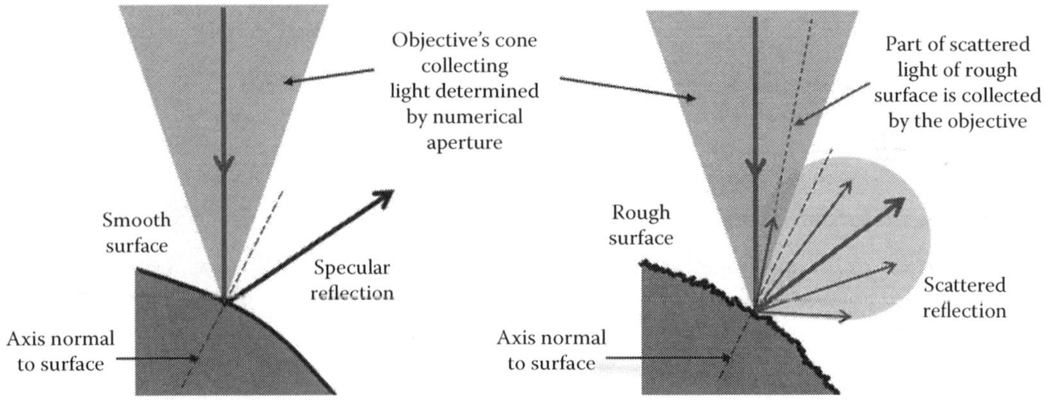

FIGURE 10.15
Scattered light reflecting from a rough surface allows for the measurement of higher slopes.

The common maximum NA for an interferometric objective is about 0.8 for magnifications 50–115× objectives. With this NA, smooth slopes up to 35° can be measured. Thus, a wide range of microlenses, Fresnel lenses, solar cells, and patterned sapphire structures for LED manufacturing can be successfully measured (see Figure 10.16).

In addition, objects with rough surfaces and higher slopes (up to 60°–70°) like some holographic films and woven materials can also be measured since rough surfaces allow some light to travel back to the objective.

WLI 3D microscopes for engineered surfaces must account for the fact that the sample often is not flat and may have surfaces at different angles. For this reason, the sample needs to be placed on a stage that can be tilted. Alternately, the objective can be tipped and tilted around the sample to accommodate the surface angles. During the relative tip and

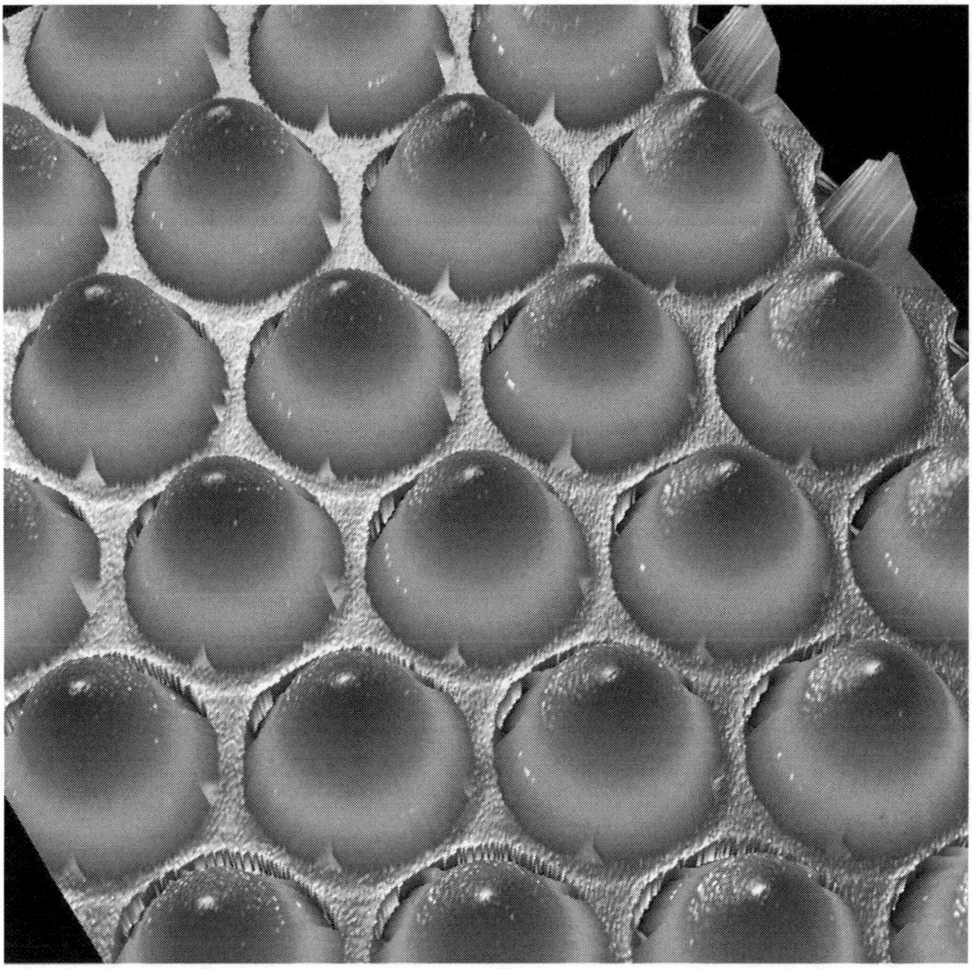

FIGURE 10.16
Patterned sapphire structure with elements about 3 μm tall and 1.5 μm wide measured with 115×, 0.8 NA objective using a special combination of fringe envelope and phase detection. Results agree very well with atomic force microscopy measurement.

tilt between the objective and sample, fringes will change density and shape as shown in Figure 10.17. For best measurement results with low noise, dense fringes should be avoided, which can often be done by changing the tip or tilt.

10.5.7 Dissimilar Materials

As long as the object's surface is comprised of a single material, the light has a consistent phase shift per wavelength upon reflection. However, when two dissimilar materials are side-by-side on the surface, they will have different phase shifts Φ upon reflection for different wavelengths (unless both of them are dielectric materials with the imaginary part of the index of refraction $k = 0$ and thus transparent), and the measured height difference at the boundary where the two meet will be incorrect (Doi et al. 1997). This difficulty can be overcome by knowing the optical constants of the different materials

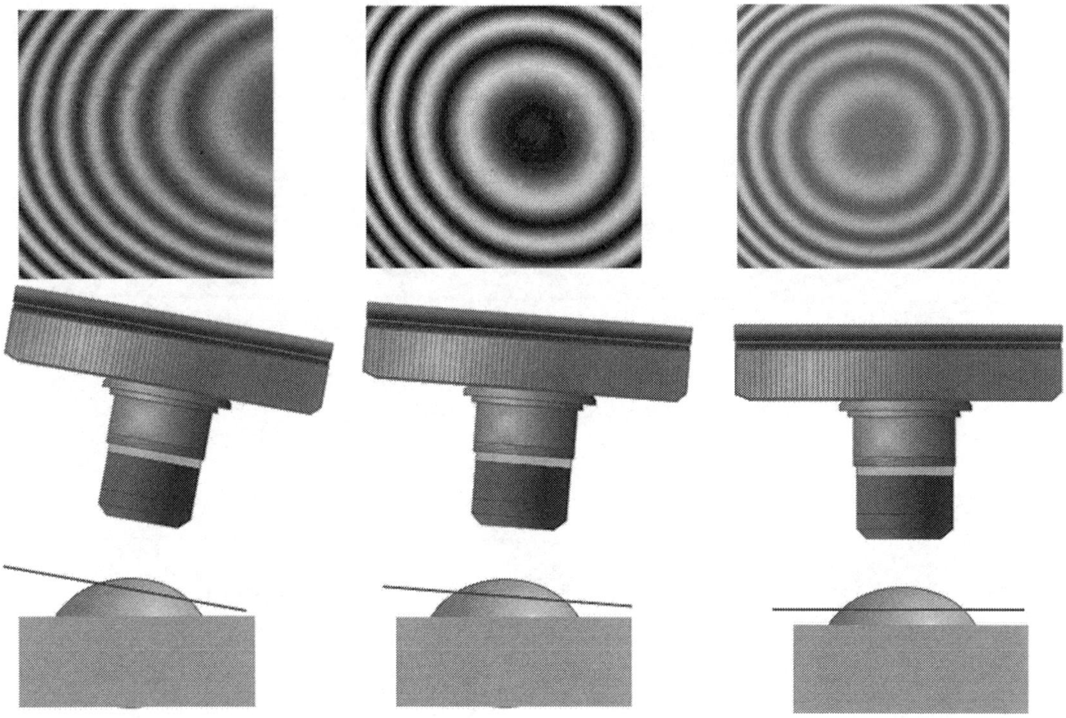

FIGURE 10.17
Tilt of interferometric objective with respect to sample changes the fringe frequency, shape, and direction. Tilt can be adjusted for better measurement of sloped surface.

for the wavelengths used in the measurement and correcting for this difference. Using monochromatic illumination, which has only single values for n and k, makes this calculation easier because the height difference can be calculated from the following equations (Born and Wolf 1975):

$$\Phi(n,k) = \arctan\left(\frac{2k}{1-k^2-n^2}\right) \tag{10.4}$$

$$\Delta h = \frac{\lambda}{4\pi}\Delta\Phi = \frac{\lambda}{4\pi}\left[\Phi(n_1,k_1) - \Phi(n_2,k_2)\right] \tag{10.5}$$

Figure 10.18 shows the phase change upon reflection effect on a plane wavefront incident on a flat sample made of two different materials.

In white-light interferometry, different materials that make up the sample will shift the peak of the envelope (metals more so than other materials) and possibly even change the shape of the envelope (for example, gold and some semiconductors). However, for most materials, this shift will not be larger than 40 nm (Harasaki et al. 2001, Park and Kim 2001). For example, a trace made of silver on glass will appear about 36 nm taller and aluminum about 13 nm taller. These offsets often may be negligible for the measurement and can be disregarded completely.

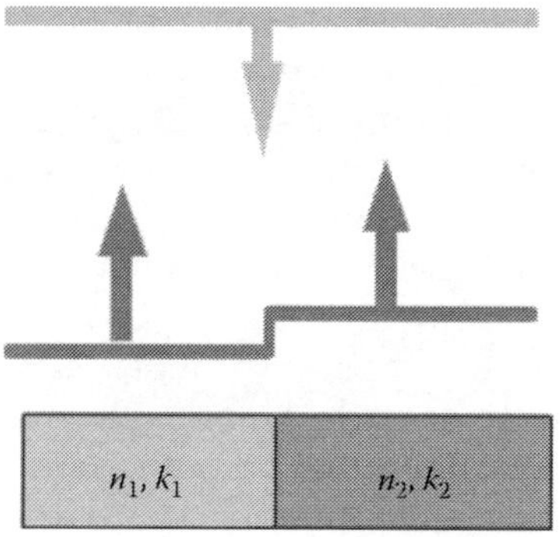

FIGURE 10.18
The phase of the wavefront reflected off materials with different n and k complex indices of refraction may be different for each material and must be accounted for when measuring surface height.

Apart from accounting for different phase shifts using constants, several alternate ways to achieve accurate measurements exist. The phase shifts can be determined by comparing the WLI result to a contact measurement like a stylus profiler (Schmit et al. 2007) and subsequently compensate for the shifts in the algorithm. Additionally, the sample may be coated with a nontransparent 100 nm layer of gold and then measured with the 3D microscope before and after coating to determine the shift. Finally, a mold of the area using silicon rubber can be used to create a replica of the sample. This replication method is also useful when the sample cannot be brought to the microscope.

10.5.8 Thick Film Measurement

If the sample is covered with a transparent film that is more than a few microns thick, two sets of localized fringes separated from each other are generated, one for each interface as shown in Figure 10.19. However, fringes for the second interface are localized below this interface by roughly half of the product of the group index of refraction for white light and the geometrical thickness of the film. A simple technique for finding the relative position

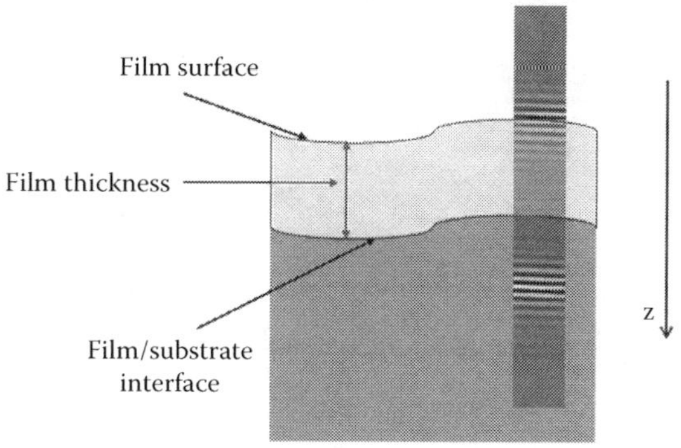

FIGURE 10.19
White-light fringes for a few micron transparent films with emphasis that the film thickness as well as its top and bottom topography can be measured.

of the peaks of the fringe envelopes can be implemented to find the thickness of the film as well as topography of the top surface and interface.

The typical range of measurable film thicknesses runs from 2 to 150 μm depending on the dispersion of the film and the NA of the objective. Typically, one sigma of repeatability for a film thickness measurement is about 6 nm. For the measurement of thicker films, objectives with lower NAs as well as illumination spectrums narrower than white light should be used to improve the contrast of fringes from the film-substrate interface. For the measurement of thinner films, a full spectrum of white light should be utilized, and a higher numerical aperture objective may be helpful. Of course it is possible to measure the thicknesses of individual layers in multiple layer coatings as long as the fringes for each layer are detected and separated. White-light fringes will also be created at defects; thus, this method can also be used for detection and localization of defects in films.

10.5.9 Measurement through the Glass Plate or Liquid Media

Some products need to be tested in their final stage, which often means the testing occurs through a protective liquid, plastic, or glass. Many engineering objects, like MEMS devices, are protected by a cover glass, and some devices in environmental chambers need to be tested under different pressures or temperatures. Biological and some mechanical samples are often immersed in liquid and require measurement through this liquid (Reed et al. 2008a,b). Because of the dispersion of the liquid, plastic, or glass, the white-light fringes may be totally washed out (see Figure 10.20). For this reason, a compensating plate needs to be placed into the reference arm of interferometer. This compensation is the most easily done for Michelson (see Figure 10.21) and Linnik-type objectives; however, compensating plates for Mirau-type objectives also exist. Compensating plates also can help with measurements of thick films as well as samples in liquid.

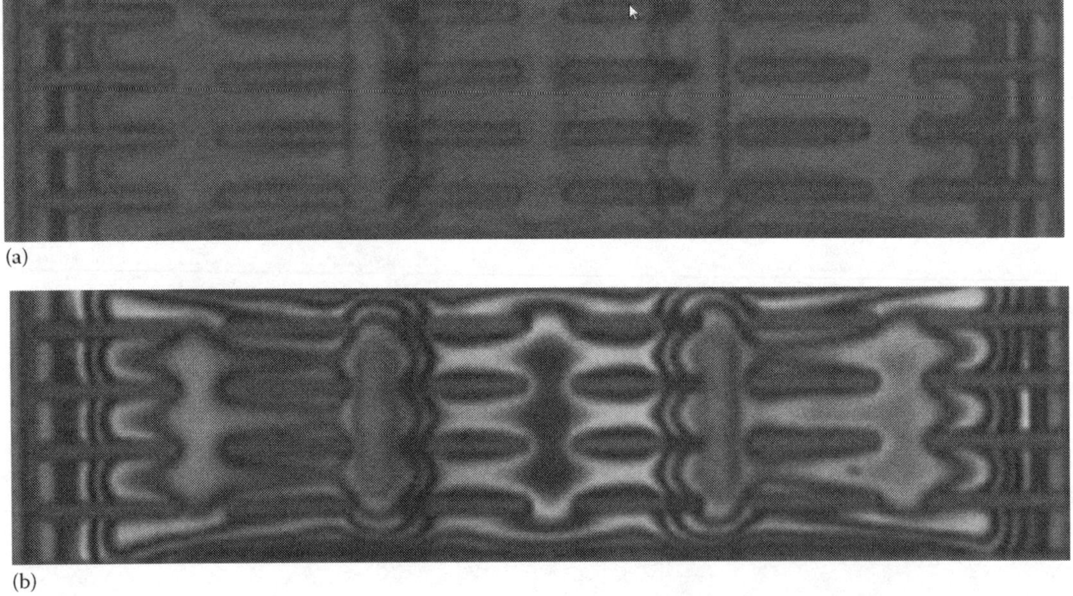

(a)

(b)

FIGURE 10.20
Image of fringes on engineering part taken using 10× interferometric objective through the 300 μm thick glass (a) without and (b) with glass compensation in reference arm of the objective.

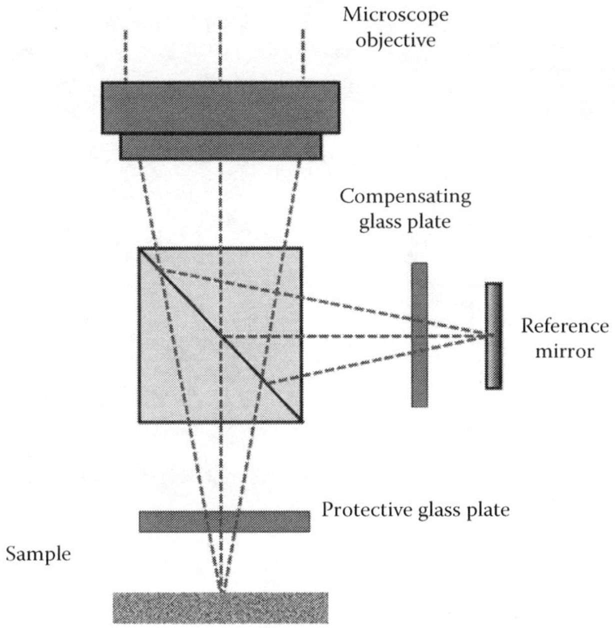

FIGURE 10.21
Michelson-type objective with a compensation plate for a measurement through the protective glass.

The effects of glass thicknesses up to 3 mm can be easily compensated for; however, for larger numerical apertures, the compensating glass thickness has to match the properties and thickness of the cover glass more precisely. Using a spectral bandwidth that is narrower than a white-light spectrum will result in better fringe contrast, and it may loosen the tolerance for incongruencies between the cover glass and the compensation glass but may be at the expense of some loss of resolution by a few nanometers.

Further improvements in thick film and through the glass measurements can be achieved by using specialized collimated illumination (see Figure 10.22) (Han and Novak 2008). This type of illumination also helps measure deep structures on a sample as shown in Figure 10.22 as well biological cells and their mechanical properties (Reed et al. 2008a,b).

10.6 3D Microscopes Based on Interference: Nomenclature

3D microscopes based on white-light interferometry may go by different names in the literature and between producers. Here is an alphabetical list of names that the reader may find. The names with their acronyms in parenthesis are most commonly used. ISO standards documents, when approved, will refer to this method as coherence scanning interferometry (CSI):

Broad-bandwidth interferometry

Coherence correlation interferometry

Coherence probe microscopy

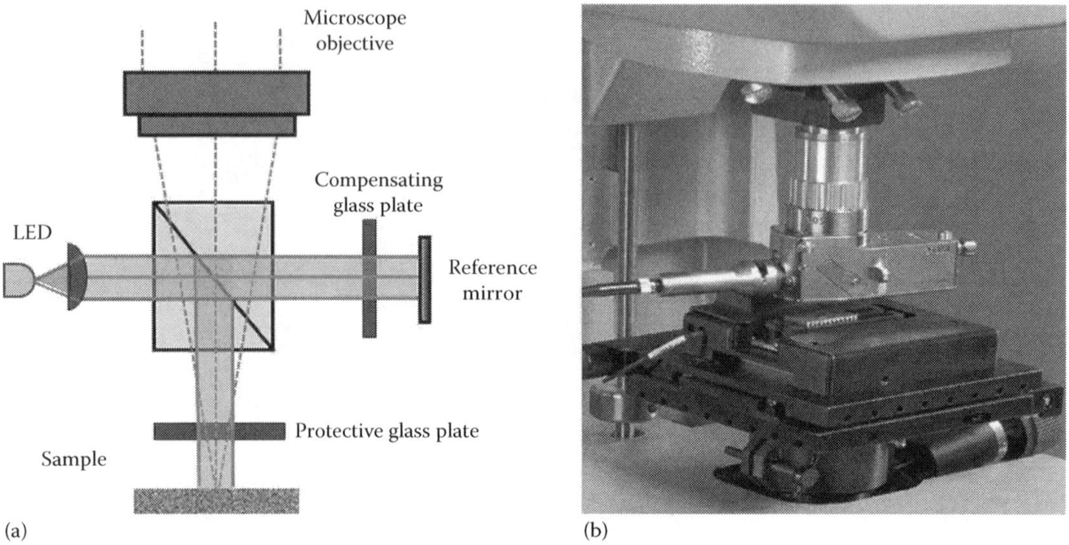

FIGURE 10.22
Michelson-type objective with a compensation plate for a measurement through the protective glass and with patented collimated illumination (shaded path) provided from under the objective for better sample penetration (a) diagram (b) photo of objective.

Coherence radar
Coherence scanning microscopy
Coherence scanning interferometry (CSI)
Mirau correlation microscopy
Fringe peak scanning interferometry
Height scanning interferometry
Interference microscope
Low coherence interferometry (LCI)
Microscopic interferometry
Optical coherence profilometry
Optical coherence microscopy
Phase correlation microscopy
Rough surface tester
Scanning white-light interferometer (SWLI)
White-light interference 3D microscope
White-light interferometry (WLI)
Vertical scanning interferometry (VSI)
White-light scanning interferometry (WLSI)
Wideband interferometry

The equivalent method for biological samples is called optical coherence tomography (OCT) but also can be called time domain OCT (TD-OCT), coherence radar, or confocal interference microscopy. This method is mainly used with near infrared illumination for better sample penetration, and point by point scanning is used. However, in some cases, systems based on full-field, white-light microscopy are used and are called full-field or wide-field OCT systems.

10.7 Summary

WLI 3D microscopy over the last two decades has become an indispensible tool both in research labs and on production floors. The exacting specifications of many high-tech products require the ability to see, sometimes on a nanometer scale, object shape and roughness. WLI 3D microscopes provide rapid, noncontact measurements with high repeatability, reproducibility, and system to system correlation; these factors facilitate excellent production-floor throughput and quality. The systems are easy to use and deliver precise measurements for a wide range of samples that require lateral resolution on the order of microns.

For larger objects, the WLI 3D microscope even has an alternate platform design to accommodate these large-scale measurements. Finally, the rather small measured area inherent to all microscopes can be extended by taking multiple overlapping measurements and stitching images together. However, if a measurement does not include roughness, small features, or high local slopes, then other methods that are not based on the microscope setup could be considered; alternate methods should shorten measurement time, but nanometer level vertical resolution will be sacrificed; look for resolutions at hundreds of nanometers or microns.

Acknowledgment

Thank you to Bruker Nano Surfaces Division for providing many of the pictures for this chapter.

References

Ai, C. and E. Novak, Centroid approach for estimation modulation peak in broad-bandwidth interferometry, U.S. Patent 5,633,715 (1997).

Born, M. and E. Wolf, *Principles of Optics: Electromagnetic Theory of Propagation, Interference and Diffraction of Light*, Cambridge University Press, Cambridge, U.K., p. 415 (1975).

Caber, P.J., Interferometric profiler for rough surfaces, *Appl. Opt.*, 32, 3438 (1993).

Danielson, B.L. and C.Y. Boisrobert, Absolute optical ranging using low coherence interferometry, *Appl. Opt.*, 30, 2975 (1991).

Deck, L. and P. de Groot, High-speed noncontact profiler based on scanning white-light interferometry, *Appl. Opt.*, 33(31), 7334–7338 (1994).

Doi, T., K. Toyoda, and Y. Tanimura, Effects of phase changes on reflection and their wavelength dependence in optical profilometry, *Appl. Opt.*, 36(28), 7157 (1997).

Dresdel, T., G. Hausler, and H. Venzke, Three dimensional sensing of rough surfaces by coherence radar, *Appl. Opt.*, 31(7), 919–925 (1992).

de Groot, P. and L. Deck, Surface profiling by analysis of white light interferograms in the spatial frequency domain, *J. Mod. Opt.*, 42, 389–401 (1995).

de Groot, P. and X.C. de Lega, Signal modeling for low coherence height-scanning interference microscopy, *Appl. Opt.*, 43(25), 4821 (2004).

de Groot, P. and X.C. de Lega, Interpreting interferometric height measurements using the instrument transfer function, *Proc. FRINGE 2005*, Osten, W. Ed., pp. 30–37, Springer Verlag, Berlin, Germany (2006).

de Groot, P., X.C. de Lega, J. Kramer et al., Determination of fringe order in white light interference microscopy, *Appl. Opt.*, 41(22), 4571–4578 (2002).

Han, S. and E. Novak, Profilometry through dispersive medium using collimated light with compensating optics, U.S. Patent 7,375,821, May 20 (2008).

Harasaki, A., J. Schmit, and J.C. Wyant, Improved vertical scanning interferometry, *Appl. Opt.*, 39(13), 2107–2115 (2000).

Harasaki, A., J. Schmit, and J.C. Wyant, Offset envelope position due to phase change on reflection, *Appl. Opt.*, 40, 2102–2106 (2001).

ISO 25178-604, Geometrical product specification (GPS)—Surface texture: Areal- part 604: Nominal characteristics of non-contact in (coherence scanning interferometry), instruments, International Organization for Standardization (2012). http://www.iso.org/iso/home/store/catalogue-tc

Kino, G.S. and S.S.C. Chim, The Mirau correlation microscope, *Appl. Opt.*, 29(26), 3775–3783 (1990).

Larkin, K.G., Efficient nonlinear algorithm for envelope detection in white light interferometry, *J. Opt. Soc. Am. A*, 13(4), 832–842 (1996).

Novak, E. and F. Munteanu, Application Note #548, AcuityXR technology significantly enhances lateral resolution of white-light optical profilers, http://www.bruker-axs.com/optical_and_stylus_profiler_application_notes.html (2011). (Last accessed on october 20, 2012).

Olszak, A.G. and J. Schmit, High stability white light interferometry with reference signal for real time correction of scanning errors, *Opt. Eng.*, 42(1), 54–59 (2003).

Park, M.-C. and S.-W. Kim, Direct quadratic polynomial fitting for fringe peak detection of white light scanning interferograms, *Opt. Eng.*, 39, 952–959 (2000).

Park, M.-C. and S.-W. Kim, Compensation of phase change on reflection in white-light interferometry for step height measurement, *Opt. Lett.*, 26(7), 420–422 (2001).

Reed, J., M. Frank, J. Troke, J. Schmit, S. Han, M. Teitell, and J.K. Gimzewski, High-throughput cell nano-mechanics with mechanical imaging interferometry, *Nanotechnology*, 19, 235101 (2008a).

Reed, J., J.J. Troke, J. Schmit, S. Han, M. Teitell, and J.K. Gimzewski, Live cell interferometry reveals cellular dynamism during force propagation, *ACS Nano*, 2, 841–846 (2008b).

Schmit, J., High speed measurements using optical profiler, *Proc. SPIE*, 5144, 46–56 (2003).

Schmit, J., K. Creath, and J.C. Wyant, Surface profilers, multiple wavelength and white light interferometry, in *Optical Shop Testing*, Malacara, D. Ed., 3rd edn., Chapter 15, pp. 667–755, John Wiley & Sons, Inc., Hoboken, NJ (2007).

Schreiber, H. and J.H. Bruning, Phase shifting interferometry, in *Optical Shop Testing*, Malacara, D. Ed., 3rd edn., Chapter 14, pp. 547–666, John Wiley & Sons, Inc., New York (2007).

11

Focus-Based Optical Metrology

Kevin Harding

CONTENTS

11.1 Introduction

A method of optical metrology perhaps not as widely used as laser triangulation, structured light, and machine vision is to use information about the focus of an optical system to provide a means of measurement. One of the simplest forms of this method is used in most autofocus cameras today. A simple analysis analyses an image based upon overall contrast from a histogram of the image, the width of edges, or more commonly the frequency spectrum derived from a fast Fourier transform of the image. That information might be used to drive a servo mechanism in the lens, moving the lens until the quantity measured on one of the earlier parameters is optimized. Moving from a fuzzy image to a sharp image is something just about anyone can do instinctively with a manual camera. Given an image like Figure 11.1, the left-hand image shows an in-focus image with distinctive surface texture evident, while the right-hand image is clearly out of focus and the surface detail is lost. The resulting FFT images shown in Figure 11.2 show more high-frequency content (Figure 11.2a) for the in-focus image versus the blurred image (Figure 11.2b).

The autofocus mechanism on cameras is intended to provide a clear image, not a measurement. For any distance subject, the depth over which an image is in focus will increase. This depth-of-field of an image, the distance over which the focus of a given size feature

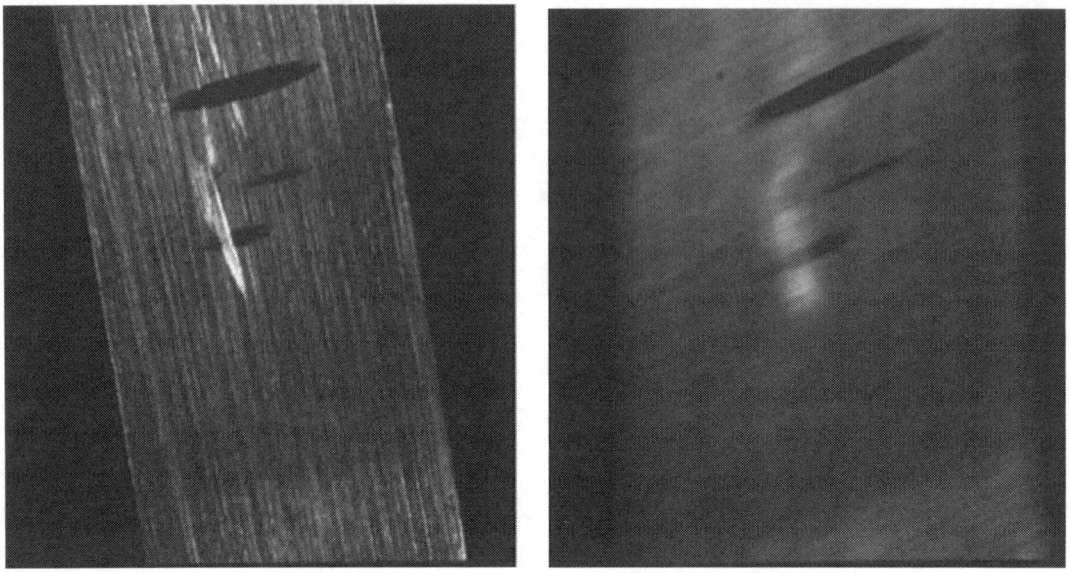

(a) (b)

FIGURE 11.1
The in-focus figure (a) shows more surface detail than the out-of-focus figure (b).

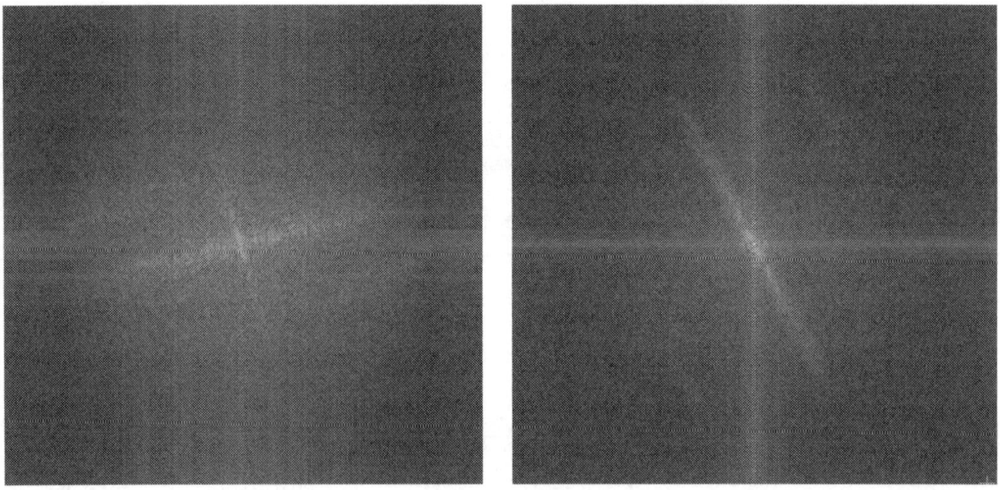

(a) (b)

FIGURE 11.2
The fast Fourier transform (FFT) of the in-focus image in Figure 11.1 (a) shows more high-frequency content than the FFT of the out-of-focus image (b).

does not appear to change, can be as large as infinity for an object sufficiently far away (see Chapter 2). For a subject close to the lens, the depth-of-field is a function of the lens F-number (cone angle of light collected) and the size of the feature. The result is approximately given by twice the F-number times the feature size (see Chapter 2). This implies the focus will only localize a 1 mm size feature within, perhaps, 22 mm or so for an f/11 system. Due to this relationship, depth from focus quality metrics has not often been used to measure larger subjects over meter-size scales. However, many enhancements have been made to focus metrics when measuring over very small areas and distances.

Focus-based systems can be considered in two categories:

1. Point-based systems
2. Area-based systems

Point-based systems are ones where only the focus attributes of a single point are used to make a measurement. This local point is typically restricted to cover only a few microns of size. Measurements of a part are made with such systems typically by scanning the point across the part. The means by which the distance measurements are made include the following:

- Comparison of the focal position relative to the lens focal length—referred to as conoscopy
- Comparison of light levels when the point is imaged through a restricting aperture—referred to as confocal imaging
- Comparison to a changing metric through focus such as chromatic shift—referred to as chromatic confocal imaging

In a like manner, the area-based methods largely fall into two approaches:

- Depth from focus—where only those features in best focus are used
- Depth from defocus—where the system analyzes how the feature focus changes

For any focus-based method, there is an inherent assumption that there is something in the image to be focused. Looking at a completely smooth surface, there may not be any edges, texture, or other features available, or the character of the light returned many not change. In these applications, focus-based methods may not be useable.

Features such as surfaced texture, grain structure, or other uniformly distributed patterns are typically best as a reference to be used in focus-based methods. In some cases, a pattern may be either imparted to the part surface, such as spray painting black and white dots on the surface, or a pattern of lines, dots, or other structure may be projected onto the part surface. Structured projection using triangulation is commonly used for 3D measurement, but in this case, there would be no triangulation angle, so the light projection and viewing axis may be the same. Active illumination patterns used with focus-based analysis allow the measurement of distance to be made in-line without shadows but typically do not provide the resolution of a triangulation-based method.

We will explore both the point-based and area-based methods in more detail.

11.2 Point-Based Distance Measurement

11.2.1 Conoscopic Imaging

The first method of focus point metrology, also referred to as conoscopic holography,[1,2] focuses a laser spot onto the subject then evaluates the wavefront shape of the beam using that focus point as a source. To understand the concept of operation, we consider first a simple interferometer arrangement such as shown in Figure 11.3. When the beam from the

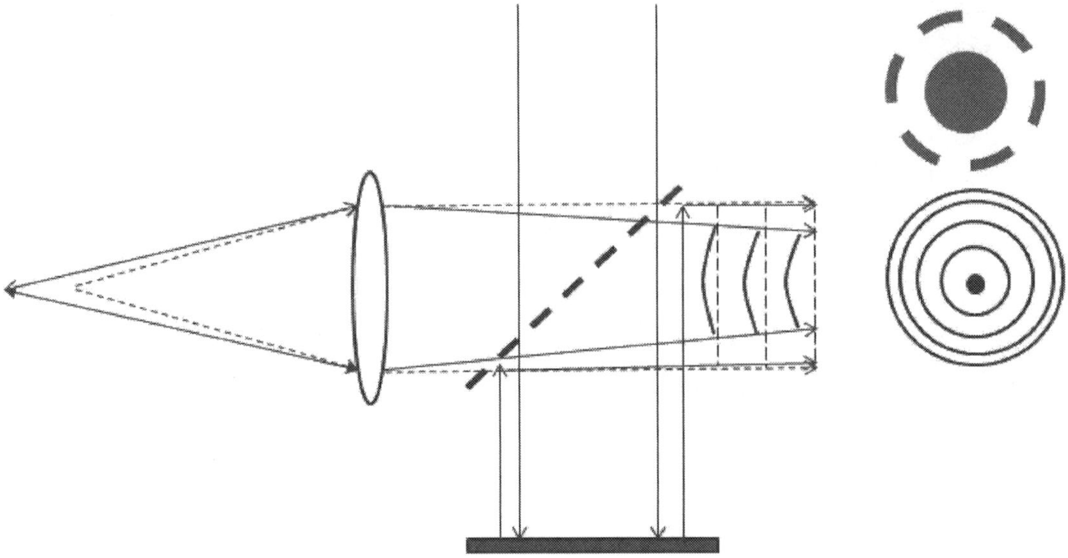

FIGURE 11.3
When the part is near one focal length from the lens, the interference pattern shows very few interference rings (upper right pattern) but increases as the wavefront becomes curved with the focus shift (lower right pattern).

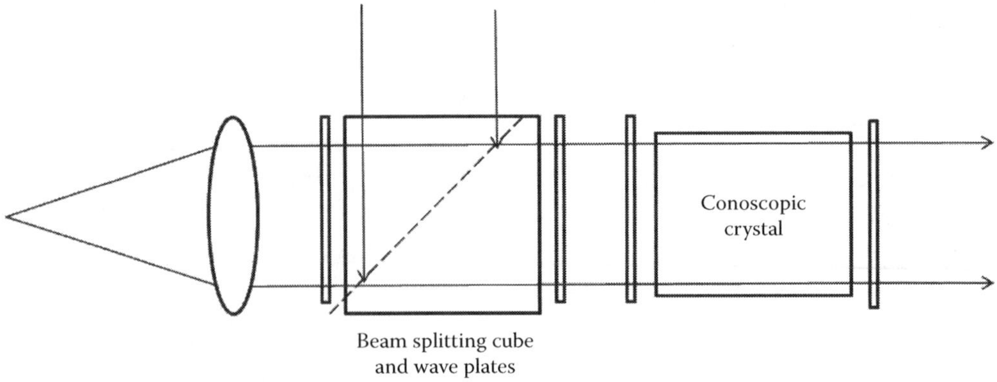

FIGURE 11.4
A common path conoscopic system for generating interference patterns that change frequency with distance.

lens is collimated (parallel rays), we know the focal point is at the focal length of the lens. As the spot moves beyond the focal length of the lens in either direction, the resulting wavefront changes from a collimated wave to a curved one. In a simple interferometer, the wavefront from the subject spot creates interference fringe in the form of circular patterns (see Figure 11.3). In practice, the conoscopic sensor uses a crystal rather than a reference mirror (see Figure 11.4). The crystal spits the beam into two wavefronts by polarization (e and o beams), slightly displaced along the light path, that interfere to form the patterns. The advantage of the conoscopic arrangement is it is a common path interferometer with very little path difference between the two beams at any distance, so very little coherence is required, and the system is very stable. By analyzing the frequency of these interference lines, the distance to the point can be measured.

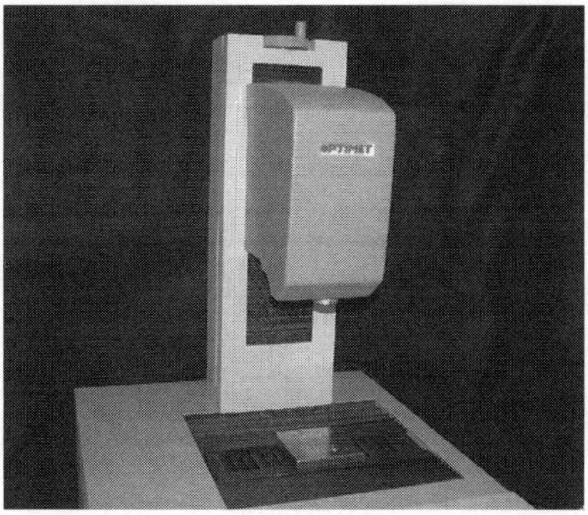

FIGURE 11.5
Scanning system using a conoscopic sensor made by Optimet.

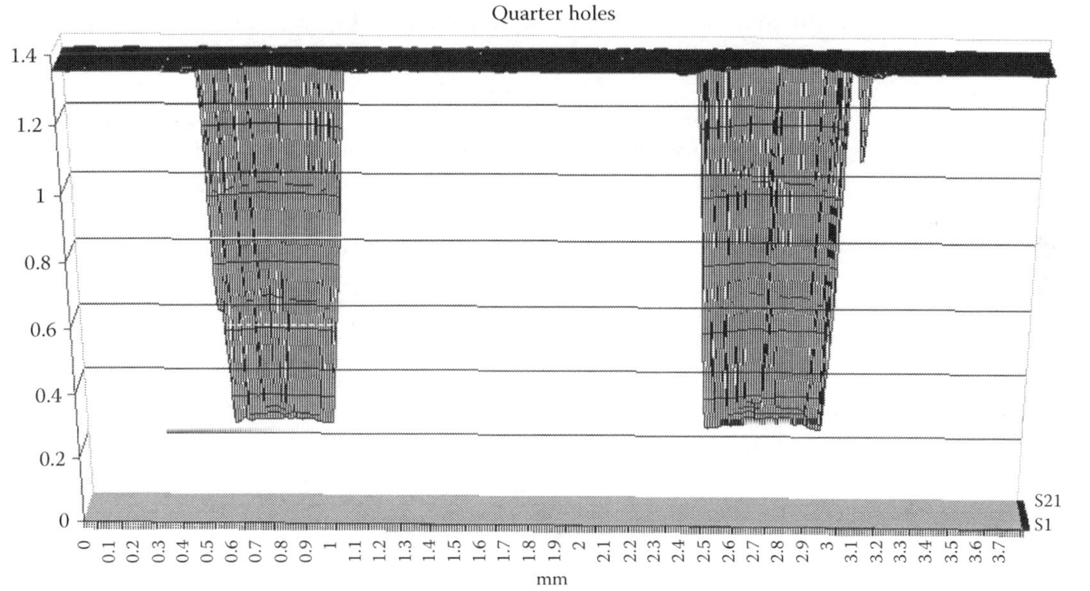

FIGURE 11.6
The ability to measure blind holes scanned by a conoscopic sensor are an advantage of the in-line measurement ability.

To measure a part, the sensor will typically be scanned over the part (see Figure 11.5). Since the measurement is in-line, requiring no triangulation angle, features such as a blind hole can be scanned (see Figure 11.6). A feature such as a corner can be scanned from the direction of the apex (Figure 11.7a), providing good data down each surface. The angle to the surface will eventually spread the spot across a range of depth, causing the depth measurement to be averaged. At incidence angles more than 45°, the spot is covering more depth than the width of the spot. A cylinder (Figure 11.7b) will begin to look parabolic

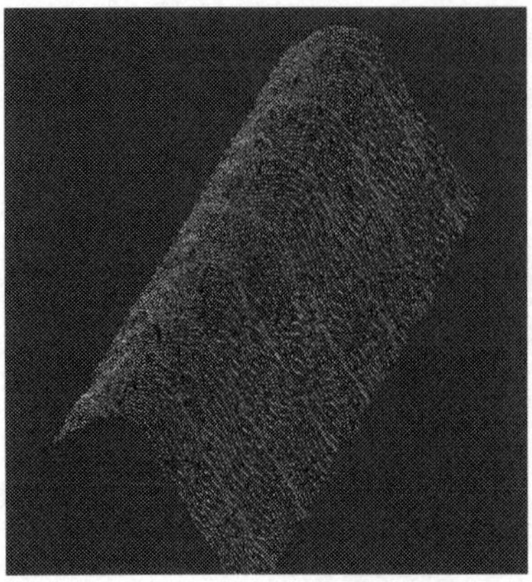

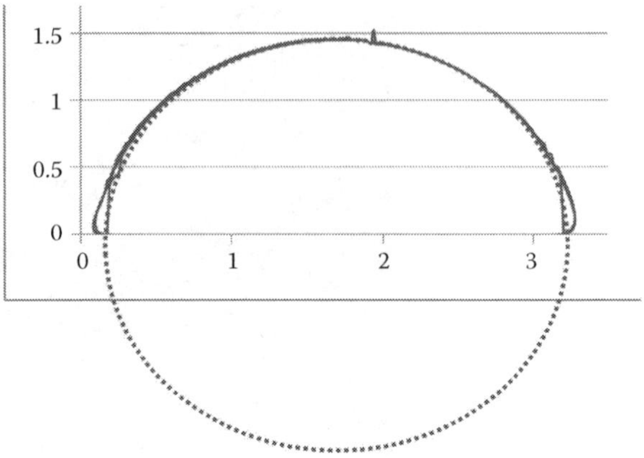

FIGURE 11.7
Multiple scans across an edge scanned with the conoscopic sensor (a) and overlaid scans of a 3 mm pin (b) showing the errors at increased angles relative to an ideal circle.

above such an angle, just based upon this geometric effect. Otherwise, as long as there is sufficient light level return, the sensor should be able to make a measurement into a high aspect ratio feature such as a hole or across an edge.

Typically, a laser beam is projected onto the part surface. The size of the laser spot will determine the in-plane resolution of the system from a practical perspective. That is, if a feature smaller than the laser focus spot is scanned, the spot will average the distance across that feature. Making small movements across the feature may provide some information about the feature, but this information is at best inferred.

For example, the rounded corner in Figure 11.7a is scanned with a probe using a focus spot only a few microns wide. The number of data points around the edge is sufficient to define the edge shape. But in Figure 11.8, a thin wire about 25 μm thick is scanned with a

FIGURE 11.8
A thin wire, scanned with a conoscopic probe with a focus spot comparable to the wire thickness. At the peak, the curvature is not correctly captured.

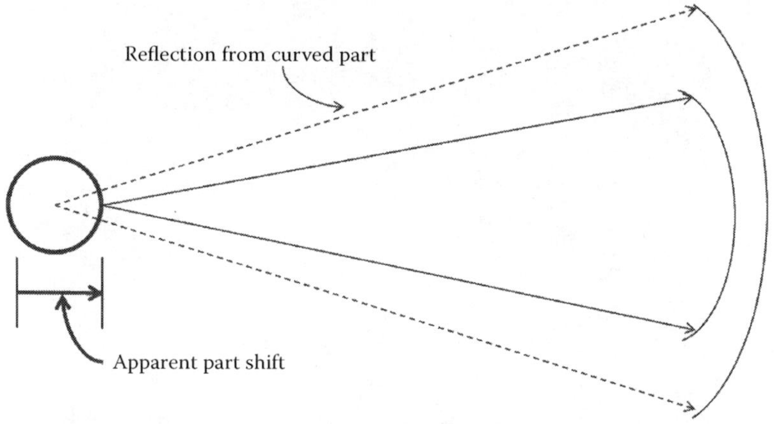

FIGURE 11.9
The shape of the wavefront may be modified by the surface of a feature the size of the laser focus, causing the distance measurement to be incorrectly interpreted.

laser spot size of about 30 μm. The width of the wire is averaged with the size of the laser spot. In addition, as the spot goes across the top of the wire, the pattern from the wire is incorrectly interpreted by the system due to insufficient sampling, or the wavefront may even be modified by the shape of the surface (see Figure 11.9), causing the system to show a flat spot or dip on the top of the wire that is not there. For a very sharp edge, the sensor will average over the edge making it appear larger than it actually is.

In order to measure a part, interference fringes are created, requiring that the light be coherent. If the surface is too scattering or translucent, the coherence of the light may be disrupted, and no measurement will be made. This is true of some plastic materials that let the light enter and scatter over multiple sites within the plastic. If the part has many machining grooves that are comparable to the focus spot size (rather than much larger than the spot), errors may also occur.

For a well-behaved surface, the data can be very fine with a high signal to noise ratio and good data quality. The scan of a back of a penny in Figure 11.10 and the end of a drill in Figure 11.11 shows such fine detail data. Overall, the conoscopic sensor has been applied to many applications ranging from dental scanning to fine groove structure measurement. Typical measurement ranges for these sensors range from a half of a millimeter to several millimeters, with depth resolutions on the range of 1–3 μm and spot sizes from 5 to 50 μm. As with any sensor, the best combination needs to be determined to fit the application.

11.2.2 Confocal Imaging

The second method of focus point-based metrology, restricting the image through an aperture, is referred to as confocal imaging.[3,4] Confocal microscopy was originally developed in 1957 to provide better throughput for fluorescence imaging microscope.[3] In fluorescence imaging the sample is flooded with an excitation light, usually in the ultraviolet, causing certain biological dyes to fluoresce. A microscope then views the wide-field image at the fluorescing wavelength. However, the microscope sees both the features of interest and a large background noise, greatly reducing the contrast of the fluorescent marked features of interest. By focusing the light to a small spot, then viewing the spot

FIGURE 11.10
The back of a penny scanned by a conoscopic sensor is able to resolve very fine detail.

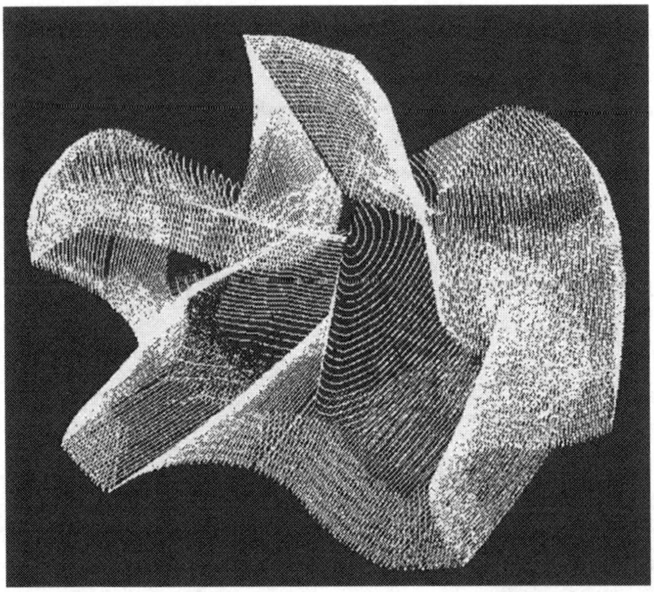

FIGURE 11.11
The back of the end of a drill scanned by a conoscopic sensor is able to resolve very fine detail.

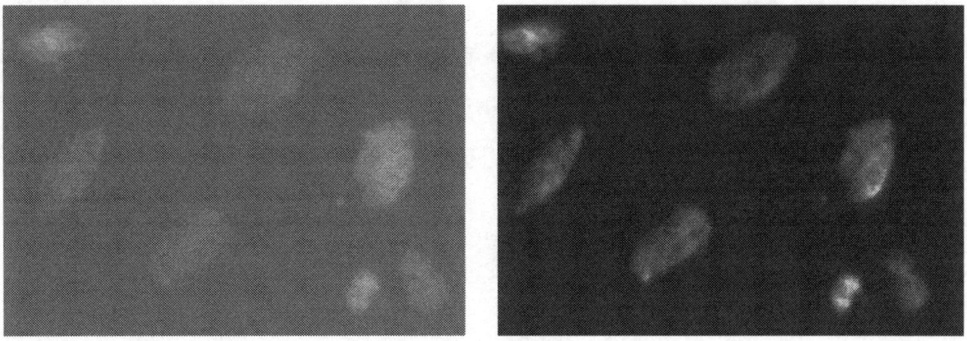

(a) (b)

FIGURE 11.12
Images of biological cells with a wide-field microscope (a) show low contrast, where a confocal scan (b) show low background noise and high contrast.

through an aperture placed in the conjugate position to the light source (effectively, an image of the light source pinhole), the background light can be almost completely eliminated (see Figure 11.12). Focusing the light to a small spot also provides higher light levels at the observation point over flood lighting over the entire sample. For the purpose of metrology, we are not interested typically in fluorescent images, but we can gain localization of a point at a particular distance from the microscope with the light source and viewing point focused together in this confocal arrangement.

In the simplest form, a white light passes through a pinhole and is focused onto a point on the part or subject. The same single-image point is imaged by a microscope objective and focused through a second pin hole (see Figure 11.13) onto a detector. When the light through the pinhole is maximized, the light point is at best focus. Confocal imaging has been in use for many years to map out the structured of biological samples. Scanning the aperture across a part creates one narrow plane of 3D information. Then the microscope is changed in depth, and another scan is made. In this manner, the 3D information is slowly built layer by layer.

Because the confocal method necessarily collects information at only a shallow depth, the separation of one layer to the next needs to be very fine, on the micron level, if a full volumetric map is desired. For a sample even only a few hundred microns thick, this requirement can mean hundreds of scan layers and long scan times. If the application only

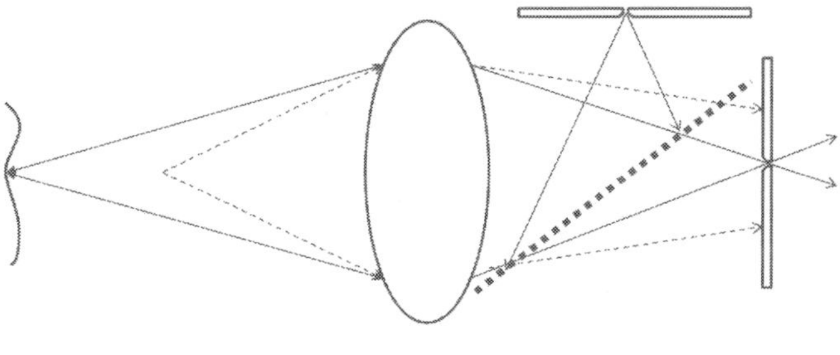

FIGURE 11.13
In confocal imaging, an aperture only passes light from a small focus spot on the part that is at best focus, blocking out-of-focus light with a pinhole.

requires sampling at fixed heights, then the system can make larger steps. An alternative is to use a vibrating pinhole that moves along the optical axis a small amount. By detecting the peak signal relative to the vibration, a limited depth range can be added to the sensor. This method has been used, for example, to measure the thickness of thin films a few microns thick. The vibrating pinhole produces two peak points at the top and bottom of the thin film.

The simplest approach to build an image is to mechanically scan the microscope across the part, move to the next height, and move the microscope again. Mechanical movements like these can take a long time even with fast stages. As an alternative, using a laser source, a small focus spot can be scanned over a sample using mirrors that scan both the light source and viewing axis. A faster scanning method referred to as Nipkow disk uses a spinning disk with an array of pinholes to raster scan across the part. The spinning disk approach typically provides faster data collection, as much as 100 frames per second. Fast data collection over the area is useful in biological imaging for observing live cells but may not be of value as an industrial metrology tool where a mirror or mechanical scan is fast enough. Biological systems also can use line confocal systems for cell analysis, sacrificing some of the 2D resolution based upon geometry assumptions, in return for greater speed. However, these methods do not lend themselves well to industrial measurement where the spatial geometry may not be known.

A limitation of many mirror-based scanning systems is they cause the standoff to the part to be larger. The spot size, and hence lateral and depth resolution, is determined primary by the wavelength used and the numerical aperture (N.A.) or cone angle of light of the imaging objective. A wide collection angle means a smaller spot and shallower depth-of-field but also will mean a shorter standoff between the objective and subject. To maintain the same collection angle or N.A. at longer standoffs requires a larger lens. Making a larger lens with a high N.A. and good performance can become very expensive, so the limitation is a practical one.

The assumption made that the pinhole well defines the focal plane of the measurement is that areas not in best focus will pass very little light through the aperture. This is a good assumption for many biological samples but may not always hold for a uniform industrial part. For example, a mirrorlike surface not at the best focus plane may produce a bright glint that returns sufficient light to be seen as a false point. For any diffuse surface, the sharply focused light point will diffuse out in all directions, very little of which would make it through the detector pinhole if it is not at best focus.

Figure 11.14 shows an image of a fine ground surface taken with a point confocal microscope. The small spot size of the microscope allows fine detail to be defined on the surface that might be missed with a larger mechanical stylus type of sensor (or the need for a very small, fragile tip). This fine surface scan detail compares well with surface measurements made with white-light interferometry and high-resolution mechanical scanning such as atomic force microscopy (AFM) down to the tens of nanometer level.

The confocal imaging method is effective at measuring fine features over small areas. However, the need to have a high numerical aperture (short standoff) and to scan the part limits the use of this method in industrial metrology applications to parts suitable for measurement on a microscope like system (see Figure 11.15).

11.2.3 Chromatic Confocal Imaging

In the third method of focus-based metrology, rather than restricting the light coming back or analyzing the light wavefront shape, the return light is imaged with a degree

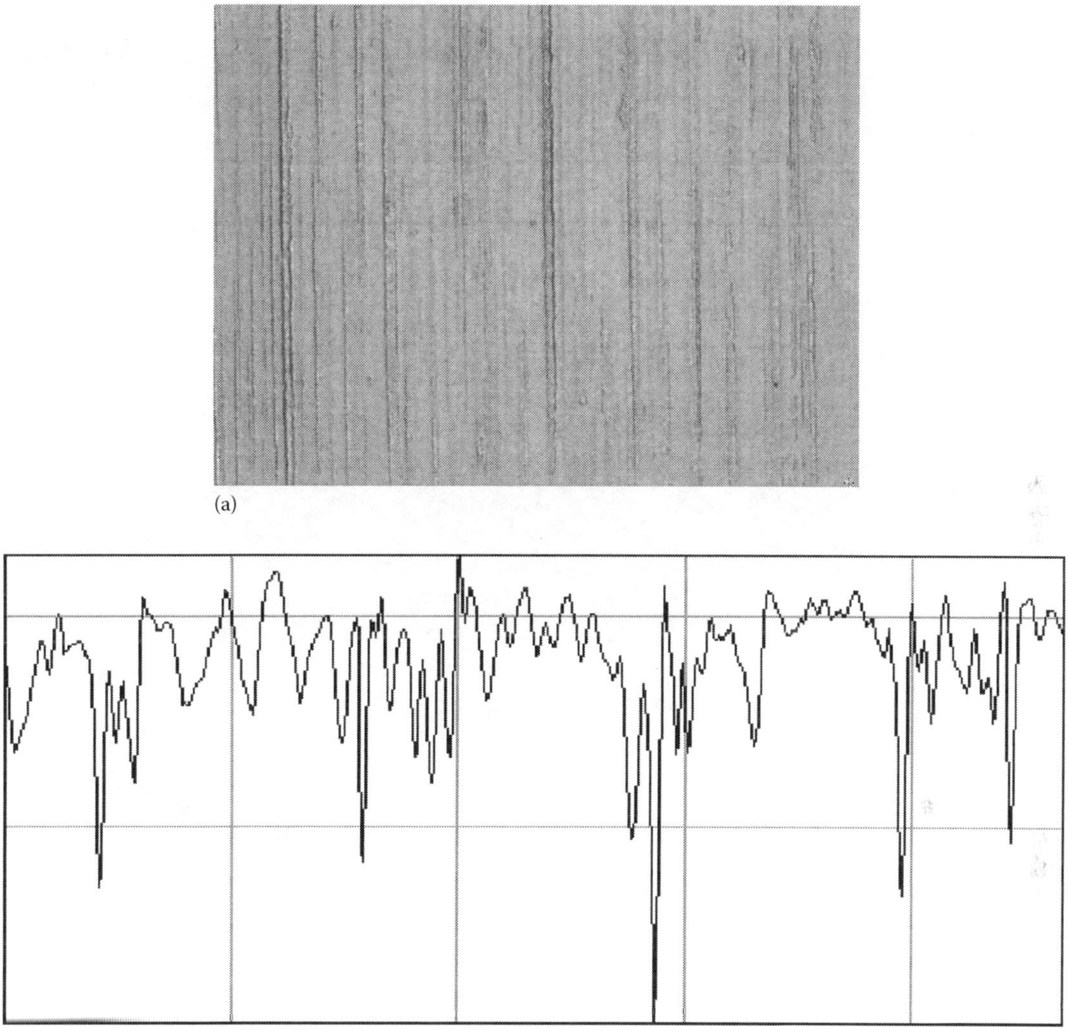

(a)

(b)

FIGURE 11.14
Confocal microscope scan of a fine ground (0.4 μm) surface showing fine texture detail in a 2D gray scale map (a) and a single line profile across the surface (b).

of chromatic aberration, causing blue, green, yellow, and red light to focus at different distances. By spectrally analyzing this return light, using a grating to split the wavelengths and directing the spectrum onto a linear detector array, the distance to the part surface can be determined (see Figure 11.16).[5–7] Made by the company STIL in France, for example, the chromatic confocal sensor approach allows for a range of measurements to be made without moving the sensor or pinhole in depth. This method provides the advantages of confocal imaging in a continuous range sensor with the potential for faster scanning of surfaces.

In chromatic confocal imaging, the spatial resolution is again determined by the spot size of the focused light on the part. However, the depth resolution is no longer completely determined by the depth-of-field and hence the numerical aperture of the focusing lens. A new parameter, the wavelength of the peak return light, drives the resolution within the

FIGURE 11.15
A commercial confocal microscope system. Samples need to fit under the microscope for scanning.

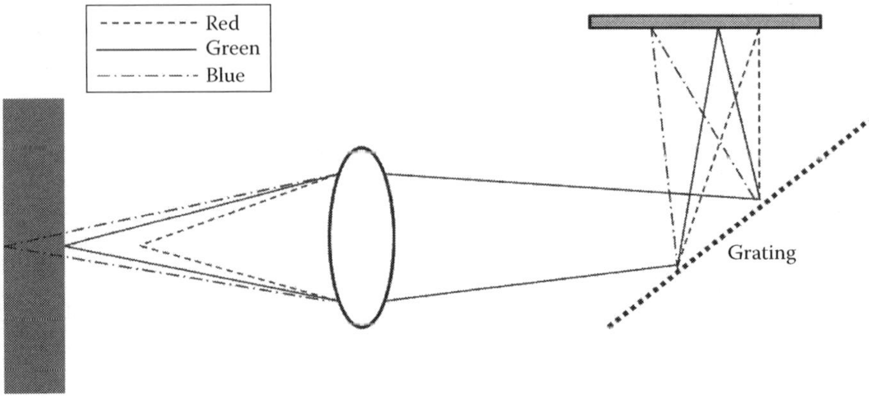

FIGURE 11.16
The chromatic confocal system uses different colors focusing at different depths that can be split with a grating onto a detector array or using a spectrometer to determine depth.

focus region of the lens. This can be thought of as a fine measurement within the focus spot range. Where in regular confocal imaging the light intensity is used to determine best focus, in this method, it is the wavelength of the light as determined by some form of spectrometer. This effect also means that the method will typically be less sensitive to local variations in surface reflectance and texture that may cause a false reading in an intensity-based confocal method due a bright glint.

Development in the area of spectral analysis has been driven by applications such as material analysis and biological applications to achieve improved spectral separation with better gratings and larger detector arrays for the analysis. Spectroscopic confocal systems can leverage these advances to achieve high-depth resolution without resorting to very high numerical aperture, shallow focus systems. This does not mean the focus spot size and depth are not important. As stated, the spatial resolution is still largely driven by the spot size, which is determined by the numerical aperture of the focusing lens. But, whereas in the previous method the focus range might need to be a couple microns or less, a focus range of even hundreds of microns can still be used while achieving submicron-level depth resolution using the chromatic confocal method. The same 100 μm range that would require a 100 images with standard confocal methods may now be measured in one scan with the chromatic confocal approach.

Being a white-light system, issues of laser scatter and speckle noise are not a big issue with chromatic confocal measurement. For this reason, this method can be used to measure such scattering features as the weave of paper (see Figure 11.17). Another potential application area often degraded by speckle noise is measuring the surface roughness of machined parts (see Figure 11.18).[8] On machined surfaces, the machining or grinding marks often create noise that create interference patterns with laser-based systems that would not be present in a white-light system. Due to the noise, and also due to the spot size in the conoscopic technology, the laser-based conoscopic system described previously is not effective at measuring fine (submicron) surface roughness, whereas the confocal methods described typically provide good agreement to stylus and even AFM methods, as stated before.

Figure 11.19 shows a map taken from a scan of the back of a penny using a commercial chromatic confocal system. The detail in the edges and fine features are evident, standing out better than for the laser-based system scan in Figure 11.10. The tradeoff to achieve the higher definition is a matter of scan time. Where Figure 11.10 took about 20 min to produce a map with a 2 μm resolution and a 20 μm spot size, Figure 11.19 took several hours to produce a map at a 0.1 μm resolution and micron level spot size.

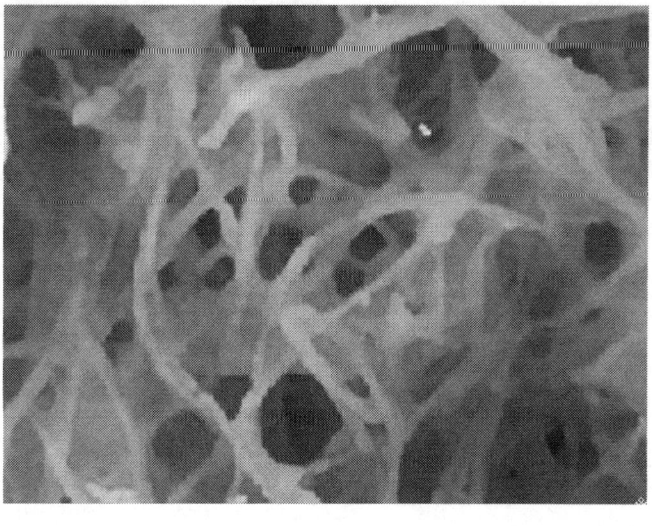

FIGURE 11.17
A chromatic confocal scan showing the weave of paper (highly scattering structure).

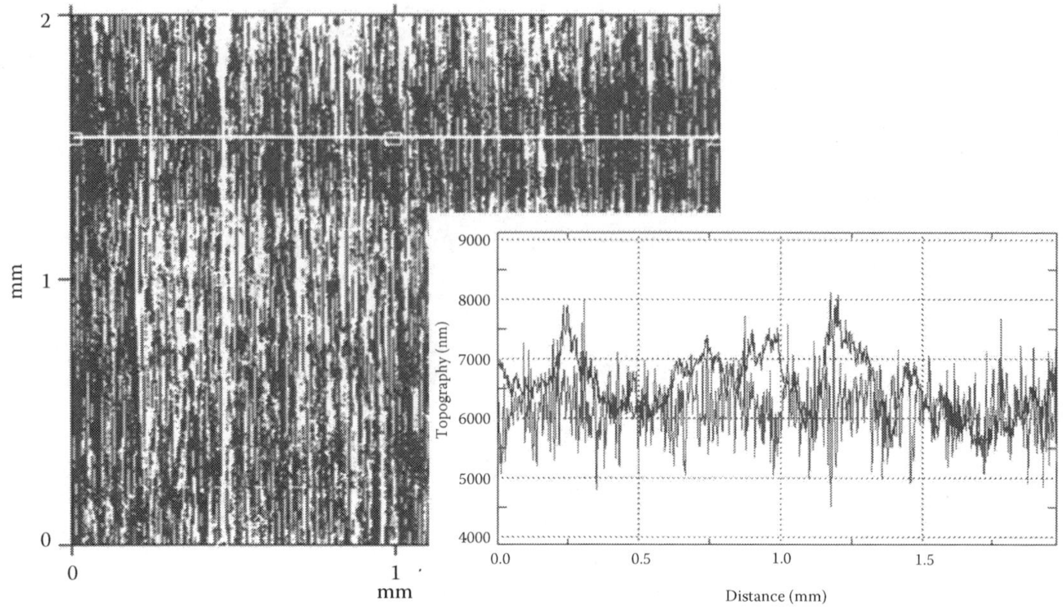

FIGURE 11.18
A 3D grayscale scan (gray is depth change) of a 0.2 μm finish machined metal surface taken with a chromatic confocal system.

FIGURE 11.19
A confocal microscope scan of a section of the back of a penny showing fine detail.

The commercial systems available typically use a fiber optic-based method to deliver the light to and from a delivery lens (see Figure 11.20). A base unit contains the spectrometer for the measurement as well as the light source. For some applications, the remote head may be useful to access tight locations that might be hard to reach with a microscope. However, the base unit still makes the unit less portable than some other gages that may be self-contained. In any case, as a point measurement device, the method will only measure at a point unless the head is scanned by some other means such as a motion stage, mirror system, or other directing means such as an active optical system.

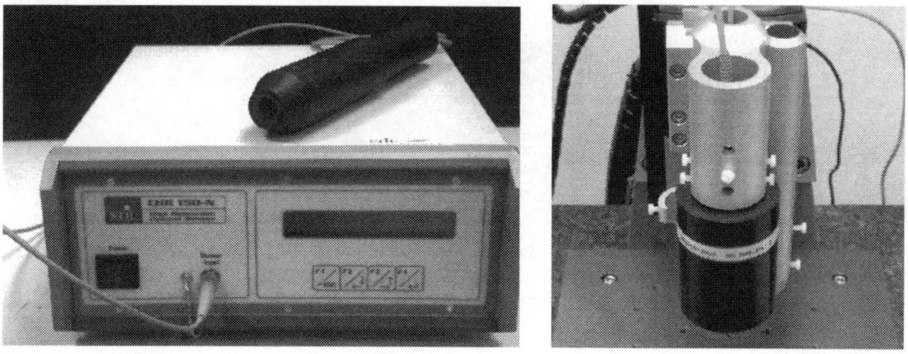

(a) (b)

FIGURE 11.20
A chromatic confocal commercial system with a fiber delivery probe showing the control unit and fiber probe (a) and the probe on a motion stage for scanning (b).

11.2.4 Area-Based Focus Metrology Methods

Area-based optical metrology methods use a video camera rather than a point detector to view a part. As with the point-based systems, a lens with a low F-number (high numerical aperture) is used to define a focus plane. In the simplest form called depth from focus, the regions that are in best focus (by contrast or other method) are segmented out from other regions. The focus plane is then stepped in depth to create a volumetric map. As with confocal imaging, to achieve a full map, the steps can be no greater than the defined focal plane depth.

A more computational method called depth from defocus uses information about how the image changes focus, based upon parameters like the lens design to calculate the depth of features both in focus and as they become more defocussed. This method uses fewer spars image planes that bound the upper and lower limits of the volume to be mapped. We will now discuss these two methods in more detail and consider the pros and cons of each approach as a metrology tool.

11.2.5 Depth from Focus

Depth from focus (DFF) uses multiple images taken at different focal depths using a configuration as shown in Figure 11.21. The simplest form of depth from focus is to take a larger number of images and search through the area within each image that has the least amount of blur. Some microscope systems use this approach to define regions within each image that is best focused and then combine those regions to build up a single, in-focus image. In the case of biological imaging, measurement is not of primary interest. In such cases, the multiple image planes of information are combined together using what is called deconvolution methods, creating one flat image from the multiple images.

By sectioning out areas of each images that are in best focus and recombining them (assumes good registration exists or can be established), a single clear, in-focus image can be generated. The effect is to collapse the volumetric image into one with all levels in focus. This is the basic premise behind convolution-based confocal imaging systems used in biological applications.

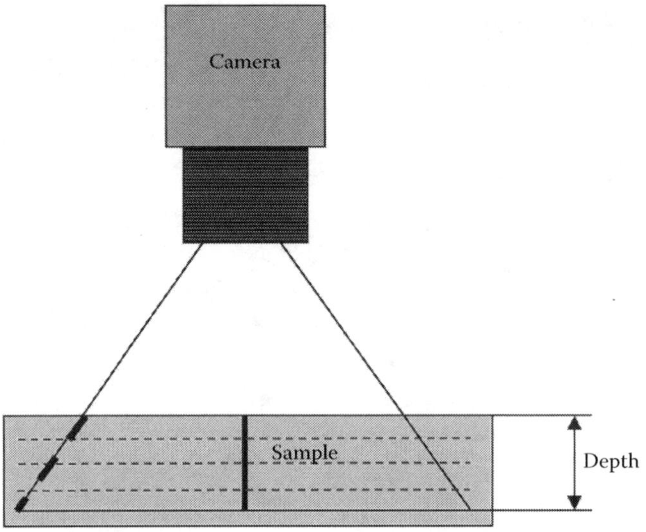

FIGURE 11.21
Multiple images taken as focus that are moved in depth by fixed steps builds a 3D map using depth from focus imaging.

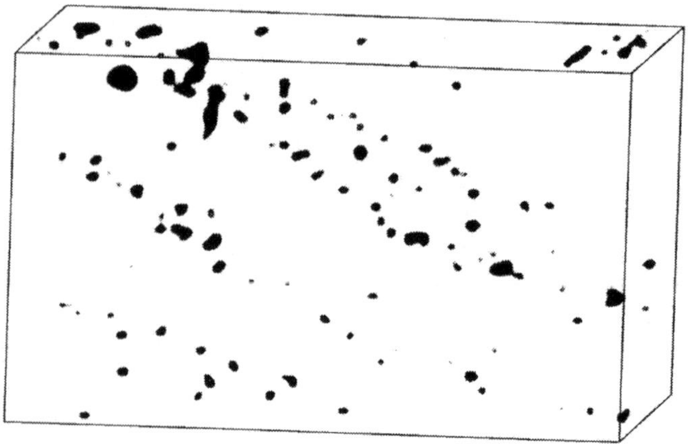

FIGURE 11.22
A map of inclusions in a media taken by stepping through the focus depth, then combining the images.

If the depth information at each focal plane is associated with each set of in-focus features, then a volumetric map can be made to within the resolution of the focus steps used. An example of a volume map in a transparent media made using up to 100 focus planes is shown in Figure 11.22.[9] The means of determining what is in or out of focus is key in this method for determining an accurate mapping.

The analysis of most DFF methods is designed around a focus quality metric. This metric generally is not based upon the physical model of the imaging system but rather relies on experimental data to determine the difference between a sharp, local region of an image versus that of a blurred (out-of-focus) one.[10] The two desirable conditions for an ideal focus quality metric is it should change uniformly through the focus region and be computational efficiency. Various focus metrics have been proposed including spatial, entropy, and frequency-based metrics.[10–13] The focus metric, F, is evaluated at each point (x, y) across K elemental images. The estimated depth is the depth at which the focus measure is at a maximum using the expression

$$\hat{z}(x,y) = \underset{i \in \{1 \ldots K\}}{\arg\max}\, F(x,y;z_i) \tag{11.1}$$

However, due to noise, the focus quality metric often exhibits local maxima which hamper accurate localization of the focus quality peak along the range of interest. In addition, the behavior of the focus metric depends heavily on the texture, frequency content, and contrast of the object.

In essence, in DFF methods, the search image is partitioned into focus zones by movement of the image plane, and a simple analysis picks the areas of best focus based on an objective function (focus metric). This combination makes DFF relatively slow in data acquisition but fast in computation, which is the characteristic of many commercial DFF-based 3D systems.

In Figure 11.23, a simple corner has lines on it (which can be texture on the part or projected lines). The region over which each image is seen as being in clear focus, first covering the apex, then extending about half way down the slope, then excluding the apex and covering the down slope area, can provide a regional estimation of depth. It is then straightforward to use this information to create a pseudo 3D map of the surface. The three images of the corner are captured at different focus positions.[13] A Sobel edge detection operator was used to find the clarity of the edges of the pattern, and the result is used as a focus metric in Equation 11.1.

Figure 11.23 shows the focus images along with the corresponding focus metric maps. Figure 11.24 shows the result of using this focus depth estimation on a lion head model.

(a)

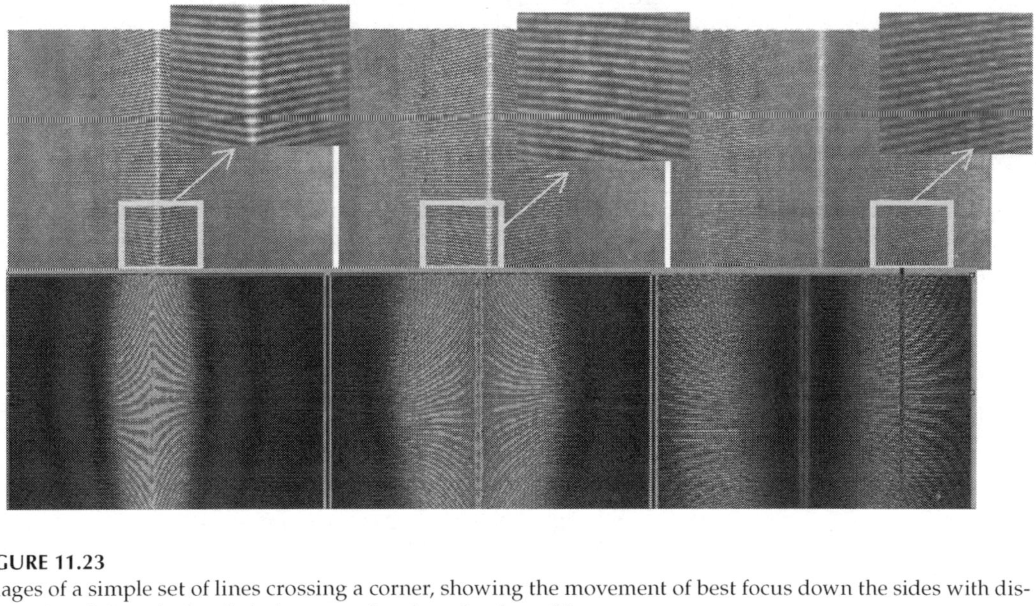

(b)

FIGURE 11.23
Images of a simple set of lines crossing a corner, showing the movement of best focus down the sides with distance (a) and the calculated clarity map of each under them (b).

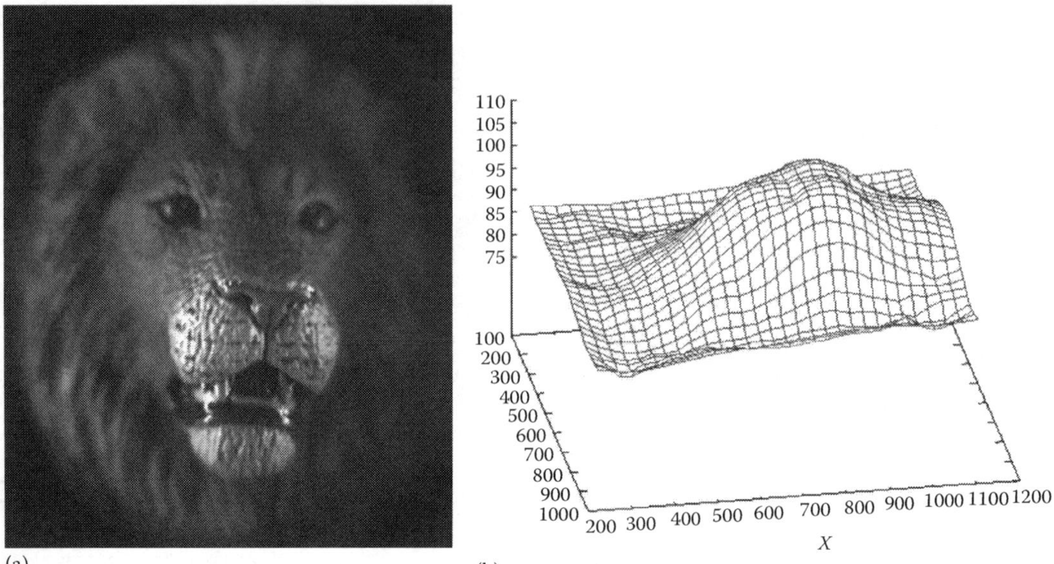

(a) (b)

URE 11.24
Image of a lion head model (a) from which the texture on the model surface is used to general a depth from focus map (b).

The measurement is coarse relative to other 3D methods and dependent upon the definition of the focus depth. Clearly, changes in the feature size will change the region seen as being in focus, as will any filters used in the edge detection operation. For systems with very sharply defined focus depths and consistent feature sizes, this method will give consistent results. So, a part with uniformly spaced futures like a grid pattern, an array of holes or machining marks might be good candidates for mapping with this simple approach. However, for a subject with a variety of feature sizes like a person's face, the determination of the depth will vary in resolution and quality with the size of the feature, causing the depth resolution to be uncertain.

11.2.5.1 Structured Pattern, Focus-Based Methods

All of the earlier suggest that there is clear edge or feature information to be analyzed for focus clarity. For a smooth part, this may not necessarily be the case. Some parts have a lot of detail such as machining grooves or surface texture, where others may be very smooth such as a painted panel. An alternative to using the inherent features on a part is to project a pattern, such as lines, onto the part such as shown in Figure 11.24. Active-structured light-based depth from focus quality uses a structured pattern that is separate from the texture.[13–15] The frequency content of the fuzziness can then be modeled as a Laplacian calculated around a narrow band of the primary frequency (spacing) of the pattern projected onto the part. This estimation can be done using a local operator over x and y of the form

$$S(x,y) = e - (x'^2 + y'^2)/2a^2 x \cos(2(\pi)/T\, x\, x' + (\phi)) \tag{11.2}$$

where

$$x' = x\cos(\theta) - y\sin(\theta) \tag{11.3}$$

$$y' = -x\sin(\theta) + y\cos(\theta) \quad (11.4)$$

where
T is the primary period of the pattern projected on the subject
a is the standard deviation of the equivalent Gaussian filter
θ is the angle of illumination to the surface normal
ϕ is the phase offset[13]

This approach assumes that the effect of blur is primarily to spread the pattern projected and a decrease in the rate of change of intensity (derivative of the contrast) of the edges. In some cases, such as autofocus systems, just the contrast of edges is considered in each area.

Clearly, one of the advantages of this approach is the period of the pattern being used to determine the focus quality is very well known and controlled. The pattern projection can be done in a very similar way as is done with structured light triangulation (Chapter 7), but there is no triangulation angle, so the method only requires line of sight access from one direction.

Using the focus quality of the pattern clearly does not define a distance at every point as with phase-shifting triangulation, only defining where the pattern is in best focus. Generally, a focus-based method like this will give resolution on the order of the feature size, where a triangulation method might provide 1/500 of the pattern size using phase-shift analysis. What drives the size of any active illumination pattern for use in depth from focus methods is the aperture of the optics needed to achieve a desired depth-of-field (DOF) of the pattern. A sharper focus plane (shallow DOF) implies a larger aperture lens.

Using the structured pattern to create the map addressed two important challenges associated with measuring a smooth object with focus clarity information. First, smooth surfaces have very few sharp lines or features on them that are usable to generate 3D data based upon a clearly focused edge. On the other extreme, some smooth parts may have a great deal of texture in some areas, such as a hole or machined edge but be very smooth in other areas, such as on a flat area. Using the structured pattern ensures that there are local patterns to analyze over the whole viewable part surface, as well as separates the 3D object's general shape information from texture variations. The analysis of a face model image taken from multiple focus depths is shown in Figure 11.25.[15] Figure 11.25a shows the structured light pattern on the face, visible on the far points like the chin and sides, then Figure 11.25b shows on the near points on the nose and forehead and the resulting calculated 3D map.

11.2.5.2 Depth-of-Field Effects

One of the key optical challenges for obtaining good 3D data from depth from focus clarity information is to control the DOF of the images used. Using the expression for DOF of approximately two times the feature size times the F-number, we can consider what might be a typical example. The optical resolution is assumed to be no better than diffraction limited given approximately by the working F-number in microns (i.e., the distance to the subject divided by the imaging lens aperture diameter). The image does not become unusable beyond this DOF range, but the edges of features of this size will begin to blur noticeably beyond this region.

To visibly see defocus, typically two times the DOF is used. To do depth from focus analysis, it is not essential that the steps cover only the part depth, only that they "exceed"

(a)

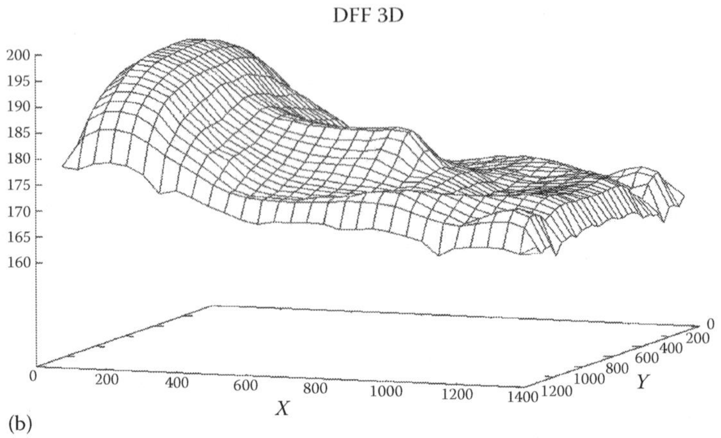

(b)

FIGURE 11.25
A face model with a projected line pattern, at best focus on the far points like the chin (a), Combines with an image at the near points like the nose and forehead resulting in a calculated 3D map (b).

the full range of the part. That is, the depth covered can be more than the depth of the part, which provides flexibility in the actual location of the part along the line of sight of the system (a radial movement of the subject relative to the optical view). To collect the highest possible resolution data on a part, it is desirable to collect a full set of clear images separated by no more than the DOF step.

Collecting a full set of images through the depth of a part imposes some practical limits of time, how big the part can be versus the size of the lens, and how many steps can be taken till the lens runs into some feature on the part. A small lens far from a part will provide a large depth-of-field but will provide very poor resolution of maybe millimeter size. A microscope objective will provide very shallow, well-defined focus regions of micron size but has very little standoff to the part and can only see a small field size of maybe a few millimeter at a time.

11.2.6 Depth from Defocus

Conversely, the amount of defocus blur can be used to estimate how far a particular image feature is from best focus.[13,16–20] In this case, the blur is typically modeled as a convolution of the in-focus image and an effective point spread function of the lens that can be calculated geometrically from

$$R = \left\{\frac{D}{2}\right\} \times \left\{\frac{1}{f} - \frac{1}{o} - \frac{1}{s}\right\} \tag{11.5}$$

where

R is the blur radius
D is the diameter of the collection aperture
f is the focal length of the lens
o is the object distance to the subject
s is the image distance to the sensor[13]

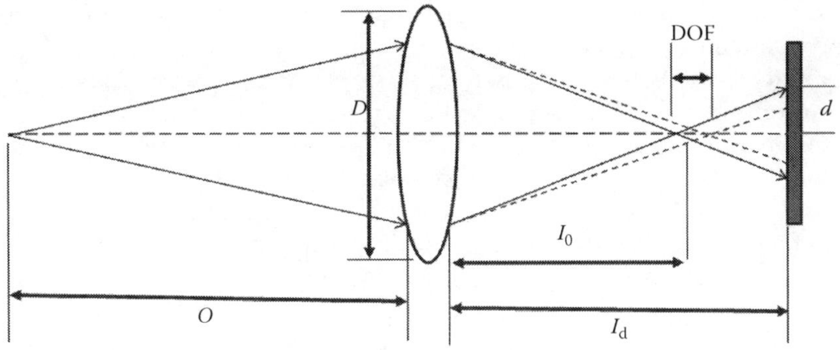

FIGURE 11.26
Parameters of a lens used to relate the defocus radius, d, to the object distance O.

Defocus is related to the amount of light spot spread on the detector which is a measurable scalar. By determining the size of defocus at each pixel, the distance of the object to the lens can be calculated based on imaging system parameters. Referring to Figure 11.26, for a lens of focal length f, and object distance of O and an image distance of I_0, for a detector at some other distance I_1, there will be a defocus radius at the detector plane of d that can be expressed as

$$d = \frac{D(I_0 - I_1)}{2I_0} \tag{11.6}$$

where the F-number of the system F is just f/D. Using the standard imaging expression

$$\frac{1}{O} + \frac{1}{I_0} = \frac{1}{f} \tag{11.7}$$

Then we can say that the defocus radius relates to the lens parameters by

$$d = \frac{\left(I_0 - fI_1 - fI_0\right)}{2FI_0} \tag{11.8}$$

Then the object distance O for any given blur radius d can be given by

$$O = \frac{fI_1}{\left(f + I_1 - 2Fd\right)} \quad \text{for } I_1 \geq I_0 \tag{11.9}$$

$$O = \frac{fI_1}{\left(f - I_1 + 2Fd\right)} \quad \text{for } I_1 < I_0 \tag{11.10}$$

For a fast (low F-number, high numerical aperture) optical system such as a microscope objective, this expression can just be reduced to

$$O = \pm\frac{fI_1}{2Fd} = \pm\frac{fI_1}{Fb} = \pm\frac{I_1 D}{b} \tag{11.11}$$

where b is the blur diameter. So the object distance is only a function of the detector distance, the diameter of the lens (the F-number effectively), and the size of the blur spot. Depth from defocus methods use this relationship as the basis of determining how far any given out-of-focus point is from the known best focus plane. With only one focus plane, this expression does not tell us whether the out-of-focus point is before our best focus plane or after it. So, at least two in-focus planes are used to determine the sign of the expression in Equation 11.11 (is the point before or after focus) by comparing the distance from the two focus planes to solve for the location between them.

Given the earlier simple analysis as the starting point, the challenge is then to determine what points in the image are out of focus and by how much to determine the defocus radius we need to find the object distance. There are quite a few methods in the reference material.[13–22] We will briefly describe two methods later to explore the strengths and weaknesses of each. More detail on the analysis can be found in the cited material.

11.2.6.1 Spectral Domain Depth from Defocus

The most common method to isolate optical defocus relies on inverse filtering.[21] This approach uses the original input image, $i_0(x, y)$, and the point spread function of the lens (PSF), $h(x,y;\theta_1)$, to calculate what an output image $i_1(x, y)$ should look like:

$$i_1(x,y) = i_0(x,y) \otimes h(x,y;\theta_1) \overset{\Im}{\Leftrightarrow} I_1(u,v) = I_0(u,v) \times H(u,v;\theta_1) \tag{11.12}$$

where the capital letters signify the Fourier-transformed version. The parameters of the imaging system may include the focal length, aperture size, and the lens to detector distance.

The ratio of resulting images captured at different depths in the Fourier domain can be expressed as

$$\frac{I_1(u,v)}{I_2(u,v)} = \frac{I_0(u,v) \times H(u,v;\theta_1)}{I_0(u,v) \times H(u,v;\theta_2)} = \frac{H(u,v;\theta_1)}{H(u,v;\theta_2)} \tag{11.13}$$

in which the unknown scene effect has been eliminated. The frequency content of the image can be evaluated over a small region (window) of the image in order to localize depth estimation. However, large defocus amounts and the windowing effect of using a small region average can induce significant errors in Equation 11.13.

By evaluating the features in the image in terms of the defocus at any given region, the analysis can explicitly relate the frequency content of the captured images to the amount of defocus seen at a feature and thereby find the distance of the object from Equation 11.11.

For most incoherent and well-corrected imaging systems, it is common to assume a uniform or a Gaussian point spread function. By expressing the geometrical optics relationship for two different parameter settings, one can find a linear relationship between the two resulting defocus positions. For example, if the aperture size is changed between the two images, the amount of defocus shift due to the aperture change can be used to correct the depth calculation. The magnification change due to changing system parameters has to be taken into account to relate the images by normalizing the images to a fixed magnification ratio.

11.2.6.2 Spatial Domain Depth from Defocus

In the spatial domain approach, the relationship between the two corresponding local regions in the focus images is expressed through a matrix in spatial coordinates. The matrix is defined by the convolution of two adjacent images in depth as outlined in Figure 11.27. The matrix $t(x, y)$ is defined such that

$$i_2(x,y) = i_1(x,y)\lfloor \otimes \rfloor t(x,y) \tag{11.14}$$

where $\lfloor \otimes \rfloor$ denotes a constrained convolution where the operator does not reach beyond the boundaries of the image $i_1(x, y)$. The map between object points can be calculated using the known point spread function of the lens or through a calibration that uses a series of images of a known target such as a progressive sign wave target used to find the modulation transfer function of a lens (see Figure 11.28).

If the patch size in the image used to define the depth steps is large, the effect of windowing the region is not very significant, allowing a relatively course map to be generated. However, if a dense depth map is needed, the windowing effect can be a significant source of error in the spectral domain-based techniques since the spectral content would need to be very small to allow a small patch to be used. In such cases, the convolution matrix in the spatial domain approach provides a more accurate representation of the relationship between corresponding local regions in elemental images.

The point spread function of the lens, found either from the lens design or by measuring it with an MTF chart, can be used to calculate a lookup table with the desired level of detail to map out a surface. This table maps each object distance to a specific convolution ratio.[22] There are other approaches to developing this mapping function, but the result is to generate a mapping between the amount of defocus seen in each region on the subject and a distance. This mapping effectively subdivides the space between in-focus images to allow the areas between best focus to be given a value of depth or distance from the system.

This approach implies the need to have an in-focus image at least at the two extremes of the range to be measured. Of course, having periodic focus planes through the range to be measured allows this vernier measurement from the defocus function to be used to get more data points. Alternately, the number of in-focus images taken by the system can be greatly reduced and still generate the same resolution mapping as one with many more images. This can be time saving in some applications, but more often is used to provide finer detail than is practical to define by the depth-of-field steps of a depth from focus system.

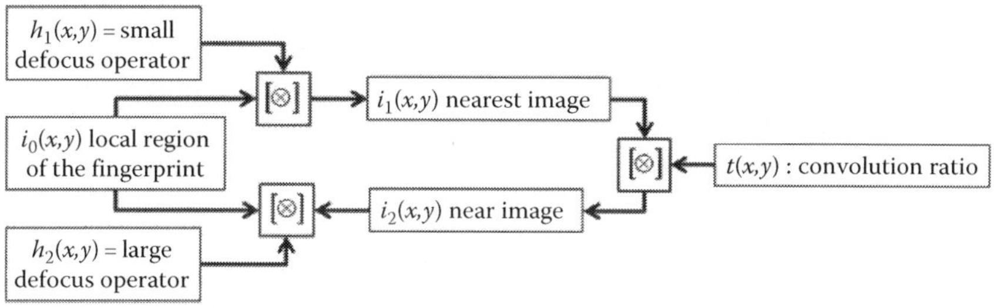

FIGURE 11.27
Flow diagram shows the process from focus image capture to the matrix-based analysis of the spatial domain DFD algorithm.

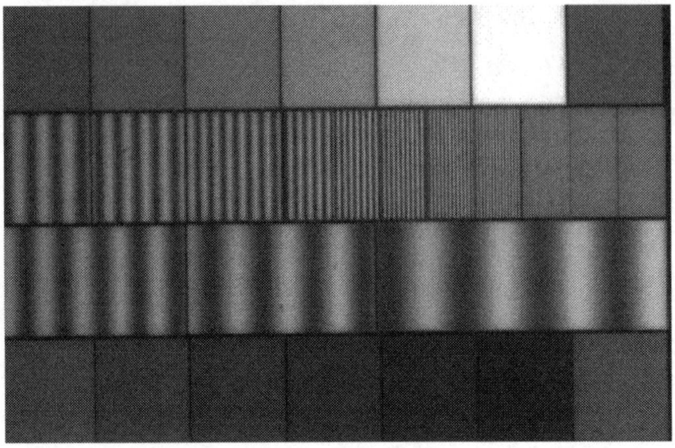

FIGURE 11.28
A standard MTF target can be used to evaluate the response of a lens using a series of spatial frequencies defined by sine wave patterns that get finer and finer.

11.2.7 Application Considerations

Although the principle of all DFD methods rely on direct or indirect measurement of defocus, the choice of algorithm and the achievable performance is heavily application dependent. Understanding the special characteristics of each method and the application at hand is essential in choosing the best algorithm for a given type of subject. The two broad DFD categories reviewed exhibit distinct features that affect their application-specific performance.

Spectral domain DFD methods (using the sizes of features on the part) capitalize on the transformation of frequency content in a local object patch and, as such, perform best when the objects of interest exhibit a balanced spatial frequency spectrum. That is, if the part has a very regular, well-defined pattern, such as machining marks, the spectral domain DFD can capitalize on that structured to generate a very consistent resolution mapping. However, in reality, the high spatial frequencies of an image are often dominated by noise and the low frequency content is often not adequate for generating a dense depth map. As a result, this category of DFD algorithms often suggests the use of a prefilter to deemphasize the two extreme spatial frequency contents. In addition, if an object happens to have a higher signal to noise ratio in a certain frequency range, then measurement is more accurate within that range but may be much less effective in regions that do not contain the right size features on the surface of the part.

The magnification of features can be critical in the spatial domain DFD methods and should be carefully compensated for using either calibration means or other correction tools. The same is true when objects in the scene are dynamic and may move between two focus images. In such cases, spectral domain DFD is typically more robust to misalignment of the images. This is not the case for spatial domain techniques that rely heavily on pixel-to-pixel correspondence between the elemental images.

Most DFD algorithms are designed with no assumption about the object properties in the scene. This level of generality is often not needed in real-world applications of these methods. The information about the part such as texture, machining marks, local holes, and surface roughness can be used to improve the performance of any of these DFD approaches.

11.2.8 Focus-Based Method Summary

There are several commercial systems on the market that use either depth from focus or some form of depth from defocus. Such systems generally take the form of a microscope, often with different power options, and a stage mechanism to move either the part or the microscope to different depths. For any given resolution of the microscope, the systems typically have a preset step size that matches the resulting depth resolution to the spatial, in-plane resolution in order to generate a complete 3D map. For some applications, it may be sufficient to have separated, well-defined planes to get a general shape or to measure the separation of key features.

For any of the focus-based methods to work, the part to be measured needs to have features of appropriate size and spacing to allow the features or locations of interest to be mapped. For some types of features viewed on the microscopic level, it is possible to get shadows, diffraction points, or other artifacts that may confuse the system and either show points at an incorrect depth or even generate phantom points. As with any method, a testing protocol to verify that the data are meaningful and correct for the application is typically a good idea.

An example of a commercial focus-based system is shown in Figure 11.29. A scan from a back of a penny is shown in Figure 11.30. The definition of the edges on the monument on the penny is not quite as well defined as in the confocal scan in Figure 11.19. There also appears to be surfaced artifacts not seen in the confocal scan. Since the confocal scan is active, it can reliably pick up points at all areas of the penny, including smooth areas, but the focus-based method in this case appears to exhibit more noise on smooth areas. This is a specific example and is not necessarily exemplary of all applications but suggests smooth areas may present a potential source of noise for focus-based methods.

FIGURE 11.29
A commercial focus-based measurement system, based upon a microscope approach for micron- to submicron-level measurements over small features.

FIGURE 11.30
The scan from a back of a penny showing fine detail from a focus-based measurement system.

A surface with a roughness of 0.2 μm is shown in Figure 11.31, taken at both 40× and 100× magnification. Note that the different magnifications used on this very fine surface do not necessarily show more structure, suggesting that the information at features less than the 0.2 μm potentially are not being fully captured by the system. In industry tests, the roughness of surfaces with 0.05, 0.1, and 0.2 μm micro-finishes do not typically show any measured differences with focus-based systems. This is in comparison to measurements done with the active confocal systems (see Figure 11.19) which did show differences at this level and had good agreement with other methods such as white-light interferometry and precision mechanical methods. This limitation of focus-based methods is just a reflection of the requirement to have something to focus upon. If the surface structure is below the optical spatial resolution of the system, which is going to generally be true for any features smaller than the visible light wavelength of a half micron and perhaps up to a micron (depending on lens numerical aperture), then the method is not likely to be effective at measuring the surface.

The scan time for images can require hundreds to thousands of images, requiring many minutes to several hours. To really achieve micron- or submicron-level measurements requires taking microscope size views, typically less than a millimeter in size, both in depth and across the part. So, for fine measurement, this method is typically going to be slow compared to other scanning methods like white-light interferometry or conoscopic laser scans but possibly faster than point confocal methods.

For larger volumes, there are currently no commercial systems really using the focus-based approach. Using a lab-based system, Figure 11.32 shows an example scan of a 25 mm diameter metal cylinder with a pattern on it for focus purposes, compared against a structured light scan of the same part. Agreement on this scan is on the half a millimeter level, but the fine structure used for the focus determination is not actually mapped. For parts on a meter scale, a custom system may provide a few millimeters of resolution using a focus-based method.

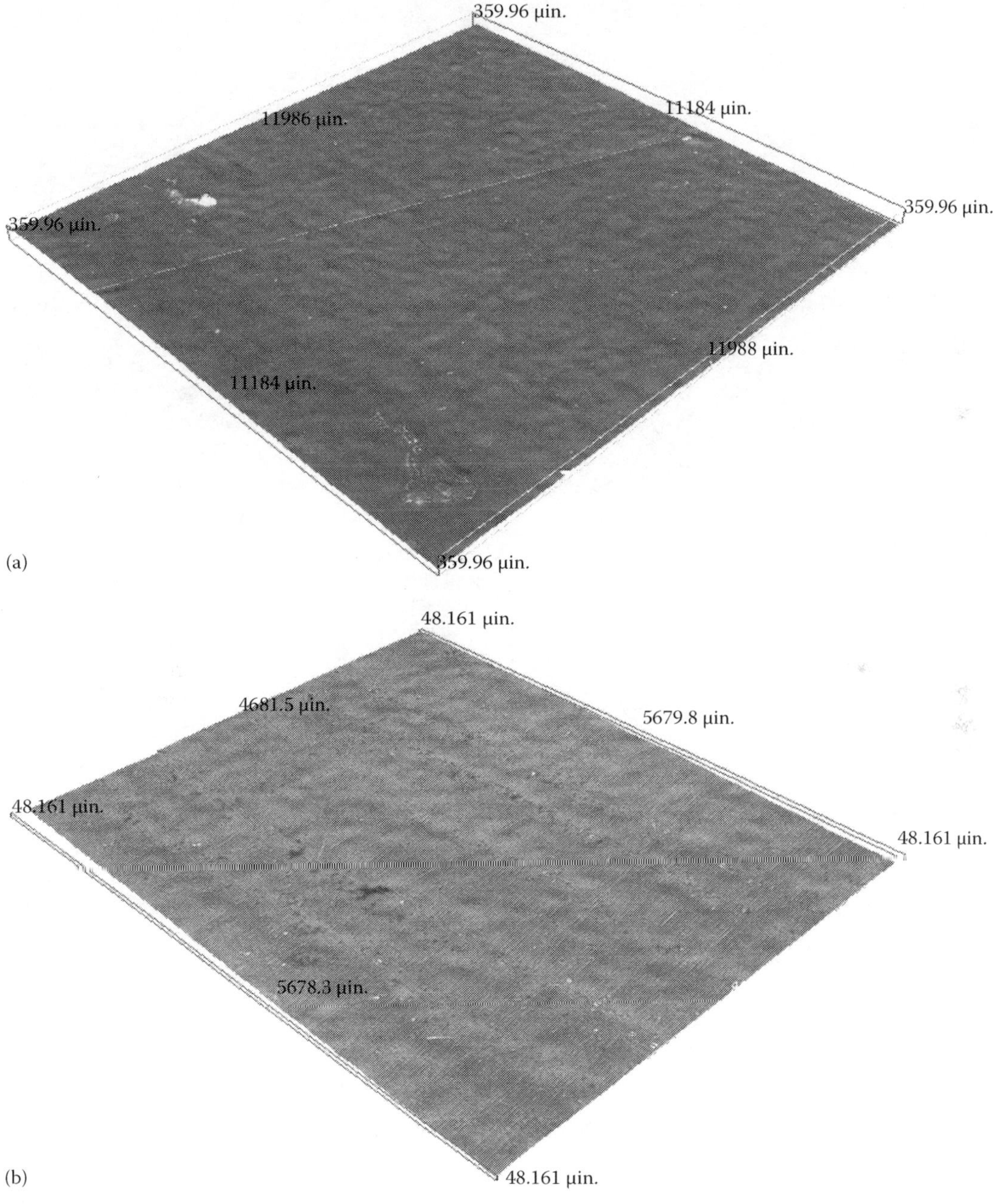

FIGURE 11.31
The scan of a 0.2 μm surface finish sample showing fine detail from a focus-based measurement system for magnifications of 40X (a) and 100X (b) showing little difference.

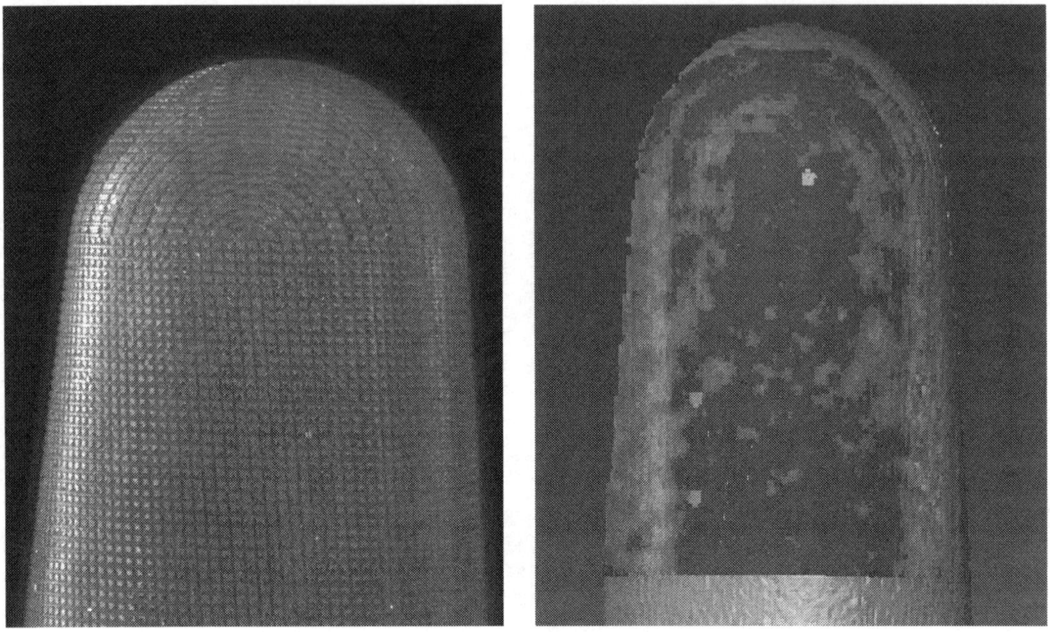

(a) (b)

FIGURE 11.32
The scan of a 25 mm diameter textured cylinder (a) and the 3D map generated by depth from defocus methods compared to a scan of the same part using a structured light method capable of 0.1 mm resolution (b). The DFD method agrees to the structured light method to within about 0.5 mm.

11.3 Summary

Optical metrology methods that use some focus as part of the measurement have the property of being in-line, requiring no triangulation angle. They can be used in point-based measurements or by capturing a series of area images. The mechanism of the method can be either active, projecting a focused point or pattern, or passive by using existing features on the part such as patterns or textures. There are many variations for determining what is at a particular focus, including

- Using the wavefront from the spot
- Using the peak intensity through a pinhole
- Analyzing the light with a spectrometer with an active chromatic dispersion projected
- Finding the areas of best focus for a series of images
- Calculating the defocus of each feature between in-focus images

Spot-based systems can be scanned, typically with data rates of kilohertz to tens of kilohertz. Image-based methods typically will require from a few to as many as thousands of images to map a part. In all the systems, the spatial, in-plane resolution is defined by the optical resolution and numerical aperture of the collection or viewing lens, along with any interpolations that might be used.

The strength of focus-based methods is the ability to look into holes, over edges, and other line of sight measurements. Submicron-level depth resolution is possible for both the point- and area-based approaches but at the expense of long data collection times of minutes to hours. Parts that have very rough or diffusing surfaces may not be able to be measured with the active methods, while parts with smooth, undecorated surfaces may be fine with active methods but may not be measurable with the passive methods. As with any measurement tool, some experimentation is likely needed to determine if focus-based methods are a good match to any particular application.

References

1. Sirat, G.Y., F. Paz, G. Agronik, and K. Wilner, Conoscopic holography, *Proc. SPIE* 5972, 597202 (2005).
2. Enguita, J.M., I. Álvarez, J. Marina, G. Ojea, J.A. Cancelas, and M. Frade, Toward extended range sub-micron conoscopic holography profilometers using multiple wavelengths and phase measurement, *Proc. SPIE* 7356, 735617 (2009).
3. Pawley, J.B. ed., *Handbook of Biological Confocal Microscopy*, 3rd edn. Berlin, Germany: Springer (2006). ISBN 0-387-25921-X.
4. Confocal Microscopy Patent, Filed in 1957 and granted 1961. U.S. Patent 3013467.
5. Boudoux, C. et al., Rapid wavelength-swept spectrally encoded confocal microscopy, *Opt. Express*, 13(20), 8214–8221 (2005).
6. Maly, M. and A. Boyde, Real-time stereoscopic confocal reflection microscopy using objective lenses with linear longitudinal chromatic dispersion, *Scanning*, 16(3), 187–192 (1994).
7. Tiziani, H.J. and H.M. Uhde, 3-dimensional image sensing by chromatic confocal microscopy, *Appl. Opt.*, 33(10), 1838–1843 (1994).
8. Lyda, W., D. Fleischle, T. Haist, and W. Osten, Chromatic confocal spectral interferometry for technical surface characterization, *Proc. SPIE*, 7432, 74320Z (2009).
9. Liao, Y., E. Heidari, G. Abramovich, C. Nafis, A. Butt, J. Czechowski, K. Harding, and J.E. Tkaczyk, Automated 3D IR defect mapping system for CZT wafer and tile inspection and characterization, *SPIE Proc.*, 8133, p. 114–124 (2011).
10. Xiong, Y. and S. Shafer, Depth from focusing and defocusing, in *Proceedings of the International Conference on Computer Vision and Pattern Recognition*, pp. 68–73 Cambridge, MA (1993).
11. Zhang, R., P.-S. Tsai, J.E. Cryer, and M. Shah, Shape-from-shading: A survey, *IEEE Trans. Pattern Anal. Mach. Intell.*, 21(8), 690–706 (1999).
12. Daneshpanah, M. and B. Javidi, Profilometry and optical slicing by passive three-dimensional imaging, *Opt. Lett.*, 34, 1105–1107 (2009).
13. Daneshpanah, M., G. Abramovich, K. Harding, and A. Vemory, Application issues in the use of depth from (de)focus analysis methods, *SPIE Proc.*, 8043 p. 80430G (2011).
14. Girod, B. and S. Scherock, Depth from defocus of structured light, *SPIE Proc.*, Vol. 1194, Optics, Illumination and Image Sensing for Machine Vision, Svetkoff, D.J. ed., Boston, MA, September, pp. 209–215 (1989).
15. Harding, K., G. Abramovich, V. Paruchura, S. Manickam, and A. Vemury, 3D imaging system for biometric applications, *Proc. SPIE*, 7690 p. 76900J (2010).
16. Lertrusdachakul, I., Y.D. Fougerolle, and O. Laligant, Dynamic (De)focused projection for three-dimensional reconstruction, *Opt. Eng.*, 50(11), 113201 (November 2011).
17. Watanabe, M., S.K. Nayar, and M. Noguchi, Real-time computation of depth from defocus, *SPIE Proc.*, Vol. 2599, 3 Dimensional and Unconventional Imaging for Industrial Inspection and Metrology, Harding, K.G. and Svetkoff, D.J. eds., pp.14–25 philadelphia (1995).

18. Xian, T. and M. Subbarao, Performance evaluation of different depth from defocus (DFD) techniques, *SPIE Proc.*, Vol. 6000, Two and Three-Dimensional Methods for Inspection and Metrology III, Harding, K.G., ed., Boston, MA p. 600009-1 to 600009-13 (2005).
19. Ghita, O., Whelan, P.F., and Mallon, J., Computational approach for depth from defocus, *J. Electron. Imag.*, 14(2), 023021 (April–June 2005).
20. Cho, S., W. Tam, F. Speranza, R. Renaud, N. Hur, and S. Lee, Depth maps created from blur information using images with focus at near and at far, *SPIE Proc.*, Vol. 6055, Stereoscopic Displays and Virtual Reality Systems XIII, Woods, A., ed., San Jose, CA p. 60551D (2006).
21. Liu, R., L. Li, and J. Jia, Image partial blur detection and classification, *Proceedings of the International Conference on Computer Vision and Pattern Recognition*, pp. 1–8 Cambridge, MA (2008).
22. Ens, J. and P. Lawrence, An investigation of methods for determining depth from focus, *IEEE Trans. Pattern Anal. Mach. Intell.* 15, 97–108 (1993).

Part V

Advanced Optical Micro-Metrology Methods

12

Parallel Multifunctional System for MEMS/ MOEMS and Microoptics Testing

Małgorzata Kujawińska, Michał Józwik, and Adam Styk

CONTENTS

12.1 Introduction

In the early chapters of this book, we took a look at the challenges and tools for measuring very large parts and structures. We then looked at medium-sized parts like engines and appliances and finally down to very small-part technologies such as white-light interferometry and microscopy. We now look at an application to some of the smallest parts made today and ones that are expected to play an increasingly important role in the future. The methods described in this chapter derive from the basic tools described in previous chapters but make use of some of the most advanced methods of diffractive optics today to achieve the very precise and difficult measurements.

MicroElectroMechanical Systems and MicroOptoElectroMechanical Systems (M(O) EMS) technology integrates mechanical, electrical, and optical elements on a common silicon or glass substrate. Each wafer may contain from tens up to several thousand devices. M(O)EMS, together with simpler examples of array microoptics (e.g., arrays of microlenses), has become one of the most important drivers for product innovation in many applications including automotive, sensing, imaging, medical, and defense. These applications of M(O)EMS cover more and more responsible tasks, so often 100% quality control is required. Thus, the M(O)EMS industry, which hardly existed two decades ago, has a huge potential for expansion with an expected Compound Annual Growth Rate (CAGR) of 10%–15% [1].

To meet the demand for increased volume and lower cost, M(O)EMS manufacturers are currently migrating from 6 to 8 in. or even 10 in. production lines. Also, more production is being outsourced to M(O)EMS foundries as process technologies become more standardized. On the other hand, microelements cover more and more responsible functional tasks, and therefore often 100% quality control is required. These are the reasons why for most M(O)EMS devices and microoptics arrays at least half of the production costs stem from packaging and testing. This is because packaging and testing are serial processes consuming a significant amount of equipment time. Wafer-level packaging and testing will therefore be necessary if the volume–cost curve is to meet market requirements.

The main challenge for the test equipment producers is the adaptation of the test systems to the wide range of applications. The M(O)EMS structures and microoptics arrays vary strongly in size, pitch, and functionality. The instruments on the market today are inflexible and expensive, and the testing process is time consuming. Thus, new inspection strategies and a generic testing platform able to adapt to the wide range of applications are required by the industry. The electrical probing cards in the semiconductor industry are a good example for such a concept. The basic setup (handling and positions the wafers) is generic, while the probe cards and the electronic equipment are customer specific [2]. Integrated circuits (IC) are thus routinely tested at wafer level using standard test equipment.

In contrast, wafer-level testing of M(O)EMS and microoptics is a relatively new concept. Besides the fact that the semiconductor industry had more time to develop scalable testing solutions, there are also differences in testing requirements that render M(O)EMS and microoptics testing even more challenging. For IC testing, the device under test is stimulated by electrical signals, while the probe cards sense the resulting electrical signals at the devices' outputs. In contrast, testing M(O)EMS involves physical stimulation (such as pressure and acceleration) and sensing mechanical, electrical, or optical outputs. Moreover, tests typically require well-controlled environments (temperature, vibration, pressure, etc.) to mimic the application conditions (see Chapter 5).

Nonelectrical testing of M(O)EMS and microoptics is mainly carried out using optical methods that allow for noncontact and noncontaminating inspection [3]. However, the optical inspection step is today one of the bottlenecks in production of M(O)EMS, as well as array microoptics. In the production process, both passive parameters (i.e., x-y-(z) dimensions) and active parameters (e.g., resonance frequencies and deformations) of microstructures need to be tested. Passive MEMS testing is often based on microscope systems connected with a camera. Image processing is utilized for the detection of the x and y dimensions. For many producers, this is the only test approach. Some suppliers offer confocal techniques or interferometry to measure the z-direction. The shape of the object can thus be measured. Active MEMS testing is typically carried out using interferometers [4] and vibrometers [5].

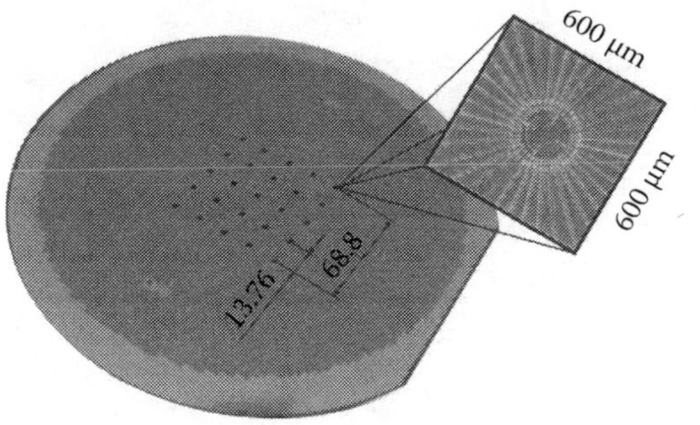

FIGURE 12.1
A wafer with MOEMS structures and exemplary microelement in an array together with exemplary dimensions of pitch between measurement channels and dimension of a single microelement.

In particular, the measurement of active parameters is time consuming and thus not suitable for testing in mass production. To remove this bottleneck, the test equipment needs to overcome the high ratio between required measurement resolution and wafer size. The diameter of standard wafers for high-volume production of M(O)EMS is today up to 10 in. (250 mm). The structures to be inspected on these wafers have a typical lateral size from 100 μm^2 to 10 mm^2 (Figure 12.1). The inspection ratio covers thus the range from about 10^{-3} to 10^{-7}. If we consider the typical defect size which is in the range of some 10 nm, this ratio would be further increased.
Today, there are four approaches to overcome this ratio:

1. A high-resolution measurement system measures one structure at a time and scans the system serially from structure to structure. These systems are suitable for spot testing. The control of the lateral dimensions is carried out in the production line. However, these systems are not suited for wafers with several thousand microstructures and for active M(O)EMS testing, because of the long measurement time per wafer.
2. Increase the field of view at high resolution (several cameras/more pixels) to inspect several structures at the same time and scan over the whole wafer area. This approach requires an improvement of the camera technology (13 Giga-pixels to cover an 8 in. wafer). Image processing of such large images is time consuming, and for in-line measurements, processor capacity requires capabilities far above today's state of the art. As early as in 2003, Aswendt et al. presented a whole wafer inspection system with rather small spatial resolution [6]. Systems that follow this approach today use several cameras to increase the spatial resolution.
3. Inspect larger areas of the wafer or the whole wafer with a low-resolution measurement system, identify suspicious areas, and then inspect these areas with a second measurement system at higher resolution. This approach was introduced by Osten et al. [7] as scaled topometry. This approach is a promising method for time-efficient measurements in particular for objects with very small pitch, that is, microlenses. The main challenge is to identify suspicious areas. This approach is the most time efficient. However, it does not cover the requirement of 100% production control.
4. Inspect the areas of interest only in parallel using an adapted array of sensors [8,9] (possibly scan the whole array). The fourth approach is very well suited for M(O)EMS and microoptics in arrays testing. It requires a priori information

of the object. The size of the active area and the pitch between the M(O)EMS structures (Figure 12.1) need to be known when parts of the inspection system are designed. Because of the high accuracy and reproducibility in micro production, the array approach has large potential to combine high-accuracy measurements with the reduction of the measuring time by a factor similar to the number of inspection channels. Furthermore, the approach is fully compatible with the requirement of 100% production control. At the same time, the systems should be multifunctional and reduce the investment and maintenance costs. We present the first implementation of this fourth approach in more detail to explore the challenges of very small-scale metrology. This system was developed within the European Commission-funded SMARTIEHS project [10].

12.2 Concept of Parallel and Multifunctional System Realized in SMARTIEHS

The most common serial approach of existing inspection systems (single-channel scanning microscope or interferometer) causes long measurement times per wafer and low throughput rates. Furthermore, the existing systems on market are based on assembled macrooptics; thus, the investment cost is rather high.

In the SMARTIEHS system, a wafer-to-wafer inspection concept is introduced in order to enable the parallel testing of several dozens of M(O)EMS or microoptic structures within one measurement cycle. To obtain this, an exchangeable microoptical probing wafer is adapted and aligned with the M(O)EMS wafer under test. The probing wafer comprises an array of microoptical testing devices. The quantities to be measured include shape, out-of-plane deformation, and resonance frequencies. The measurements of these quantities are typically carried out using interferometers. Some first approaches for an array interferometric sensor principle had been developed within microoptical microphone technology in the form of a point scanning interferometer array [11,12] but at the time of this writing are at a concept state only.

The configuration, spacing, and resolution of the interferometer array are designed for each specific application. The illumination, imaging, and excitation modules are modular and can be moved from one interferometer array to the other. The modules can thus be interchanged if the spatial distribution of the M(O)EMS structures or changed functionality requires this. The array configuration can be nonregular and optimized for time-efficient inspection strategies. More than 100 interferometers can thus be arranged on an 8 in. (or bigger) wafer and decrease the inspection time of a wafer by a corresponding factor.

The wafer-to-wafer concept addresses the production of complete interferometers in an array arrangement by standard micro-fabrication technologies. All the interferometers in one array have to be precisely spaced within micrometer tolerances. Additionally, as each interferometer consists of multiple functional optical layers, the layers need to be arranged at a certain distance with micrometer accuracy. Possible approaches to solve this challenge are stacking of different microoptical wafers to a "semi-monolithic" block or applying double-sided fabrication of elements on one wafer.

The multifunctionality of the system is demonstrated by two different probing wafer configurations:

- A refractive Mirau-type interferometer. This interferometer is used for low-coherence interferometry to measure the shape and deformation of the M(O)EMS structures.
- A diffractive Twyman–Green-type interferometer. This interferometer is applied for vibration analysis using laser interferometry (LI) to find the resonance frequency and spatial mode distribution. Furthermore, the LI will be applied to measure shape and deformations on smooth surfaces.

The inspection system is integrated in a commercially available probing station (e.g., SUSS prober PA 200 [13]). Making use of commercial tools developed and proven on a related field can be of significant savings in an effort to develop a new metrology system. Wafer handling and measurement routines can thus easily be automated and integrated in the production line as the M(O)EMS wafer is mounted and positioned using the wafer chuck of the prober. Figure 12.2 shows a 3D CAD representation of the prober with the inspection system integrated.

The optical configuration of the instrument is shown in Figure 12.3. It comprises the two different 5 × 5 interferometer arrays. The left side of the image shows the low-coherent interferometer (LCI) and the right side the laser interferometer (LI) array. The light sources are arranged in an array and positioned on each side of each interferometer unit. The light is guided by a beam splitter toward the probing wafer.

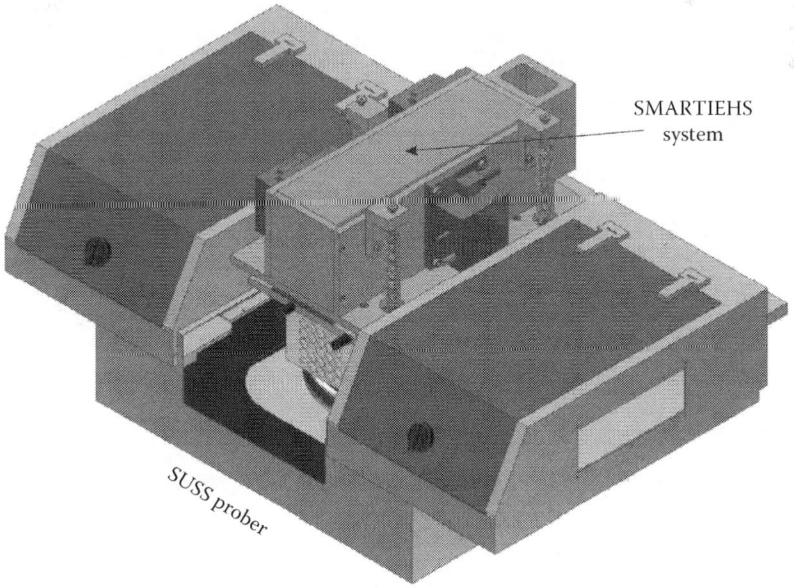

FIGURE 12.2
CAD 3D representation of the SUSS test station (PA200) for M(O)EMS testing, with the optical inspection system attached at the scope mount of the prober.

A glass wafer containing mini-lenses produced by standard micro-fabrication processes is used for the imaging of the interference fringes toward the camera. A distributed array of 5×5 smart pixel imagers detects the interferometric signals. The signal processing is based on the "on pixel" processing capacity of the smart pixel camera array [14], which can be utilized for phase shifting or envelope maximum determination (see Chapter 7). Each micro-interferometer image will thus be detected by a NxM pixels sub-arrays distributed in the imaging plane.

An excitation system for the M(O)EMS structures is needed for active testing. A glass wafer consisting of indium tin oxide (ITO) electrodes is applied for electrostatic excitation of the resonance frequency of the structures [15]. For the measurement of deformation, the active area on the M(O)EMS is excited either electrostatic by the ITO wafer or using a tailor-made pressure chuck.

The heart of the inspection system is the innovative interferometer (probing) arrays, which allow implementation of the concept of parallel measurement of several M(O)EMS structures. Therefore, the next two sections explore in more detail the LCI and LI arrays design and technology.

12.3 Low-Coherence Interferometry Probing Wafer

12.3.1 Concept

The LCI is realized in a Mirau interferometer configuration [3]. The illumination is provided by an array of light-emitting diodes (LED) (Figure 12.3). Each illumination system consists of a LED, a lens, and an aperture stop. The LED is slightly defocused versus the imaging lens. Camera and illumination aperture are, seen from the object side, "virtually" positioned in the same vertical and horizontal position.

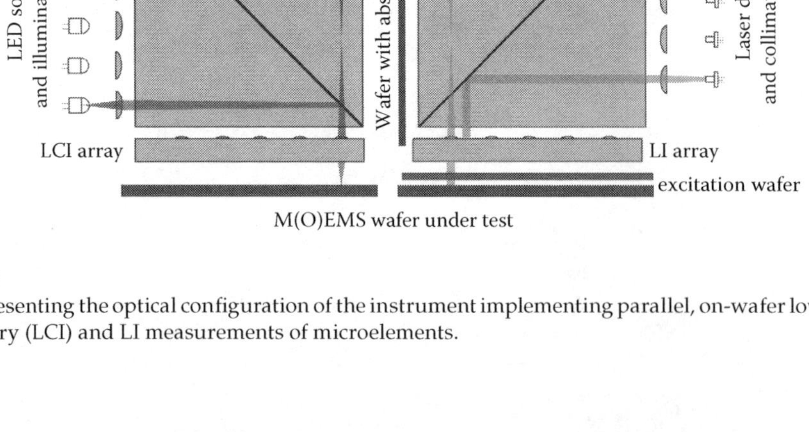

FIGURE 12.3
A scheme presenting the optical configuration of the instrument implementing parallel, on-wafer low-coherence interferometry (LCI) and LI measurements of microelements.

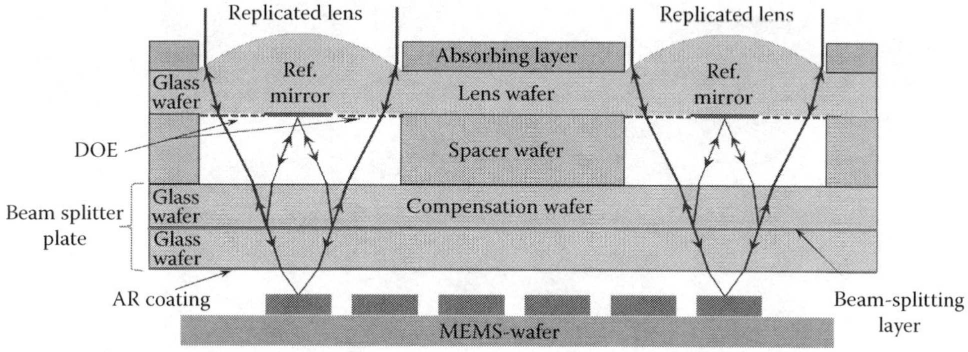

FIGURE 12.4
Optical configuration of the LCI.

The design of the LCI array is shown in Figure 12.4. It consists of two glass substrates separated by a spacer. The upper wafer is carrying the plano-convex imaging lenses replicated onto the top surface and the reference mirror on the backside. The beam splitter wafer is composed from a glass wafer with a partly reflective coating on the upper side and the compensation wafer. The beam splitter wafer divides the incoming light into a reference beam and an object beam. The reference mirrors are placed in order to reflect the light coming from the beam splitter wafer and thus form the reference beam of the Mirau interferometer. The object beam is reflected back from the M(O)EMS structure under test. Both beams are interfering in the beam splitter wafer. The interference fringes are imaged by the imaging lens onto the camera, where the camera detects the interference signal.

12.3.2 Technology

The LCI matrix is fabricated on a 4 in. basis within a square of nearly 70×70 mm^2, where the pitch between the channels is 13.76 mm. The details of LCI technology development are described in [16].

First, the 700×700 μm^2 reference mirrors are created by a liftoff process over a surface of a 500 μm thick glass substrate. In order to reduce the light reflected back toward the camera, an additional amorphous silicon (a-Si) layer is placed between the glass substrate and the aluminum (Al) layer. A 230 nm thick layer of a-Si is created by low-temperature plasma-enhanced chemical vapor deposition (PECVD). A 150 nm thick Al layer is evaporated onto the a-Si, forming the mirror layer. The aluminum evaporation is performed with a high-energy process in order to achieve low-roughness layers and compensate as much as possible for the roughness induced by the a-Si layer.

The refractive microlens used in LCI has a spherical shape, a diameter of 2.5 mm, and a sag of 162 μm in order to achieve a numerical aperture (NA) of 0.135 and an effective focal length of 9.3 mm. The correction of the chromatic and spherical aberrations induced by the microlens is achieved by a diffractive optical element (DOE) placed on the same surface as reference mirror. The replication of the nearly planar DOE needs to be done prior to the microlens. The design was realized by using ZEMAX, a commercial ray trace software package, considering a diameter of 2.5 mm with a nonstructured 700×700 μm^2 at the center to fit the area occupied by the micro mirrors.

The DOE is representing a phase function of radial symmetry. They are fabricated by variable dose laser lithography in a photoresist layer [17], with a DWL400 FF system from

FIGURE 12.5
Image of the lens array replicated into a UV curable polymer.

Heidelberg Instruments. The DOE master is obtained from selective exposition of a 3 μm thick layer of photoresist by an intensity-modulated laser beam. Then, the structures are replicated over the glass substrate with a hybrid polymer resin ORMOCOMP HA497 [18]. The use of this hybrid polymer improves homogeneity of the layer thickness.

Once the DOE is created around the micro mirror, the microlens is formed directly onto the opposite surface of the glass substrate. The photolithography and resist reflow are basic technologies in lens mastering process. The obtained resist-based spherical lenses are then replicated in a PDMS in order to create the master to be used in the last replication process. The final microlenses (Figure 12.5) are obtained by UV molding of ORMOCOMP polymer [19] in a SUSS MicroTec contact mask aligner MA6 with special UV-molding tools and software [20]. The lateral positioning accuracy on the wafer is achieved by using lithographic techniques to generate the structures.

The last optical component required for the Mirau interferometer is a beam splitter plate (BSP). It divides the illumination beam into an object beam and reference beam and later recombines both beams on their return paths. The BSP should achieve a ratio-transmitted light/reflected light of 50/50 under normal incidence. In order to equalize the optical paths of the reflected and transmitted light, the BSP, made of a dielectric multilayer stack, is sandwiched between two similar Borofloat 33 substrates. Within the Mirau configuration, the beam splitter plate is located at approximately half of the microlens focal length. This distance is adjusted by spacers that also equalizes the optical path lengths of the reference and the measurement arms of the interferometer and are used to reduce the crosstalk between channels. The stray light can be eliminated also by an additional absorbing layer deposit on the top of glass wafer with the lenses. The final LCI matrix, shown in Figure 12.6, is produced with antireflective coatings on the lenses and at the bottom surface of BSP.

12.4 LI Probing Wafer

12.4.1 Concept

The design of the LI is based on a Twyman–Green interferometer configuration. Its architecture requires custom-designed DOE. Those DOEs are diffraction gratings optimized

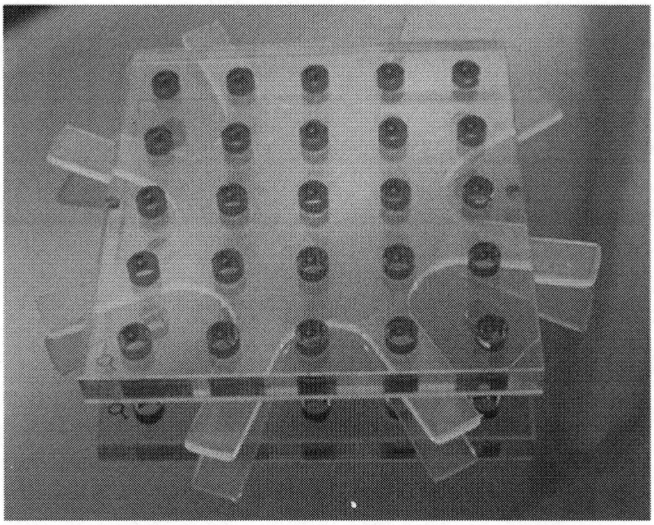

FIGURE 12.6
An image of the assembled LCI array.

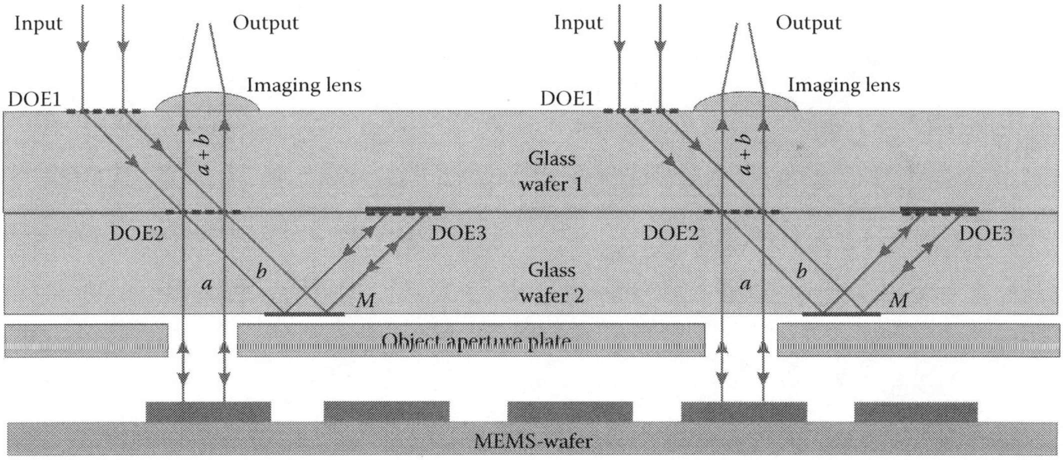

FIGURE 12.7
Optical configuration of the LI.

in order to obtain maximum output intensities and maximum interference contrast for an operating wavelength of 650 nm.

The LI probing wafer is composed of two glass substrates, as shown in Figure 12.7 (the arrows indicate the directions of beams propagation in LI). The first diffractive grating (DOE1) is etched at the top surface of the first substrate. DOE1 allows changing the direction of an incident, collimated beam in order to illuminate the beam splitting diffraction grating (DOE2) fabricated at the top surface of the second substrate. This surface also contains the third diffraction grating (DOE3) and bottom surface of the second substrate that has a mirror. The transmitted light of DOE2 forms the measurement beam, which goes through the holes in the object aperture plate and is reflected back by the investigated M(O)EMS structure. The light reflected of DOE2 is directed through the mirror M to DOE3 acting as the reference mirror. The mirror M enables to equalize an optical path difference in reference arm (distance between DOE2 and DOE3) and object arm (distance between DOE2 and object) in order to maximize contrast of interferometric fringes. Both beams are after reflections recombined by DOE2 and imaged through an imaging optics at a camera. As imaging optics, the generation of the 5 × 5 microlens array was realized using a standard replication process on the top surface of the first substrate. The alignment of the lenses with respect to the gratings was realized via alignment marks (fiducials) using the backside alignment microscope.

The theoretical optimization of gratings profiles was achieved by simulations with rigorous coupled wave approach (RCWA) [20]. Because of its deflection function for the incidence beam, DOE1 is optimized for the first diffraction order at a normal incidence. The angle of the first diffraction order is governed by the grating equation. For illumination wavelength $\lambda = 658$ nm and air-SiO_2 interface (nSiO2 = 1.457 [21]), it equals 26.85°. DOE1 (Figure 12.8) comprises two binary grating bars within a period of 1 μm. The smaller bar

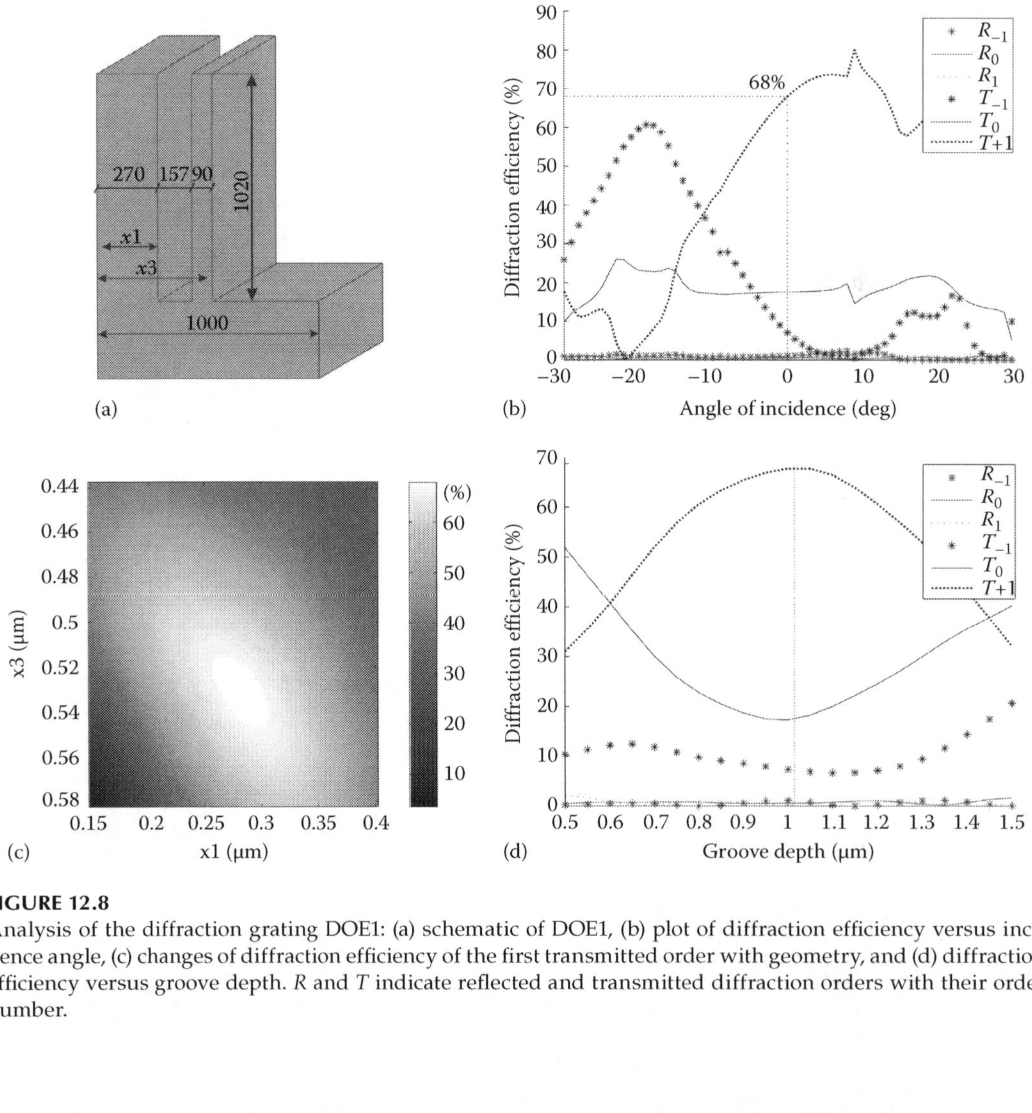

FIGURE 12.8
Analysis of the diffraction grating DOE1: (a) schematic of DOE1, (b) plot of diffraction efficiency versus incidence angle, (c) changes of diffraction efficiency of the first transmitted order with geometry, and (d) diffraction efficiency versus groove depth. *R* and *T* indicate reflected and transmitted diffraction orders with their order number.

is only 90 nm wide. Moreover, the distance between the two bars is 157 nm. Figure 12.8c shows the analysis of the diffraction efficiency of the first transmitted order at normal incidence with respect to changes of bar width ×1 and the distance ×3 (width of the both segments and the spacer). The diffraction efficiency versus incidence angle is shown in Figure 12.8b. At normal incidence, the efficiency of the first transmitted order (T_{+1}) is close to 68% for an optimal grating depth of 1020 nm (Figure 12.8d). The incoupling grating is the most demanding one from a technological point of view because of the high-aspect-ratio structures.

The diffraction grating DOE2 (Figure 12.9a) works as a beam splitter. The illumination beam from DOE1 is impinging at DOE2 at the incidence angle of 26.85°, and it is diffracted creating two useful orders: 0th with efficiency of 52% and +1 with efficiency of 26% (Figure 12.9b). The 0th order beam is transmitted to a reference, and +1 one illuminates an object. Then after reflections, DOE2 recombines beams and directs them toward the camera. The measurement beam after reflection from an object is at normal incidence and is transmitted as the 0th order with an efficiency of 52%. The reference

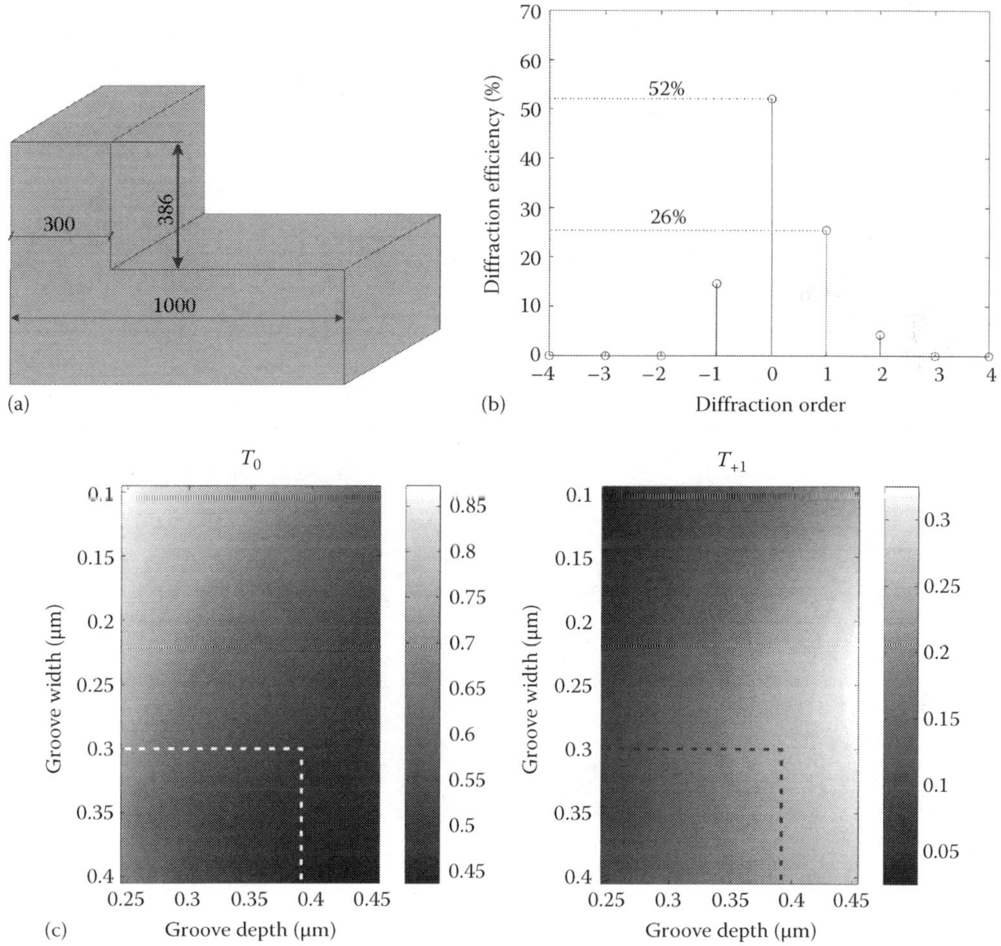

FIGURE 12.9
Analysis of diffraction grating DOE2: (a) schematic of DOE2, (b) diffraction efficiency of the transmitted diffraction orders, and (c) changes of diffraction efficiency of the first transmitted orders with DOE2 geometry.

beam hits DOE2 under an angle of 26.85° and becomes diffracted as +1 order with an efficiency of 23%. Assuming the same reflection coefficient in both interferometer arms, the total budget of DOE2 efficiencies gives almost equal output in the reference and measurement arms (approximately 12%). DOE2, due to its multifunctionality, is more sensitive to fabrication tolerances, as shown in Figure 12.9c. Variations in depth and duty cycle affect the efficiency of different orders used in the interferometric setup in different ways, causing a decrease of an interferogram contrast. From a technological point of view, the structure of the DOE2 is rather uncritical with respect to lateral feature size, duty cycle, and depth.

The diffraction grating DOE3 is a retro reflecting grating. The main task of DOE3 is to reflect back the incident light into the same direction. Thus, the grating is acting as a reference mirror of the laser interferometer. To accomplish this optical function, a fused silica-etched grating is covered with silver. The high reflectivity obtained in the −1st diffraction order is strongly dependent on the silver coating quality. As can be seen from the design in Figure 12.10a, the grating groove has a width of only 160 nm. Figure 12.10b shows the simulation of the efficiency of the optimized grating structure in the function of DOE3 geometry.

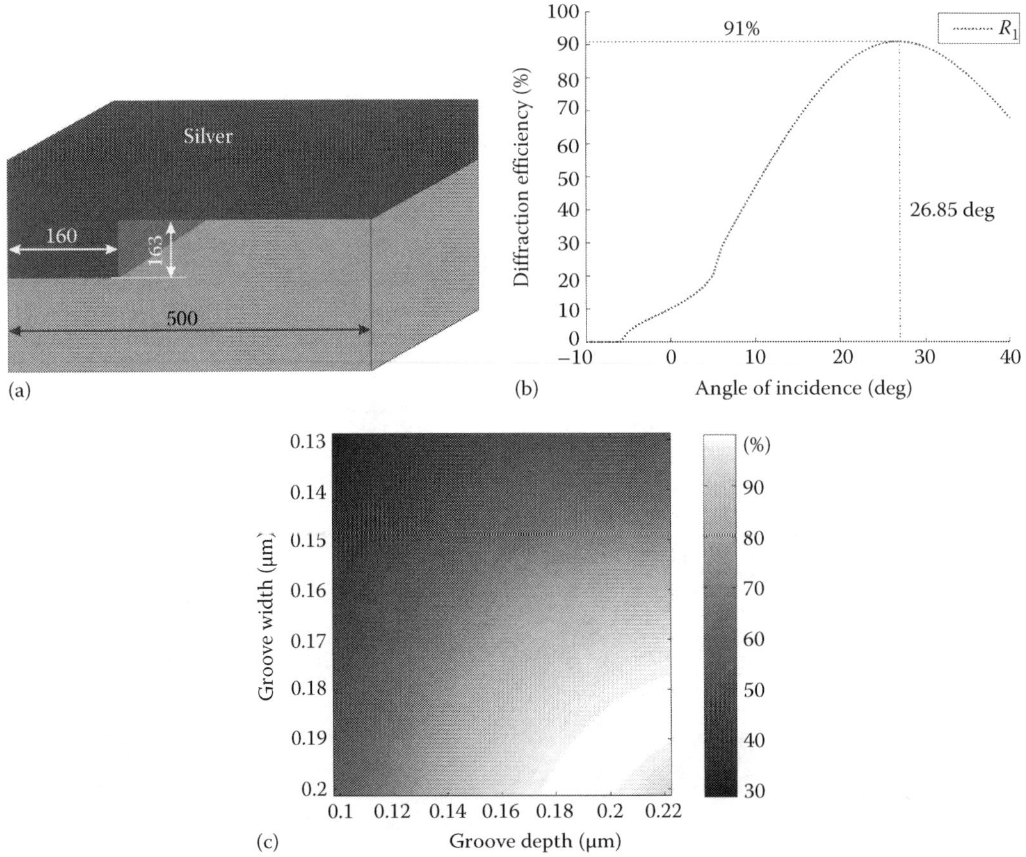

FIGURE 12.10
Analysis of diffraction grating DOE3: (a) schematic of DOE3, (b) diffraction efficiency versus incidence angle, and (c) changes of diffraction efficiency with DOE3 geometry.

12.4.2 Technology

The fabrication process of the integrated laser interferometer can be divided into four main stages. In the first phase, the grating arrays are fabricated on two 6.35 mm thick glass substrates (fused silica mask blank). All diffraction gratings used for creation of LI have a binary profile, that is, the profile is made of subwavelength ridges etched periodically into the substrate [22]. In particular, DOE2 and DOE3 are simply made of one ridge within a period; DOE1 has two different ridges in a single period. The fabrication of the gratings was carried out by a standard binary optics fabrication process [23]. The resist pattern is first transferred into the chromium layer by reactive ion etching (RIE); then it is used as a hard mask for the final RIE etching of the grating into the fused silica substrate. The second phase is the replication of the microlens on the top of the first glass substrate. The third phase is dedicated to creation of mirrors on the bottom surface of second glass substrate. In the last phase, the two glass substrates are aligned and bonded together. The details of the whole fabrication process and its results have been discussed in Ref. [24].

Theoretical results of diffraction efficiency are a bit higher but are still in good agreement with the measured values after fabrication. The assembly procedure was developed especially for the LI array considering the best order and compatibility of the individual technological steps. This procedure uses elastomer as a bonding material. The elastomer layer was generated on the backside of the upper substrate in a process where a plane reference glass substrate treated with an anti-sticking layer was used as the replication tool. The assembled LI array with 5×5 channels is shown in Figure 12.11.

FIGURE 12.11
Image of the assembled LI array.

12.5 Smart Camera and Data Analysis

12.5.1 Smart Pixel Camera

Interferometric signals obtained in each channel are analyzed by smart pixel imagers placed in a 5×5 matrix at the camera module (Figure 12.12) that transfers the images via a high-speed frame grabber to the PC. The pitch between the imager chips is adapted to the interferometer pitch. Every imager features 140×140 smart pixels that allows for the direct demodulation of the time-dependent interferometric signal and increases the parallelism to a total resolution of about half a million interferometric sub-channels. The smart pixel camera modules can be individually focused. This enables the compensation of deviations of the nominal focal lengths that may occur in the production of the mini-lenses. The imaging optics has a magnification of 9×. The imaging system is designed without an intermediate image. The field of view is $600 \times 600\ \mu m^2$.

The demodulation of incoming signals is performed on the detection level with demodulation frequencies (ω_D) up to 100 kHz [14]. Every pixel of the detection module creates two output signals: I (in-phase) and Q (quadrature-phase). They are the result of internal processing of the collected interferometric signal, which is integrated and sampled. The signal samples are further multiplied by a reference signal of demodulation frequency and averaged on two paths, then phase shifted by $\pi/2$ with respect to each other (see Figure 12.13). A sample and hold stage in both circuit branches allows simultaneous demodulation and readout, which means that while the stored values are read out of the imager pixel field, the input signal is demodulating and generating the next values. These two branches create I and Q channels of the imager. Each of the output I and Q values may be created as an average of the sampled signal over N_{avg} demodulation periods before they are read out. The signal averaged over N_{avg} demodulation periods is denoted as one frame, with frame number m. A total number of M frames, depending on the application, are acquired. Hence, the total number of demodulation periods is MN_{avg} (or $M(N_{avg} + 1)$, depending on camera settings), and the total number of sample periods is $4MN_{avg}$ $(4M(N_{avg} + 1))$, since every demodulation period is sampled with four samples. The signals I and Q from the sampled detector intensity $I_D(n_{sh})$ in the frame m may be described as

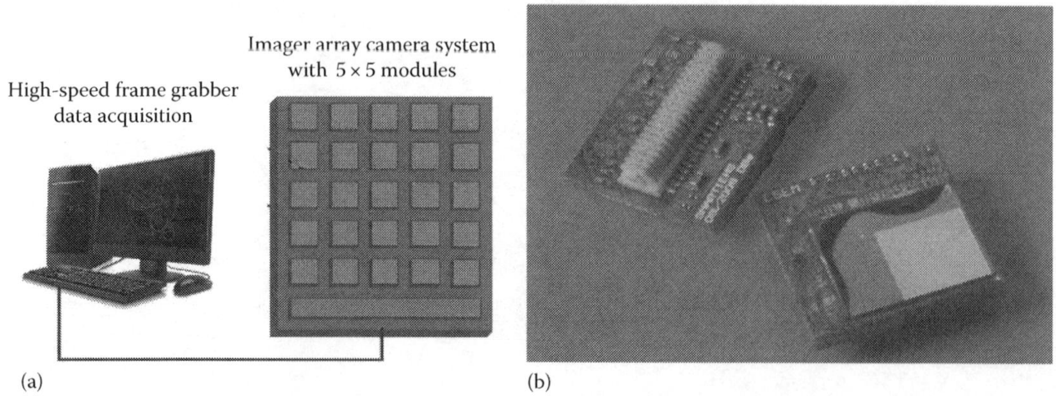

FIGURE 12.12
Smart pixel camera: (a) an array of 5×5 detection modules and (b) image of a single detection module.

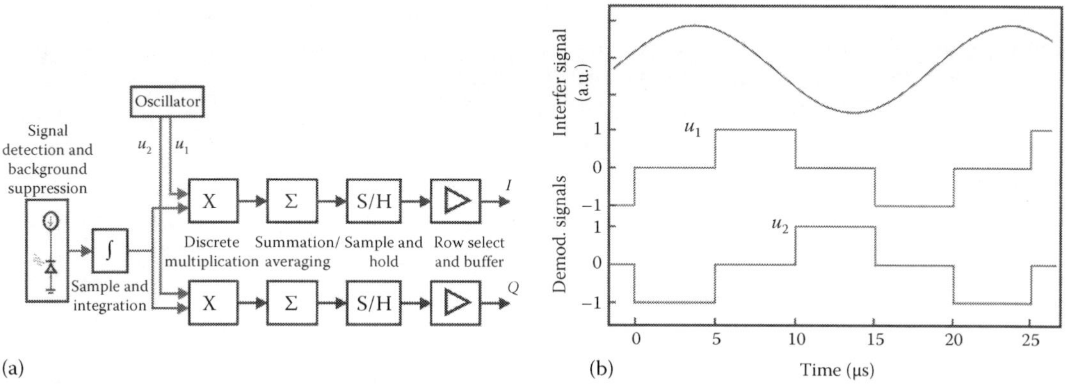

FIGURE 12.13
Smart pixel camera working principle: (a) electronic signal flow and (b) I/Q signals creation; u_1 and u_2 are the reference signals $\pi/2$ phase shifted.

$$I(m) = \sum_{n_{avg}=1}^{N_{avg}} I_D(4(m-1)N_{avg} + 4(n_{avg}-1)+1) - I_D(4(m-1)N_{avg} + 4(n_{avg}-1)+3) \quad (12.1)$$

$$Q(m) = \sum_{n_{avg}=1}^{N_{avg}} I_D(4(m-1)(N_{avg}+1) + 4(n_{avg}-1)+2) - I_D(4(m-1)N_{avg} + 4(n_{avg}-1)+4) \quad (12.2)$$

Each pixel has a background suppression circuit to avoid saturation and small input signal contrast. These features allow for demodulation of signals with a high dynamic range [25]. To do so, a detected interferometric signal needs to be modulated over time. In most of the cases, this modulation is performed by linear scanning (linear change of optical path difference) in one of the interferometer arms. The speed of scanning has to be suited exactly to the demodulation frequency set in the camera. The scan length in one demodulation (lock-in) period is $\lambda/2$.

12.5.2 Signal Processing in LCI

In the SMARTIEHS, instrument measurements performed by means of low-coherence interferometry (LCI—see Chapter 10) are applied for topography and deformation investigations. The smart pixel camera with demodulation mode is well suited for signal analysis in LCI. LCI data capture requires continuous shift (with constant and well-defined velocity) of the object wafer with respect to a reference surface. The camera then produces tomographic 3D data. This means that a cube of data is returned with the information of the reflected light intensity within every (m, n_x, n_y) "voxel," where m is the acquired slice number and n_x and n_y are the imager pixel coordinate. For an opaque object such as most M(O)EMS, a few voxels only, which correspond to the surface of an object, will have relevant values. So, for any vertical scan of the LCI, only one depth point will be found for each pixel. In biological applications with translucent objects line the eye, multiple points may be captured, defining a volume rather than just a surface. In this application, the goal of the processing algorithms is to extract the information of the position of the object surface within the cube of data. This means extracting the topographic image from the tomographic 3D data. For the data processing, two algorithms have been implemented: (a) convolution and max search and (b) minimize energy.

The first algorithm allows extracting every single surface point from the corresponding z-scan. It means that the value of the envelope modulating carrier interferometric signal is tracked:

$$S(i,m,n_x,n_y) = E^2(i,m,n_x,n_y) = I^2(i,m,n_x,n_y) + Q^2(i,m,n_x,n_y) \quad (12.3)$$

where

I and Q are the two signals obtained from the camera
i is the number of imager
m is the number of acquired slice (frame)
n_x n_y are the pixel coordinates in each imager
S is proportional to the reflectivity of the object in the coherence plane

For simplicity, we present here the treatment of data obtained from ith imaginer. The "cube" of data (Figure 12.14a) is made of a stack (in the z-direction) of 2D (x, y) images. However, since the object is opaque, the optical-reflected signal will be relevant only at the surface of the object. This means that most of the values in the "cube" will be zeros except those that lie on the part surface. Only the points corresponding to the surface of the object will have a significant intensity (Figure 12.14b). An object surface is represented by "topographic" data where every value at pixel (n_x, n_y) gives the height of the surface.

Several algorithms may be used to extract the topographic information from the 3D cube of data. The most straightforward algorithm searches along the z-direction to find the position of the point with the highest intensity. However, as the signal might contain a certain amount of noise, it is not an accurate approach. Therefore, a convolution is previously applied to the signal, as a low-pass filter (along z) (Figure 12.14c) and the maximum of the modified signal are found. However, in order to get a subframe resolution, a parabola can be fitted between the points in the neighborhood of the maximum. Another possibility is to calculate the center of gravity of these points.

Unfortunately, some z-scans may have too weak of a surface signal to be able to show the surface point, especially close to edges or particular "dark" spots. If this happens, then the height value is randomly found from any noise spike. These particular bad points appear as "outlier" points on the surface. To get a correct value for these outlier points, the minimum energy algorithm is used. It takes into consideration the outlier points' neighborhood. In a first step, the surface is calculated with the convolution and max-search algorithm to get a first surface $h_0(n_x, n_y)$. Then a quality value, Q_y, is calculated:

$$\begin{aligned} Q_Y &= aS\big(n_x,n_y,h_0(n_x,n_y)\big) \\ &\quad + b\big|\big(4h_0(n_x,n_y) - h_0(n_x-1,n_y) - h_0(n_x+1,n_y) - h_0(n_x,n_y-1) - h_0(n_x,n_y+1)\big)\big| \end{aligned} \quad (12.4)$$

where a and b are the parameters established during the calibration process. This quality factor is a balance between the intensity of the signal at the point (n_x, n_y) and the height difference from this point to its neighbors. This value is maximized by letting $h_0(n_x, n_y)$ vary. A new value $h_1(n_x, n_y)$ is obtained. This value is calculated for every point of the surface. This process is iterated with the new values found.

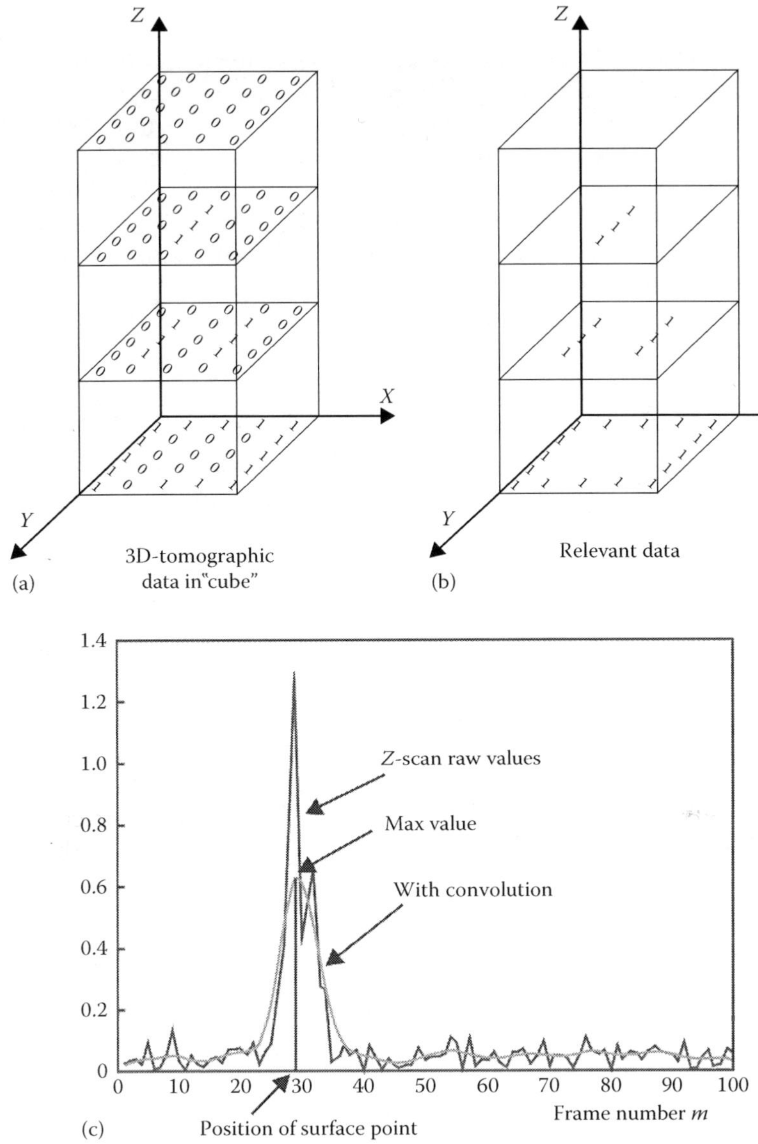

FIGURE 12.14
3D tomographic data cube returned by the camera: (a) all data, (b) relevant data, and (c) the principle of the convolution and max-search algorithm.

The difference between this algorithm and a simple low-pass filter is that we still take into account the volume data. If the $S\ (n_x, n_y, m)$ scan has a small local maximum at the surface height, it will be preferred to the absolute maximum which can be quite far from the surface. The direct neighbors will be much less influenced in comparison to a simple low-pass filter on the surface.

Former described algorithms (convolution and max search) are characterized by high-speed processing but are more subject to noise influence producing erroneous "spikes" in the evaluated surface topography, while the latter one (minimum energy) is much more resistant to noise but requires a longer calculation time.

12.5.3 Signal Processing in LI

The LI module in the SMARTIEHS system is responsible for determination of resonance frequencies of the parts and their vibration modes characterization. It also employs the smart pixel camera working in the demodulation mode. In order to investigate resonant frequencies of the object under test, new quasi-heterodyned measuring technique has been suggested. This technique is based on the usage of the sinusoidal modulation of the light source with a carefully chosen frequency ω_L. The light modulation frequency has to fulfill the $\omega_L = \omega_V - \omega_D$ condition, where ω_V is the frequency of the investigated vibration (excitation signal frequency) and ω_D is the camera demodulation frequency. If the investigated object vibrates at ω_V and the aforementioned condition is fulfilled, the signals from the camera become

$$I = \left(\frac{2\pi\gamma\chi A_V}{\lambda}\right) I_0 \sin(\phi_0 + \delta)\cos(\phi_V - \phi_L) \tag{12.5}$$

$$Q = \left(\frac{2\pi\gamma\chi A_V}{\lambda}\right) I_0 \sin(\phi_0 + \delta)\sin(\phi_V - \phi_L) \tag{12.6}$$

where
- I_0 is the bias
- γ is the contrast of the interferogram
- ϕ_0 is the phase of the interferogram (depends on the shape of the object)
- δ is an additional phase shift that may be introduced intentionally
- ϕ_V and ϕ_L are the initial phases of vibration and light modulation, respectively
- λ is the wavelength used
- A_V is the amplitude of vibration

If the object vibration amplitude is small, that is, $A_V \ll \lambda$ (M(O)EMS are usually actuated with an amplitude ranging between 40 and 80 nm or even smaller), it can be calculated from the given signals:

$$MD = \sqrt{(I)^2 + (Q)^2} = \left|\left(\frac{2\pi\gamma\chi A_V}{\lambda}\right) I_0 \sin(\phi_0 + \delta)\right| \tag{12.7}$$

Unfortunately, the signal given by Equation 12.7 is a proportional measure for A_V only for interferogram points that fulfill the condition $\phi_0 + \delta = \pi/2$ or $-\pi/2$ (so-called working point condition) or are in the very close neighborhood of this working point. In every other interferogram point, the evaluated amplitude will be incorrect (scaled with sinus function). In order to assure the correctly evaluated values for each point of the interferogram, additional frames, with a mutual phase shift of well-defined steps between them, need to be captured. Further details of the presented technique may be found in Refs. [26,27].

An amplitude determination is the most crucial step during dynamic characterization of M(O)EMS devices. The scheme of the full procedure is presented in Figure 12.15. For the resonant frequency determination, one need to acquire a scan of the amplitude behavior over a certain range of frequencies. The collection of the objects' amplitude values for each sampled frequency is called the objects' frequency response function, and the previously shown procedure leads to its determination. The maximum amplitude in the frequency response function shows the resonant frequency of the object.

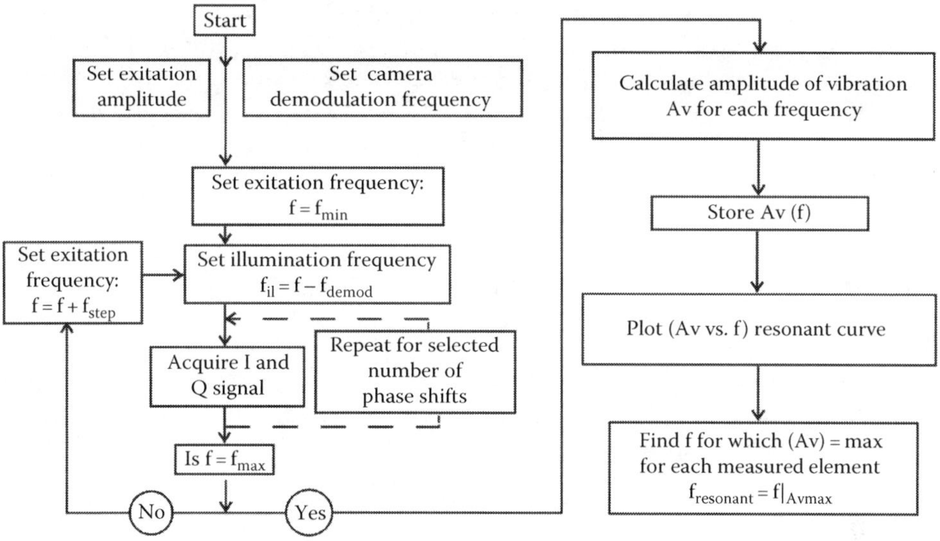

FIGURE 12.15
Procedure for the measurement of M(O)EMS frequency response function.

It should be noted that for the resonant frequency determination a single value of amplitude or its estimator for each imager for each applied frequency (in the scanning range) is sufficient. This reduces the data flow significantly. Since the camera output is in the form of two signals (I and Q) to evaluate the searched value for each imager for each sampled frequency, two calculation methods have been proposed, namely, the *max method* and the *envelope method*. The max method requires at least four phase shifts of the interferometer unit (see Chapter 7) with respect to the measured objects to fulfill the working point condition in every pixel of every imager. The method is simple yet provides limited accuracy of the vibration amplitude measurement. Therefore, the preferred approach is the envelope method that calculates the amplitude value for each pixel of each imager for every sampled frequency along phase-shift dimension using

$$\sqrt{\left(MD_{\delta=0}\right)^2+\left(MD_{\delta=\pi/2}\right)^2}=\left|\left(\frac{2\pi\gamma A_V}{\lambda}\right)I_0\right| \tag{12.8}$$

This method requires only one phase shift of $\pi/2$ value. The evaluated "true" vibration values for each sampled frequency are stored. In the result, the information on I_0 and λ is also present; however, this dependence does not destroy the evaluated frequency response function. As the last stage of processing the calculation, an average value for each imager for every sampled frequency along two dimensions of imager size is performed. The obtained measurement time for 25 channels for 100 frequency samples is approximately 20 s, mostly limited by the time of control of driving signals in the measurement system.

In general, the laser interferometer unit in the SMARTIEHS instrument is employed for dynamic measurements. However, as an additional feature, it may also be employed for topography and deformation measurements if specific conditions on object topography are fulfilled. The interferogram phase (which is directly related to the objects' topography) can be measured using the smart pixel camera in intensity mode. For the phase measurement in this mode, the temporal phase shifting (TPS) method of automatic fringe pattern analysis has been implemented. It is the most accurate method of interferogram analysis, and since the SMARTIEHS instrument has the ability to perform accurate phase shifts (changes of the optical path difference in the interferometer), its implementation is straightforward. In the system, the algorithms requiring from 4 to 8 frames in two classes A and B (A, standard $\pi/2$ phase shift class algorithms; B, extended averaging class algorithms) [28] have been implemented.

12.6 M(O)EMS Inspection System Platform

The design of the M(O)EMS inspection system platform is motivated, on the one hand, by industrial requirements and, on the other hand, by the assumption of modular design and flexibility. The opto-mechatronics design of the platform is determined by the three basic modules:

- A commercial prober system used as system platform. The prober system performs a chuck movement in x, y, z and rotation around z as well as a scope movement in x, y, and z. These features are used to realize a semiautomatic test system.
- The optical system consisting of interferometer wafers, light sources with the corresponding lenses, beam splitters, and a camera unit. The mechanical mounting of these components has to realize the required optical path lengths including various adjustment possibilities to handle tolerances of the custom made and commercial elements.
- A high-precision z-drive needed for focusing of the measurement interferometers versus the M(O)EMS wafer. Furthermore, the z-drive has to realize a highly uniform movement and stable positioning of the optical system in order to perform the measurement tasks.

The mechanical unit was manufactured and is assembled at SUSS probe station (Suss prober PA 200 [13]). The main aim of the unit is to assure mechanical stability and protection for the optical measurement head. The optical head consists of two interferometer types. For each interferometer, special modules holding an array of interferometers and light sources as well as beam splitters were designed. The final SMARTIEHS measurement system mounted on the probe station and its control unit are presented in Figure 12.16.

In order to perform measurements, each interferometer type requires scanning or steplike changing of the optical path difference. For this purpose, the high-precision drive consists of three voice coil drives. The selection of the voice coils was motivated by the large required travel range of 1 mm with a linearity of better than 1%. The position signals in the z-axis (scanning axis) are measured by three commercial interferometers made by *SIOS*, and their measuring heads are fixed at the carrier frame. Based on the *SIOS* interferometer signal, the resolution of 3 nm and the positioning accuracy of 10 nm were achieved.

Besides the z-scan, the platform enables pitch and roll motions (r_x- and r_y-directions) for the parallel alignment of the interferometer array to the M(O)EMS wafer. To realize a straight-lined and uniform z-motion, the platform is weight compensated by pull springs and guided with star-shaped leaf springs which provide a ratio of horizontal to vertical stiffness of over 10,000.

FIGURE 12.16
Image of the assembled SMARTIEHS system and its control unit.

The system includes an illumination module made of a matrix of LEDs operating at a wavelength of 470 nm for LCI and a matrix of collimated laser diodes operating at the wavelength of 658 nm for LI. Illumination is directed down to optical probing wafer via the beam splitter. Optical wafer is placed under the cube beam splitter at a distance of 2 mm. The light sources, the beam splitter cube, and the smart pixel camera are assembled by mechanical means on the complete architecture of the measurement system. The coupling between the optical unit, the high-precision *z*-drive, and other hardware components like the frequency generator for the excitation unit is operated by the multifunctional system software.

M(O)EMS (which frequency ranges up to several MHz) are dynamically excited by electrostatic forces using electrodes. To implement excitation for an array of microelements, a glass wafer structured with transparent ITO electrodes was especially developed by the

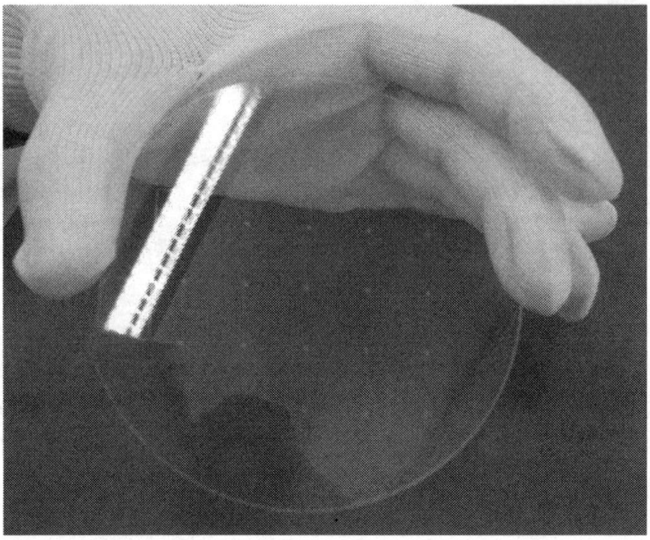

FIGURE 12.17
Excitation wafer with ITO electrodes.

Fraunhofer Institute for Reliability and Microintegration IZM (Chemnitz, Germany). The transparent electrodes permit introduction of a large electrode surface without disturbing the optical measurements. The excitation wafer with ITO electrodes (Figure 12.17) is implemented between the optical LI wafer and object. The result of the dynamic measurements is a frequency response function of the tested M(O)EMS, which is obtained by sweeping the excitation frequencies. The excitation signal given by the signal generator is sinusoidal. The inspection system software synchronizes the measurement and excitation process according to the selected range of excitation frequency.

12.7 Example Measurement Results

12.7.1 Measurement Objects

The validation test of the inspection system was performed on reference wafers specially designed and produced by Institute FEMTO-ST (Besancon, France). The reference wafer contains 25 groups of micro-machined silicon structures, that is, membranes and step objects (depth of 5 μm), as shown in Figure 12.18a. The groups are spaced in the form of a 5×5 matrix with a pitch of 13.76 mm, equal to the pitch in LI and LCI interferometer wafers. Each group, referred to one measurement channel, is composed of nine structures (Figure 12.18b):

- One step object marked as S1 ($400 \times 400\ \mu m^2$)
- Eight square membranes, marked as M1–M8, with lateral dimensions from 400×400 to $750 \times 750\ \mu m^2$

The structures are arranged in 3×3 matrix with a pitch of 3 mm.

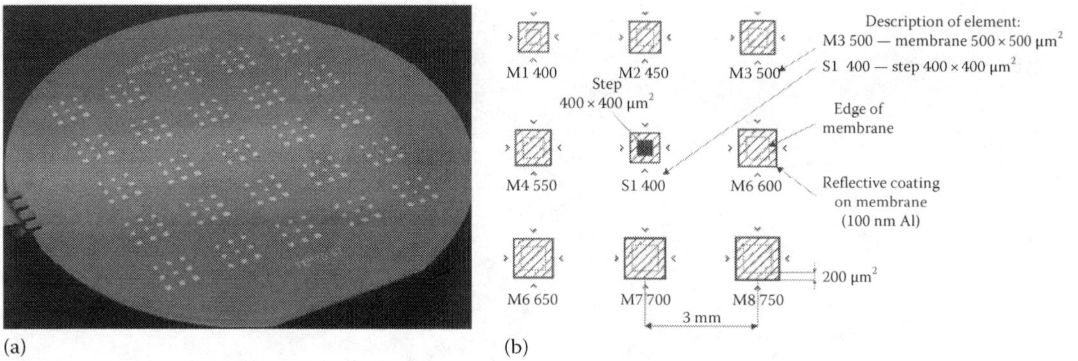

FIGURE 12.18
Reference wafer: (a) image of the fabricated wafer and (b) technical drawing of the group referred to one measurement channel.

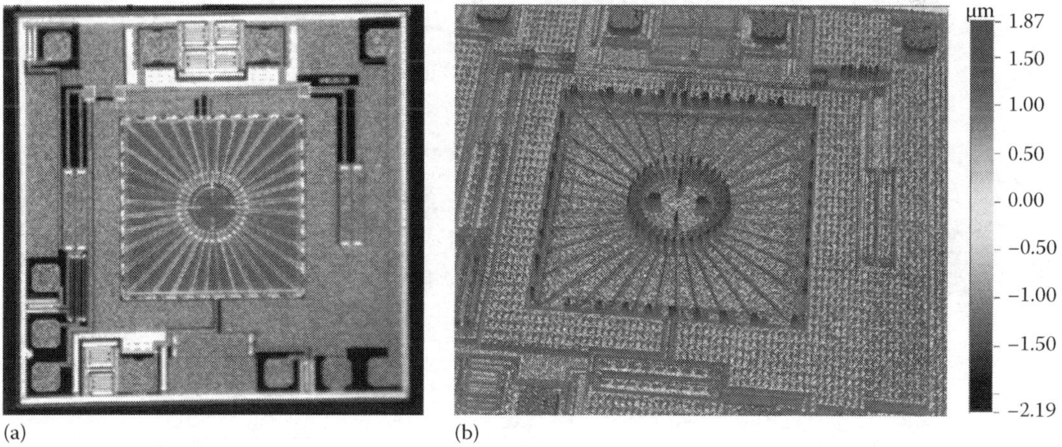

FIGURE 12.19
A single IR sensor: (a) microscope image and (b) topography obtained by Veeco optical profiler.

The functional test of the inspection system has been performed on an IR sensor wafer produced by Melexis [29]. The elements on the wafer consist of a thin silicon nitride membrane with small silicon on oxide pillars with a height of about 2 μm. A photo of the object and the result of measurement using a Veeco white-light optical profiler are shown in Figure 12.19a and b, respectively. The pitch between two sensor structures is 1.72 mm. The pitch of the interferometer array is 13.76 mm and matches the 8th structure in the sensor matrix. The size of the membrane is about 750 × 750 μm². The most important criteria for dynamic testing is the resonance frequency.

12.7.2 LCI Results

The final tests of LCI probing of the wafer showed that there are 19 correct measurement channels. The other six channels had misfunctioned due to camera failure (four channels indicated in Figures 12.20 and 12.22 by black squares) and too much distortion in the image, which was impossible to analyze (two channels).

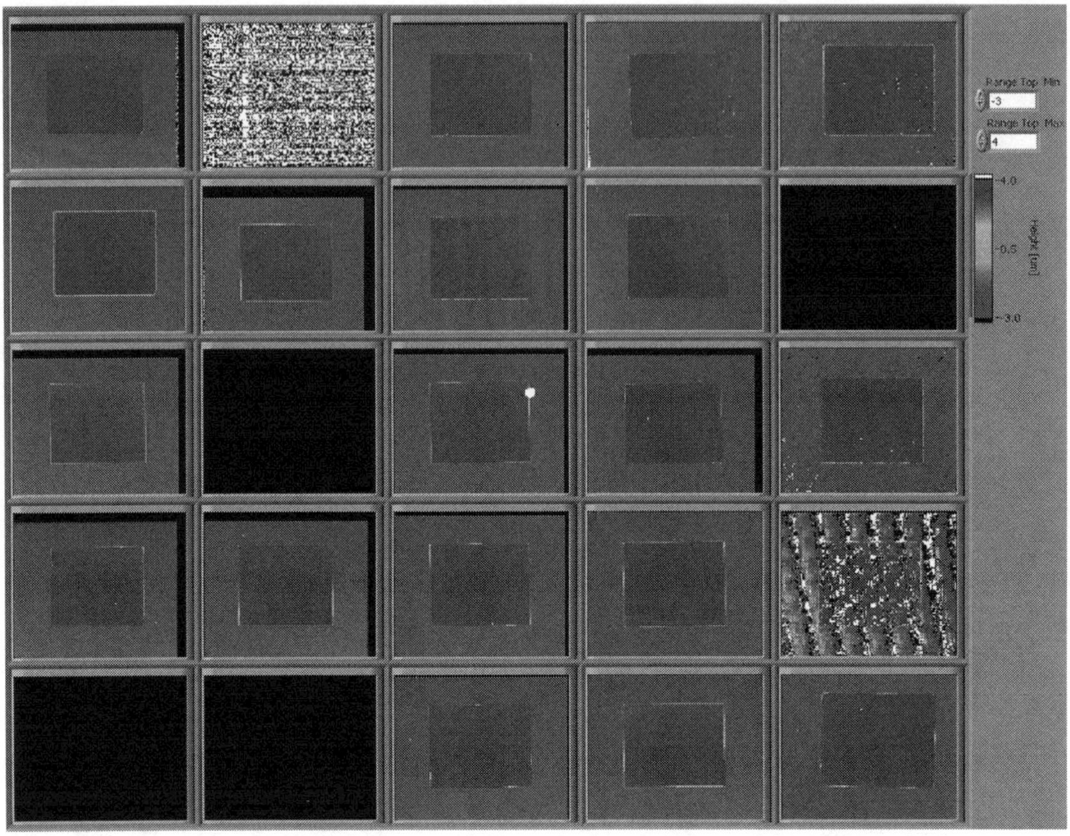

FIGURE 12.20
Results of step objects measurement performed with LCI.

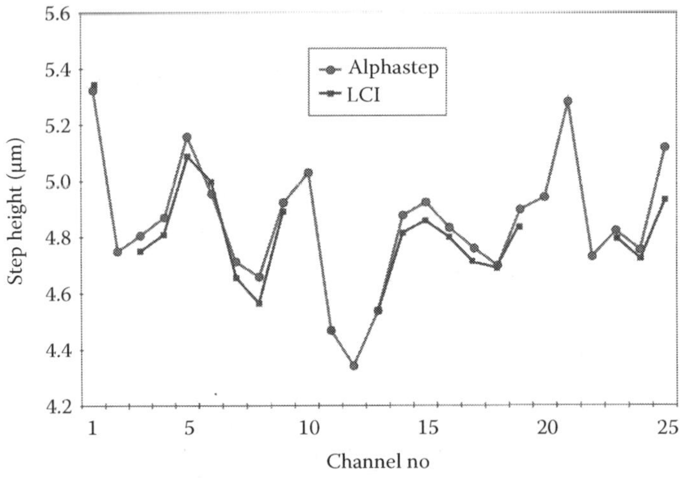

FIGURE 12.21
Comparison of the measurements of the height of step objects in all channels performed with LCI to Alphastep profilometer.

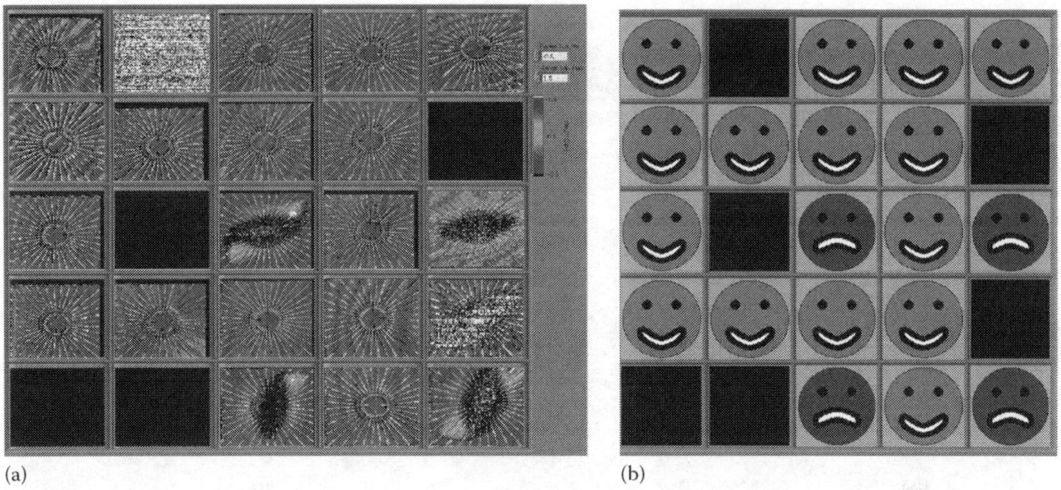

FIGURE 12.22
Results of measurements of IR sensors by LCI probing wafer: (a) their topography and (b) visualization of the fault detection.

The step object height measurements performed on a reference wafer give a good indication of the precision of the LCI instrument (Figure 12.20). These step measurements show that the SMARTIEHS system can guarantee a high absolute precision, thanks to the interferometers used for the scanning. This absolute precision is comparable to the accuracy of the commercial instruments and is better than 100 nm.

Figure 12.21 shows the differences in the results of measurements obtained by the SMARTIEHS LCI and the mechanical profilometer made by Tencor Alphastep. This is a systematic difference between the LCI and the Alphastep profilometer of approximately −43 nm. The repeatability (sequential in time measurements) for the LCI was 21 nm. The evaluated LCI parameters are in good agreement with the assumed instrument specification.

The results of IR sensor topography measurement are presented in Figure 12.22a. In four working channels, we can observe broken membranes in the sensor. These defects were introduced intentionally in order to present an example automatic procedure which recognizes and indicates broken elements as shown in Figure12.22b.

12.7.3 LI Results

The final tests of LI probing of the reference wafer were done on the 19 working LI channels. Four channels were without camera signal, while the next two had distorted and noisy optical images. The main task of the LI measurements is to determine the dynamic characteristic of the M(O)EMS. First, the frequency response function (FRF) of the microelements is investigated using the approach presented in Section 5.3 referred to as the quasi-heterodyned method. The measurements were performed at the reference objects—the membranes of dimensions: 750×750 μm^2. The results of a frequency scan from 20 to 270 kHz are shown in Figure 12.23. From these data, the resonance frequency of the 1st mode was found. In the selected FRF curves, additional peaks around 250–270 kHz are also visible.

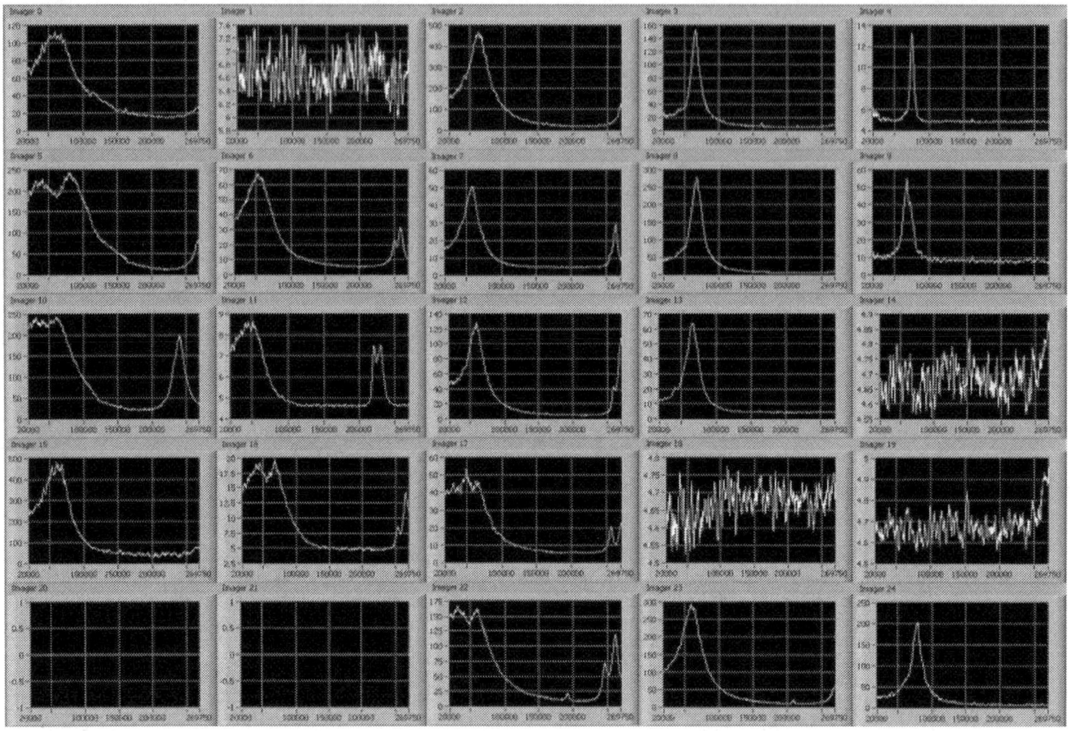

FIGURE 12.23
FRF curves for square membranes of 750 × 750 μm² size obtained for the 20–270 kHz frequency range.

These peaks correspond to the membrane's higher vibration modes and are plotted together with the reference data (see Figure 12.24). The comparison between the 1st mode resonance frequencies obtained in the SMARTIEHS system and Polytec laser vibrometer reference system shows the measurements provided by LI are correct (Figure 12.24). The data show very good agreement, keeping in mind that the FWHM of the resonant mode is large (around 15 kHz). It is worthwhile to mention that the height of the peaks in each channel does not depend on the amplitude of vibration only but also takes into account the quality of the interferogram in each channel. That is, the signal amplitude depends on the fringe pattern contrast as well as on the illumination conditions and imaging quality. This factor may explain the differences in plots height achieved in each channel. Also, the amplitude of vibration in each channel is slightly different due to differences in the electrostatic actuation that was employed. All objects were actuated at once with a specially designed excitation wafer containing ITO electrodes. The excitation wafer nonflatness caused differences in distance between object and electrode and influences the local excitation force.

Further investigations concerning FRFs of the IR sensors at the wafer were delivered by Melexis. In this case, the same wafer was tested as by LCI; however, a different set of sensors was measured. The results of these investigations within the 100–300 kHz frequency range are presented in Figure 12.25. As shown in the figure, the 1st resonance frequency for most of the objects under investigations appears around 225 kHz. A distinct shift of the frequency may be observed in the third imager in the first row in which the 1st resonant frequency is determined to be approximately 160 kHz. The frequency shift is caused by

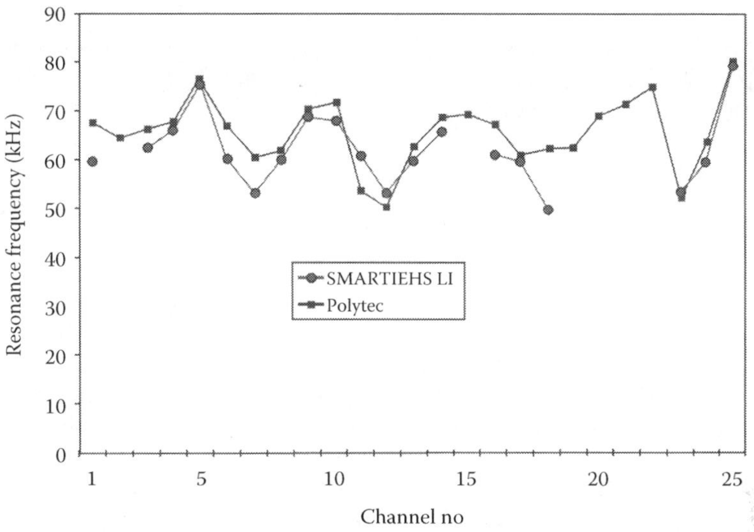

FIGURE 12.24
Comparison of the resonance frequencies of the square membranes measured with SMARTIEHS LI matrix and reference system Polytec MSA-500.

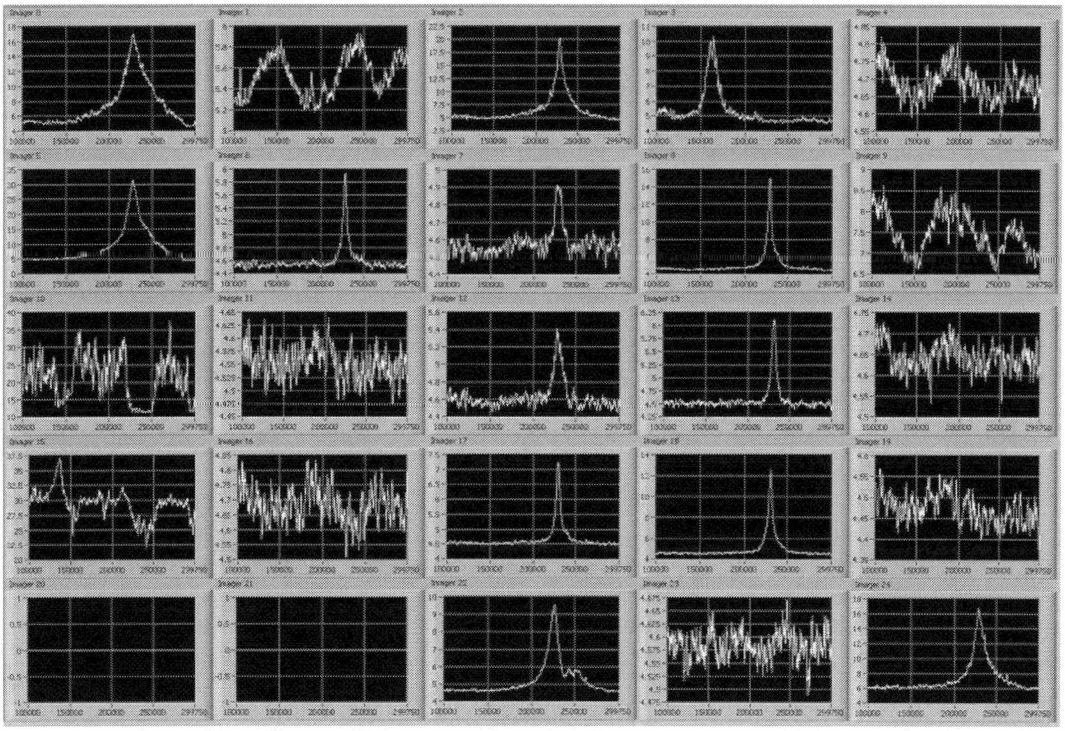

FIGURE 12.25
FRS curves of the IR sensors obtained for the 100–300 kHz frequency range.

damage (a crack) on the sensor in this channel. Note that there are significantly lower values of the peaks in the FRF curves with respect to the ones obtained while measuring square membranes. These lower values are caused by insufficient force of excitation of the IR sensor induced through the ITO layer.

12.8 Conclusions and Future Prospects

The next-generation test equipment for micro production must be more time and cost efficient and multifunctional than the state-of-the-art equipment today. The concepts introduced in this chapter exemplify novel ways being developed to fill this metrology tool gap. The approach described is based on a parallel inspection approach and measurement made by the systems that are comparable in size with the measured objects. Effectively, an array of very small metrology sensors was produced to measure an array of very small devices. Specifically, an array of microoptical sensors were integrated onto a probing wafer that was adapted to measure a M(O)EMS wafer. In practice, this approach is expected to be able to inspect more than 100 structures within one measurement cycle on an 8 in. wafer. An exchangeable probing wafer enables cost-effective testing of several different parameters in the same inspection platform.

The mechanical system consists of a weight-compensated scanning platform with three voice coils. This allows for parallel alignment of the interferometer array to the MEMS wafer and realizes a straight-lined and highly uniform z-motion to measure with a LCI. The microoptical components of the probing wafers were produced using standard micro-fabrication processes borrowed from the semiconductor industry. The smart pixel imager array provided 5×5 imagers with a resolution of 140×140 pixels each. The proof of principle of the functionality of a probing wafer containing interferometers was demonstrated. The measurements performed using LCI and direct LI demonstrators have shown the proper beam forming and object imaging in both interferometers. The measurement error in both interferometers was below $\lambda/10$.

The developed inspection system can be improved with regard to the spatial resolution, the numbers of channels, the range of measured parameters, and the measurement speed. The main limitation for a smaller footprint of the channels is the size of the smart pixel imager. The footprint of the microoptical interferometers is only a few mm^2. The next-generation smart pixel cameras will be smaller and have increased spatial resolution and measurement speed. The variety of probing wafers can be extended, enabling the measurement of additional parameters such as in-plane displacement/strain distributions or in-depth layers visualization.

The measurement speed for both interferometer configurations is lower than expected mostly by the time of control of driving signals in the measurement system. This parameter is not connected with the technological issues of the interferometer arrays and may be enhanced by further development of controlling devices and their protocols.

In general, this chapter has demonstrated the concept of measuring larger numbers of very small devices using an equal number of very small sensors that can be configured to the task. The combination of multiple metrology techniques in an adaptable platform opens up possibilities for a wide range of applications for the measurement of the microscale and even nanoscale parts and devices of the future.

Acknowledgments

The system reported had been developed within SMARTIEHS, which is a collaborative project funded under the Grant Agreement 223935 to the 7th Framework Program Objective 2007-3.6. Several people have contributed to the work presented in this chapter. Kay Gastinger (in SINTEF up to 2011, actually, NTNU—Trondheim Norwegian University of Science and Technology) has been the initiator and coordinator of the SMARTIEHS project. Odd Løvhaugen and Kari Anne H Bakke (SINTEF) provided optical design and analysis of LCI. Christoph Schaeffel, Steffen Michael, Roman Paris, and Norbert Zeike (Institut für Mikroelektronik- und Mechatronik-Systeme gemeinnützige GmbH, IMMS) have given input of great value to the mechanical design and assembly of the inspection system. Uwe Zeitner, Dirk Michaelis, Peter Dannberg, Maria Oliva, Tino Benkenstein, and Torsten Harzendorf (Fraunhofer Institute for Applied Optics and Precision Engineering, IOF) took care of the technological realization of LCI and LI arrays. Christophe Gorecki, Jorge Albero, and Sylwester Bargiel (Institute of FEMTO-ST, Université de Franche-Comté) have developed technology of microlenses and the main technological components of LCI. Stephan Beer (CSEM S.A.) has developed the smart pixel camera. Patrick Lambelet and Rudolf Moosburger (Heliotis AG) had worked on LCI data treatment and instrument exploitation. Kamil Liżewski, Krzysztof Wielgo (Warsaw University of Technology), and Karl Henrik Haugholt (SINTEF) provided their input to the mechanical design and tests of LI and LCI demonstrators. The authors want to thank these people for their valuable contributions.

References

1. Yole Développement, Status of the MEMS Industry, Yole Développement, Lyon (2008).
2. enableMNT Industry reviews, Test and measurement equipment and services for MST/MEMS-worldwide, December 2004.
3. Osten, W. (ed.), *Optical Inspection of Microsystems*, Taylor & Francis Group, New York (2006).
4. Petitgrand, S., Yahiaoui, R., Danaie, K., Bosseboeuf A., and Gilles, J.P, 3D measurement of micromechanical devices vibration mode shapes with a stroboscopic interferometric microscope, *Opt. Lasers Eng.*, 36, 77–101 (2001).
5. Michael, S. et al., *MEMS Parameter Identification on Wafer Level using Laser Doppler Vibrometer*, Smart Systems Integration, Paris, France, 321–328 (2007).
6. Aswendt, P., Schmidt, C.-D., Zielke, D., and Schubert, S., ESPI solution for non-contacting MEMS-on-wafer testing, *Opt. Lasers Eng.*, 40, 501–515 (2003).
7. Osten, W., Some answers to new challenges in optical metrology, *Proc. SPIE*, 7155, 715503 (2008).
8. Gastinger, K., Løvhaugen, P., Skotheim, Ø., and Hunderi, O., Multi-technique platform for dynamic and static MEMS-characterisation, *Proc. SPIE*, 6616, 66163K (2007).
9. Gastinger, K., Haugholt, K.H., Kujawinska, M., and Jozwik M., Optical, mechanical and electro-optical design of an interferometric test station for massive parallel inspection of MEMS and MOEMS, *Proc. SPIE*, 7389, 73891 (2009).
10. SMARTIEHS—SMART Inspection system for high speed and multifunctional testing of MEMS and MOEMS, www.ict-smartiehs.eu
11. Kim, B., Schmittdiel, M.C., Degertekin, F.L., and Kurfess, T.R., Scanning grating microinterferometer for MEMS metrology, *J. Manuf. Sci. Eng.*, 126, 807–812 (2004).

12. Johansen, I-R. et al., Optical displacement sensor element, International patent publication no. WO 03/043377 A1.
13. Cascade Microtech, Inc., http://www.cmicro.com/products/probe-systems/200mm-wafer/pa200/pa200-semi-automatic-probe-system (accessed 2012)
14. Beer, S., Zeller, P., Blanc, N., Lustenberger, F., and Seitz, P., Smart pixels for real-time optical coherence tomography, *Proc. SPIE*, 5302, 21–32 (2004).
15. Albero, J., Bargiel, S., Passilly, N., Dannberg, P., Stumpf, M., Zeitner, U.D., Rousselot, C., Gastinger, K., and Gorecki, C., Micromachined array-type Mirau interferometer for parallel inspection of MEMS, *J. Micromech. Microeng.*, 21, 065005 (2011).
16. http://www.schott.com/hometech/english/download/brochure_borofloat_e.pdf
17. Gale, M.T., Direct writing of continuous-relief micro-optics micro-optics in *Elements, Systems and Applications*, Herzig, H.P. (ed.), Taylor & Francis Group, London, U.K., pp. 87–126 (1997).
18. Micro Resist Technology GmbH, Hybrid Polymers-OrmoComp®, http://www.microresist.de/products/ormocers/ormocomp_en.htm (accessed 2012)
19. Dannberg, P., Mann, G., Wagner, L., and Brauer, A., Polymer UV-moulding for micro-optical systems and O/E-integration, *Proc. SPIE*, 4179, 137–45 (2000).
20. Moharam, M.G. and Gaylord, T.K., Rigorous coupled-wave analysis of planar-grating diffraction, *J. Opt. Soc. Am.*, 71, 811–818 (1981).
21. Weber, M.J., *Handbook of Optical Materials*, CRC Press, Boca Raton, FL (2003).
22. Astilean, S., Lalanne, P., and Chavel, P., High-efficiency subwavelength diffractive element patterned in a high-refractive-index material for 633 nm, *Opt. Lett.*, 23, 552–554 (1998).
23. Stern, M.B., Binary optics fabrication, in *Microoptics: Elements, Systems and Application*, Herzig, H.P. (ed.), Taylor & Francis Group, London, U.K., pp. 53–58 (1997).
24. Oliva, M. et al., Twyman–Green-type integrated laser interferometer array for parallel MEMS testing, *J. Micromech. Microeng.* 22, 015018 (2012).
25. Beer, S., Real-time photon-noise limited optical coherence tomography based on pixel-level analog signal processing, PhD dissertation, University of Neuchâtel, Neuchâtel (2006).
26. Styk, A., Kujawińska, M., Lambelet, P., Røyset, A., and Beer S., Microelements vibration measurement using quasi-heterodyning method and smart-pixel camera, *Fringe*, Osten, W. and Kujawińska, M. (eds.), Springer-Verlag, Berlin, Heidelberg, pp. 523–527 (2009).
27. Styk, A., Lambelet, P., Røyset A., Kujawińska M., and Gastinger K., Smart pixel camera based signal processing in an interferometric test station for massive parallel inspection of MEMS and MOEMS, *Proc. SPIE*, 7387, 73870M, (2010).
28. Schmit, J. and Creath, K., Extended averaging technique for derivation of error-compensating algorithms in phase-shifting interferometry, *Appl. Opt.* 34, 3610–3619 (1995).
29. Melexis, S.A., Microelectronic integrated systems, www.melexis.com

Index

Q

R

S

Made in the USA
Monee, IL
12 November 2026

d0bb40f2-0222-4549-9939-c87f9c9df231R01